Eddy Covariance

Springer Atmospheric Sciences

For further volumes:
http://www.springer.com/series/10176

Marc Aubinet • Timo Vesala • Dario Papale
Editors

Eddy Covariance

A Practical Guide to Measurement and Data Analysis

Springer

Editors
Prof. Marc Aubinet
Université de Liège
Gembloux Agro-Bio Tech
Unit of Biosystem Physics
Passage des Deportés 2
5030 Gembloux
Belgium
Marc.Aubinet@ulg.ac.be

Prof. Timo Vesala
Department of Physics
University of Helsinki
PO Box 54
00014 Helsinki
Finland
tvesala@mappi.helsinki.fi

Dr. Dario Papale
University of Tuscia
Department for Innovation in Biological,
Agro-food and Forest systems (DIBAF)
Via Camillo de Lellis
01100 Viterbo
Italy
darpap@unitus.it

ISBN 978-94-007-2350-4 e-ISBN 978-94-007-2351-1
DOI 10.1007/978-94-007-2351-1
Springer Dordrecht Heidelberg London New York

Library of Congress Control Number: 2011942320

Printed on acid-free paper

Springer is part of Springer Science+Business Media (www.springer.com)

Preface

As soon as the first eddy covariance networks developed, in the mid-1990s, the need for standardisation became clear. Standardisation concerned not only material but also data treatment, corrections, computation. In order to harmonise these procedures in the frame of the EUROFLUX network, some software intercomparison exercises were developed: "golden files" were circulated between teams and treated with different software packages with the aim to compare the computation results. It rapidly appeared that beyond some bugs that appeared in new software and were corrected immediately, important differences remained between different computations that were due to the use of different hypotheses. The necessity to clarify these choices, and to propose a standardised (even if perfectible) eddy covariance flux computation procedure, led us to publish a first methodological paper (Aubinet et al. 2000). Eleven years later, this paper remains an often cited reference in the field.

However, as the theory and measurement techniques progressed since, and since the eddy covariance techniques is becoming also a monitoring exercise and not more only a purely scientific activity, the necessity of an update of this paper and of creating something that could help to install an eddy covariance site and manage it correctly grew. In December 2008, during a meeting at the Hyytiälä Forestry Field Station (Finland) celebrating the tenth anniversary of the EUROFLUX network constitution, the idea was launched (originally by Samuli Launiainen) to produce such an update. However, it appeared rapidly that if we wanted to produce a self-standing document, useful to eddy flux practisers, we could not limit its size to those of a paper.

We thus decided to tackle the edition of a book with the general objective to give to eddy flux practisers the theoretical and practical information necessary in order to develop eddy covariance measurements, from site installation to data treatment. After preparing a book plan, structured in 17 chapters, we chose different first authors, known for their skills in the field and asked them to constitute a team of co-authors and prepare their chapters. The present book is the result of the two and half year long work that followed.

After a first chapter recalling the theoretical bases on which eddy covariance method relies, Chap. 2 describes technical requirements of the eddy covariance set-up: tower positioning and dimensioning (height, position, system positioning on the tower), sonic and gas analysers, dimensioning, calibration and maintenance.

Chapter 3 describes the general procedure used in order to get "uncorrected" fluxes and to discuss the pros and cons of different computation alternatives. This implied especially a description of the data acquisition set-ups, and a detailed discussion on flux computation (fluctuation computation, first quality control on raw data, time lagging, rotation and flux computation).

Chapter 4 concentrates on the different corrections procedures necessary in order to get good quality fluxes and on the quality tests on these fluxes.

Chapter 5 focuses on the problem of night flux underestimation, its causes and its impact on flux measurements. It described different screening or correction procedures and discussed their pros and cons.

Chapter 6 specifies the conditions when data gap filling is necessary and which precautions should be taken when performing data gap filling. It presented and compared the different data gap filling procedures and their (dis)advantages.

Chapter 7 identifies and quantifies the different causes of uncertainty in flux measurements and analyses how they combine during scaling up.

Chapter 8 describes the main footprint models and the way they could be combined with vegetation cover maps (in order to identify the sources/sinks of flux) or with quality tests (in order to evaluate the general quality of data).

Chapter 9 presents the different possibilities to partition eddy flux into ecosystem respiration and gross ecosystem photosynthesis. Different approaches based on night-time or on day-time data were described.

Chapter 10 focuses on disjuncted eddy covariance technique, which is especially adapted to capture tracer gas.

Chapters 11–16 describe the specific requirements for flux measurements in specific ecosystems like forests, grasslands, croplands, wetlands, lakes or urban environment.

Finally, Chap. 17 describes the objectives of a data base, the way it should be maintained and managed. In addition, it proposes some policies for data use, exchange and publication.

The editors would like to thank the co-authors of the chapters for their enthusiasm and their involvement in this long (but, hopefully, useful) work that we hope can contribute to reinforce the links between the different eddy covariance networks in the world.

Dario Papale and Timo Vesala, although editors of this book, would like to thank M. Aubinet for taking care of the lion's share of the editing job.

The book idea and preparation has been also supported by the IMECC EU project and the ABBA Cost Action.

This book is dedicated to all the field (often anonymous) technicians whose continuous system care, maintenance and follow up constitute an inestimable contribution to ecosystem studies and to the Ph.D. students that decide to base their work on these unique measurements.

Folks, mark already in your calendars "the 20th Anniversary of EUROFLUX" to be held around 10 December in 2018, once again in Hyytiälä. We do not know yet what will be the main product of the meeting then.

Marc Aubinet
Dario Papale
Timo Vesala

Contents

Contributors

Deborah A. Agarwal Lawrence Berkeley National Laboratory, Berkeley, CA, USA, DAAgarwal@lbl.gov

Christof Ammann Agrosocope Reckenholz Tanikon Res Stn ART, CH-8046 Zurich, Switzerland, christof.ammann@art.admin.ch

Nicola Arriga DIBAF, University of Tuscia, Viterbo, Italy, arriga@unitus.it

Marc Aubinet Unit of Biosystem Physics, Gembloux Agro-Bio Tech, University of Liege, 5030 Gembloux, Belgium, Marc.Aubinet@ulg.ac.be

Mika Aurela Finnish Meteorological Institute, P.O. Box 503, FI-00101 Helsinki, Finland, Mika.Aurela@fmi.fi

Dennis Baldocchi Department of Environmental Science, Policy and Management, University of California, Berkeley, CA, USA, baldocchi@berkeley.edu

Alan G. Barr Environment Canada, 11 Innovation Blvd, Saskatoon, SK S7N 3H5 Canada, Alan.Barr@ec.gc.ca

Pierre Béziat Centre d'Etudes Spatiales de la BIOsphère (CESBIO), Toulouse, France

Eric Ceschia Centre d'Etudes Spatiales de la BIOsphère (CESBIO), Toulouse, France, Eric.ceschia@cesbio.cnes.fr

Andreas Christen Department of Geography and Atmospheric Science Program, University of British Columbia, Vancouver, Canada, andreas.christen@ubc.ca

Robert B. Cook Oak Ridge National Laboratory, Oak Ridge, TN, USA, cookrb@ornl.gov

Ankur R. Desai Atmospheric and Oceanic Sciences, University of Wisconsin, Madison, USA, desai@aos.wisc.edu

Werner Eugster Department of Agricultural and Food Sciences, Institute of Agricultural Sciences, ETH Zurich, Zurich, Switzerland, werner.eugster@agrl.ethz.ch

Christian Feigenwinter Institute of Meteorology, Climatology and Remote Sensing, University of Basel, Basel, Switzerland, feigenwinter@metinform.ch

Joshua B. Fisher Jet Propulsion Laboratory, California Institute of Technology, Pasadena, CA, USA

Thomas Foken Department of Micrometeorology, University of Bayreuth, 95440 Bayreuth, Germany, thomas.foken@uni-bayreuth.de

Mathias Göckede Department of Forest Ecosystems & Society, Oregon State University, Corvallis, OR, USA, mathias.goeckede@oregonstate.edu

André Granier UMR1137 Ecologie et Ecophysiologie Forestières, Centre de Nancy, INRA, F-54280 Champenoux, France, agranier@nancy.inra.fr

Bernard Heinesch Unit of Biosystem Physics, Gembloux Agro-Bio Tech, University of Liege, 5030 Gembloux, Belgium, bernard.heinesch@ulg.ac.be

David Y. Hollinger USDA Forest Service, Northern Research Station, 271 Mast Road, Durham, NH 03824 USA, dhollinger@fs.fed.us

Andreas Ibrom Risø National Laboratory, Biosystems Department, Technical University of Denmark (DTU), Roskilde, Denmark

Risø National Laboratory for Sustainable Energy, Technical University of Denmark (DTU), Frederiksborgvej 399, 4000, Roskilde, Denmark, anib@risoe.dtu.dk

Catharine van Ingen Microsoft Research, San Francisco, CA, USA, vaningen@microsoft.com

Natascha Kljun Department of Geography, Swansea University, Swansea, UK, n.kljun@swansea.ac.uk

Katja Klumpp INRA, Grassland Ecosystem Research (UREP), Clermont-Ferrand, France, katja.klumpp@clermont.inra.fr

Olaf Kolle Max-Planck Institute for Biogeochemistry, Jena, Germany, olaf.kolle@bgc-jena.mpg.de

Werner L. Kutsch Institute for Agricultural Climate Research, Johann Heinrich von Thünen Institute (vTI), Braunschweig, Germany, werner.kutsch@vti.bund.de

Quentin Laffineur Unit of Biosystem Physics, Gembloux Agro-Bio Tech, University of Liege, 5030 Gembloux, Belgium

Gitta Lasslop Max-Planck Institute for Biogeochemistry, 07745 Jena, Germany, gitta.lasslop@zmaw.de

Tuomas Laurila Finnish Meteorological Institute, P.O. Box 503, FI-00101 Helsinki, Finland, tuomas.laurila@fmi.fi

Monique Y. Leclerc Laboratory for Environmental Physics, The University of Georgia, Griffin, GA, USA, mleclerc@uga.edu

Ray Leuning Marine and Atmospheric Research, CSIRO, PO Box 3023, Canberra, ACT 2601 Australia, ray.leuning@csiro.au

Henry W. Loescher National Ecological Observatory Network, Boulder, CO 80301, USA

Institute of Alpine and Arctic Research (INSTAAR), University of Colorado, Boulder, CO 80303, USA, hloescher@neoninc.org

Bernard Longdoz UMR1137 Ecologie et Ecophysiologie Forestières, Centre de Nancy, INRA, F-54280 Champenoux, France, longdoz@nancy.inra.fr

Hongyan Luo National Ecological Observatory Network, Boulder, CO 80301, USA

Institute of Alpine and Arctic Research (INSTAAR), University of Colorado, Boulder, CO 80303, USA, hluo@neoninc.org

Matthias Mauder Institute for Meteorology and Climate Research, Atmospheric Environmental Research, Karlsruhe Institute of Technology, Kreuzeckbahnstr. 19, 82467 Garmisch-Partenkirchen, Germany, matthias.mauder@kit.edu

Christine Moureaux Gembloux Agro-Bio Tech, University of Liege, Gembloux, Belgium, christine.moureaux@ulg.ac.be

J. William Munger School of Engineering and Applied Science, and Department of Earth and Planetary Sciences, Harvard University, Cambridge, MA, USA, jwmunger@seas.harvard.edu

Anne Ojala Department of Environmental Sciences, University of Helsinki, Helsinki, Finland, Anne.Ojala@helsinki.fi

Steven R. Oncley Earth Observing Laboratory, NCAR, P.O. Box 3000, Boulder, CO, 80307-3000, USA, oncley@ucar.edu

Dario Papale DIBAF, University of Tuscia, Viterbo, Italy, darpap@unitus.it

Elizabeth Pattey ECORC, Agriculture and Agri-Food Canada, Ottawa, Canada, elizabeth.pattey@agr.gc.ca

Ronald Queck Department of Meteorology, Institute of Hydrology and Meteorology, TU Dresden (TUD), Dresden, Germany, ronald.queck@tu-dresden.de

Üllar Rannik Department of Physics, University of Helsinki, Helsinki, Finland, ullar.rannik@heuristica.ee

Corinna Rebmann Department Computational Hydrosystems, Helmholtz Centre for Environmental Research – UFZ, 04318 Leipzig, Germany, corinna.rebmann@ufz.de

Markus Reichstein Max Planck Institute für Biogeochemistry, Jena, Germany, mreichstein@bgc-jena.mpg.de

Andrew D. Richardson Department of Organismic and Evolutionary Biology, Harvard University Herbaria, 22 Divinity Avenue, Cambridge, MA, 02138 USA, arichardson@oeb.harvard.edu

Janne Rinne Department of Physics, University of Helsinki, FI-00014, Helsinki, Finland, Janne.Rinne@helsinki.fi

Andrey Sogachev Risø National Laboratory for Sustainable Energy, Technical University of Denmark, Roskilde, Denmark, anso@risoe.dtu.dk

Jean-François Soussana INRA, Grassland Ecosystem Research (UREP), Clermont-Ferrand, France, Jean-Francois.Soussana@clermont.inra.fr

Paul C. Stoy Department of Land Resources and Environmental Sciences, Montana State University, P.O. Box 173120, Bozeman, MT, 59717-3120 USA, paul.stoy@montana.edu

Juha-Pekka Tuovinen Finnish Meteorological Institute, P.O. Box 503, FI-00101 Helsinki, Finland, Juha-Pekka.Tuovinen@fmi.fi

Timo Vesala Department of Physics, University of Helsinki, Helsinki, Finland, timo.vesala@helsinki.fi

Eva Van Gorsel CSIRO, Canberra, Australia

Roland Vogt Institute of Meteorology, Climatology and Remote Sensing, University of Basel, Basel, Switzerland, Roland.Vogt@unibas.ch

Georg Wohlfahrt Institute of Ecology, University of Innsbruck, Innsbruck, Austria, Georg.Wohlfahrt@uibk.ac.at

Chapter 1
The Eddy Covariance Method

Thomas Foken, Marc Aubinet, and Ray Leuning

1.1 History

The eddy covariance method for measuring exchanges of heat, mass, and momentum between a flat, horizontally homogeneous surface and the overlying atmosphere was proposed by Montgomery (1948), Swinbank (1951), and Obukhov (1951). Under these conditions, net transport between the surface and atmosphere is one-dimensional and the vertical flux density can be calculated by the covariance between turbulent fluctuations of the vertical wind and the quantity of interest.

Instrumentation limitations hampered early implementation of this approach. In 1949, Konstantinonov (Obukhov 1951) developed a wind vane with two hot wire anemometers to measure the shear stress but the full potential of the eddy covariance method only emerged after the development of sonic anemometers, for which the basic equations were given by Schotland (1955). After the development of the first sonic thermometer (Barrett and Suomi 1949), a vertical sonic anemometer with a 1 m path length (Suomi 1957) was used during the O'Neill experiment in 1953 (Lettau and Davidson 1957). The design of today's anemometers was developed by Bovscheverov and Voronov (1960) and later by Kaimal and Businger (1963) and

T. Foken (✉)
Department of Micrometeorology, University of Bayreuth, 95440 Bayreuth, Germany
e-mail: thomas.foken@uni-bayreuth.de

M. Aubinet
Gembloux Agro-Bio Tech, Unit of Biosystem Physics, University of Liege, 5030 Gembloux, Belgium
e-mail: Marc.Aubinet@ulg.ac.be

R. Leuning
Marine and Atmospheric Research, CSIRO, PO Box 3023, Canberra, ACT 2601, Australia
e-mail: ray.leuning@csiro.au

M. Aubinet et al. (eds.), *Eddy Covariance: A Practical Guide to Measurement and Data Analysis*, Springer Atmospheric Sciences, DOI 10.1007/978-94-007-2351-1_1,

Mitsuta (1966). These phase shift anemometers have now been replaced by running time anemometers with delay time measurements (Hanafusa et al. 1982; Coppin and Taylor 1983).

Early micrometeorological experiments from the 1950s to 1970s were designed to study fundamental aspects of atmospheric turbulence over homogeneous surfaces, whereas studies in the 1980s investigated the turbulent fluxes of momentum, sensible, and latent heat over heterogeneous surfaces. Similar experiments were conducted in the United States (FIFE, Sellers et al. 1988), in France (HAPEX, André et al. 1990), and in Russia (KUREX, Tsvang et al. 1991). These experiments were to become the basis of many further micrometeorological experiments (Foken 2008) that needed researchers who were highly experienced in micrometeorology and sensor handling.

The possibility of continuous eddy flux measurements arose in the 1990s with the development of a new generation of sonic anemometers (see reviews by Zhang et al. 1986; Foken and Oncley 1995) and infrared gas analyzers for water vapor and carbon dioxide, together with the first comprehensive software packages for the eddy covariance method (McMillen 1988). In the early 1990s, the eddy covariance method became more and more widely used by the ecological community for the measurement of the carbon dioxide and water exchange between an ecosystem and the atmosphere. The first measuring towers of what later became the international FLUXNET network (Baldocchi et al. 2001) were installed, and introductions into techniques new for nonmicrometeorologists were written (Aubinet et al. 2000; Moncrieff et al. 1997a, b). In parallel, the development of new analyzer types allowed an extension of the investigated trace gas spectrum. In particular, Tunable Diode Laser and Quantum Cascade Laser spectrometers were used for the measurement of methane and nitrous oxide (Smith et al. 1994; Laville et al. 1999; Hargreaves et al. 2001; Kroon et al. 2010), Proton Transfer Reaction Mass Spectrometers for volatile organic compounds (Karl et al. 2002; Spirig et al. 2005), and Chemiluminescent sensors for Ozone (Güsten and Heinrich 1996; Gerosa et al. 2003; Lamaud et al. 1994, a.o.).

Some milestones in the development of the eddy covariance method are given in Table 1.1 with the reference to the Chapters of this book.

1.2 Preliminaries

1.2.1 Context of Eddy Covariance Measurements

Eddy covariance measurements are typically made in the surface boundary layer, which is approximately 20–50 m high in the case of unstable stratification and a few tens of meters in stable stratification (see Stull 1988; Garratt 1992; Foken 2008; for complete definitions of layers in the atmosphere). Fluxes are approximately constant with height in the surface layer; hence measurements taken in this layer

Table 1.1 History of the development of the eddy covariance method

Historical milestone	References	See chapter/ section
Theoretical basis of the eddy covariance method	Montgomery (1948), Swinbank (1951), Obukhov (1951)	Section 1.2
Three-dimensional sonic anemometer	Bovscheverov and Voronov (1960), Kaimal and Businger (1963), Mitsuta (1966)	Chapter 2
Instrumental requirements	McBean (1972)	Chapter 2
Gas analyzer for water vapor (UV)	Buck (1973), Kretschmer and Karpovitsch (1973), Martini et al. (1973)	
Gas analyzer for water vapor (IR)	Elagina (1962), Hyson and Hicks (1975), Raupach (1978)	Chapter 2
Correction of the effect of the air density	Webb et al. (1980)	Section 4.1
Gas analyzer for carbon dioxide (IR)	Ohtaki and Matsui (1982), Elagina and Lazarev (1984)	Chapter 2
Transformation of buoyancy flux into sensible heat flux	Schotanus et al. (1983)	Section 4.1
System of transfer functions for spectral correction	Moore (1986)	Section 4.1
Fetch conditions	Gash (1986)	Chapter 8
Real-time data processing software	McMillen (1988)	Chapter 3
Source regions for fluxes (footprint), based on Gash (1986)	Schmid and Oke (1990), Schuepp et al. (1990)	Chapter 8
Relaxed eddy accumulation method, based on Desjardins (1977)	Businger and Oncley (1990)	
Influence of tubing of closed path sensors	Leuning and Moncrieff (1990)	Section 4.1.3 Chapter 3
Theoretical basis for flux footprints and sampling strategies	Horst and Weil (1994), Lenschow et al. (1994)	Chapter 8
Addressing the problem of the unclosed energy balance at the surface	Foken and Oncley (1995)	Section 4.2
Quality tests for eddy covariance data	Foken and Wichura (1996), Vickers and Mahrt (1997)	Section 4.3
Addressing the problem of vertical advection	Lee (1998) and many others	Section 1.3, Chapter 5
Methodology for FLUXNET network (EuroFlux)	Aubinet et al. (2000)	All chapters
Gap filling in the FLUXNET network	Falge et al. (2001a, b)	Chapter 6
Organization of an international network (FLUXNET)	Baldocchi et al. (2001)	All chapters

Foken et al. (1995), Foken (2008), Moncrieff (2004), modified

are representative of the fluxes from the underlying surfaces which are desired to be known. Here atmospheric turbulence is the dominant transport mechanism, justifying the use of the eddy covariance approach to measure the fluxes.

Some preliminary definitions are necessary before discussing the eddy covariance approach in detail.

1.2.2 Reynolds Decomposition

The description of turbulent motions in the following theory sections requires the decomposition of the time-series of each variable ζ into a time-mean part, $\bar{\zeta}$, and a fluctuating part, ζ', the so-called Reynolds decomposition (Fig. 1.1). This can be written as:

$$\zeta = \bar{\zeta} + \zeta' \tag{1.1a}$$

where:

$$\bar{\zeta} = \frac{1}{T} \int_{t}^{t+T} \zeta(t) dt \tag{1.1b}$$

The application of Reynolds decomposition requires some averaging rules for the turbulent value ζ' which are termed Reynolds postulates:

$$\begin{aligned}
&\mathrm{I} \quad \overline{\zeta'} = 0 \\
&II \quad \overline{\zeta\xi} = \bar{\zeta}\,\bar{\xi} + \overline{\zeta'\xi'} \\
&III \quad \overline{\bar{\zeta}\xi} = \bar{\zeta}\,\bar{\xi} \\
&IV \quad \overline{a\zeta} = a\bar{\zeta} \\
&V \quad \overline{\zeta + \xi} = \bar{\zeta} + \bar{\xi}
\end{aligned} \tag{1.2}$$

where a is a constant.

Stricto sensu, these relations are valid only when averages are by "ensemble" averaging (i.e., averaging over many realizations under identical conditions, Kaimal and Finnigan 1994). However, this is never possible in atmospheric measurements, so averages are most often computed on the basis of time series of statistical quantities by making use of the ergodic hypothesis which states that time averages are equivalent to ensemble averages (Brutsaert 1982; Kaimal and Finnigan 1994). To fulfil this assumption, the fluctuations have to be statistically stationary during the averaging time chosen (see Chap. 4).

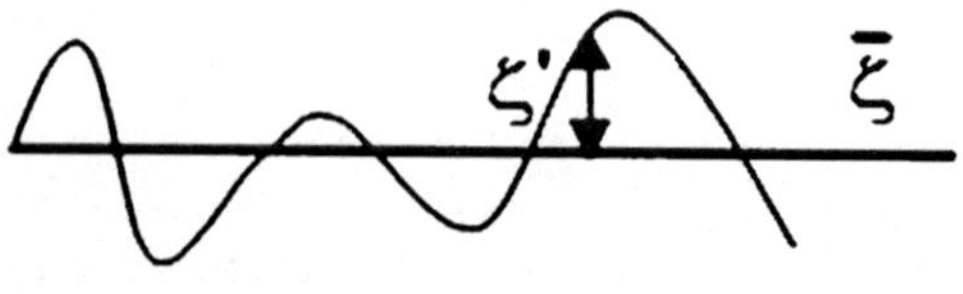

Fig. 1.1 Schematic presentation of Reynolds decomposition of the value ζ (Foken 2008)

1.2.3 Scalar Definition

The following variables are commonly used in the literature (and throughout this book) to define the scalar intensity of an atmospheric constituent *s*: *density* (ρ_s, kg m^{-3}) and *molar concentration* (c_s mol m^{-3}) represent the mass and the number of moles of *s* per volume of air, respectively. The *mole fraction* (mole mole^{-1}) is the ratio of the moles of *s* divided by the total number in the mixture (also equal to the ratio of the constituent partial pressure to the total pressure), the *molar mixing ratio* ($\chi_{s,m}$, mole mole^{-1}) is the ratio of the constituent mole number to those of dry air, and the *mass mixing ratio* (χ_s, kg kg^{-1}) is the ratio of the mass of the constituent to the mass of dry air. These variables are related by the perfect gas and the Dalton laws.

However, among these variables, only the molar and mass mixing ratios are conserved quantities in the presence of changes in temperature, pressure, and water vapor content (see Kowalski and Serrano-Ortiz (2007) for a more complete discussion). Unfortunately, the variables that are directly measured in the field by infrared gas analyzers are rather density and molar concentration, quantities that are not conserved during heat conduction, air compression/expansion or evaporation, and water vapor diffusion. Therefore, variations in these quantities may appear even in the absence of production, absorption, or transport of the component. The corrections that are necessary to take these effects into account were extensively discussed by Webb et al. (1980) and reexamined by Leuning (2003, 2007). They will be presented in Sect. 4.1.4.

The conservation equations developed in the section below are written using the mass mixing ratio but, for convenience, the other variables will also appear in this book. Conversion factors of one variable into another are given in Table 1.2.

Table 1.2 Conversion factors between different variables characterizing scalar intensity

Conversion factor	Molar mixing Ratio, $\chi_s=$	Mass mixing Ratio, $\chi_{sm}=$	Molar concentration, $c_s=$	Density, $\rho_s=$
Molar mixing ratio, χ_s X	1	$\frac{m_s}{m_d}$	$\frac{p_d}{R\overline{\theta}}$	$\frac{m_s p_d}{R\overline{\theta}}$
Mass mixing Ratio, χ_{sm} X	$\frac{m_d}{m_s}$	1	$\frac{m_d p_d}{m_s R\overline{\theta}}$	$\frac{m_d p_d}{R\overline{\theta}}$
Molar concentration, c_s X	$\frac{R\overline{\theta}}{p_d}$	$\frac{m_s R\overline{\theta}}{m_d p_d}$	1	m_s
Density, ρ_sX	$\frac{R\overline{\theta}}{m_s p_d}$	$\frac{R\overline{\theta}}{m_d p_d}$	$\frac{1}{m_s}$	1

Note that p_d corresponds to the dry air pressure (namely $p - p_v$). As a result, the exact conversion of mass or molar mixing ratio into concentration or density needs the knowledge of water vapor pressure (for details see list of symbols)

1.3 One Point Conservation Equations

The equation describing the conservation of any scalar or vector quantity ζ in the atmosphere may be written as

$$\underbrace{\frac{\partial \rho_d \zeta}{\partial t}}_{I} + \underbrace{\vec{\nabla}(\vec{u}\rho_\mathrm{d}\zeta)}_{II} + \underbrace{K_\zeta \Delta(\rho_d \zeta)}_{III} = \underbrace{S_\zeta}_{IV} \tag{1.3}$$

where $\vec{u}$ is the wind velocity vector, $\vec{\nabla}$ and Δ represent the divergence $\left(\frac{\partial}{\partial x}, \frac{\partial}{\partial y}, \frac{\partial}{\partial z}\right)$ and Laplacian $\left(\frac{\partial^2}{\partial x^2} + \frac{\partial^2}{\partial y^2} + \frac{\partial^2}{\partial z^2}\right)$ operators, ρ_d is the dry air density, K_ζ is the molecular diffusivity of the quantity ζ, and S_ζ represents its source/sink strength. This equation is instantaneous and applies to an infinitesimal volume of air. It states that the *rate of change of the quantity* (I) can be due to its *atmospheric transport* (II) to *molecular diffusion* (III) or to its *production by a source/absorption by a sink* into the infinitesimal volume (IV). It can be applied to any scalar or vector quantity provided source terms are defined accordingly. In particular, if ζ is 1, Eq. 1.3 is the continuity equation, if ζ is air enthalpy, it is the enthalpy conservation equation, and if ζ is the mixing ratio of an atmospheric component (water vapor, carbon dioxide, etc.), it is the scalar conservation equation. If the quantity is a component of the velocity vector in one given direction, Eq. 1.3 expresses the conservation of the momentum component in this direction. The three equations describing the momentum conservation in the three directions constitute the Navier Stokes equations.

Application of these equations to the surface boundary layer requires application of the Reynolds decomposition rules: the variables ζ, $\rho_\mathrm{d}, \vec{u}$, and S_ζ should each be decomposed into a mean and a fluctuating part according to Eq. 1.1, followed by application of the averaging operator, and appropriate rearrangement and simplification. This procedure will be applied to each equation below.

1.3.1 Dry Air Mass Conservation (Continuity) Equation

By replacing ζ by 1 in Eq. 1.3, one obtains

$$\frac{\partial \rho_\mathrm{d}}{\partial t} + \vec{\nabla}(\vec{u}\rho_\mathrm{d}) = 0 \tag{1.4}$$

as there is neither a source nor sink of dry air in the atmosphere. Application of the time- averaging operator gives immediately:

$$\frac{\overline{\partial \rho_\mathrm{d}}}{\partial t} + \vec{\nabla}(\overline{\vec{u}\rho_\mathrm{d}}) = 0 \tag{1.5}$$

1.3.2 Momentum Conservation Equation

By replacing ζ in Eq. 1.3 with the component of wind velocity in one given direction, u_i, one obtains the momentum conservation equation in this direction:

$$\frac{\partial \rho_\mathrm{d} u_\mathrm{i}}{\partial t} + \vec{\nabla}(\vec{u}\rho_\mathrm{d}\, u_\mathrm{i}) = S_\mathrm{i} \tag{1.6}$$

In Eq. 1.6, the source/sink terms correspond to momentum source/sink, namely to forces. Forces that can act on air parcels in the atmospheric boundary layer are drag, pressure gradient, Coriolis forces, viscous forces, or buoyancy. The first three forces are considered negligible for a flat, horizontally homogeneous surface boundary layer above the roughness elements (i.e. not including vegetation) (Businger 1982; Foken 2008; Stull 1988). Buoyancy appears only in the equation for vertical momentum. The horizontal component of momentum parallel to the mean wind is dominant in the surface boundary layer and thus the buoyancy term is not considered. In a Cartesian coordinate system (x, y, z) where x corresponds to the horizontal, parallel to the average wind velocity, y to the horizontal, perpendicular to the average velocity, and z to the vertical; u, v, w are the x, y, and z components of velocity, respectively, and this equation is written as

$$\frac{\partial \rho_\mathrm{d} u}{\partial t} + \frac{\partial \rho_\mathrm{d} u^2}{\partial x} + \frac{\partial \rho_\mathrm{d} v u}{\partial y} + \frac{\partial \rho_\mathrm{d} w u}{\partial z} = 0 \tag{1.7}$$

Application of the Reynolds decomposition to Eq. 1.7 and use of the following simplifications (Businger 1982; Stull 1988):

$$\begin{aligned} &I \;\; \left|p'/\bar{p}\right| \ll \left|\rho'_\mathrm{d}/\overline{\rho_\mathrm{d}}\right| \\ &II \;\; \left|p'/\bar{p}\right| \ll \left|\theta'/\bar{\theta}\right|, \\ &III \;\; \left|\rho'_\mathrm{d}/\overline{\rho_\mathrm{d}}\right| \ll 1 \\ &IV \;\; \left|\theta'/\bar{\theta}\right| \ll 1 \end{aligned} \tag{1.8}$$

where p is the pressure and θ the air temperature, leads to

$$\frac{\overline{\partial u}}{\partial t} + \bar{u}\frac{\partial \bar{u}}{\partial x} + \bar{v}\frac{\partial \bar{u}}{\partial y} + \bar{w}\frac{\partial \bar{u}}{\partial z} + \frac{\partial \overline{u'^2}}{\partial x} + \frac{\partial \overline{v'u'}}{\partial y} + \frac{\partial \overline{w'u'}}{\partial z} = 0 \tag{1.9}$$

Equation 1.8, III corresponds to the *Boussinesq-approximation* (Boussinesq 1877), which neglects density fluctuations except in the buoyancy (gravitation) term, because the acceleration of gravity is relatively large in comparison with the other accelerations in the momentum equation. By choosing a coordinate system such that

$\bar{v}$ and $\bar{w}$ are zero and assuming *horizontal homogeneity* (horizontal gradients nullify) and *steady state conditions* (time derivative nullifies) we obtain finally

$$\frac{\partial \overline{w'u'}}{\partial z} = 0 \tag{1.10}$$

Where $\overline{w'u'}$ is the *eddy covariance* term. Equation 1.10 suggests that, under the preceding assumptions, this flux is constant with height and that it is representative of the vertical flux of momentum through a horizontal plane above the surface roughness elements. This approach is called the *eddy covariance method*.

Neglecting the *pressure gradient*, *molecular/viscous transport*, *gravity*, and *Coriolis* terms to derive Eq. 1.10 does not have significant impact on the eddy covariance method over flat, homogeneous surfaces. These conditions are however rare in ecosystems located in a patchy landscape or undulating topography. *Steady-state conditions* are also rare in the surface layer because of diurnal variations in atmospheric stability. It is then necessary to measure the change in storage term using an array of sensors (Sect. 2.5) or to assume quasi-steady conditions. Methods to estimate errors caused by neglecting the storage term are included in the data quality procedures discussed in Sect. 4.3.

1.3.3 Scalar Conservation Equation

By replacing ζ in Eq. 1.3 by χ_s, the mixing ratio of one atmospheric component, one obtains

$$\frac{\partial \rho_d \chi_s}{\partial t} + \vec{\nabla}(\vec{u}\rho_d\,\chi_s) = S_s \tag{1.11}$$

Through application of the Reynolds decomposition and the continuity Eq. 1.5, Leuning (2003) showed that Eq. 1.11 can be written as

$$\overline{\rho_d}\overline{\frac{\partial \chi_s}{\partial t}} + \overline{\rho_d \vec{u}}\;\vec{\nabla}\,(\overline{\chi_s}) + \vec{\nabla}\left[\overline{\rho_d}\;\overline{\vec{u}'\chi'_s}\right] = \overline{S_s} \tag{1.12}$$

This equation states that the source term $\overline{S_s}$ is given by the sum of the rate of change of the mixing ratio χ_s, advection due to spatial gradients in χ_s, and to divergences in the eddy fluxes.

Expanding this in terms of spatial derivatives and assuming constant dry air density give the one point conservation equation of a scalar:

$$\overline{\rho_d}\frac{\overline{\partial \chi_s}}{\partial t} + \overline{\rho_d u}\frac{\partial \overline{\chi_s}}{\partial x} + \overline{\rho_d v}\frac{\partial \overline{\chi_s}}{\partial y} + \overline{\rho_d w}\frac{\partial \overline{\chi_s}}{\partial z} + \frac{\partial \overline{\rho_d}\overline{u'\chi'_s}}{\partial x} + \frac{\partial \overline{\rho_d}\overline{v'\chi'_s}}{\partial y} + \frac{\partial \overline{\rho_d}\overline{w'\chi'_s}}{\partial z} = \bar{S}_s \tag{1.13}$$

Considering that $\bar{v}$ and $\bar{w}$ are zero, due to axis choice (Sect. 3.2.4) and assuming *horizontal homogeneity* (horizontal gradients nullify) and *steady state conditions* (time derivative nullifies) we get, similar to Eq. 1.10:

$$\frac{\partial \overline{\rho_d} \overline{w' \chi'_s}}{\partial z} = \bar{S}_s \tag{1.14}$$

expressing that the vertical gradient of eddy covariance is equal to the tracer source/sink term in the volume element. In the case of passive tracers (water vapor, CO_2), this term is zero. In the case of active tracers (ozone, VOCs, $NO_x, \ldots$), $\bar{S}_s$ corresponds to the rate of chemical production/destruction of the component in the volume element.

1.3.4 Enthalpy Equation

By replacing ζ by $c_p\theta$, the air enthalpy, one obtains

$$\frac{\partial \rho c_p \theta}{\partial t} + \vec{\nabla}(\vec{u} \rho c_p \theta) = S_\theta \tag{1.15}$$

where c_p is the specific heat of the air and ρ is moist air density. The same development as before leads to

$$\frac{\overline{\partial \theta}}{\partial t} + \bar{u}\frac{\partial \bar{\theta}}{\partial x} + \bar{v}\frac{\partial \bar{\theta}}{\partial y} + \bar{w}\frac{\partial \bar{\theta}}{\partial z} + \frac{\partial \overline{u'\theta'}}{\partial x} + \frac{\partial \overline{v'\theta'}}{\partial y} + \frac{\partial \overline{w'\theta'}}{\partial z} = \frac{1}{\bar{\rho} c_p}\left[\frac{\partial R}{\partial z}\right] \tag{1.16}$$

and

$$\frac{\partial \overline{w'\theta'}}{\partial z} = \frac{1}{\bar{\rho} c_p}\left[\frac{\partial R}{\partial z}\right] \tag{1.17}$$

Where $\frac{\partial R}{\partial z}$ is the vertical radiative flux divergence, which is close to zero in clear surface layers (no fog, rain, smoke, etc.).

1.4 Integrated Relations

Eddy covariance measurements can be used as a tool to estimate fluxes exchanged by ecosystems. To this end, preceding equations may be integrated both horizontally over the area of interest, *A* $(2L \times 2L)$, and vertically, from soil to the measurement height h_m (Fig. 1.2).

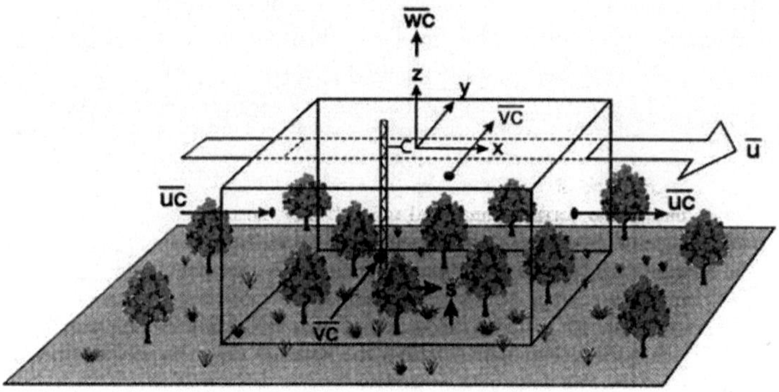

Fig. 1.2 Schematic image of integration of Eq. 1.15 on a control volume in homogeneous terrain (Finnigan et al. 2003)

1.4.1 Dry Air Budget Equation

Integrating Eq. 1.5 on the control volume and assuming horizontal homogeneity gives:

$$\int_0^{h_m} \frac{\partial \overline{\rho_d}}{\partial t} dz + \bar{w}\overline{\rho_d}|_{h_m} + \overline{w' \rho'_d}|_{h_m} = 0 \tag{1.18}$$

where assumptions of zero flux of dry air at the ground and no net sources or sinks of dry air in the layer of air below h_m are made. Slight imbalances between molar fluxes of CO_2 and O_2 during photosynthesis or respiration or fluxes of nitrogen or volatile organic compounds are extremely small and do not invalidate Eq. 1.18.

1.4.2 Scalar Budget Equation (Generalized Eddy Covariance Method)

Integrating (1.13) in the control volume gives

$$\frac{1}{4L^2}\int_{-L}^{L}\int_{-L}^{L}\int_0^{h_m}\left[\underbrace{\overline{\rho_d}\frac{\partial \overline{\chi_s}}{\partial t}}_{I} + \underbrace{\overline{\rho_d u}\frac{\partial \overline{\chi_s}}{\partial x} + \overline{\rho_d v}\frac{\partial \overline{\chi_s}}{\partial y} + \overline{\rho_d w}\frac{\partial \overline{\chi_s}}{\partial z}}_{II} + \underbrace{\frac{\partial \overline{\rho_d}\overline{u'\chi'_s}}{\partial x} + \frac{\partial \overline{\rho_d}\overline{v'\chi'_s}}{\partial y}}_{III} + \underbrace{\frac{\partial \overline{\rho_d}\overline{w'\chi'_s}}{\partial z}}_{IV}\right] dz\,dx\,dy$$

$$= \frac{1}{4L^2}\int_{-L}^{L}\int_{-L}^{L}\int_0^{h_m}\underbrace{\overline{S_s}\,dz\,dx\,dy}_{V} \tag{1.19}$$

Equation 1.19 represents the complete budget equation of the component *s*. It shows that the component produced by the source or absorbed by the sink (*V*) may be either stored in the control volume (*I*), or transported by advection (*II*), or by turbulence (*III* and *IV*). In these conditions, the source/sink term represents both the sources/sinks inside the air volume and those at the lower limit of the volume (soil, litter).

This equation may be simplified in several ways using different hypotheses. The most common simplification, thoroughly discussed by Finnigan et al. (2003), supposes that the measurement system is placed in a horizontally homogeneous equilibrium layer where all horizontal gradients in Eq. 1.19 are negligible and the mixing ratios and turbulent fluxes measured on the tower are assumed representative of the whole volume. In these conditions, horizontal integration is unnecessary and a simplified one-dimensional mass balance can then be deduced as

$$\underbrace{\int_0^{h_m} \overline{\rho}_d \overline{\frac{\partial \chi_s}{\partial t}} dz}_{I} + \underbrace{\int_0^{h_m} \overline{\rho_d w \frac{\partial \chi_s}{\partial z}} dz}_{II} + \underbrace{\overline{\rho_d \overline{w' \chi'_s}}\big|_{h_m}}_{IV} = \underbrace{F_s}_{V} \tag{1.20}$$

where $\overline{w' \chi'_s}\big|_{h_m}$ represents the vertical turbulent flux at the top of the control volume and F_s the averaged source/sink strength in the whole control volume, that is, the *net ecosystem exchange for the component s*. Term *II* represents the vertical advection at the top of the control volume that results from dry air density change with time in the air layer below h_m. By application of the dry air conservation Eq. 1.18, this term may be rewritten as

$$\int_0^{h_m} \overline{\rho_d w} \overline{\frac{\partial \chi_s}{\partial z}} dz = -\int_0^{h_m} \left[\int_0^z \overline{\frac{\partial \rho_d}{\partial t}} dz' \right] \overline{\frac{\partial \chi_s}{\partial z}} dz \tag{1.21}$$

and after integration by parts as

$$\int_0^{h_m} \overline{\rho_d w} \overline{\frac{\partial \chi_s}{\partial z}} dz = \int_0^{h_m} [\overline{\chi_s}(z) - \overline{\chi_s}(h)] \overline{\frac{\partial \rho_d}{\partial t}} dz \tag{1.22}$$

Equation 1.20 may thus be rewritten as

$$\int_0^{h_m} \overline{\rho}_d \overline{\frac{\partial \chi_s}{\partial t}} dz + \int_0^{h_m} [\overline{\chi_s}(z) - \overline{\chi_s}(h)] \overline{\frac{\partial \rho_d}{\partial t}} dz + \overline{\rho_d} \overline{w' \chi'_s}\big|_{h_m} = F_s \tag{1.23}$$

However, most often, term II is negligible so that Eq. 1.20 may be written more simply as

$$\underbrace{\int_0^{h_m} \overline{\rho}_d \overline{\frac{\partial \chi_s}{\partial t}} dz}_{I} + \underbrace{\overline{\rho_d} \overline{w' \chi'_s}\big|_{h_m}}_{IV} = \underbrace{F_s}_{V} \tag{1.24a}$$

This equation is at the basis of the *generalized eddy covariance method*: it suggests indeed that the flux of a scalar exchanged by an ecosystem (F_s, term V) can be estimated as the sum of the vertical eddy covariance $\overline{w'\chi'_s}|$ at height h_m (F_s^{EC}, term IV) and of the change of storage of the scalar between the soil and this height (F_s^{STO}, term I), namely:

$$F_s = F_s^{EC} + F_s^{STO} \tag{1.24b}$$

The above hypotheses are known to work fairly well in daytime conditions when turbulence is fully developed, but they appear to be too restrictive to describe completely nighttime conditions. It is then necessary to include the horizontal and vertical advection terms in the conservation equation, with the additional assumption that the vertical integral of $\overline{\rho_d w}\, \partial\overline{\chi}_s/\partial z$ as measured on a single tower is representative of the whole volume. Equation 1.19 then becomes

$$\underbrace{\int_0^{h_m} \bar{\rho}_d \frac{\partial \overline{\chi_s}}{\partial t} dz}_{I} + \underbrace{\int_0^{h_m} \left[\overline{\rho_d w} \frac{\partial \overline{\chi_s}}{\partial z} dz \right]}_{IIa} + \underbrace{\int_0^{h_m} \left[\overline{\rho_d u} \frac{\Delta \bar{\chi}_{s,x}}{\Delta x} + \overline{\rho_d v} \frac{\Delta \bar{\chi}_{s,y}}{\Delta y} \right] dz}_{IIb} + \underbrace{\overline{\rho_d w' \chi'_s}|_{h_m}}_{IV} = \underbrace{F_s}_{V} \tag{1.25a}$$

in which $\Delta\overline{\chi}_{s,x} = \overline{\chi}_{s,x=L} - \overline{\chi}_{s,x=-L}$ is the difference in mixing ratios at height z between the downwind ($+L$) and upwind ($-L$) vertical planes normal to the x-direction, with a similar definition for $\Delta\overline{\chi}_{s,y}$ in the y-direction. Equation 1.25a may be rewritten as

$$F_s = F_s^{EC} + F_s^{VA} + F_s^{HA} + F_s^{STO} \tag{1.25b}$$

where F_s^{VA} and F_s^{HA} represent vertical (Term IIa) and horizontal (Term IIb) advection of component s. One problem is that these terms cannot be measured on a single tower, and full use of this equation requires a three-dimensional array of instrumentation. The importance of the different terms of this equation will be discussed in Chap. 5.

1.5 Spectral Analysis

Thorough eddy covariance analysis, application of data quality criteria, or correct assessment of some correction factors require a spectral analysis of the (co)variances. The aim of this section is to give the necessary information about spectral analysis of a signal, atmospheric turbulence (co)spectra and the effects of measurement on these (co)spectra to allow the reader to perform these analyses. More details on spectral analysis can be found in the textbooks of Stull (1988), Kaimal and Finnigan (1994), or Foken (2008).

1.5.1 Spectral Analysis of Turbulence

Any turbulent flow may be thought of as a superposition of eddies over a wide range of sizes. As a result, the fluctuation with time of the signals (velocity components, temperature, scalar densities) measured by sensors placed in such flow vary over a wide range of frequencies. The relation between spatial and temporal scale can be established thanks to Taylor's frozen turbulence hypothesis (Taylor 1938) which assumes that eddies do not change significantly in size when convected by the mean wind past a fixed observer. Spectral analysis uses signal frequency decomposition. It is performed by applying an integral transformation which converts a function of time into a function of frequency (f [Hz]):

$$\mathcal{F}_s(f) = \frac{1}{\sqrt{2}} \int_{-\infty}^{\infty} \chi_s(t) e^{ift} dt \tag{1.26}$$

$\mathcal{F}(f)$ is called the Fourier transform of the signal. Of special interest in eddy covariance are the power spectrum C_{ss} of a signal χ_s and the cospectrum C_{ws} of two signals w and χ_s. The first one is defined as

$$C_{ss}(f) = \mathcal{F}_s(f) . \mathcal{F}_s{}^*(f) \tag{1.27}$$

and the second as the real part of the cross-spectrum, defined as

$$C_{ws}(f) = \mathcal{F}_w(f) \cdot \mathcal{F}_s{}^*(f) \tag{1.28}$$

where $\mathcal{F}_s{}^*(f)$ is the complex conjugate of $\mathcal{F}_w(f)$. $C_{ss}(f)$ and $C_{ws}(f)$ are the spectral and cospectral density, respectively. The main interest of (co) spectra is that their integral over the whole frequency range equals the (co)variance of the signals:

$$\int_0^{\infty} C_{ss}(f) df = \sigma_s^2$$

$$\int_0^{\infty} C_{ws}(f) df = \overline{w' \chi'_s} \tag{1.29}$$

so that (co) spectra may be thought of as a distribution of (co)variances into the different frequency bands of width df.

1.5.2 Spectral Analysis of Atmospheric Turbulence

In the frequency range of interest to micrometeorology, turbulence spectra can be divided into three major spectral regions: (1) at low frequencies (typically 10^{-4} Hz) is the energy containing range, where turbulent energy is produced; (2) at intermediate

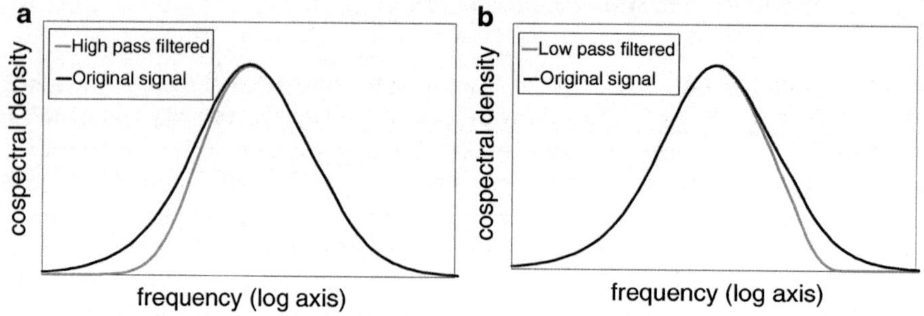

Fig. 1.3 Typical atmospheric cospectrum (*black curve*) with effects of high pass filtering (**a**, *gray curve*) and low pass filtering (**b**, *gray curve*)

frequencies is the inertial subrange, where energy is neither produced nor dissipated but is transformed to smaller and smaller eddies due to an "energy cascade" process (see, e.g. Stull 1988); and (3) at higher frequencies is the dissipation range where turbulent energy is dissipated through viscosity. By considering similarity arguments, one can deduce that the shapes of suitably normalized atmospheric (co)spectra are repeatable and can be described by universal relations. Parameterizations of the momentum and sensible heat cospectra proposed by Kaimal et al. (1972) are given by

$$-\frac{f\,C_{\mathrm{uw}}(f)}{u_*^2} = \frac{12n}{(1+9.6n)^{\frac{7}{3}}} \tag{1.30a}$$

$$-\frac{f\,C_{w\theta}(f)}{u_*\,\theta_*} = \frac{11n}{(1+13.3n)^{\frac{7}{4}}} \quad \text{for} \quad n \le 1$$

$$-\frac{f\,C_{w\theta}(f)}{u_*\,\theta_*} = \frac{4n}{(1+3.8n)^{\frac{7}{3}}} \quad \text{for} \quad n \ge 1 \tag{1.30b}$$

where n is a dimensionless frequency defined as: $n = f(h_{\mathrm{m}} - d)/\overline{u}$ and d is the zero-plane displacement height. The $\overline{u'w'}$ and $\overline{w'\theta'}$ covariances are normalized by u_*^2 and $u_*\theta_*$, respectively, where $u_*$is the friction velocity and θ_* is the dynamic temperature. An illustration of Eq. 1.30 is given in Fig. 1.3 (black curve).

1.5.3 *Sensor Filtering*

Eddy covariance systems, like any sensor, act as frequency filters in dampening high and low frequencies. The reasons for this may be diverse and will be discussed in detail in Sect. 4.1.3. In this chapter, we limit ourselves to the description of the impact of low or high frequency filtering on the (co)spectra shape and on the resulting error that affects fluxes. The evolution of this impact with measurement height and wind velocity will also be discussed.

To represent high or low frequency dampening by a measurement system, signal theory generally use sigmoidal transfer functions which are equal to 1 in the frequency range where the signal is not attenuated and that decays to zero in the range where signal is attenuated. The shapes of these functions vary according to the processes that are responsible of the dampening (Sect. 4.1.3).

As an example, Fig. 1.3 depict the impact of filtering on a typical cospectrum, Fig. 1.3a showing the effect of a low pass filtering and Fig. 1.3b the effect of a high pass filtering.

The relative error on the fluxes due to frequency losses, $\frac{\delta_s}{F_s^{EC}}$, may be computed according to:

$$\frac{\delta_s}{F_s^{EC}} = 1 - \frac{\int_0^\infty C_{ws}(f)T_{ws}(f)df}{\int_0^\infty C_{ws}(f)df} \tag{1.31}$$

where $C_{ws}(f)$ is the ideal cospectral density and $T_{ws}(f)$ is the apparatus transfer function. The ratio in the integrals on the right hand side of Eq. 1.31 is represented in Fig. 1.3 by the ratio of areas below the gray and black curves. Figure 1.3a shows clearly that low pass filtering causes a loss of covariance and always induces a systematic error. Figure 1.3b suggests the same thing for high pass filtering. However, it could be deceptive as the low frequency spectral range (energy containing range) is not so well defined as it could also depend on mesoscale atmospheric movements. In some conditions, it is possible to observe cospectral densities of different signs at low and high frequencies. In these conditions, the impact of the low pass filtering is not necessarily systematic.

1.5.4 *Impacts of Measurement Height and Wind Velocity*

The preceding observations allow predicting the impact of measurement height and of wind velocity on errors due to frequency losses, which is synthesized in Fig. 1.4.

Equations 1.30 describe universal cospectra as functions of the nondimensional frequency $n = \frac{f(h_m-d)}{\bar{u}}$. This implies that the decrease in $h_m - d$ shifts the cospectrum toward higher frequencies (Fig. 1.4). However, as the apparatus transfer function does not depend on measurement height, systems placed at lower heights would be more sensitive to high-frequency losses (Fig. 1.4a) while systems placed at higher heights would be more sensitive to low-frequency losses (Fig. 1.4b). The first ones would require set ups able to capture fluctuations at higher frequencies while the second would need longer averaging times (the main cause of low-frequency losses). Further details about frequency losses and their correction are presented in Sect. 4.1.3.

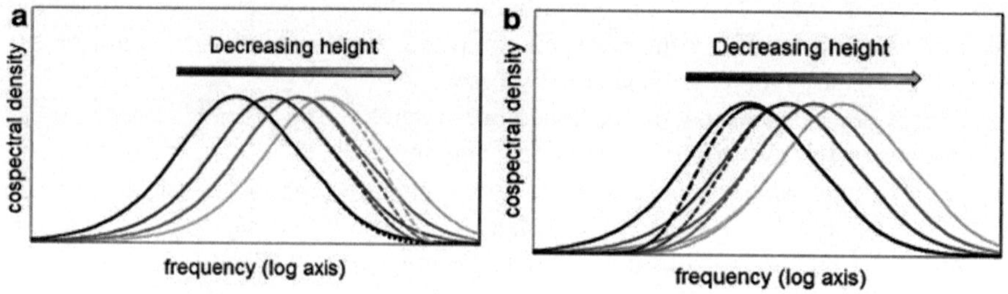

Fig. 1.4 (**a**) Undamped (*full line*) and low pass filtered (*dotted line*) cospectra; (**b**) Undamped (*full line*) and high pass filtered (*dotted line*) cospectra

Acknowledgments MA acknowledges financial support by the European Union (FP 5, 6, and 7), the Belgian Fonds de la recherche Scientifique (FNRS-FRS), the Belgian Federal Science Policy Office (BELSPO), and the Communauté française de Belgique (Action de Recherche Concertée).

References

André J-C, Bougeault P, Goutorbe J-P (1990) Regional estimates of heat and evaporation fluxes over non-homogeneous terrain, Examples from the HAPEX-MOBILHY programme. Bound Layer Meteorol 50:77–108

Aubinet M, Grelle A, Ibrom A, Rannik Ü, Moncrieff J, Foken T, Kowalski AS, Martin PH, Berbigier P, Bernhofer C, Clement R, Elbers J, Granier A, Grünwald T, Morgenstern K, Pilegaard K, Rebmann C, Snijders W, Valentini R, Vesala T (2000) Estimates of the annual net carbon and water exchange of forests: the EUROFLUX methodology. Adv Ecol Res 30: 113–175

Baldocchi D, Falge E, Gu L, Olson R, Hollinger D, Running S, Anthoni P, Bernhofer C, Davis K, Evans R, Fuentes J, Goldstein A, Katul G, Law B, Lee X, Malhi Y, Meyers T, Munger W, Oechel W, PawU KT, Pilegaard K, Schmid HP, Valentini R, Verma S, Vesala T (2001) FLUXNET: a new tool to study the temporal and spatial variability of ecosystem-scale carbon dioxide, water vapor, and energy flux densities. Bull Am Meteorol Soc 82:2415–2434

Barrett EW, Suomi VE (1949) Preliminary report on temperature measurement by sonic means. J Meteorol 6:273–276

Boussinesq J (1877) Essai sur la théorie des eaux courantes. Mem Savants Etrange 23, 46 pp

Bovscheverov VM, Voronov VP (1960) Akustitscheskii fljuger (Acoustic rotor). Izv AN SSSR Ser Geofiz 6:882–885

Brutsaert W (1982) Evaporation into the atmosphere. D. Reidel Publ. Co., Dordrecht, 299 pp

Buck AL (1973) Development of an improved Lyman-alpha hygrometer. Atmos Technol 2:213–240

Businger JA (1982) Equations and concepts. In: Nieuwstadt FTM, Van Dop H (eds) Atmospheric turbulence and air pollution modelling: a course held in The Hague, 21–25 September 1981. D. Reidel Publ. Co., Dordrecht, pp 1–36

Businger JA, Oncley SP (1990) Flux measurement with conditional sampling. J Atmos Ocean Technol 7:349–352

Coppin PA, Taylor KJ (1983) A three component sonic anemometer/thermometer system for general micrometeorological research. Bound Layer Meteorol 27:27–42

Desjardins RL (1977) Description and evaluation of a sensible heat flux detector. Bound Layer Meteorol 11:147–154

Elagina LG (1962) Optitscheskij pribor dlja izmerenija turbulentnych pulsacii vlaschnosti (Optical sensor for the measurement of turbulent humidity fluctuations). Izv AN SSSR, ser Geofiz. 12:1100–1107

Elagina LG, Lazarev AI (1984) Izmerenija tschastotnych spektrov turbulentnych pulsacij CO_2 v prizemnom sloje atmosphery (Measurement of the turbulence spectra of CO_2 in the atmospheric surface layer). Izv AN SSSR Fiz Atm Okeana 20:536–540

Falge E, Baldocchi D, Olson R, Anthoni P, Aubinet M, Bernhofer C, Burba G, Ceulemans R, Clement R, Dolman H, Granier A, Gross P, Grunwald T, Hollinger D, Jensen NO, Katul G, Keronen P, Kowalski A, Lai CT, Law BE, Meyers T, Moncrieff H, Moors E, Munger JW, Pilegaard K, Rannik U, Rebmann C, Suyker A, Tenhunen J, Tu K, Verma S, Vesala T, Wilson K, Wofsy S (2001a) Gap filling strategies for long term energy flux data sets. Agric For Meteorol 107:71–77

Falge E, Baldocchi D, Olson R, Anthoni P, Aubinet M, Bernhofer C, Burba G, Ceulemans R, Clement R, Dolman H, Granier A, Gross P, Grunwald T, Hollinger D, Jensen NO, Katul G, Keronen P, Kowalski A, Lai CT, Law BE, Meyers T, Moncrieff H, Moors E, Munger JW, Pilegaard K, Rannik U, Rebmann C, Suyker A, Tenhunen J, Tu K, Verma S, Vesala T, Wilson K, Wofsy S (2001b) Gap filling strategies for defensible annual sums of net ecosystem exchange. Agric For Meteorol 107:43–69

Finnigan JJ, Clement R, Malhi Y, Leuning R, Cleugh HA (2003) A re-evaluation of long-term flux measurement techniques, Part I: averaging and coordinate rotation. Bound Layer Meteorol 107:1–48

Foken T (2008) Micrometeorology. Springer, Berlin/Heidelberg, 308 pp

Foken T, Dlugi R, Kramm G (1995) On the determination of dry deposition and emission of gaseous compounds at the biosphere-atmosphere interface. Meteorol Z 4:91–118

Foken T, Oncley SP (1995) Results of the workshop 'Instrumental and methodical problems of land surface flux measurements'. Bull Am Meteorol Soc 76:1191–1193

Foken T, Wichura B (1996) Tools for quality assessment of surface-based flux measurements. Agric For Meteorol 78:83–105

Garratt JR (1992) The atmospheric boundary layer. Cambridge University Press, Cambridge, 316 pp

Gash JHC (1986) A note on estimating the effect of a limited fetch on micrometeorological evaporation measurements. Bound Layer Meteorol 35:409–414

Gerosa G, Cieslik S, Ballarin-Denti A (2003) Ozone dose to a wheat field determined by the micrometeorological approach. Atmos Environ 37:777–788

Güsten H, Heinrich G (1996) On-line measurements of ozone surface fluxes: Part I: methodology and instrumentation. Atmos Environ 30:897–909

Hanafusa T, Fujitana T, Kobori Y, Mitsuta Y (1982) A new type sonic anemometer-thermometer for field operation. Pap Meteorol Geophys 33:1–19

Hargreaves KJ, Fowler D, Pitcairn CER, Aurela M (2001) Annual methane emission from Finnish mires estimated form eddy co-variance campaign measurements. Theor Appl Climatol 70: 203–213

Horst TW, Weil JC (1994) How far is far enough?: The fetch requirements for micrometeorological measurement of surface fluxes. J Atmos Ocean Technol 11:1018–1025

Hyson P, Hicks BB (1975) A single-beam infrared hygrometer for evaporation measurement. J Appl Meteorol 14:301–307

Kaimal JC, Businger JA (1963) A continuous wave sonic anemometer-thermometer. J Clim Appl Meteorol 2:156–164

Kaimal JC, Finnigan JJ (1994) Atmospheric boundary layer flows: their structure and measurement. Oxford University Press, Oxford, 289 pp

Kaimal JC, Wyngaard JC, Izumi Y, Coté OR (1972) Spectral characteristics of surface layer turbulence. Q J R Meteorol Soc 98:563–589

Karl TG, Spirig C, Rinne J, Stroud C, Prevost P, Greenberg J, Fall R, Guenther A (2002) Virtual disjunct eddy covariance measurements of organic compound fluxes from a subalpine forest using proton transfer reaction mass spectrometry. Atmos Chem Phys 2:279–291

Kowalski AS, Serrano-Ortiz P (2007) On the relationship between the eddy covariance, the turbulent flux, and surface exchange for a trace gas such as CO_2. Bound Layer Meteorol 124:129–141

Kretschmer SI, Karpovitsch JV (1973) Maloinercionnyj ultrafioletovyj vlagometer (Sensitive ultraviolet hygrometer). Izv AN SSSR Fiz Atm Okeana 9:642–645

Kroon PS, Hensen A, Jonker HJJ, Ouwersloot HG, Vermeulen AT, Bosveld FC (2010) Uncertainties in eddy covariance flux measurements assessed from CH4 and N2O observations. Agric For Meteorol 150:806–816

Lamaud E, Brunet Y, Labatut A, Lopez A, Fontan J, Druilhet A (1994) The Landes experiment: biosphere–atmosphere exchanges of ozone and aerosol particles, above a pine forest. J Geophys Res 99:16511–16521

Laville P, Jambert C, Cellier P, Delmas R (1999) Nitrous oxide fluxes from a fertilized maize crop using micrometeorological and chamber methods. Agric For Meteorol 96:19–38

Lee X (1998) On micrometeorological observations of surface-air exchange over tall vegetation. Agric For Meteorol 91:39–49

Lenschow DH, Mann J, Kristensen L (1994) How long is long enough when measuring fluxes and other turbulence statistics? J Atmos Ocean Technol 11:661–673

Lettau HH, Davidson B (eds) (1957) Exploring the atmosphere's first mile, vol 1. Pergamon Press, London/New York, 376 pp

Leuning RL (2003) Measurements of trace gas fluxes in the atmosphere using eddy covariance: WPL corrections revisited. In: Lee X et al (eds) Handbook of micrometeorology. Kluwer Academic, Dordrecht/Boston/London, pp 119–132

Leuning R (2007) The correct form of the Webb, Pearman and Leuning equation for eddy fluxes of trace gases in steady and non-steady state, horizontally homogeneous flows. Bound Layer Meteorol 123:263–267

Leuning RL, Moncrieff JB (1990) Eddy covariance CO_2 flux measurements using open and closed path CO_2 analysers: correction for analyser water vapour sensitivity and damping of fluctuations in air sampling tubes. Bound Layer Meteorol 53:63–76

Martini L, Stark B, Hunsalz G (1973) Elektronisches Lyman-Alpha-Feuchtigkeitsmessgerät. Z Meteorol 23:313–322

McBean GA (1972) Instrument requirements for eddy correlation measurements. J Appl Meteorol 11:1078–1084

McMillen RT (1988) An eddy correlation technique with extended applicability to non-simple terrain. Bound Layer Meteorol 43:231–245

Mitsuta Y (1966) Sonic anemometer-thermometer for general use. J Meteorol Soc Jpn Ser II 44:12–24

Moncrieff J (2004) Surface turbulent fluxes. In: Kabat P et al (eds) Vegetation, water, humans and the climate. A new perspective on an interactive system. Springer, Berlin/Heidelberg, pp 173–182

Moncrieff JB, Massheder JM, DeBruin H, Elbers J, Friborg T, Heusinkveld B, Kabat P, Scott S, Søgaard H, Verhoef A (1997a) A system to measure surface fluxes of momentum, sensible heat, water vapor and carbon dioxide. J Hydrol 188–189:589–611

Moncrieff JB, Valentini R, Greco S, Seufert G, Ciccioli P (1997b) Trace gas exchange over terrestrial ecosystems: methods and perspectives in micrometeorology. J Exp Bot 48:1133–1142

Montgomery RB (1948) Vertical eddy flux of heat in the atmosphere. J Meteorol 5:265–274

Moore CJ (1986) Frequency response corrections for eddy correlation systems. Bound Layer Meteorol 37:17–35

Obukhov AM (1951) Charakteristiki mikrostruktury vetra v prizemnom sloje atmosfery (Characteristics of the micro-structure of the wind in the surface layer of the atmosphere). Izv AN SSSR ser Geofiz 3:49–68

Ohtaki E, Matsui T (1982) Infrared device for simultaneous measurement of fluctuations of atmospheric carbon dioxide and water vapor. Bound Layer Meteorol 24:109–119

Raupach MR (1978) Infrared fluctuation hygrometer in the atmospheric surface layer. Q J R Meteorol Soc 104:309–322

Schmid HP, Oke TR (1990) A model to estimate the source area contributing to turbulent exchange in the surface layer over patchy terrain. Q J R Meteorol Soc 116:965–988

Schotanus P, Nieuwstadt FTM, DeBruin HAR (1983) Temperature measurement with a sonic anemometer and its application to heat and moisture fluctuations. Bound Layer Meteorol 26:81–93

Schotland RM (1955) The measurement of wind velocity by sonic waves. J Meteorol 12: 386–390

Schuepp PH, Leclerc MY, MacPherson JI, Desjardins RL (1990) Footprint prediction of scalar fluxes from analytical solutions of the diffusion equation. Bound Layer Meteorol 50:355–373

Sellers PJ, Hall FG, Asrar G, Strebel DE, Murphy RE (1988) The first ISLSCP field experiment (FIFE). Bull Am Meteorol Soc 69:22–27

Smith KA, Clayton H, Arah JRM, Christensen S, Ambus P, Fowler D, Hargreaves KJ, Skiba U, Harris GW, Wienhold FG, Klemedtsson L, Galle B (1994) Micrometeorological and chamber methods for measurement of nitrous oxide fluxes between soils and the atmosphere: overview and conclusions. J Geophys Res 99:16541–16548

Spirig C, Neftel A, Ammann C, Dommen J, Grabmer W, Thielmann A, Schaub A, Beauchamp J, Wisthaler A, Hansel A (2005) Eddy covariance flux measurements of biogenic VOCs during ECHO 2003 using proton transfer reaction mass spectrometry. Atmos Chem Phys 5:465–481

Stull RB (1988) An Introduction to boundary layer meteorology. Kluwer Academic, Dordrecht/Boston/London, 666 pp

Suomi VE (1957) Sonic anemometer – University of Wisconsin. In: Lettau HH, Davidson B (eds) Exploring the atmosphere's first mile. Pergamon Press, London/New York, pp 256–266

Swinbank WC (1951) The measurement of vertical transfer of heat and water vapor by eddies in the lower atmosphere. J Meteorol 8:135–145

Taylor GI (1938) The spectrum of turbulence. Proc R Soc Lond A164(919):476–490

Tsvang LR, Fedorov MM, Kader BA, Zubkovskii SL, Foken T, Richter SH, Zelený J (1991) Turbulent exchange over a surface with chessboard-type inhomogeneities. Bound Layer Meteorol 55:141–160

Vickers D, Mahrt L (1997) Quality control and flux sampling problems for tower and aircraft data. J Atmos Ocean Technol 14:512–526

Webb EK, Pearman GI, Leuning R (1980) Correction of the flux measurements for density effects due to heat and water vapour transfer. Q J R Meteorol Soc 106:85–100

Zhang SF, Wyngaard JC, Businger JA, Oncley SP (1986) Response characteristics of the U.W. sonic anemometer. J Atmos Ocean Technol 2:548–558

Chapter 2
Measurement, Tower, and Site Design Considerations

J. William Munger, Henry W. Loescher, and Hongyan Luo

2.1 Introduction

Although the number of sites making eddy-covariance (EC) CO_2 flux measurements throughout the world has increased rapidly over the last two decades it is still a challenge to define and build a new system. There are myriad options for tower design and placement and a steadily growing range of instrument options and configurations. Selecting among these options is based on finding an optimal solution that best achieves the precision and accuracy required to satisfy the scientific objectives for a site and often for the lowest installation and operational costs. Site design is only the first step to ensuring accuracy and precision of the results. Site operation must also include a program of quality assurance tests that verify whether the measurement system *as installed* is operating within the accuracy and precision goals over time. This calibration and validation is essential in knowing its overall performance, associated uncertainties, and for confidence in comparing data among different sites and within the same site over the duration of its operation. In this chapter, we provide theoretical basis and practical guidance for tower location and design and advice on instrument selection, installation, and operation.

J.W. Munger (✉)
School of Engineering and Applied Sciences, and Department of Earth and Planetary Sciences, Harvard University, Cambridge, MA, USA

H.W. Loescher • H. Luo
National Ecological Observatory Network, Boulder, CO 80301, USA

Institute of Alpine and Arctic Research (INSTAAR), University of Colorado, Boulder, CO 80303, USA
e-mail: hloescher@neoninc.org; hluo@neoninc.org

M. Aubinet et al. (eds.), *Eddy Covariance: A Practical Guide to Measurement and Data Analysis*, Springer Atmospheric Sciences, DOI 10.1007/978-94-007-2351-1_2,

2.2 Tower Considerations

One of the most important decisions is where to put a tower and the tower design and measurement locations. Inevitably there will be compromises between science requirements, engineering standards, cost and practicality. Although there is no ideal tower design or location, we point here to some guidelines based on theory and practice, and collected wisdom from the micrometeorology/biometeorological/surface-layer research communities. Installing a new tower site becomes a task of how to best minimize the systematic biases caused by a large suite of potential flow distortions and maintain the ecologic integrity of the site. Here we describe the source of these biases and present guidelines to optimize the scientific integrity of a research program employing tower-based measurements of eddy covariance (turbulent and gradient approaches) and micrometeorology. Understanding how the presence of tower and tower design affect the surrounding flows (wind) and microclimate, and identifying the issues to optimize the tower size, height, placement, physical properties, and orientation in order to minimize these effects are also the subject of this chapter section.

2.2.1 Theoretical Considerations for Tower Design

2.2.1.1 Diverse Ecosystems and Environments

Measuring an ecosystem microclimate and scalar exchanges above the canopy presents a unique suite of challenges. Ecosystems around the globe are structurally and functionally diverse as well as they are found in all the environmental extremes. Towers need to be designed to best capture the ecological drivers and processes from complex forest ecosystems to relatively simple grasslands. Towers, supporting infrastructure and instrumentation, have to be robust enough to withstand the expected environmental extremes over their lifetimes. In addition, the tower needs to provide year-round, safe access to instrumentation by technicians in extreme conditions, such as high temperatures, 0–100% relative humidity (RH), ice and snow loading, high winds, lightning, and nesting birds and insects; all of which provide unique design challenges.

2.2.1.2 Physical Effects on Surrounding Flows Due to the Presence of Tower Structure

The need to both make measurements through a plant canopy and to access the well-mixed surface layer above canopy presents the challenging requirement for a stable nonmoving platform that is inside a flexible and moving plant canopy. Though this is most physically challenging for measurements over forest canopies, short stature crops and grasslands are not free from these issues. Several types of distortion

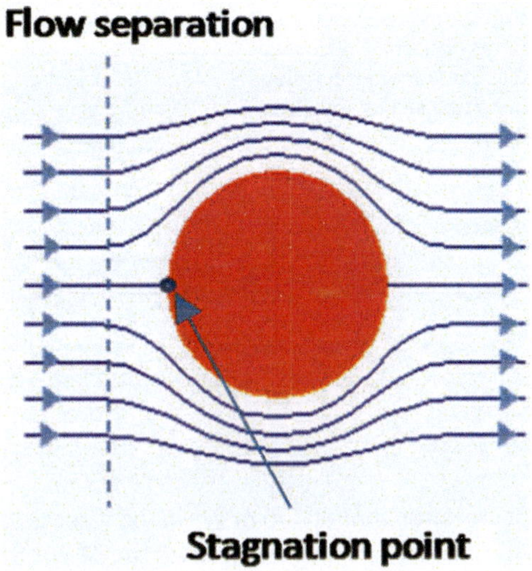

Fig. 2.1 Conceptual diagram of flow streamlines around an obstacle as viewed perpendicular to the flow (i.e., from the top of a vertical post). The point upstream where streamlines start to diverge is indicated by a *dashed line*. The stagnation point where flow velocity reaches zero is indicated by a *dot*

(streamline, wake, and chimney effects, etc.) can affect tower-based measurements. Each of these has to be evaluated for specific ecosystem types (structure and the environmental conditions) that a tower is to be placed into.

Wind Streamline Flow Distortion on the Windward Side of the Tower

The tower structure presents an obstacle to airflow and distorts the wind velocity and direction nearby. A schematic view of flow around an obstacle (Fig. 2.1) shows how the streamlines separate upstream of the obstacle. A stagnation point (windspeed = 0) forms in the upwind side of the obstacle due to the increased pressure field by the wind striking the obstacle. The flow is distorted for some distance downstream of the obstacle as well. Note the upstream distance affected by flow distortion increases with the size of the obstacle (Akabayashi et al. 1986). Wind speed decreases were observed at a distance of 1 tower diameter in front of and on the windward side of the tower at a constant 9.2 m s^{-1} wind speed inside a wind tunnel (Cermak and Horn 1968). In the windward side of an ocean-deployed tower platform, wind speed was decelerated by up to 30% within 1 tower diameter away from the tower (Thornthwaite et al. 1962, 1965). Recirculation flows were also observed within the region of separation downstream (Davies and Miller 1982) (Fig. 2.2).

Wind Flow Distortion on the Lee Side of the Tower

Wind speeds are attenuated on the lee side of a tower, that is, the wake area. Moses and Daubek (1961) reported up to 50% decrease in wind speed for light winds and a

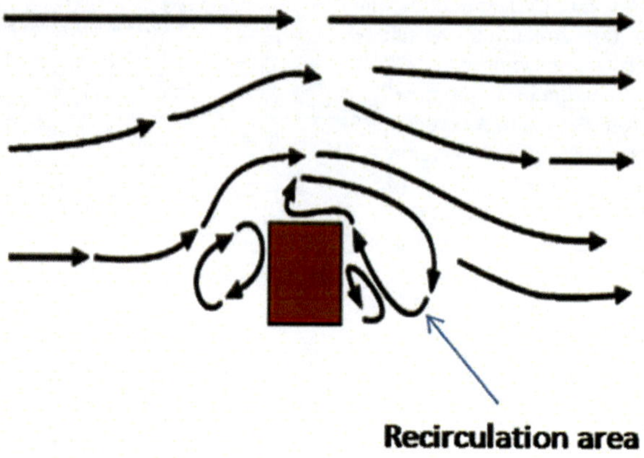

Fig. 2.2 View from the side of the obstacle. Conceptual diagram for windward flow distortion and lee-side flow distortion

decrease of 25% at 4–6 m s^{-1} in the wake area from a 2 m × 2 m cross section tower. Similarly, wind speed decreased by 7% for 8–12 m s^{-1} in another report (Shinohara 1958). Wake effects behind an obstacle apparently reduce in size and magnitude as wind speed increased because of more rapid restructuring of turbulent flow. In a wind tunnel experiment, a decrease of 40% at a distance within 2 tower diameters at a 9.2 m s^{-1} wind speed was also observed (Cermak and Horn 1968) (Fig. 2.3). This wake effect occurred in a well-defined, constrained ±30° sector from centerline, downwind. The wake effects are affected by the size (both length and width) of the obstacle.

Wind Flow Distortion on the Sides of the Tower

As air flows around a tower, the tower changes the flow streamline, and results in the acceleration of wind around the sides of tower. This acceleration is due to decreased pressure on both tower sides and act as a jet (Munson et al. 1998). The maximum accelerated wind speed within both jets was 18% on an in situ "boxlike" platform (Thornthwaite et al. 1962, 1965). Moreover, the increase in flow along the tower sides, up to 19%, was observed in another experiment (Dabberdt 1968). In a wind tunnel experiment, flow around the tower sides was increased up to 6% at distance between 1 and 2 obstacle diameters (Cermak and Horn 1968).

Wind Flow Distortion at the Top of the Tower and Chimney Effects

When air flow passes a vertical obstacle (tower in our case), it separates and accelerates around the obstacle sides and top. The separated flows also accelerate vertically along the wall of the obstacle (Fig. 2.4). The upward deflection and acceleration of winds in the windward side of a tower were observed (Fig. 2.5, Sanuki and Tsuda 1957). Wind speed acceleration at the upwind leading edge of a boxlike platform was observed up to 40% (Thornthwaite et al. 1962, 1965).

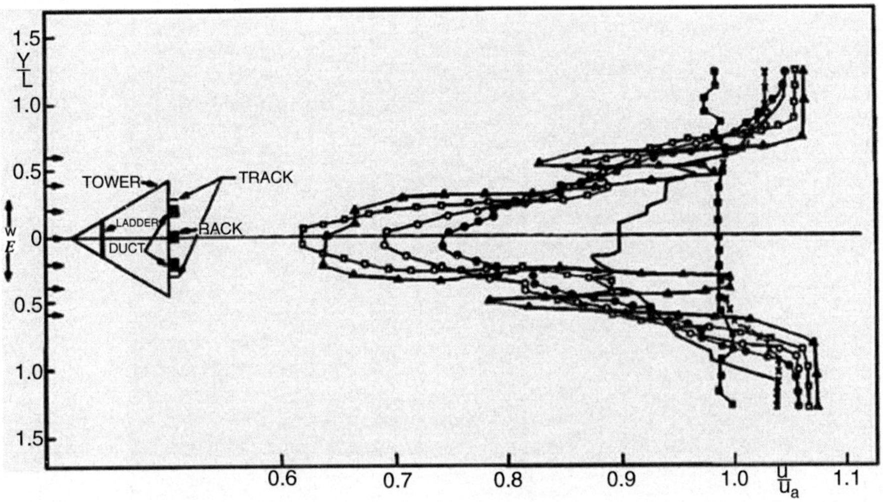

Fig. 2.3 Transverse velocity profiles, tower model, showing lateral variation across the wake for several downstream positions made in a wind tunnel. Additional Note for Figure 2.3: Units for the axis are normalized according to engineering nondimensional analyses, where the *y*-axis is the measured crosswind component (v) and normalized by the physical length scale (L), and the *x*-axis is the measured longitudinal wind speed (u) and normalized by the constant, controlled wind speed in the tunnel (u_a), where u_a was 9.2 m s^{-1} and mounting was at 0°. *Solid square*, station −1 (−37.5 cm); *cross*, station 0, (0.00 cm); *solid triangle*, station 1 (37.5 cm); *open square*, station 2 (75 cm); *open circle*, station 3 (150 cm); *solid circles*, station 4 (300 cm); *solid line*, station 5 (750 cm) (Reproduced from Cermak and Horn (1968))

Fig. 2.4 View from the side of the obstacle. Conceptual diagram to show the vertical deflection of airflows (chimney effects)

Heating of the tower base and strucuture induces convective circulation that may reinforce the vertical deflection, leading to a strong ‘chimney’ effect that preferentially move air from the near the ground up to the top of tower. This type of effect is a function of how much tower (and foundation) mass is present, its spatial distribution, heat capacity of the tower and foundation, structural shape of the tower,

Fig. 2.5 Flow around an anemometer tower with a roof at a height equal to a tower diameter. Upward deflection and acceleration of winds were observed in the windward side of a tower (Sanuki and Tsuda 1957)

degree of disturbance to the existing plant canopy (openings or clearings made in the ecosystem during tower construction), and the amount of input net radiation to the ecosystem. Any factor in site design or implementing tower-based measurements that alter the natural conditions enhances these effects.

Site disturbance can alter the localized convection around the tower by removing plant material, leaf litter, and disturbing the soil. If the ground heats up more (or foundation for that matter) than prior to site disturbance, the amount of net radiation and albedo will change, generating local circulations around the tower. Similarly, large tower structure, large tower foundations, increased concrete mass, and increased disturbance will all generate additional convection and enforce chimney effects. Concrete foundations of the tower as well as the metal structure heat up faster due to their lower heat storage capacity than the surrounding soils. To reduce this effect, wherever possible, smaller tower structure and smaller concrete foundation are preferred for the tower establishment.

2.2.1.3 Size of Horizontal Supporting Boom

At the 1976 International Turbulence Comparison Experiment, some participants reported mean upflow wind speed of $0.1\ \mathrm{m\,s^{-1}}$ caused by the 0.05 m diameter horizontal support structure (Dyer 1981), which was large enough to invalidate eddy covariance measurements. Therefore, the size of the mounting boom for anemometer also needs to be minimized, and should only be sized to provide a secure and stable measurement platform.

2.2.1.4 Tower Deflection and Oscillations

The physical stability of the tower can affect the measurement of winds and turbulent structure (Barthlott and Fiedler 2003). Because the eddy covariance technique utilizes the covariance between wind speed and scalar concentrations, that is, turbulent fluctuations of temperature, CO_2, H_2O, etc., any movement in the tower that covaries with either the turbulent fluctuations of wind speed or scalar that is in interest, for example, towers that sway with the wind, or wind-induced harmonic motion or vibrations, contributes toward uncertainty in the estimates. The current ability to measure wind speed accurately is 0.02 m s^{-1}, consequently, tower or boom movement are required to be below this threshold and shall not have moments that covary with the wind between 1 and 20 Hz (harmonic effect). Fast response accelerometers can be used to quantify this motion. Movement due to personnel working on the tower can be discounted, because the movement they generate does not covary with wind or scalar exchange and data recording may be suspended anyway while personnel are on the tower servicing instruments. It should also be noted that tower sway makes it uncomfortable for personnel to work on the tower structure.

2.2.1.5 Recirculation Zone at the Opening in a Tall Canopy

After flow passes an obstacle, wake effects due to the pressure gradient form a recirculating flow (Arya 1988) (Fig. 2.2). This flow can be caused by canopy edges (Chen et al. 1990, 1992, 1993a, b), by the creation of openings in the canopy for tower access, or by other structures such as buildings (Fig. 2.6, note the return flow pointed by the lower wind vane, which is opposite to the top wind vane, (Vaucher et al. 2004)). Recirculation areas form in canopy openings with a horizontal length scale (distance) equal to 2–5 canopy heights (i.e., the vertical length scale, Fig. 2.7, Detto et al. 2008). The size of the recirculation area can vary from 1 to 15 canopy heights, depending on the width to height and length to width ratios of the contributing obstacles (Arya 1988). This is still a concern even in nonforest ecosystems, though the affected areas are smaller and the sensors are closer to the ground. The larger the obstacle (tower) size, the larger the tendency to have larger recirculation areas. The returning flow also increases the propensity of up-flows and reinforces the chimney effects, which could significantly bias wind measurements as well as perturb mixing ratio gradients. To avoid the man-made formation of recirculation areas, the size of the openings in canopy should be minimized during the construction and tower placements. Also, the removal of trees and branches, which provide resistance (drag) against the formation of these recirculation areas, should be minimized. Flow recirculation is most obvious in tall stature, forest canopies, but it must still be considered even for short grass and crop canopies. Note that in a short canopy, the scales of support structures and sensors will be larger relative to the canopy height.

Fig. 2.6 An experiment to show the cavity flow with westerly flow (10 m above ground) over the top of a building and easterly return flow at 2 m above ground (Reproduced from Vaucher et al. (2004))

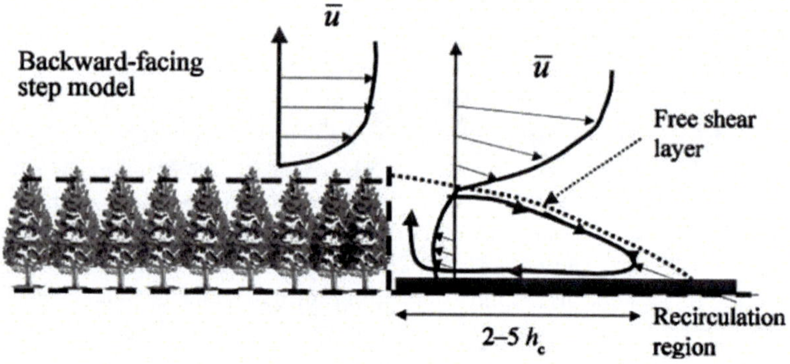

Fig. 2.7 Conceptual model for the structure of turbulence near forest edge (Detto et al. 2008). Recirculation areas form in forest openings with a length (distance) equal to 2–5 canopy height (h_c)

2.2.2 *Tower Design and Science Requirements*

2.2.2.1 Tower Location Requirements

The tower should be located in a representative ecosystem of interest. Micrometeorological requirements include adequate fetch for all desired wind directions and atmospheric stabilities, and should be centered in or on the downwind side of a spatially homogeneous and structurally uniform vegetative canopy, which in practice is often difficult to achieve. The tower (and associated boom orientation) should also be positioned to maximize the exposure time for winds blowing from the desired land cover type, and with the longest upwind fetch attainable. Because some ecosystems do not have a uniform cover type, the prevailing winds, land cover type, and topography should be analyzed to determine the source area under different stabilities, wind speeds, and direction, and will provide valuable guidance for appropriate tower placement (see Chap. 8 and Foken and Leclerc 2004; Horst 2001; Horst and Weil 1992, 1994, 1995; Kormann and Meixner 2001; Schmid 1994; Schmid and Lloyd 1999; Schuepp et al. 1990). In complex terrain, placement of the tower should be situated to minimize flows toward or away from the site and to minimize horizontal flux divergence, advective motions, and drainage in the airshed (Lee 1998; Loescher et al. 2006a; Paw et al. 2000).

Discontinuities in ecosystem structure, which can also affect local circulations, flows, and subsequent measurements, should be avoided in tower site selection. Plant canopies are dynamic and can also alter the structure and increase surface heating through natural disturbance even after a tower has been erected, for example, tree falls, windthrow, and manmade disturbances, for example, logging, harvests, clearings, roads, development, and even the gap created to allow a tower to pass through a plant canopy. Even in micrometeorologically ideal locations (uniform source/sink strength, flat and even terrain, plant canopy, and short roughness lengths), studies have demonstrated that small clear-cuts changed the local circulations during periods of convective turbulence and changed the flow statistics at a tower site (Leclerc et al. 2003; Loescher et al. 2006a). Anomalous flows can also occur and affect tower-based measurements even in short-stature vegetation when the surface conditions are modified from the desired conditions (e.g., homogeneous-managed conditions perturbed by harvest, grazing, mowing, etc.). Even these seemingly small, microscale discontinuities in stature and conditions can perturb wind fields and patterns of latent and sensible heat, or be unrepresentative of the local microclimate, for example, temperatures, long- and short-wave radiation, reflected photosynthetically active radiation (PAR), etc. In another example, low-level jets can descend through the boundary-layer and can alter flows around a tower (Karipot et al. 2008, 2009). These flows, however, are temporally uncommon and can easily be removed from datasets once they are identified. Standing waves can be quite common, and occur with conditions of mechanical turbulence, high wind speeds, and over short stature ecosystems with undulating topography, such as dunes, grass fields, prairie, and tundra. These anomalous flows can cause directional systematic bias in datasets. Even after a site has been chosen and tower has been established, datasets should be examined with rigor periodically. Relocation of tower should be considered if these flows are detected.

Criteria to site a tower are also contingent on the scientific requirements for the research study or program in interest. Minimizing the flows induced by the tower infrastructure toward or away from a site becomes important in reducing uncertainties in EC and local-scale micrometeorological estimates at the seasonal-to-interannual time scales rather than process-based studies. This is because the relative magnitude of uncertainties can change diurnally or alter with synoptic-scale changes in climate.

While there are no uniformly accepted criteria, a general guideline to site a tower is to have adequate fetch to measure the representative ecosystem in question among the expected environmental conditions, that is, ≥80% contribution by the representative ecosystem with the design goal of 90% contribution. Annual and interannual net ecosystem exchange (NEE) estimates benefit from contiguous time series over as much of the source area as possible making the ≥80% contribution of the source area to these measurements paramount. Whereas process-based studies or campaign-based studies can constrain measurements to a predefined suite of environmental conditions, for example, select time period, summer light response curves in unstable atmospheres, or nighttime NEE with $u* > 0.25$ m s^{-1}, and limit the number of averaging periods only during times when the ≥80% criteria is met. Estimating uncertainties during the postprocessing of eddy-covariance data can also be used as a robust diagnostic tool to assess proper tower location (Göckede et al. 2004, 2006, 2008), where footprint and directional analyses, topography, and vegetative mapping can be used in concert to diagnose data quality.

Other ecological criteria may also be important when considering the location of the tower, such as avoidance of wildlife migrations path (e.g., corridor of seasonal caribou movement), or breeding grounds for endangered species. All final decision-making and criteria for specific tower site locations should be documented, archived, linked to best available practice, and attached as metadata to the datasets collected from the tower.

2.2.2.2 Tower Structure Requirements

Most available commercial towers are commonly made of steel or aluminum. Safety and access issues and the need to comply with appropriate regulations and design criteria are not discussed here. Local tower erection companies should be consulted. The material itself, however, is less important than its ability to meet site-specific science requirements, which does include some tradeoffs between size and stability, and flow-distortions and thermally induced chimney effects. Tower with a large projected footprint (>6 m^{-2}) may be very stable and appropriate for large forest structure, but inappropriate for short-stature forested canopies with high stem density. The structural integrity of the tower should have minimum sway, harmonic motion, or vibrations. The tower movement should be <1.0 mm per 1 m in height when subjected to personnel on the tower or with windspeeds equal to of less than 20 m s^{-1}. The tower should not oscillate within the 1–20 Hz frequency. It is not uncommon for towers to fail to operate correctly once weight loading by instruments, ice, and personnel exceeds the structural requirements. Towers must also be capable of supporting measurements made within the range of expected environmental conditions, for example, winds to 40 m s^{-1}, humidity ranging 0–100%, temperature range of –50°C to +50°C, salt air, ice (12.7 mm of accumulated ice), snow (weight, 5.1 m year^{-1} in depth), and rain (0–6.35 m year^{-1}). Hence, safety issues aside, a tower system should be site specifically designed to have sufficient strength and stability to simultaneously withstand the weight loading, applied temperature and/or other accompanying environmental phenomena without experiencing yielding, failure, or detrimental deformation. More specific guidelines and requirements may be mandated by local zoning and permitting procedures, which must be followed. Tower size should be optimized to meet safety and regulatory requirements while also supporting the necessary instrumentation. Excessive tower size increases the local impact on the nearby environment by perturbing the local microclimate and inducing flow distortions noted above.

Tower design and materials should attempt to minimize the thermal mass and reflective surfaces that can alter the radiation environment, that is, short wave, long wave, ultraviolet, infrared, albedo, and temperature spectra. Reducing the exposed thermal mass and minimizing changes to the radiation environment will reduce the propensity of localized tower-induced convection and chimney effects. The tower structure should also not create safe harbor for stinging insects or dangerous animals, for example, snakes, scorpions, raccoons, bears, and hunters. Fences climbing prevention hardware, and appropriate warning signs are generally needed to prevent unauthorized human access.

2.2.2.3 Tower Height Requirements

Towers must have the ability to access the environment through and above the plant canopy. The tower shall be high enough to place the sensors at top layer well above the surrounding plant canopy in the well-mixed surface layer, but not so high that the footprint during stable night-time conditions extends beyond the boundary layer above the ecosystem of interest. Nor shall the tower height be too low such that the tower top measurements are influenced by the roughness layer or individual canopies close to the tower.

Because of the wide range of structural and functional diversity of ecosystems, two separate criteria will be applied to determine the tower height for the top measurement level: (1) a fixed tower-measurement height (h_m) of 6 m above all grasslands (or shrublands) where $h_m > [d + 4(h_c - d)]$ and $h_c \leq 1.75$ m, h_c is the mean canopy height and d is the zero plane displacement height (Monteith and Unsworth 2008), and (2) $h_m \approx d + 4(h_c - d)$ over forested or more structurally complex ecosystems. Both criteria are founded on the work from multiple studies (e.g. Dyer and Hicks 1970; Hicks 1976; Lemon 1960; Monin and Obukhov 1954). If a research program has many towers among several ecosystem types, the criteria used should maintain consistency and uniformity among tower measurement heights within any particular ecosystem type to provide regional and spatial comparable data with similar uncertainties.

There will be some instances when the canopy height will change over time (e.g., actively growing young forests and crops). In these cases, the tower design needs to incorporate the capability of changing tower height and moving sensors over time. Eddy covariance measurements should be maintained in the same turbulent structure at $d + 4\ (h_c - d)$ to maintain the same relationship with ecosystem structure for any gradient measurements, and maintain the same source area for any downward facing sensors. For sites that are actively accruing canopy height, plan to construct the tower at least 5 $(h_c - d)$, but mount the sensors at $d + 4(h_c - d)$ at inception. The tower and measurement height shall be changed at a convenient time of the year when the height of the sensors (h_m) is $< d + 3.6(h_c - d)$. For crops with changing canopy heights that remain below 3 m, 8 m is the suggested measurement height. Horizontal wind profiles measured at the tower sites can be used to determine d.

2.2.2.4 Tower Size Requirements

Tower size (horizontal dimensions, not height) shall be large enough to be safe and secure for many years of operation, but should also limit the impact on the surrounding environment and scientific measurements of interest and minimize the flow distortion as described earlier in Sect. 2.2.1.2.

Large tower structures create larger canopy openings, which promote wind recirculation (see Fig. 2.7), enforcing chimney effect, changing the local microclimate, and, for the biological concerns, introducing opportunistic plant species, which will locally alter ecosystem structure around tower due to the radiation and temperature changes at the opening (edge effects). Minimizing the tower foundation and canopy opening shall be implemented to limit disturbance and to mimic the natural environment. For the same reason, adjacent vegetation shall not be removed or disturbed unless absolutely necessary.

There are site-specific interactions between the tower presence, ecosystem type and structure, and local microclimate. As a general guideline for tower design and establishment, the spacing between tower and nearby trees should mimic the existing mean distance between trees (i.e., mimicking the existing natural ecosystem structures and openings).

In order to minimize tower-based uncertainties, a custom-tailored tower design to site-specific environmental conditions is ideal, but not practical. To reduce the uncertainties of tower effects on the same suite of measurements made among multiple tower sites within a certain research program, the projected tower base is recommended to be no larger than 4 m^2, for example, 2 m $\times$ 2 m, because (1) sources of uncertainty caused by the tower design among all sites will be similar and (2) this will enhance the interchangeability of all sensors and supporting hardware, and uniform tower health and safety training. Special consideration can be made to increase the tower structural members (and size of the projected base) for closed plant canopies if the average projected tree crown area is $>6\times$ the tower projected area (4 m^2), that is, >24 m^2, canopy cranes. On the other hand, at nonforested sites, structural elements should be minimized because there is no canopy to mask the tower.

Tower shape also has some bearing on the ability to meet site-specific criteria and scientific requirements. Some researchers prefer triangular climb-up towers (antennae-style) because they are light weight and can be easily transported into remote areas, small foundations are needed, small holes through the plant canopy can be created, and minimal impact on the surrounding microclimate can be achieved. They do, however, have limited expansion room for additional science and instrumentation. It is important to properly account for the weight and especially surface area of instrumentation on the tower to not exceed the design specifications. Furthermore, since the tower structure serves as the climbing structure, design and placement of instruments, booms, and cables needs to be considered to avoid interfering with safe tower access. Walk-up scaffolding-style towers are larger with likely more flow distortions, create large canopy access holes, and require larger foundations. They do, however, allow simpler instrument-mounting options without interfering with tower access (up to the limits imposed by not having sensors interfere with one another), and personnel may be more comfortable using stairs rather than ladder-style climbing. In either case, appropriate fall-restraint systems are necessary. Selecting a tower style is partly a matter of taste and optimization of construction and operation costs against scientific returns, and matching the tower design to the specifics of local ecosystem structure. The considerations noted above provide guidance on measurement issues to be considered.

2.2.2.5 Instrument Orientation Requirements

The tower, instrument placement, and overall design should minimize any disturbance to the radiation and other microclimatic environment of interest (Culf et al. 1995, 1996). Challenges to measure the surface layer (and microclimate within a plant canopy, Sect. 2.5) occur when, by necessity, the tower and supporting structure have to be fixed, stable, and surrounded by a flexible plant canopy. The tower and booms that extend horizontally out from the tower superstructure have to be positioned within, close to, and able to assess the ecological strata of interest (ecosystem structure, microclimate, etc.). This will be partially dependent on minimizing the gap size in the plant canopy created by the tower. All the meteorological measurements (with the exception of radiation) should be mounted on a stable horizontal boom with a minimum distance that is no less than the length of 2x the face-width of the tower. Anemometers should not be mounted on the tower sides, or in the wind wake area, and should be placed on the windward side of tower on a stable horizontal boom. Gill et al. (1967) suggested that, to achieve wind speed measurements, accurate to within 5%, anemometers should located no less than 2 tower diameters from an open lattice cylindrical obstacle. This finding can also be applied to the face-width of square, rectangular, or triangular towers. The windward side is the one facing the site-specific predominant wind direction, which is where booms should be mounted for EC measurements. However, for sites with distinct differences between daytime and nighttime wind directions, the preferred orientation for anemometers should be optimized to measure the daytime winds. Wind measurements made when the anemometer is not facing the streamline need to be inspected for biases and distortions. If anemometers are mounted directly above the tower top (not recommended), they must be mounted at least 5 tower diameters above the tower top (Perrin et al. 2007).

2.2.2.6 Tower Installation and Site Impact Requirements

During tower construction, installation, and operation, extreme care shall be given to minimize any impacts on the surrounding environment in order to minimize perturbations to the ecological variables that we wish to measure, for example, small canopy openings, reduced thermal mass of tower structure and foundation, boardwalks not in the view shed of sensors, etc. Extensive clearing around the base of the tower or use of large construction equipment should not be permitted. Boardwalks should be considered if impact to the surround soil and plants is expected to increase seasonally and over time.

Although the tower structure can be robust, special consideration is needed to address site-specific conditions, for example, sites with marine salts or areas with sand storms may need more frequent painting and protection from rusting. Higher winds in alpine environments may require a stronger, well-guyed tower to maintain the stability requirements, while other sites may not allow guy wires because they interfere with migrating birds. Regularly scheduled tower inspection and preventative maintenance according to tower manufacturer recommendations are essential for assuring site reliability and safety of personnel. Guy wires may be applied wherever permitted to secure tower stability and to withstand high winds, though guy wires can become a source of tower failure in tall-stature ecosystems if trees or branches frequently fall and are massive enough to break a guy. Cross hatched guys, or cross braced, and free of any contact with trees or branches are recommended to ensure that the tower is stable and safe.

In many locations lightning is common and measures need to be taken to minimize the potential for instrument damage and loss of data. Proper grounding of the tower, guy wires, anchors, and buildings is an essential component of construction. Induced voltages could occur in long signal wires from sensors to data acquisition systems. It is inexpensive insurance to place surge voltage protectors (varistors, suppression diodes, gas-discharge tubes) or optical isolators on each signal or control line, including serial and network communications. An excellent connection to earth ground is essential for diverting surges and avoiding buildup of stray voltages. Obviously, site personnel should never risk their lives by working on or around a tower when lightning storms are nearby.

Many towers require a shelter near the base of the tower to house instrumentation, gas cylinders, and supporting equipment. Its structure and placement should not affect the local biotic and abiotic environment of interest. Placement of the shelter can be adjacent to the tower if there are closed-canopy conditions which can shield the tower measurements from changes in microclimate caused by the shelter, for example, reflected radiation, changes in turbulent structure, heat, traffic, etc. Otherwise the shelter should be located away from the tower. In order to rigorously minimize impacts of the shelter on measurements, the location of the instrument shelter shall be on the prevailing leeward side of the tower (site specific). The distance between the tower structure and the instrument shelter shall apply a 5:1 ratio of tower-shelter horizontal separation to shelter height (optimum for grasslands) or 3:1 ratio (minimum for closed forest canopies). Exterior color should mimic the color and environmental reflectivity (albedo) of the surrounding landscape, and the roof design, type, and slope should minimize perturbations to air flows affecting the tower measurements.

2.3 Sonic Anemometer

2.3.1 General Principles

Eddy covariance flux measurements are based on determining the correlation between changes in vertical wind velocity and deviations in a scalar quantity such as mixing ratio of a trace gas or air temperature (see Chap. 1). The measurements must be frequent enough to capture the variability due to atmospheric turbulence, which is typically >1–10 Hz depending on the surface characteristics as discussed in Chap. 4. The principle of sonic anemometry-thermometry (SAT) was demonstrated prior to the 1960s (Kaimal and Businger 1963a, b), and was developed into more robust field deployable instruments in the 1970s (Campbell and Unsworth 1979). Availability of reliable and relatively inexpensive three-dimensional SAT was a key technology allowing the extensive networks of CO_2 flux measurements that exist today. The basic principle of a SAT is to measure the difference in transit time for an ultrasound pulse between pairs of transducers arranged at a known distance apart (d_{pl}). Transit time (t) is dependent on the speed of sound and velocity of air in its path, hence the difference in the inverse of transit times for sound pulses traveling in opposite directions along the same path depends on the wind velocity (u_{pl}) along the transducer axis and the speed of sound (c) can be derived from the sum of the inverse of transit times:

$$u_{pl} = \frac{d_{pl}}{2}\left(\frac{1}{t_{1,2}} - \frac{1}{t_{2,1}}\right) \tag{2.1}$$

$$c = \frac{d_{pl}}{2}\left(\frac{1}{t_{1,2}} + \frac{1}{t_{2,1}}\right) \tag{2.2}$$

where d_{pl} is the path length and $t_{1,2}$ and $t_{2,1}$ are the transit times from transducer 1 to 2 and 2 to 1, respectively.

Speed of sound is a function of air density, which depends on temperature and the mixing ratio of other gases, especially water vapor. The equations relating speed of sound to temperature are presented in Sect. 3.2.1.1.

2.3.2 Problems and Corrections

Although SAT measurements are grounded in physical principles, several critical issues affect the measurement when applied to field practice. Nearly all these issues are dealt with in the SAT software by theory-based and empirical correction terms. In most cases, users don't need to consider them explicitly, but should be sufficiently aware of these underlying issues to recognize when results might not be valid. The fundamental data measured by SAT is the delay time for a sonic pulse. However, there is finite delay between applying an excitation voltage to a transducer and generation of a sonic pulse. The delay is affected by the transducer temperature and must be accounted for by factory calibration and built-in corrections.

The path that a sonic pulse takes between a pair of transducers is distorted by winds oriented across its axis, giving rise to crosswind contamination of sonic temperature measurements. They must thus be corrected for this effect. Current sonic anemometers include this correction in their firmware, but this is not the case for the older Solent R2 models and the METEK USA1 without a turbulence processor. The correction was first given for anemometers with Cartesian coordinate systems (Schotanus et al. 1983) and recalculated for the omnidirectional probes (Liu et al. 2001):

$$\left(\overline{\omega'\theta}\right)_{\text{corrected}} = \left(\overline{\omega'\theta}\right)_{\text{uncorrected}} + \frac{2\overline{\theta}}{\overline{c^2}}\left(\overline{u}\,\overline{u'\omega'}A + \overline{v}\,\overline{v'\omega'}B\right) \tag{2.3}$$

The coefficients A and B are given in Table 2.1.

Finally, the supporting structure of the SAT can also perturb the flow by blocking portions of the measurement volume or generating small-scale turbulent eddies and wake effects, and as discussed above, the measurement platform itself (boom and tower) obstructs the wind (discussed below). The effects of flow distortion and shadowing by the transducers have been extensively analyzed, (Dyer 1981; Kaimal et al. 1990; Miller et al. 1999; Wyngaard 1981), and SAT manufacturers have incorporated these results into probe design, calibration, and data processing firmware. This point is discussed more in details in Sect. 4.1.5.1. Users don't need to make corrections but these considerations impose limits on the data range that is acceptable for the SAT they are using. In particular, the attack angle (deviation of the wind streamline from horizontal) should be considered and checked that it

Table 2.1 Coefficients for Eq. 2.3 according to Liu et al. (2001), φ: angle between the measuring axis and the horizontal line for different sonic anemometer types currently in use. For most of the recent sonic anemometers, the correction is included in the firmware (except older R2 and USA1 without turbulence processor)

Factor	CSAT3	USA-1	Solent (all other types)	Solent-R2
A	7/8	¾	$1 - 1/2\cos 2\varphi$	½
B	7/8	¾	$1 - 1/2\cos 2\varphi$	1

is within the range that the SAT specifications indicate as valid. Large errors can arise for winds outside that range (Gash and Dolman 2003). Early versions of three dimensional sonic anemometers used an array of transducers arranged on orthogonal axes to measure the three components of wind velocity to simplify construction and directly provide velocities in an orthogonal coordinate system. This turns out to not be an optimal geometry, in part because self-shadowing is large for winds aligned with one of the transducer axes. Most SAT now available employ nonorthogonal configurations. Trigonometric axis transformations are made in signal processing to derive the orthogonal components of each wind vector using results from all the transducer pairs collecting data from the same physical volume (Sect. 3.2.4). Improvements in the design and fabrication of ultrasonic transducers have made them smaller and more reliable.

2.3.3 Requirements for Sonic Choice, Positioning, and Use

Today, several manufacturers provide fast sonic anemometers suitable for flux measurements. Typically SATs cycle through the measurement axes more rapidly than the data reporting rate and the output data are the result of averaging several separate measurements for each axis to reduce noise. Signal processing algorithms can generate data quality flags to identify potential errors. Results from SAT intercomparisons in a controlled environment highlight some distinctions between them and point to key design attributes to consider in selecting an anemometer (Loescher et al. 2005). In general "yoke" style SAT, with transducers arrayed at the end of a horizontal boom, are preferable to "post" style units in which the transducers are arrayed above the support structure creating asymmetrical flow distortion. Deviations were observed even within the manufacturers specified acceptance angle. Even after accounting for the influence of water vapor on temperature inferred from SAT (see Sect. 3.2.1.1), it is not recommended as an accurate absolute temperature measurement, but it can be calibrated against a collocated reliable absolute temperature measurement to correct for any offset that could be imparted by uncertainty in the transducer delay time for instance. After calibration, the absolute temperature from SAT is suitable for inclusion in flux calculations such as computation of molar volume. In most cases, the temperature errors are a constant offset and do not impact the computed variances and covariances. Some models of SAT, however, exhibit a nonlinear dependence of sonic temperature on absolute air temperature so that the fluctuations in temperature ($\theta' = \theta - \overline{\theta}$) are neither symmetric about the mean nor constant as the magnitude of θ changes, resulting in incorrect temperature covariances, making them unreliable for heat-flux measurements. Transformation from buoyancy flux – which is what the SAT delivers – to sensible heat flux is described in Sect. 4.1.2.

SAT specifications are continually evolving, making it impractical to recommend a specific manufacturer and model as ideal. Instead, we provide a set of overall attributes to consider in selecting a new SAT. First of all, a 3-axis SAT is required to make eddy covariance measurements. Two-axis versions that are intended for measuring horizontal wind speed and direction only are not suitable. Measurement accuracy and precision is affected by the quality of the transducers electronics and calibration. Research grade SATs that are suitable for eddy covariance will use better transducers with minimal temperature sensitivity and better electronics components with improved accuracy to measure the very small differences in signal transit time at very low wind speeds. Research-grade SATs will typically have resolution of 0.01 $\mathrm{m\,s^{-1}}$ and 0.01°C, or better, for wind and temperature, respectively, that are required to measure the respective turbulent fluctuations. SAT measurements are affected by local environmental conditions. Because transducer performance is affected by temperature, the data processing and internal calibration tables must account for this temperature dependence in the pulse transit times. Be sure to select a model that has an operating temperature that spans the expected local range, or select optional shifted range as appropriate. Raindrops and ice, which block the transducer path and attenuate the sound pulse, degrade SAT measurements. All SATs will fail in very heavy rain or under icing conditions when the sound pulses are attenuated too much. In light rain conditions (e.g., <0.5 mm $\mathrm{h^{-1}}$), transducer geometry, selection of materials, wicking, and proprietary signal processing algorithms are solutions to minimize data loss. Angled transducer surfaces are less prone to accumulate water droplets that block sound transmission, which is an advantage for nonorthogonal versions compared to orthogonal heads where the transducer face of the vertical axis is horizontal. Additionally, hydrophobic materials and physical design of the transducer to wick droplets away improve performance or speed recovery after the end of rain event. Finally, SAT manufacturers have developed proprietary internal software to improve performance in the face of some signal degradation. In cold environments, optional heating elements to prevent icing on the transducers may be necessary. Measurement accuracy depends on knowing the path length between transducers. The sonic array must be handled with care to avoid bending the support arms, and returned for recalibration if any accident changes the alignment.

The mode of data output and power requirements are additional considerations; SAT data are typically generated in digital form and optionally available as analog output via some internal digital to analog conversion. Other analog signals can be optionally sent to the SAT and digitized by on-board electronics to be included in the SAT output stream providing a way to merge data from other sensors with the wind data. Data logging time stamping and clock maintenance options are discussed in more detail in Sect. 2.1.

Although the SAT provides wind speed and direction, an independent measurement of wind speed and direction is desirable for comparison and redundancy. The traditional wind vane and spinning cup anemometer is a typical approach for wind measurement. Alternative configurations use an integrated vane and propeller configuration. Two-axis sonic anemometers that are priced competitively with the best quality mechanical anemometers are available. All mechanical anemometers need a minimal wind speed (typically 0.1–0.2 $\mathrm{ms^{-1}}$) to overcome inertia of the vane and the cup (or propeller). Minimizing the mass of the components and friction in the moving parts reduces the threshold, though it can increase fragility of the device. Optimizing this tradeoff depends on the range of wind speeds likely to be encountered at a site, and how dependent the science questions are to the accuracy of low wind speed measurements. Wind sensors need to be located upwind of supporting structures to minimize artifacts and errors due to flow distortion and wake effects. A single measurement at an arbitrary height above the canopy has limited value as a network measurement. Multiple measurements that observe the wind speed profile near the canopy interface can be analyzed to define drag coefficients, roughness lengths, and zero plane displacement height based on the assumption of logarithmic wind profile. In order to provide a cross-check on SAT measurements, it is essential to perform sensor maintenance according to manufacturer guidelines and monitor the data itself to check for changes in the low-speed threshold. Moving parts, such as bearings, will wear over time and must be replaced periodically to ensure consistent data.

Corroboration of sonic temperature would require a secondary measurement of ambient temperature. Precise and accurate temperature measurements are possible using any of the typical temperature-sensitive devices (thermistors, thermocouples, platinum resistance thermometers) as incorporated in commercially available temperature probes in conjunction with appropriate signal conditioning and data logger connections that will be described in sensor documentation. Proper shielding of the measurement sensor from solar heating and radiative cooling is essential for unbiased temperature measurements. Fan-aspirated radiation shields provide the most effective radiation shielding and will reduce heating errors to 0.1° or less, independent of wind speed and radiation intensity.

In addition to comparing horizontal wind and sonic temperature against independent wind and air temperature measurements, tracking the ratios of variance and means of the wind components helps to identify sudden changes in performance or wind sectors with anomalous data that should be investigated (Tropea et al. 2007). Some SATs include a "zeroing chamber" that can fit over the transducer array without interfering with the signals. This chamber is used to verify that none of the transducer pairs have a zero-offset when measuring in a zero-wind condition, and to adjust the internal constants if necessary. Data quality checks are discussed in more detail in Sect. 4.3.

The characteristics of sonic anemometers, as well as of gas analyzers (see below), and their location make them act as filters that remove high- and low-frequency components of atmospheric signals and reduce the magnitude of the measured flux. Several correction procedures to account for lost flux components are presented in Sect. 4.1.3. Here we discuss design considerations to minimize the magnitude of these corrections. Spatial averaging along the path length of the sonic anemometer or a gas analyzer and spatial averaging due to the separation between velocity and scalar sensors are important causes of low-pass filtering. Therefore, path length and separation should be always small in relation to the size of the turbulent eddies. Because eddy size scales with height above the surface, anemometers with short path lengths are required close to the ground. A "rule-of-thumb" to avoid significant high-cut frequency corrections is for the path length to be 1/20 times the measurement height (e.g., using an anemometer with a 15 cm path length only at heights greater than 3 m), though Kristensen and Fitzjarrald (1984) show reasonable flux measurements down to only a few times the height using single-axis (vertical) sonic anemometers. Van Dijk (2002) later modified this conclusion to show that three-dimensional sonic anemometers (currently in use) have more path averaging and need to be deployed at higher position than had been previously recommended. Similarly, the path length in open-path CO_2-H_2O sensors (see below) and the separation between anemometer and scalar sensor or inlet dictate minimum height-above-surface specifications.

2.4 Eddy CO_2/H_2O Analyzer

2.4.1 General Description

The second component of a system to determine CO_2 and water-vapor fluxes is a fast-response analyzer for measuring turbulent fluctuations in CO_2 and H_2O molar concentrations at high frequency. Currently, most sites use a nondispersive infrared absorption analyzer (commonly referred to as infrared gas analyzer – IRGA), in either an open- or closed-path configuration. The relative advantages and disadvantages of these systems will be discussed in Sect. 2.4.4. For either system, the measurement scheme consists of a broadband IR light source, band-pass filters (rather than a monochromator or other *dispersive* device to select wavelength) to select a wavelength range that spans absorption lines for CO_2 and water vapor, and a detector. Light is absorbed by CO_2 and H_2O in the light path, and the reduced intensity observed by the detector is a nonlinear function of the molar concentration of CO_2 and H_2O. A closed-path analyzer has an internal sample cell (optical bench) that is flushed by sampled air while in open-path sensors the sample cell is in the open air. In order to account for variations in the light source intensity and detector response, light absorption is evaluated by comparing the detector signal with a reference signal. In the closed-path analyzer, the reference signal is measured using a second cell purged by a small flow of air with known (can be zero) CO_2 and H_2O molar concentration. In the open-path sensor, intensity of light at an adjacent nonabsorbing wavelength is used as reference signal.

Detector signals are converted to mixing ratios using a calibration equation and constants (see Sect. 3.2.1.2) and accounting for density of air at pressure and temperature in the sample cell. In open-path analyzers, temperature and pressure vary with ambient conditions, so their fluctuation has to be accounted for through the so-called Webb-Pearman-Leuning (WPL) density corrections (Webb et al. 1980, Sect. 4.1.4.2). In closed paths, temperature and pressure within the sample cell are different from the ambient conditions but can be precisely controlled to constant values, reducing the need to account for temperature fluctuations in the WPL density correction. In both systems, however, the dilution correction should take water vapor fluctuations into account (Sect. 4.1.4.4).

In addition to dilution effects, a correction for spectral interference should also be considered. The underlying spectroscopic details of this are beyond the scope of this chapter, but the proportionality between light absorbance and density depends on the temperature, the pressure and the composition of the sample matrix, especially its water content. At higher mixing ratios, the gain (change in absorbance for a unit change in density of the analyzed gas) tends to decline (in part due to band broadening). McDermitt et al. (1993) derived a calibration function for IRGAs based on nonoverlapping line approximation that includes the influence of pressure, temperature, and water vapor on the CO_2 signal and assuming broadening coefficients for dry air. This correction is generally incorporated in factory calibrations and should not be introduced in the standard data treatment. If very accurate absolute CO_2 mixing ratios are required, the application of this equation to raw signals would be recommended (Sect. 2.4.2.3). In any case, it is important to realize that temperature and pressure fluctuations affect the computed mixing ratio so that they are held to a minimum. Water vapor fluctuations affect computed mixing ratios as well and need to be quantified both to compute the water-vapor flux (latent heat) and to accurately account for the water vapor affect on CO_2 mixing ratio by the spectral corrections included in factory calibration and the dilution corrections applied in data processing.

2.4.2 *Closed-Path System*

2.4.2.1 Absolute and Differential Mode

Closed-path system may run in either *absolute mode*, if the reference has zero CO_2 and H_2O concentrations, or *differential mode***,** if the reference has constant molar concentrations near ambient conditions. In absolute mode, dry CO_2-free, purge gas is achieved using a cylinder of compressed nitrogen or CO_2-free air, a purge-gas generator, or chemical scrubbers in line with a compressor pump. In the latter case, this requires some attention to replenish as they are used up. In differential mode, a cylinder of compressed gas of required mixing ratio is needed. Operationally, absolute mode is simpler, but if the analyzer zero is being recorded as part of routine calibration (see below), the data recording must span a wide range from zero to >400 ppm CO_2. For either mode of operation, the flow rate necessary to purge the reference cell is typically only a few $cm^3 min^{-1}$, so gas cylinders can last a long time or power requirements for compressors or zero-air generators are modest. But, if routine calibration includes a zero check, the capacity of the purge-gas source to deliver adequate flow rates to flush the sample cell must be considered. In differential mode the data only span a narrow range centered on the ambient mean mixing ratio; for applications using analog-to-digital converters for data logging this allows better signal resolution.

2.4.2.2 Tubing Requirements for Closed-Path Sensors

A closed-path sensor needs to have sample air brought to it. The presence of tubing introduces possibility for chemical and physical alteration of the air as well as attenuation of high-frequency variation. The following requirements are given to minimize these artifacts and limit the magnitude of correction terms:

1. Minimize attenuation of high-frequency variability throughout the sampling system
2. Avoid water condensation within tubes and analyzer
3. Avoid pressure fluctuations and air contamination caused by the pump
4. Comply with the analyzer range of operational parameters
5. Stabilize and monitor the air flow
6. Keep the analyzer chamber clean
7. Avoid generation or loss or the analytes of interest (artifacts)

Air transport through the tube has two main consequences on the measurements: First, the sampling of CO_2 concentration lags that of wind velocity, which has to be accounted for in covariance computation (Sect. 3.2.3.2); secondly, diffusion and physical mixing of the sample stream as air passes through the inlet tubing attenuate high-frequency fluctuations in mixing ratios. In laminar flow conditions ($Re < 2{,}100$, where $Re = \frac{2Q}{\pi r_t \nu'}$ is the Reynolds number, Q is air flow in the tube, r_t is tube radius, and ν is kinematic viscosity of air), a parabolic velocity profile is established with a maximum velocity at the center of the tube that is twice the average velocity of air through the tube. The sample arriving at the CO_2 analyzer is thus a mixture of air that entered the tubing at different times and atmospheric fluctuations are smeared out. The velocity profile for turbulent flow is more constant across the tubing cross section with only a very thin boundary layer at low velocity adjacent to the tubing wall; so there is less physical mixing due to the velocity shear. Thus, choosing the inlet diameter and flow rate to maintain the Reynolds number above a threshold of 3,000–3,500 for maintaining turbulent flow is desirable (Lenschow and Raupach 1991; Leuning and King 1992) to minimize the loss of high-frequency fluctuations by passing through an inlet line (Leuning and King 1992). However, power limitations or site configurations sometimes may preclude achieving a high enough flow rate to maintain turbulent flow. In these conditions, specific corrections for high-frequency losses are required (Sect. 4.1.3.2) and data quality may suffer if the correction magnitude is too large. Even though turbulent flow reduces the physical mixing of sample air, other mechanisms including adsorption of analyte on the tubing wall and mixing induced by fittings and bends in the tubing may still attenuate high-frequency fluctuations; so it is always essential to evaluate the spectra and cospectra to detect anomalies at high frequency and, when necessary, apply appropriate corrections (Sect. 4.1.3.2).

Recommendations for Set Up

The ideal set up consists in an inlet placed as close as possible to the sonic averaging volume, a sampling tube, a mass flow controller, the analyzer, and the pump (Fig. 2.8). The pump needs to be downstream of the analyzer to preserve the variability in mixing ratio, but pressure pulsing from the pump operation must be avoided. This is easily accomplished by including a ballast volume (B) between the analyzer and pump. Operating the analyzer at negative pressure, ~25 kPa below ambient is recommended. Filters are necessary to protect the detector cell from damage by debris or particles and reduce accumulation of material on the tubing walls that may absorb/desorb CO_2 or water vapor. It is recommended to place two filters, a first one at the inlet and a second one near the analyzer for additional protection. For the inlet filter, Teflon© membranes (e.g., Pall Zefluor 47 mm diam, 2 μm pore size, Gelman ACRO 50, 1 μm pore size) in an open-face filter holder are a good option. Because Teflon© is hydrophobic it makes a good barrier to liquid water and does not itself interact with the water vapor in the atmosphere. Accumulated dirt on the filter may interact with water, so changing filters regularly is required even if they are not becoming clogged. The second filter (e.g., Gelman inline Teflon©disk filter) is installed close to the analyzer as a final protection against dirt or liquid water entering the detector cell.

Dead volumes, sharp bends, and restrictions in the tubing between the inlet and the analyzer promote mixing and will further attenuate high-frequency fluctuations. They should be minimized as much as practical by selecting properly sized fittings and configuring plumbing so that smooth curves rather than 90° elbows are possible.

In addition, the tubing material must be considered. Aside from the obvious need to be impervious to damage from UV radiation and possible exposure to extreme temperatures (e.g., polyethylene becomes brittle in cold temperature), the tubing must not interact with the compounds being analyzed. CO_2 is fairly inert on dry surfaces, but H_2O is particularly reactive and tends to equilibrate with surfaces so that ambient fluctuations in its mixing ratio are attenuated by passing through a tube. Hydrophobic materials such as Teflon,© polyethylene, and Synflex are the best choices to minimize wall absorption, but accumulated dust and coatings of semivolatile organics can also absorb water; so keeping the tubing clean by using a filter at the inlet and cleaning or replacing the tubing periodically should be planned.

Finally, when using an inlet tube it is important to avoid conditions that allow condensation anywhere along the path to the analyzer. For example, water vapor may condense in the tubing if it passes through an air conditioned room and outdoor temperature and relative humidity are high. Or in another common example, when the above canopy environment is warm and humid and the analyzer is at the base of the tower with dew-point temperatures. Heating the sample line to maintain temperatures above the ambient dew point at all points or reducing the pressure in the tubing so that the partial pressure of water vapor does not exceed saturation will prevent condensation.

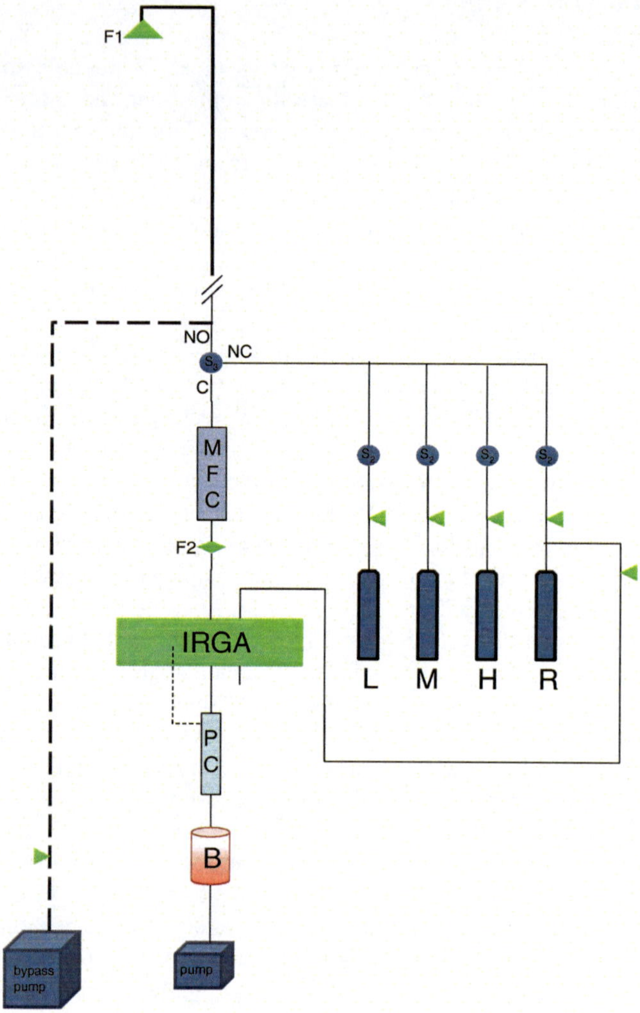

Constraints on Tube Dimensions and Mass Flow

Mass flow (Q), tube length (L_t), and radius (r_t) should be dimensioned in order to reach the best compromise providing the highest Reynolds number, Re and the shortest lag time, $t_l = \frac{L_t \pi r_t^2}{Q}$ possible, while minimizing the pressure drop in the analyzer chamber, $\Delta p \lesssim 8\frac{Q L_t \rho \nu}{\pi r_t^4}$ (using laminar flow as a lower limit).

This implies conflicting constraints on Q and r_t (higher mass flows and lower radius will lead to larger Reynolds numbers and lower lag time but also to higher pressure drops and the need for larger pumps). This clearly indicates the need to reduce the tube length and thus to place the analyzer as close as possible, that is, a few meters, from the tube inlet. When the sampling point is above a tall canopy (e.g., forests), this would mean placement on the tower and would require a more elaborate environmental enclosure to protect the instrument from weather and environmental variability. However, in this case, the system to access would be more difficult, which would impede maintenance and calibration. An alternative would be to place the analyzer at ground level, to flow the air from the canopy top to the ground at high mass flow (e.g., $>10\,\mathrm{l\,min^{-1}}$) through a large radius tube, in order to maximize *Re* without unduly increasing Δp, and to divert a portion of this flow through the analyzer from a junction close to the analyzer at a mass flow sufficient to flush the detector cell. Manually adjustable flow valves or active pressure and flow controllers are required to balance the bypass and sample flows and to maintain cell pressure at the desired values (Fig. 2.8).Whatever the set up that is chosen, a careful analysis of flux cospectra is necessary in order to apply the most relevant high-frequency correction to the flux (Sect. 4.1.3).

To avoid changes in cell pressure and lag times as conditions change (temperature, accumulating dirt on filters, and aging of pumps affect pressure and flow) inclusion of active pressure and flow control elements (mass flow controller) is recommended rather than using manually adjusted restrictions (e.g., needle valves) to set the desired pressure and flow (Fig. 2.8).

Fig. 2.8 Schematic of a CO_2 analyzer configured for a 3-point calibration with zeroing check. Option for bypass flow is indicated by *dashed line* connecting to Bypass pump. Inlet and in-line filter are designated by F1 and F2. S_3 indicates a 3-way switching valve with the Common, Normally Open, and Normally Closed ports marked as C, NO, and NC that is used to select whether the analyzer gets sample or calibration standard. Two-way valves (shutoff) on the calibration gases, indicated by S_2, select which calibration gas is in use. Solenoid valves controlled by a data logger would allow the calibrations to be automated, or they could be performed manually. The cylinder indicated by R would be filled with a gas mixture having approximately ambient CO_2 to operate in differential mode or CO_2-free air (optionally N_2) from a cylinder or produced by a CO_2 scrubber to operate in absolute mode. Flow adjusters/restrictors are designated by *green triangles*. The restrictors on the calibration gases limit the flow so that the mass flow controller (*MFC*) does not have to overcome a sudden change from subambient pressure during sampling to high pressure from the compressed gas standards. Frits, capillaries, or needle valves are suitable devices for the control elements. The *line* from cylinder R to the IRGA is connected to the reference cell and has a restriction to limit its flow to the minimum needed to purge the cell. A pressure controller downstream of the IRGA is used to maintain constant pressure in the sample cell. Either an integrated pressure controller that combines pressure transducer, electronics, and controlling valve, or a separate controller with input from a transducer on the detector cell could be used. A ballast (B) between analyzer and pump damps the pressure and flow oscillations induced by the pump. A control element is shown upstream of the bypass pump to adjust its flow as necessary to achieve the desired overall flow and pressure

2.4.2.3 Calibration for CO_2

Individual instrument calibration constants are derived by fitting the calibration function (McDermitt et al. 1993) to a series of known and traceable standards. The approximations underlying the calibration fit are best met at CO_2 mixing ratios below 1,100 ppm and the temperature and pressure corrections are most accurate below 500 ppm, which is adequate for typical ambient concentrations. In fact, the resolution of the instrument can be enhanced if the calibration is constrained to the range of expected ambient values (350–650 ppm), rather than the full range of the analyzer, which is often 0–3,000 ppm. For eddy covariance fluxes, the mean mixing ratios are subtracted from the observations rendering accuracy of absolute mixing ratios less important than accurate determination of the turbulent fluctuations. Hence, the slope of the calibration curve near the observed mixing ratio rather than the intercept term is most critical. However, the nonlinear response must still be correctly accounted for unless fluxes will be biased because fluctuations in mixing ratio above and below the mean will cause disproportionate instrument response. Even so, increasing the absolute accuracy of mixing ratios adds value to the EC estimates through supplemental activities that can provide additional process-level understanding, for example, advection and transport studies, and enhance scaling activities, for example, inverse modeling; so it is beneficial to make the best calibration that can be achieved with available resources.

The cell pressure and temperature of the gas being analyzed are part of the calibration equation. While these are measured by the analyzer, it is important to realize that the pressure and temperature sensors only measure one point in the analyzer, not necessarily the sample gas that is in the cell. It is essential to minimize temperature and pressure differences between sample and calibration modes and to reduce fluctuations in T and p that induce gradients in the analyzer such that measured T and the conditions within the cell diverge. Enhancing the thermal management and pressure control of the instrument beyond factory defaults provides better analytical results. Protecting the analyzer from solar heating or rapid temperature cycling by air conditioning and heating is a simple minimal step. Ideally, placing the analyzer in a temperature-controlled enclosure that maintains the instrument housing at a relatively stable temperature is the best option. If the analyzer is well calibrated and operated according to the design specifications, the results from internally computed mixing ratio are generally quite accurate. However, there is no substitute for periodic measurement of field calibration standards with traceable known mixing ratios to confirm the accuracy of the instrument's calibration curve and detect problems. A multipoint calibration with at least three points spanning the range of ambient mixing ratio is necessary to verify that sensor nonlinearity is correctly compensated and assure absolute accuracy. Over the typical range of ambient concentrations in the atmosphere, a third-order polynomial, constrained to that range, is adequate to compute CO_2 mixing ratios to better than 0.1 ppm accuracy (Ocheltree and Loescher 2007, Fig. 2.3). If an automated field calibration is being used for closed-path sensors, it is also prudent to have it cycle on intervals that are not even fractions of a day so that the time series is not biased by always removing data at the same time each day.

If a highest absolute accuracy in CO_2 mixing ratios is required, it is best to record raw signals and compute mixing ratios based on calibration against known standards. One important practical consideration is that the dilution gas for CO_2 standards must be air. Using other gases such as N_2 or even synthetic air that has an O_2/N_2 ratio very different from ambient air affects the shape of absorbance bands and violates the simplifying assumptions behind the McDermitt et al. (1993) calibration equation. Although in practice the deviation is small, it should be noted that the IR absorption lines selected in broadband absorption gas analyzers primarily cover the $^{12}CO_2$ lines. Ambient measurements at ^{13}C: ^{12}C isotopic ratios very far from typical ambient levels or calibration against CO_2 standards with isotopic ratios very different from ambient (say, CO_2 from fossil fuel sources) can lead to mixing ratio errors of a few 0.1 ppm. For flux measurements this uncertainty will be inconsequential, but it is a consideration for accurate absolute mixing ratio measurements.

2.4.2.4 Water Vapor Calibration

Water vapor calibration cannot rely on compressed gas standards because gas mixtures with stable water vapor mixing ratios are not available. Dew-point generators, which bubble a flow of air through a temperature-controlled volume of water deliver air with a known water vapor pressure and are used to manually calibrate the H_2O channel of IRGAs (Loescher et al. 2009). It is important to operate the dew-point generator within flow conditions at which the air stream will achieve thermal equilibrium with the water chamber, and avoid deviations in pressure that would affect the resulting saturation vapor pressure of water. An alternative to direct calibration of H_2O is to provide a secondary measurement of absolute humidity (e.g., using a chilled mirror (Loescher et al. 2009)) or compute absolute humidity from ambient temperature and relative humidity measured independently.

The accuracy of water vapor calibration affects the accuracy of measured CO_2 mixing ratios and fluxes through the dilution corrections and WPL term (Sect. 4.1.4) Water vapor corrections are obviously critical for accurate CO_2 flux measurements because H_2O and CO_2 fluxes are typically correlated.

2.4.3 *Open-Path Systems*

2.4.3.1 Installation and Maintenance

The larger physical size of an open-path sensor than the inlet filter for closed-path analyzer presents a flow obstruction. The open-path sensor needs to be far enough away from the SAT that its flow distortion does not interfere with the wind measurement, but not so far that the sensor separation exceeds criteria for minimizing spatial-averaging problems (see Sects. 2.3.3 and 4.1.3.2). As noted above, the separation distance needs to be smaller than the turbulent eddies, and the minimum physical separation that can be achieved sets a lower limit on the height the sensor can be used above the surface without unreasonable loss of high-frequency covariance with the wind. The cospectra should be examined for evidence of high-frequency loss and correction terms evaluated to ensure that their magnitude is reasonable. Accumulated dirt, precipitation, or ice on the sensor windows prevents its operation. To reduce instrument down time, the sensor should be tilted from vertical orientation to promote rapid runoff of droplets. The sensor windows need to be wiped periodically to remove accumulated dust following manufacturer-recommended protocols.

2.4.3.2 Calibration

For open-path sensors, automated routine calibrations are not practical. Periodic manual calibration by purging a chamber placed over the sensor path with gas mixtures having a known CO_2 is possible. It is challenging to ensure a good seal that prevents mixing of outside air with the calibration standard without causing pressure perturbations inside the housing. Alternatively, a secondary calibration by comparison to simultaneous measurements by a second well-calibrated analyzer (e.g., profile CO_2 analyzer, see below) is a reasonable solution that could meet the accuracy requirements for a flux measurement. Regular manual calibrations of open-path sensors using a consistent protocol can help to assure reliable data and detect instrument problems, but the accuracy of open-path calibrations will not match what can be achieved for a closed-path sensor.

2.4.4 Open and Closed Path Advantages and Disadvantages

Open- and closed-path sensors each have advantages and disadvantages. A closed-path sensor can be configured to precisely control temperature and pressure of the sample gas, reducing a potential source of imprecision and avoiding the need to account for covariance in air density and water vapor by including large density correction terms (see Sect. 4.1.4). Secondly, it is straightforward to implement routine automated calibrations supplying known standards to the analyzer to verify the measurement accuracy and precision. The performance of the closed-path sensor is not degraded by adverse weather conditions. A drawback to closed-path sensors is that the necessity of an inlet line induces an attenuation of high-frequency variations and also introduces some delay between when a parcel of air enters the inlet and when it reaches the analyzer. The high-frequency attenuation was discussed in Sect. 2.4.2.2. Corrections for this effect are presented in Sect. 4.1.3. The delay, lag time, must be accounted for when computing covariance between the mixing ratios and vertical wind velocities (see Sect. 3.2.3.2), and is dependent on flow rates and pressure, though it can be determined quite accurately by computing the lagged covariances between CO_2 or H_2O and vertical wind speed or temperature and selecting the time offset that gives the maximum correlation coefficient. The lag for CO_2 and H_2O would be identical if they had no wall interaction. In practice, H_2O tends to stick to the tubing walls more, but large differences in lag are evidence that the tubing or filter is contaminated by hydrophilic material and should be replaced or cleaned. Computing lagged correlations over time to detect changes is a simple and effective measurement quality check.

If high-frequency response or power requirement were the only considerations, open-path sensors would be an ideal solution. However, the gain in high-frequency response and reduced power is made at the expense of increased down time (from rain and inclement weather), and the need to include heat fluxes in the calculation of CO_2 and H_2O fluxes (Chap. 4) (Leuning 2007; Webb et al. 1980; Massman 2004), Leuning 2004), which add additional uncertainties (Chap. 7). Sensor self-heating (or radiational cooling) (Sect. 4.1.5.2, Burba et al. 2008,) may require a correction term and adds additional uncertainty to CO_2 fluxes measured by an open-path sensor that is oriented vertically. Haslwanter et al. (2009) found in a long-term comparison of collocated open and closed-path sensors that there was little overall difference in flux uncertainty. However, when using an open-path sensor for actual fluxes close to zero (no flux), the WPL and Burba corrections can sometimes be several orders of magnitude larger than the flux making estimates of uncertainty (1) among temporal scales, (2) among sites with contrasting conditions, and (3) across different technologies difficult to quantify. Finally, spatial averaging due to sensor path length and separation from the anemometer introduces unacceptable uncertainties in flux for measurement heights too close ($h_m < \sim 3$ m) to the surface. Selection between these two technologies should be based on logistical considerations, individual research requirements, and site characteristics. At sites with frequent precipitation, or large heat fluxes, the improved frequency response by open-path analyzers may not be an acceptable tradeoff. Some research objectives, such as process-based studies as opposed to those needing annual averages, can be accommodated by constraining the collection period to times when the uncertainties imposed by open-path environment are well understood and acceptably small.

The drawbacks of closed-path sensors are minimized by placing the analyzer near the inlet with very short lengths of tubing, but this requires additional engineering to provide adequate protection of the analyzer from weather and temperature variation in the harsh outdoor environment on top of a tower, which would not be needed for analyzers inside a building. On triangular towers, engineering requirements include secure mounting, but it has been successfully accomplished even in the hot, humid tropics. On scaffold towers, mounting is simplified and some protection from precipitation and solar radiation is provided by placing analyzer boxes below the top stage.

A novel CO_2 sensor that was first released in 2010 provides an integrated package with a CO_2 sensor based on open-path technology but enclosing the path may be an excellent alternative to either existing open or closed-path analyzers. The enclosure includes fast temperature and pressure sensors and is flushed by an integrated low-power flow module. This hybrid allows measurements with a minimal inlet and provides electronics intended for outdoor installation without the need for additional environmental control or user modification. The instrument specifications appear to be ideal for flux measurement, but so far it has not been in use long enough to evaluate its actual performance. The short (1–2 m) inlet tube separates the inlet located near the sonic path from the analyzer cell and electronics and is also intended to attenuate ambient temperature fluctuations, thus reducing the density correction term associated with temperature. Additionally, this inlet provides an ideal point to introduce calibration gases by supplying them through a "tee" in excess of the sample volume required.

2.4.5 *Narrow-Band Spectroscopic CO_2 Sensors*

An emerging alternative to broadband absorbance instruments is a new class of analyzers based on laser spectroscopy. Lasers provide light that can be tuned to very narrow frequencies and modulated to scan across individual lines in the IR absorbance spectra of CO_2 and H_2O. Fitting the observed spectra to well-known line strength data bases provides a signal that is nearly linear with respect to the density of CO_2 and H_2O. In practice this method still requires some calibration to account for nonideality and drift in the laser frequency output, but with significantly reduced frequency and complexity. These laser-based spectrometers achieve very long path length by employing multipass cells (Herriott) or very high reflectivity mirrors that keep most of the light inside the cell. Cavity ring down injects a short laser pulse into a high reflectivity cell and observes the decay of every small fraction of light that "leaks" out through a small aperture. The signal of interest is the duration of light pulse coming from the cell rather than its absolute intensity, τ being the transit time for light at each wavelength, rather than the absorptance. Rapid firing of the laser provides many individual measurements that are signal averaged to reduce noise and the laser frequency is modulated to scan across the absorbance spectra (Richman et al. 2004). Another alternative is an integrated cavity, which also uses very high reflectivity (but not 100%) mirrors but does not have a defined exit for the light to reach the detector. Instead the very small fraction of light transmitted through one of the mirrors is collected and its intensity is measured as the wavelength is modulated across the absorbance feature. The light that is detected at any instant in time spans the range of wavelengths that have been injected since the scan started. Sophisticated data processing algorithms incorporated in the analyzer's software deal with this. These technologies are also pioneering a wide range of new applications and measurement capability for other gas species (including isotopic discrimination) with better accuracy, precision, and resolution than can be obtained by traditional broadband IR absorbance. In principle, these systems when configured with small detector cells flushed by high volumetric flow would be ideal sensors for eddy covariance flux measurements of CO_2 as well as other species. By selection of nearby absorbance bands for different species accessed in single scan or using dual lasers these spectrometers can simultaneously measure more than one constituent. Possibilities are limited only by availability of lasers at the wavelengths of interest and time delays for tuning between different frequency regions or multiplexing different lasers. An optimum combination would include high-frequency measurement of both H_2O and CO_2 (or another scalar) because water vapor covariance is essential to account for density fluctuations. The laser spectrometer-based analyzers are also closed-path instruments, so the same considerations about tubing materials, flow rates and pressures that were described in Sect 2.4.2.2 will apply.

2.5 Profile Measurement

Net ecosystem exchange cannot be determined by eddy flux alone, but requires measurement of the storage term, (Sect. 1.4, Eq. 1.19, see also Loescher et al. 2006b). From Eq. 1.19, term I, the storage term, $\int_0^{h_m} \overline{\rho_d \frac{\partial \chi_s}{\partial t}} dz$ appeared as the vertical integral of concentrations time derivatives. Note that the vertical integral of concentrations is equivalent to finding the column-average concentration and the storage could be quantified by continually measuring from perfect mixing inlet manifold that drew air equally from all heights, or using a long-path instrument that observed total CO_2 density between the ground and sensor height instantaneously. However, it is difficult to ensure perfectly balanced sampling, and there is often useful ecological information in the shape of and changes in the concentration profile. Instruments with open paths a few to 60 m long are not commercially available, and would be difficult to deploy in most canopy situations where vegetation would obstruct the path. The number of measurement levels required to accurately quantify the mixing ratio profile depends on canopy complexity and height. Profile accuracy for observations in a mature deciduous forest was evaluated using data records from the Harvard Forest Main tower. Figure 2.9 shows the difference in CO_2 profile when one of the 8 measurement heights is deleted. Papale et al. (2006) also showed that the difference between storage estimates based on a single point and a complete profile could induce, at forested sites, differences up to 25 gC m^{-2} year^{-1} on NEE, to 80 gC m^{-2} year^{-1} on total ecosystem respiration (TER) and to 100 gC m^{-2} year^{-1} on gross primary production (GPP). Measurement levels below the canopy top are thus essential to accurately fit a profile and correctly evaluate the storage term. Removing one above-canopy measurement level produced $<1\%$ uncertainty in annual NEE estimates. There was much larger uncertainty, however, when removing a measurement level below the canopy, that is, 20–60%, in the ability to detect individual events at the 30-min timescale, for example, sweeps and ejections. The timescale of profile measurements used to quantify canopy storage needs to match the integration time for the eddy covariance fluxes, that is, 30-min. What has to be determined is how many complete column samples are needed to best characterize the column estimate for any particular 30-min average. Often the differences in concentration from the EC measurement

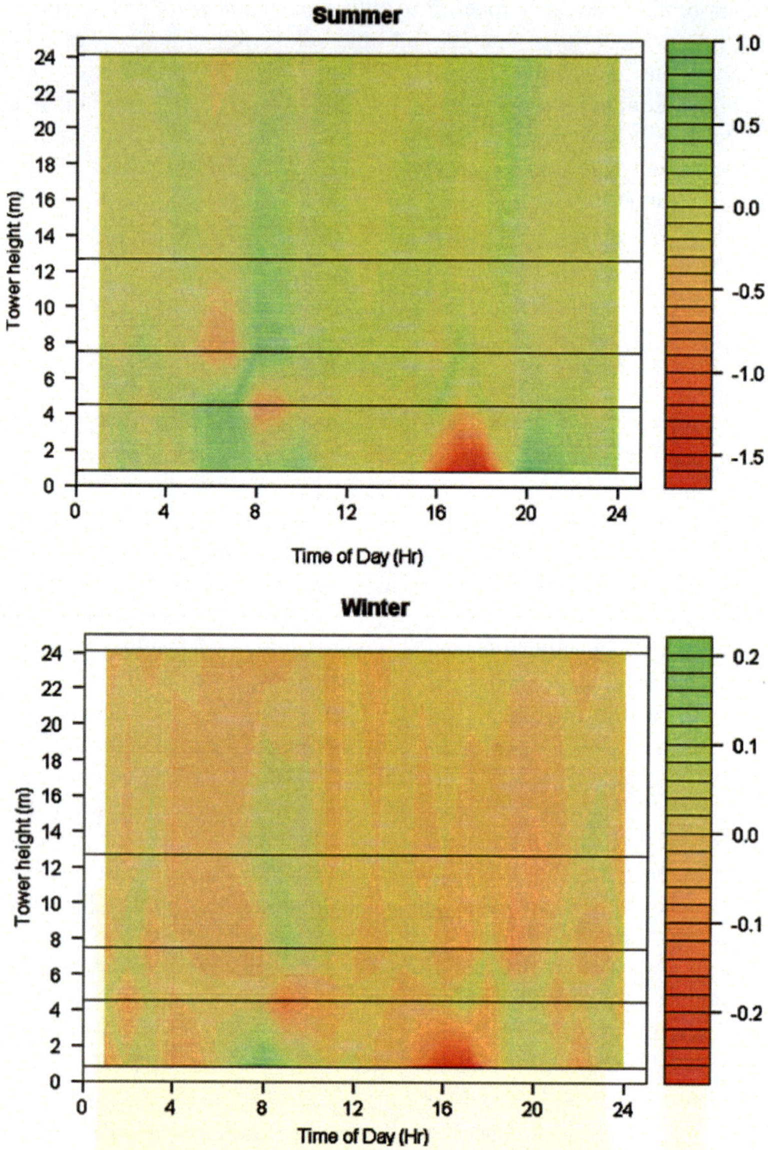

Fig. 2.9 Differences in CO_2 concentration profiles when one measurement level is removed. Data are from 2004–2008, Harvard Forest, main tower at 30-min increments at 0.3, 0.8, 4.5, 7.5, 12.7, 18.3, 24.1, and 28 m heights, with $n = 31,142$ for summer and winter seasons, respectively. We assessed the impact of removing the 4.5, 7.5, 12.7, and 24.1 m measurement levels on the CO_2 profile (0.3, 18 and 28 m heights were always fixed for both observed and predicted estimates). For each case, a generalized boosting model (gbm) was fit to the data from the tower heights, while one measurement level was excluded. Then the observed measurements from all heights were assessed against the gbm model fits. The difference between the observed data and the prediction from the model fit was computed using the following statistic; $loss\ of\ fit = ((observed - predicted)^2)^{0.5}$. CO_2 units are $\mu mol\ m^{-2}\ s^{-1}$, and we assumed normal distribution

height and the ecosystem floor can be very small under well-mixed conditions or very large when a plant canopy is decoupled from the above-canopy environment. Accurately detecting these differences is best done with a single analyzer, hence removing any among-sensor biases. When profile levels are sampled sequentially, the time offset between measurements at each level will need to be accounted for in computing the storage term, usually by some averaging or interpolation to arrive at some estimate of the average concentration profiles during successive intervals. There is typically a settling time for the analyzer to equilibrate after switching from one sample to another, which can be estimated using Allan's variance techniques (Allan 1966). Reducing the transit time and switching delay for profile sampling can be achieved by consistently pulling all the profile inlets, each with the same volume (and resistance) through a common large volume manifold, and then subsampling each profile measurement through the analyzer. The profile analyzer requires the same considerations about condensation and environmental control as the eddy analyzer.

The measurement of the concentration profile, because it can operate with reduced flow rates compared to EC, may operationally be easier to calibrate as an absolute measurement of the mixing ratio without consuming large volumes of expensive standards. For systems with frequent and easy operator access manual approaches based on supplying an excess flow of calibration standard at an inlet are an alternative to built-in calibration systems. An advantage of having good absolute calibration of a profile system is that it provides a secondary calibration of the eddy system by selecting the data from periods when the profile system takes sample from the same location as the eddy system.

2.5.1 Requirements for Measurement Levels

The number of profile levels required is dictated by the need to adequately resolve the vertical gradients in scalar quantities and adequately represent the shape of the vertical profile. Increasing height and complexity of the canopy requires more levels. Dense canopies impede vertical mixing, allowing larger concentration gradients. Vegetation strata will affect the magnitude and shape of the vertical profile and need to be considered in placing the sampling heights.

The number of measurement levels on a tower will differ among different types of ecosystem structure. There should be at least 4 measurement levels over short-stature ecosystems, grasslands, croplands, etc., and where possible, the bottom level (closest to the ground, Level 1) measuring within the canopy environment. The location of the remaining measurement levels (the distance between Level 1 and the tower top) should be mounted equal distance apart (arithmetic not logarithmic scale) over these short stature ecosystems. For shrublands, and open- and closed-canopy forests, there should be a minimum of 2 measurement levels above the canopy, which includes measurements at the top of the tower. There are no absolute criteria to determine the level closest to the canopy (but above the canopy). It will be dependent on local scalar source and sink status of individual canopies, surface roughness, and topography to best capture the vertical divergence between this measurement level and the well-mixed layer at the top of the tower. The next lower level should be associated with mean canopy height and region with the highest leaf area density. Placement of other measurement levels below-canopy should capture other ecologically significant strata, for example, established understory plant canopy.

Determining the measurement height of lowest level will be challenging at sites with significant snow accumulation. The ideal measurement height for growing season may be buried by winter snow, requiring some adjustment of sensor height and careful documentation.

2.5.2 *Requirements for Profile Mixing Ratio Measurement*

Analysis of profile mixing ratios is subject to the same concerns given above for the eddy analyzer. Instead of single sample line, multiple inlets, each with an inlet filter, would be brought to a manifold or stream select valve. The analyzer is connected to the outlet of inlet manifold and the profile inlets are opened one at a time. Analyzer output immediately after switching to a new level will need to be discarded due to pressure transients from the valve switching and to allow the inlet to be flushed with air from the selected inlet. If power is not limiting, the selection hub can be configured to allow the inlets not in use to be continually flushed by pulling on them through a bypass pump. The switching time between levels is reduced by maintaining as high a flow through the inlet as practical. A bypass flow with small subsample to the analyzer can be used to rapidly flush the inlet without needing a high flow through the analyzer itself. Furthermore, operating the inlet at a higher flow provides the reduced pressure to prevent condensation inside the tubing. To avoid artifacts due to pressure differences for each inlet height, a pressure controller should be used on the profile system

Acknowledgments J. W. Munger was supported by Office of Science (BER) U. S. Dept of Energy DE-SC0004985, and H.W. Loescher and H. Luo were supported by National Science Foundation DBI-0752017. The authors wish to thank P. Duffy for statistical support on the profile analyses.

References

Akabayashi S, Murakami S, Kato S, Chirifu S (1986) Visualization of air flow around obstacles in laminar flow type clean room with laser light sheet. Paper presented at 8th international symposium on contamination control, Milan, Italy, Sept 9–12

Allan DW (1966) Statistics of atomic frequency standards. Proc IEEE 54:221–231

Arya SPS (1988) Introduction to micrometeorology. Academic, San Diego

Barthlott C, Fiedler F (2003) Turbulence structure in the wake region of a meteorological tower. Bound Layer Meteorol 108:175–190

Burba GG, Anderson DJ, Xu L, McDermitt DK (2008) Correcting apparent off-season CO_2uptake due to surface heating of an open path gas analyzer: progress report of an ongoing study. LI-COR Biosciences, Lincoln

Campbell GS, Unsworth MH (1979) Inexpensive sonic anemometer for eddy correlation. J Appl Meteorol 18(8):1072–1077

Cermak JE, Horn JD (1968) Tower shadow effects. J Geophys Res 73(6):1869–1876

Chen J, Franklin JF, Spies TA (1990) Microclimatic pattern and basic biological responses at the edges of old-growth Douglas-fir stands. Northwest Environ J 6(2):424–425

Chen J, Franklin JF, Spies TA (1992) Vegetation responses to edge environments in old-growth Douglas-fir forests. Ecol Appl 2(4):387–396

Chen J, Franklin JF, Spies TA (1993a) An empirical model for predicting diurnal air-temperature gradients from clearcut-forest edge into old-growth Douglas-fir forest. Ecol Model 67:179–198

Chen J, Franklin JF, Spies TA (1993b) Contrasting microclimatic patterns among clearcut, edge, and interior area of old-growth Douglas-fir forest. Agric For Meteorol 63(3–4):219–237

Culf AD, Fisch G, Hodnett MG (1995) The albedo of Amazonian forest and ranchland. J Clim 8:1544–1554

Culf AD, Esteves JL, Marques Filho ADO, da Rocha HR (1996) Radiation, temperature and humidity over forest and pasture in Amazonia. In: Gash JHC, Nobre CA, Roberts JM, Victoria RL (eds) Amazonian deforestation and climate. Wiley, Chichester, pp 175–191

Dabberdt WF (1968) Wind disturbance by a vertical cylinder in atmospheric surface layer. Bull Am Meteorol Soc 49(7):767–771

Davies ME, Miller BL (1982) Wind effects on offshore platforms – a summary of wind tunnel studies. Rep., National Maritime Institute, Feltham

Detto M, Katul GG, Siqueira M, Juang J-Y, Stoy P (2008) The backward-facing step flow analogy revisited. Ecol Appl 18:1420–1435

Dyer AJ (1981) Flow distortion by supporting structures. Bound Layer Meteorol 20(2):243–251

Dyer AJ, Hicks BB (1970) Flux gradient relationships in the constant flux layer. Q J R Meteorol Soc 96:715–721

Foken T, Leclerc MY (2004) Methods and limitations in validation of footprint models special issue on footprints of fluxes and concentrations. Agric For Meteorol 127(3–4):223–234

Gash JHC, Dolman AJ (2003) Sonic anemometer (co)sine response and flux measurement I. The potential for (co)sine error to affect sonic anemometer-based flux measurements. Agric For Meteorol 119(3–4):195–207

Gill GC, Olsson LE, Sela J, Suda M (1967) Accuracy of wind measurements on towers or stacks. Bull Am Meteorol Soc 48(9):665–674

Göckede M et al (2008) Quality control of CarboEurope flux data – Part 1: coupling footprint analyses with flux data quality assessment to evaluate sites in forest ecosystems. Biogeosciences 5(2):433–450

Göckede M, Rebmann C, Foken T (2004) A combination of quality assessment tools for eddy covariance measurements with footprint modeling for the characterization of complex sites. Agric For Meteorol 127:175–188

Göckede M, Markkanen T, Hasager CB, Foken T (2006) Update of a footprint-based approach for the characterization of complex measurement sites. Bound Layer Meteorol 118:635–655

Haslwanter A, Hammerle A, Wohlfahrt G (2009) Open-path vs. closed-path eddy covariance measurements of the net ecosystem carbon dioxide and water vapour exchange: a long-term perspective. Agric For Meteorol 149(2):291–302

Hicks BB (1976) Wind profile relationships from wangara experiment. Q J R Meteorol Soc 102(433):535–551

Horst TW (2001) Comment on 'footprint analysis: a closed analytical solution based on height-dependent profiles of wind speed and eddy viscosity'. Bound Layer Meteorol 101(3):435–447

Horst TW, Weil JC (1992) Footprint estimation for scalar flux measurements in the atmospheric surface layer. Bound Layer Meteorol 59(3):279–296

Horst TW, Weil JC (1994) How far is enough – the fetch requirements for micrometeorological measurement of surface fluxes. J Atmos Ocean Technol 11(4):1018–1025
Horst TW, Weil JC (1995) How far is enough – the fetch requirements for micrometeorological measurement of surface fluxes (VOL 11, PG 1018, 1994). J Atmos Ocean Technol 12(2):447–447
Kaimal JC, Businger JA (1963a) A continuous wave sonic anemometer-thermometer. J Appl Meteorol 2(1):156–164
Kaimal JC, Businger JA (1963b) Preliminary results obtained with a sonic anemometer-thermometer. J Appl Meteorol 2(1):180–186
Kaimal JC, Gaynor JE, Zimmerman HA, Zimmerman GA (1990) Minimizing flow distortion errors in a sonic anemometer. Bound Layer Meteorol 53(1–2):103–115
Karipot A, Leclerc MY, Zhang GS, Lewin KF, Nagy J, Hendrey GR Starr G (2008) Influence of nocturnal low-level jet on turbulence structure and CO_2 flux measurements over a forest canopy. J Geophys Res 113, D10102, doi:10.1029/2007JD009149
Karipot A, Leclerc MY, Zhang GS (2009) Characteristics of nocturnal low-level jets observed in the north Florida area. Mon Weather Rev 137(8):2605–2621
Kormann R, Meixner FX (2001) An analytical footprint model for non-neutral stratification. Bound Layer Meteorol 99(2):207–224
Kristensen L, Fitzjarrald DR (1984) The effect of line averaging on scalar flux measurement with a sonic anemometer near the surface. J Atmos Ocean Technol 1:138–146
Leclerc MY, Karipot A, Prabha T, Allwine G, Lamb B, Gholz HL (2003) Impact of non-local advection on flux footprints over a tall forest canopy: a tracer flux experiment. Agric For Meteorol 115(1–2):19–30
Lee XH (1998) On micrometeorological observations of surface-air exchange over tall vegetation. Agric For Meteorol 91(1–2):39–49
Lemon ER (1960) Photosynthesis under field conditions. II. An aerodynamic method for determining the turbulent carbon dioxide exchange between the atmosphere and a corn field. Agron J 52(12):697–703
Lenschow DH, Raupach MR (1991) The attenuation of fluctuations in scalar concentrations through sampling tubes. J Geophys Res Atmos 96(D8):15259–15268
Leuning R (2004) Measurements of trace gas fluxes in the atmosphere using eddy covariance: WPL corrections revisited. In: Lee X, Massman W, Law B (eds) Handbook of micrometeorology: a guide for surface flux measurements and analysis. Kluwer, Dordrecht, pp 119–132
Leuning R (2007) The correct form of the Webb, Pearman and Leuning equation for eddy fluxes of trace gases in steady and non-steady state, horizontally homogeneous flows. Bound Layer Meteorol 123(2):263–267
Leuning R, King KM (1992) Comparison of eddy covariance measurements of CO2 fluxes by open path and closed path CO2 analysers. Bound Layer Meteorol 59(3):297–311
Liu HP, Peters G, Foken T (2001) New equations for sonic temperature variance and buoyancy heat flux with an omnidirectional sonic anemometer. Bound Layer Meteorol 100(3):459–468
Loescher HW (2007) Enhancing the precision and accuracy within and among AmeriFlux site measurements, expanded annual report, May 2007, DOE report
Loescher HW, Ocheltree T, Tanner B, Swiatek E, Dano B, Wong J, Zimmerman G, Campbell J, Stock C, Jacobsen L, Shiga Y, Kollas J, Liburdy J, Law BE (2005) Comparison of temperature and wind statistics in contrasting environments among different sonic anemometer-thermometers. Agric For Meteorol 133(1–4):119–139
Loescher HW, Starr G, Martin TA, Binford M, Gholz HL (2006a) The effect of local atmospheric circulations on daytime carbon dioxide flux measurements over a Pinus elliottii canopy. J Appl Meteorol Climatol 45(8):1127–1140
Loescher HW, Law BE, Mahrt L, Hollinger DY, Campbell J, Wofsy SC (2006b) Uncertainties in, and interpretation of, carbon flux estimates using the eddy covariance technique, J Geophys Res 111, D21S90, doi:10.1029/2005JD006932

Loescher HW, Hanson CV, Ocheltree TW (2009) The psychrometric constant is not constant: a novel approach to enhance the accuracy and precision of latent energy fluxes through automated water vapor calibrations. J Hydrometeorol 10(5):1271–1284

Massman W (2004) Concerning the measurement of atmospheric trace gas fluxes with open- and closed-path eddy covariance system: the WPL terms and spectral attenuation. In: Lee X,Massman W, Law B (eds) Handbook of micrometeorology: a guide for surface flux measurements and analysis. Kluwer, Dordrecht, pp 133–160

McDermitt DK, Welles JM, Eckles RD (1993) Effects of temperature, pressure, and water vapor on gas phase infrared absorption by CO_2. Rep., LI-COR, Inc, Lincoln

Miller DO, Tong CN, Wyngaard JC (1999) The effects of probe-induced flow distortion on velocity covariances: field observations. Bound Layer Meteorol 91(3):483–493

Monin AS, Obukhov AM (1954) Osnovnye zakono- mernosti turbulentnogo peremesivanija v prizemnom sloe atmosfery. Trudy Geofiz Inst AN SSSR 24(151):163–187

Monteith JL, Unsworth MH (2008) Principles of environmental physics, 3 edn. Elsevier, Amsterdam/Boston, xxi, 418 pp

Moses H, Daubek HG (1961) Errors in wind measurements on the towers associated with tower-mounted anemometers. Bull Am Meteorol Soc 42:190–194

Munson BR, Young DF, Okiishi TH (1998) Fundamentals of fluid mechanics, 3 edn. Wiley, New York, xvii, 877 pp

Ocheltree TO, Loescher HW (2007) Design of the AmeriFlux portable eddy-covariance system and uncertainty analysis of carbon measurements. J Atmos Ocean Tech 24:1389–1409

Papale D, Reichstein M, Aubinet M, Canfora E, Bernhofer C, Kutsch W, Longdoz B, Rambal S, Valentini R, Vesala T, Yaki D (2006) Towards a standardized processing of Net Ecosystem Exchange measured with eddy covariance technique: algorithms and uncertainty estimation. Biogeosciences 3:571–583

Paw UKT, Baldocchi DD, Meyers TP, Wilson KB (2000) Correction of eddy-covariance measurements incorporating both advective effects and density fluxes. Bound Layer Meteorol 97(3):487–511

Perrin D, McMahon N et al (2007) The effect of a meteorological tower on its top-mounted anemometer. Appl Energy 84(4):413–424

Richman B, Rella C, Crosson E, Paldus B (2004) CRDS measures atmospheric CO2. Laser Focus World 40(11):S5–S7

Sanuki M, Tsuda N (1957) What are we measuring on the top of a tower? Pap Meteorol Geophys 8(1):98–101

Schmid HP (1994) Source areas for scalars and scalar fluxes. Bound Layer Meteorol 67(3):293–318

Schmid HP, Lloyd CR (1999) Spatial representativeness and the location bias of flux footprints over inhomogeneous areas. Agric For Meteorol 93(3):195–209

Schuepp PH, Leclerc MY, Macpherson JI, Desjardins RL (1990) Footprint prediction of scalar fluxes from analytical solutions of the diffusion equation. Bound Layer Meteorol 50(1–4):353–373

Schotanus P, Nieuwstadt FTM, DeBruin HAR (1983) Temperature measurement with a sonic anemometer and its application to heat and moisture fluctuations, Boundary-Layer Meteorol 26:81–93

Shinohara T (1958) On results of an experiment for practical use of the three-cup anemometers. J Meteorol Res 10:587–590

Thornthwaite CW, Superior WJ, Field RT (1962) Evaluation of an ocean tower for measurement of climatic fluxes. Publ Climatol Lab Climatol 15(3):289–351

Thornthwaite CW, Superior WJ, Field RT (1965) Disturbance of airflow around Argus island tower near Bermuda. J Geophys Res 70(24):6047–6052

Tropea C, Yarin A, Foss JF (2007) Handbook of experimental fluid dynamics. Springer, Berlin/Heidelberg, p 1557

Vaucher G, Cionoco R, Bustillos M (2004) Forecasting stability transitions and air flow around an urban building – Phase I. Paper presented at symposium on planning, nowcasting, and forecasting in the Urban Zone, 84th AMS Meeting, Seattle, 11–15 Jan 2004, American Meteorological Society, Boston, MA

van Dijk A (2002) Extension to 3D of "The effect of line averaging on scalar flux measurements with a sonic anemometer near the surface" by Kristensen and Fitzjarrald. J Atmos Ocean Tech 19:80–82

Webb EK, Pearman GI, Leuning R (1980) Correction of flux measurements for density effects due to heat and water-vapor transfer. Q J R Meteorol Soc 106(447):85–100

Whitaker S (1984) Introduction to fluid mechanics. Krieger Publishing, Malabar, 457 pp

Wyngaard JC (1981) The effects of probe induced flow distortion on atmospheric-turbulence measurements. J Appl Meteorol 20(7):784–794

Chapter 3
Data Acquisition and Flux Calculations

Corinna Rebmann, Olaf Kolle, Bernard Heinesch, Ronald Queck, Andreas Ibrom, and Marc Aubinet

In this chapter, the basic theory and the procedures used to obtain turbulent fluxes of energy, mass, and momentum with the eddy covariance technique will be detailed. This includes a description of data acquisition, pretreatment of high-frequency data and flux calculation.

3.1 Data Transfer and Acquisition

The data transfer and acquisition mainly depend on the output data types and measuring frequency of the measuring devices. Different methods are distinguished

C. Rebmann (✉)
Department Computational Hydrosystems, Helmholtz Centre for Environmental Research – UFZ, 04318 Leipzig, Germany

O. Kolle
Max-Planck Institute for Biogeochemistry, Jena, Germany
e-mail: olaf.kolle@bgc-jena.mpg.de

B. Heinesch • M. Aubinet
Unit of Biosystem Physics, Gembloux Agro-Bio Tech, University of Liege, 5030 Gembloux, Belgium
e-mail: bernard.heinesch@ulg.ac.be; Marc.Aubinet@ulg.ac.be

R. Queck
Department of Meteorology, Institute of Hydrology and Meteorology, TU Dresden (TUD), Dresden, Germany
e-mail: ronald.queck@tu-dresden.de

A. Ibrom
Risø National Laboratory, Biosystems Department, Technical University of Denmark (DTU), Roskilde, Denmark

Risø National Laboratory for Sustainable Energy, Technical University of Denmark (DTU), Frederiksborgvej 399, 4000, Roskilde, Denmark
e-mail: anib@risoe.dtu.dk

M. Aubinet et al. (eds.), *Eddy Covariance: A Practical Guide to Measurement and Data Analysis*, Springer Atmospheric Sciences, DOI 10.1007/978-94-007-2351-1_3,

with respect to digital or analog output signals from the sonic anemometer, the analyzer, or any other additional device. The main requirements for instruments and data acquisition systems used for eddy covariance data are their response time to solve fluctuations up to 10 Hz. This means that the sampling frequency has to be high enough to cover the full range of frequencies carrying the turbulent flux, leading usually to a sampling rate of 10–20 Hz. Data acquisition in general should be flexible with respect to sampling frequency and may depend on the devices used (data logger versus personal computer; type of sonic anemometer or gas analyzer).

One needs to distinguish between two major groups of data acquisition systems, namely data loggers or computers. Explicit advantages when using data loggers are robustness, compactness, behavior in difficult conditions (low temperature, high humidity), and, above all, low power consumption, which makes such a system the preferred choice for a solar-powered eddy covariance site, especially in remote places where line power is not available. In this case, however, open-path gas analyzers would be preferred compared to closed-path gas analyzers, the latter needing a pump that consumes significantly more energy. If frequent station supervision and data collection are not feasible, an immediate processing of mean data by a logger may be advisable. In this case, the user should ensure to log not only the corrected fluxes but also the raw means for a separate postprocessing. Another convenience when using data loggers is that sensors, instruments, or devices with various output signals can be used simultaneously. The data logger can handle analog output signals, data sent through RS232 serial interface, or, in case of Campbell Scientific data loggers, according to a Synchronous Device for Measurement (SDM) protocol. Figures 3.1, 3.2, 3.3 and 3.4 are examples for LiCor and ADC gas analyzers. Other analyzers such as the Los Gatos CH_4-analyzer or the Picarro CH_4/CO_2/H_2O analyzer suit also to some of the following schemes.

Disadvantages of data logger-based systems are that graphical representation of raw or calculated data is much more complicated to realize, of much less quality, or even impossible and that raw data are usually stored in one large file which later needs to be split into files of convenient length, for example, 30 min. This could also be done online if the data logger is connected to a computer, but in this case the benefit of low power consumption may be canceled out.

Data collection with a computer equipped with one of the numerous eddy covariance software packages requires generally a connection to main power. Nevertheless, small cap rail computers with low power consumption are available on the market, which may be used with an affordable solar power supply. Except for some laptops, computers and most of the data acquisition software packages (e.g., EddySoft) can be configured to restart after a power down event so that no operator's action is required. There are several advantages to use a computer to acquire eddy covariance data: raw data are stored in files of desired length and format; visual user interfaces allow the operator to carry out or modify any program settings in an easy way; raw data as well as processed data like fluxes can be represented in colorful graphs and tables; with some software packages, more than one instance of the data acquisition program can run on the same computer allowing simultaneous acquisition from several eddy systems (e.g., useful for flux profiles). Finally, the

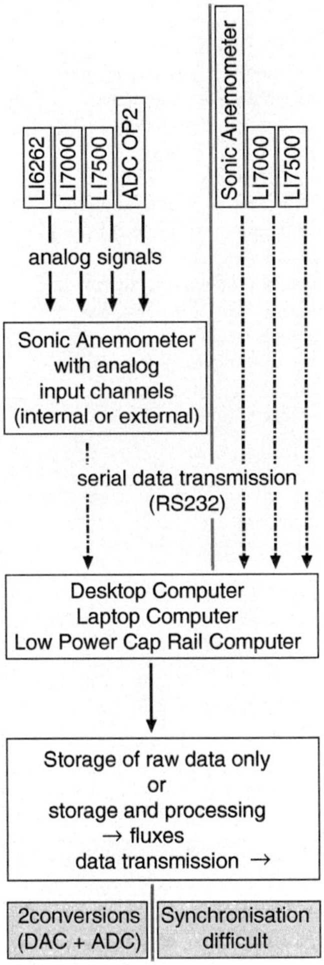

Fig. 3.1 Examples for data acquisition of sonic anemometer and various gas analyzers with a computer for analog and digital data streams. The *left path* describes the digitization of the analog signals by the sonic anemometer; the *right path* describes the direct digital data transmission from all devices to a computer. The *gray boxes* at the *bottom* show the disadvantages of each of the configurations

computer can be used for other tasks besides collecting the eddy covariance data such as flux data postprocessing; communication with a data logger to archive meteorological data; communication with a web camera capturing phenological images; data, images, graphs transmission via modem or network; data backup on peripheral storage devices; collection of other data such as status information from the eddy covariance instruments.

If a computer is used and particularly if it is running under Windows® it is recommended to transmit all data of one eddy covariance system through only one data stream. Since Windows® is not a real time operating system, it is impossible to ensure data being transmitted to different input lines of the computer, for example, several COM-ports will be synchronized throughout long periods. In practice, this means that the data should be sent to the computer via one physical or virtual COM-port (RS232, USB or Ethernet). This implies that there must be an instrument or a device in front of the computer which merges the data from the different components

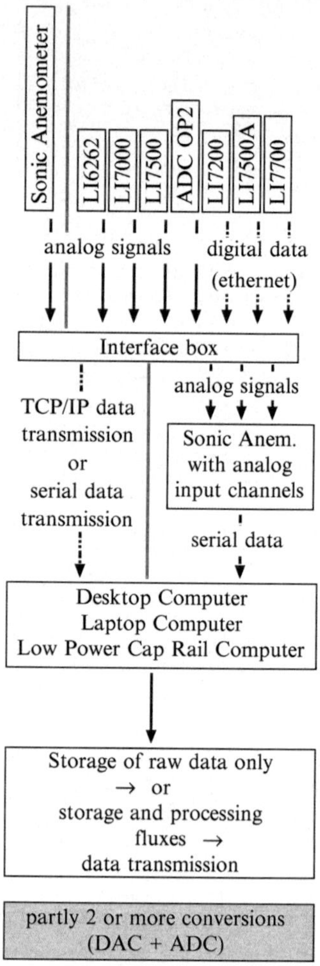

Fig. 3.2 Examples for data acquisition of sonic anemometer and various gas analyzers with a computer for analog and digital data streams. An intelligent interface box (LI-7550, LiCor Biosciences) is able to merge different signal inputs (analog or ethernet) and data output can be realized via ethernet or RS232 (*left path*) or as analog signals (*right path*). The *gray box* at the *bottom* shows the disadvantages of the configuration

of the eddy instrumentation into one data stream. In many cases, this merging device is the sonic anemometer (Figs. 3.1 and 3.2) because many of the producers of sonic anemometers provide their instruments with analog input channels (Gill, Metek, Thies, Young). This is realized by embedded analog to digital converters (ADCs) or by optional external analog input boxes. The quality and resolution of the ADCs can vary significantly. With this solution, analog output signals of gas analyzers are being digitized by the ADCs and the digital data are then merged with the sonic data and sent to the computer. The general drawback of this procedure is that in many cases two conversions are carried out: modern gas analyzers operate internally on a digital basis. To produce an analog output signal, a first conversion from digital to analog (DAC) is required. Then the data must be converted back to a digital representation to be transmitted to the computer. Those two conversions may of course reduce the signal quality (see further Chap. 4).

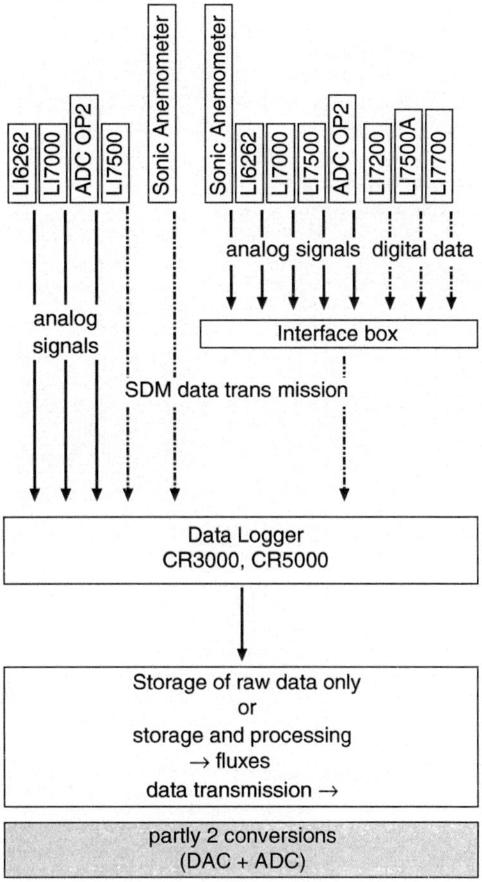

Fig. 3.3 Examples for data acquisition of sonic anemometer and various gas analyzers with a data logger for analog and digital data streams. The same interface box (LI-7550, LiCor Biosciences) as described in Fig. 3.2 might be used (*right path*) or various devices can be connected to the data logger directly (*left path*). Both paths can be used simultaneously. The *gray box* at the *bottom* shows the disadvantages of the configuration

Another option is offered by LiCor Biosciences in conjunction with their new gas analyzers. There is an interface box available to which the digital signals of the gas analyzers can be connected to as well as additional analog signals. In this case, the idea is to connect analog output signals from sonic anemometers which are digitized and merged with the digital data stream of one or more gas analyzers and digitally transmitted to the computer (Fig. 3.2) or data logger (Fig. 3.3). Again the drawback is the double conversion; in this case, the DACs of some sonic anemometers are quite limited in resolution. On the other hand, the same interface box is able to convert the digital data of the gas analyzers to analog signals which then might be connected to the analog input channels of a sonic anemometer (Fig. 3.2).

Two system configurations which are less common should also be mentioned:

- A system described by Eugster and Pluss (2010) is operating "fully digital", which means that digital data from all instruments of the system are transmitted to a computer via independent COM-ports (Fig. 3.1, path on right hand side). The problem of synchronization is controlled by a Linux operating system.

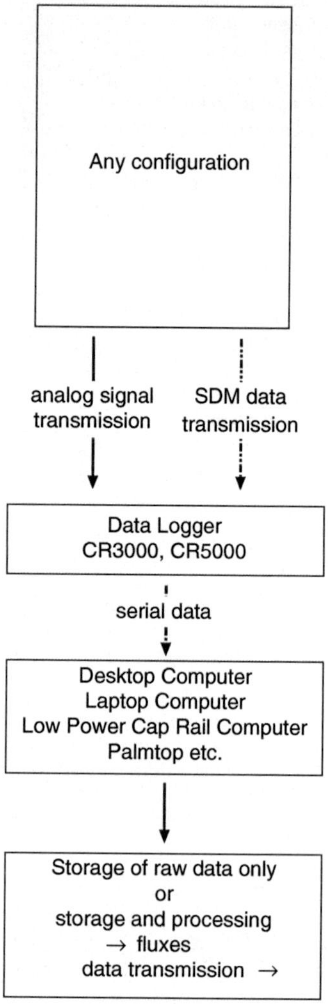

Fig. 3.4 Examples for data acquisition of sonic anemometer and various gas analyzers with a computer for analog and digital data streams with a data logger being a merging and synchronizing device

- A system where a data logger acts as merging device to which the sonic anemometer and gas analyzers are connected and the raw data are transmitted to a computer via RS232 at high frequency. This system is pretty flexible because instruments with analog and digital outputs can be mixed in various combinations (Fig. 3.4).

Before running any of the possible data acquisition software tools, it has to be ensured that hardware settings of the sonic anemometer and the analyzer are appropriately introduced into the software settings. Depending on the type of software used, also measurement frequency, number, and order of additional analog or serial input channels have to be set. The sonic anemometer azimuth alignment has to be fixed in the acquisition software to get horizontal wind components directly as an output especially when real time calculation of fluxes and wind direction is required but also to enable correct post-processing.

For closed-path CO_2/H_2O analyzers such as the LI 6262 or LI 7000 it may be possible to choose linearized or nonlinearized output signals. In this latter case also pressure and temperature signals from the analyzer have to be sampled in high-frequency resolution. For any of the signals sampled, it has to be ensured that in case of voltage signals the ranges and units are set correspondingly in the analyzer output and in the data acquisition software.

Wind components together and speed of sound are determined by any of the different types of 3D-ultrasonic anemometers, such as Campbell CSAT3, Gill R2, R3, HS, or WindMaster(Pro), METEK, or Young (see also Sect. 3.2.1.1). Each of the different anemometers has specific characteristics that have to be considered with respect to data acquisition: number of analog inputs, azimuth alignment, angle adjustment, tone settings (for Gill R3/HS), heating settings (for Metek USA-1), analog output full scale deflection (for Gill R3/HS and Windmaster Pro), sensor head correction (for Metek USA-1), and analyzer type.

3.2 Flux Calculation from Raw Data

The transformation of high-frequency signals into means, variances, and covariances requires different steps that will be detailed below. First, the sensor output signals have to be transformed in order to represent micrometeorological variables (Sect. 3.2.1). Secondly, a series of quality tests have to be applied in order to flag and/or eliminate spikes and brutal shifts that could appear in the raw signals due to electronic noise (Sect. 3.2.2). After that, variable averages, variances, and covariances have to be computed (Sect. 3.2.3). Variances and covariances require the computation of variable fluctuations, which in some cases could require some detrending (Sect. 3.2.3.1). Covariances require, in addition, a determination of the lag between the two variables that covary (Sect. 3.2.3.2). These procedures provide estimates of means, variances, and covariances expressed in an axis system that is associated with the sonic anemometer. A rotation is then needed in order to express these variables in a coordinate frame that is linked to the ecosystem under study (Sect. 3.2.4).

3.2.1 Signal Transformation in Meteorological Units

3.2.1.1 Wind Components and Speed of Sound from the Sonic Anemometer

The operating principles of sonic anemometers are described in Sect. 2.3 and in several publications and textbooks (Cuerva et al. 2003; Kaimal and Businger 1963; Kaimal and Finnigan 1994; Schotanus et al. 1983; Vogt 1995). The sonic anemometer output provides three wind components in an orthogonal axis system associated with the sonic anemometer, and the sound velocity, *c*, respectively. This variable depends on air density and thus on atmospheric pressure (*p*), vapor pressure (*e*), and absolute air temperature θ:

$$c = \sqrt{\gamma \cdot R/m_d \cdot \theta \cdot (1 + 0.32 \cdot e/p)} \tag{3.1}$$

where $R = 8.314\ \mathrm{J\,K^{-1}mol^{-1}}$ is the universal gas constant, $m_\mathrm{d} = 28.96\ 10^{-3}\ \mathrm{kg\ mol^{-1}}$ is the dry air molar mass, and $\gamma = 1.4$ the ratio of constant pressure and constant volume heat capacities. In practice, the sonic anemometer software computes the sonic temperature as (Aubinet et al. 2000; Schotanus et al. 1983):

$$\theta_S = \frac{m_d}{\gamma R}\frac{\left(c_1^2 + c_2^2 + c_3^2\right)}{3} = \frac{1}{403}\frac{\left(c_1^2 + c_2^2 + c_3^2\right)}{3} \tag{3.2}$$

where: c_1, c_2 and c_3 correspond to the speed of sound measured along each sonic anemometer axis.

However, this temperature strays from real absolute temperature (θ) by 1–2% as it does not take the dependence of sound velocity on vapor pressure (*e*) into account. The relation between sonic temperature and absolute real temperature is given by Kaimal and Gaynor (1991):

$$\theta_s = \theta \cdot (1 + 0.32 \cdot e/p) \tag{3.3}$$

This is almost equal to the virtual temperature θ_v, defined as:

$$\theta_v = \theta \cdot (1 + 0.38 \cdot e/p) \tag{3.4}$$

As a result, θ_s can be directly used to estimate the buoyancy flux and, thus the stability parameter $(h_\mathrm{m}-d)/L$. However, for sensible heat flux estimates, a correction, based on Eq. 3.3 and needing independent vapor pressure measurement (SND correction), is necessary. It is described in detail in Sect. 4.1.2.

3.2.1.2 Concentration from a Gas Analyzer

The scalar intensity of an atmospheric constituent must be expressed in the conservation equation (e.g., Eqs. 1.19–1.25) in terms of mixing ratio. Infrared gas analyzers measure either density or molar concentrations and may convert them to mole fractions either with or without correction for water vapor (Sect. 2.4.1). The signal conversion to mixing ratios requires the knowledge of high-frequency air density fluctuations and therefore an estimate of high-frequency air temperature and humidity fluctuations. In the closed-path system, the former is neglected, considering that temperature fluctuations are damped due to the air passage through the tube (for more detail, see Sect. 4.1.2.3), and the latter is taken into account by the analyzer if this option is available in the analyzer software (among others, LI-COR 6262) and chosen by the user. If this is not the case (among others, LI-COR 7000), the signal conversion must be done during data postprocessing. In the open-path system (among others, LI-COR 7500), none of these corrections are accounted for and they must be applied during data post-processing.

In case of linearized analog output mode, the output of a gas analyzer is thus a voltage signal V_χ (V_ρ) related to molar mixing ratio (density). The relation has to be determined from the settings in the data acquisition software, namely the maximum voltage output $V_{\max}$, which relates to a maximum mixing ratio (density) of the trace gas χ_{smax} (ρ_{smax}) and a zero voltage, which corresponds to a minimum mixing ratio (density) $\chi_{\text{smin}}(\rho_{\text{smin}})$:

$$\chi_s = \chi_{smin} + \frac{\chi_{smax} - \chi_{smin}}{V_{\max}} V_\chi \tag{3.5a}$$

$$\rho_s = \rho_{smin} + \frac{\rho_{smax} - \rho_{smin}}{V_{\max}} V_\rho \tag{3.5b}$$

$\chi_{\text{s max}}$ ($\rho_{\text{s max}}$) as well as $\chi_{\text{s min}}(\rho_{\text{s min}})$ have to be set according to expected values of mixing ratios (densities) at the site to optimize the analyzer's resolution and calibration gases with mixing ratios within this range should be used (see Sect. 2.4.2.3). This equation for the determination of trace gas mixing ratios (densities) applies to many gas analyzers in use with analog output. Newer sensors' output provide mixing ratios as digital signals.

3.2.2 *Quality Control of Raw Data*

Quality control of flux data is the second step of processing. High-frequency raw data often contain impulse noise, that is, spikes, dropouts, constant values, and noise. Spikes in raw data can be caused by instrumental problems, such as imprecise adjustment of the transducers of ultrasonic anemometers, insufficient electric power supply, and electronic noise, as well as by water contamination of the transducers, bird droppings, cobwebs, etc., or rain drops and snowflakes in the path of the sonic anemometer. Some instruments issue error flags in case of suspect data (e.g., USA-1, CSAT, LI7500).

Spikes can usually be detected because of their amplitude, duration, or abruptness of occurrence. Besides checks for exceeding of physical limits and standard deviations, Hojstrup (1993) suggested a procedure which defines thresholds by a point-to-point autocorrelation. Further, Vickers and Mahrt (1997) developed test criteria for quality control of turbulent time series independent of the statistical distribution with a focus on instrument malfunctions.

Any spike detection and elimination modifies the data. Especially, means and variances of an averaging interval that are used as test criteria are changing. As a consequence, the quality assessment is an iterative process (e.g., Schmid et al. 2000). However, the change in the measured data implies also that each test is site-specific, has to be applied carefully, and should not mean a simple removal of single samples or complete averaging intervals, but an application of meaningful flags. As introduced by Vickers and Mahrt (1997), commonly, hard flags are used to identify artifacts introduced by instrumental or data recording problems and soft flags are used to identify statistical abnormal behaviors which are apparently physical but do disturb the further statistical evaluation or indicate nonstationary time series (Sect. 4.3.2.1). Detected hard spikes should be checked visually either to affirm a setup artifact and to discard data or to switch to a soft spike. Data flagged with a soft flag indicating limited data quality can be used for some purposes but not for standard data analyses.

The first step of data quality flagging includes checks for physical limits (wind velocity range, temperature range, and realistic trace gas concentrations, respectively). The thresholds should be chosen not too close but can include the seasonal cycle (especially for temperature) to avoid any truncation of the measuring signal. Examples are:

Horizontal wind velocity: $|u| < 30\ \mathrm{m\,s^{-1}}$
Vertical wind velocity: $|w| < 5\ \mathrm{m\,s^{-1}}$ (close to the surface)
Sonic temperature: $|\theta_s - \theta_m| < 20$ K (θ_m: monthly mean temperature)

Site and instrument-specific thresholds can be derived from typical frequency distributions of time series that are representative for the majority of meteorological conditions. The thresholds must be corroborated by direct inspection of the time series where unusual ranges were detected. Spikes detected with these thresholds are marked with a hard flag.

In a second step, the data could be checked relative to the standard deviation σ of the average interval. Schmid et al. (2000) proposed that each value χ_i within a time series which deviates more than the product of a discrimination factor (e.g., $D = 3.5$) and the standard deviation σ_j from the mean value $\overline{\chi}_j$ should be characterized as a spike. For more selective filtering, subintervals (*j*) of the average interval are used to define the standard deviation and mean.

$$\left|\chi_i - \bar{\chi}_j\right| \geq D \cdot \sigma_j \quad \rightarrow \quad spike \qquad (3.6)$$

These data windows should comprise most of the variance of the variable in a local scale. Schmid et al. (2000) applied 15 min windows, whereas Vickers and Mahrt (1997) used moving windows of 5 min length.

As the standard deviation decreases with the elimination of spikes, the tests should be repeated several times, until either there are no more new spikes or the maximum of iterations is completed. The discrimination factor should be increased with each iterative step (*k*) by a constant term (e.g. $D_k = 3.5 + 0.3k$).

A soft spike is registered if the fluctuation from the mean is larger than the threshold value. As a second condition, the duration of the deviation can be used, for example, a spike should be shorter than 0.3 s (Schmid et al. 2000).

More complex approaches perform despiking with respect to the difference between consecutive data points. Hojstrup (1993) applied a point-to-point auto-correlation method using an exponential filter function. Each individual value χ_i is compared with a test value $\chi_{t,i}$ calculated from the course of the previous time series, according to:

$$\chi_{t,i} = \chi_{i-1} R_{M,i} + X_{M,i}(1 - R_{M,i}) \tag{3.7a}$$

where the mean ($X_{M,i}$) is computed as

$$X_{M,i} = X_{M,i-1}(1 - 1/M) + \chi_i/M \tag{3.7b}$$

the auto correlation coefficient ($R_{M,i}$) as

$$R_{M,i} = \frac{R_{M,i-1}(1 - 1/M)\sqrt{\sigma^2{}_{M,i-1}\sigma^2{}_{M,i-2}} + [(x_i - X_{M,i})(x_{i-1} - X_{M,i-1})]/M}{\sqrt{\sigma^2{}_{M,i}\sigma^2{}_{M,i-1}}} \tag{3.7c}$$

and the standard deviation ($\sigma_{M,i}$) as

$$\sigma^2{}_{M,i} = \sigma^2{}_{M,i-1}(1 - 1/M) + (\chi_i - X_{M,i})^2/M \tag{3.7d}$$

The memory of the filter is characterized by a number of points M, however, it is rather a filter constant, as the influence of previous points on following test values reduces with time distance but is theoretically infinite. During the process, the filter memory is adjusted to the varying auto correlation R_M:

$$M = \frac{-230}{\ln(|R_{M,i}|)} \quad with \quad 0.1 < |R_{M,i}| < 0.99 \tag{3.7e}$$

The comparison is made according to:

$$|\chi_i - \chi_{t,i}| \geq D \cdot \sigma_{\chi-\chi t} \quad \rightarrow \quad spike$$

where $\sigma_{\chi-\chi t}$ is the standard deviation of the differences between the test values and the actual data points. The discrimination factor D is set from 3.3 to 4.9, depending on the probability of exceeding the threshold, $D\sigma_{\chi-\chi t}$.

Clement (2004) proposed a similar approach based on the difference between consecutive data points $\Delta\chi_i = |\chi_i - \chi_{i-1}|$, which further regards dropouts (a corresponding subroutine is implemented in the flux calculation software EDIRE, University of Edinburgh, Institute of Atmospheric and Environmental Science). The threshold for the deviation of $\Delta\chi$ from the mean of the differences $\overline{\Delta\chi}$ is set relative to the standard deviation $\sigma_{\Delta\chi}$ of $\Delta\chi$ of the whole averaging interval.

$$|\Delta\chi_i - \overline{\Delta\chi}| \geq D \cdot \sigma_{\Delta\chi} \rightarrow spike$$

Detected differences are suggested as an upward or downward leg of a spike. The procedure searches within a predefined window around a detected difference for the corresponding leg. The interval between the two legs is then corrected by an offset function including the slope of the interval.

The last two methods need to be parameterized very carefully to avoid false data exclusion. Indeed, parameters are so sensitive that physically valuable data could inappropriately be eliminated. Despite this, they are very helpful to detect dropouts and spikes automatically, which are not found by the previous methods.

Further tests aim to detect variances outside a defined valid range (a variance that is either too small or too large is flagged). Unusually large skewnesses or kurtosis and large discontinuities can be detected using the Haar transform (Vickers et al. 2009). Large kurtosis in time series of the sonic temperature can for example indicate water on the transducers (Foken et al. 2004). These tests are preferably applied to moving windows of width 10–15 min.

The eliminated spikes leave gaps in the time series that need to be filled, especially when spectral analysis has to be performed on the data. For short gaps, this is frequently done by interpolating using Gaussian random numbers depending on mean and standard deviation or by the model of Hojstrup (1993). Linear interpolation can lead to a systematic error and is not recommended. Time series with more than 1% spikes should be excluded from further statistical analysis (Foken 2008).

However, the application of the methods needs to be examined carefully. It is known, that physically plausible behavior and instrument problems overlap in parameter space. This underscores the importance of the visual inspection either to confirm or deny flags raised by the automated set of tests.

3.2.3 *Variance and Covariance Computation*

3.2.3.1 Mean and Fluctuation Computations

The variance of any variable χ_s is computed as

$$\overline{\chi_s'^2} = \frac{1}{N-1} \sum_{j=1}^{N} (\chi_{s\,j} - \overline{\chi_s})^2 \tag{3.8}$$

where N is the number of samples, χ_s the scalar of interest, χ'_s its fluctuating part, and $\overline{\chi_s}$ its nonfluctuating part, that is, that part of the time series that does not represent turbulence, for example, the arithmetic mean.

The covariance of any wind component u_k or scalar χ_s with another wind component u_i is calculated as

$$\overline{\chi'_s u'_i} = \frac{1}{N}\sum_{j=1}^{N}\left[\left(\chi_{sj} - \overline{\chi_s}\right)\left(u_{i,j} - \overline{u_i}\right)\right] = \frac{1}{N}\sum_{j=1}^{N}\chi'_{sj}\, u'_{i,j} \tag{3.9a}$$

$$\overline{u'_k u'_i} = \frac{1}{N}\sum_{j=1}^{N}\left[\left(u_{k,j} - \overline{u_k}\right)\left(u_{i,j} - \overline{u_i}\right)\right] = \frac{1}{N}\sum_{j=1}^{N}u'_{k,j}u'_{i,j} \tag{3.9b}$$

where u_k, with $k = 1, 2, 3$, represent wind components u_j, v_j, or w_j.

In practice, the averages of χ_s and u_i may be computed in several ways. The first approach, referred to as block averaging (BA), is

$$\overline{\chi}_{sBA} = \frac{1}{N}\sum_{j=1}^{N}\chi_s \tag{3.10}$$

It has the advantage over the alternatives that it dampens low-frequency parts of the turbulence signal to the least degree. However, when there is a need to remove a trend in the time series, due to instrumental drift or synoptic change in atmospheric conditions, block averaging is not sufficient to calculate fluctuations from turbulence data. To remove these undesired contributions from the time series, mainly two other types of high-pass filtering are being used, namely linear detrending, where the nonfluctuating part is calculated as

$$\chi_{s\,LDj} = \beta_1 t_1 + \beta_0 \tag{3.11}$$

where β_0 and β_1 are intercept and slope of a linear regression of χ_s with time (e.g., Draper and Smith, 1998). Another way of defining the nonfluctuating term is to calculate the auto regressively filtered time series, which is sometimes falsely called running mean:

$$\overline{\chi}_{sAF,j} = \alpha\chi_{s,j} + (1-\alpha)\cdot\chi_{s,j-1} \tag{3.12}$$

where α is the constant of the filter, related to cut-off, f_c, and sampling, f_s, frequencies as:

$$\alpha = 1 - e^{-2\pi(f_c/f_s)} \tag{3.13}$$

The different detrending algorithms were compared by Rannik and Vesala (1999), Culf (2000) and Moncrieff et al. (2004). Due to the nature of turbulence, that is, varying over several orders of frequency domains, high-pass filtering does not only remove undesired contributions to the covariance but also low-frequency contributions to the flux happening at the same time scales, which must be corrected. The theoretical work by Lenschow et al. (1994) and Kristensen (1998) provided spectral transfer functions for each of the three detrending methods (cf. Rannik and Vesala (1999)). To calculate unbiased and complete fluxes, any covariance, irrespective of the detrending method used, must be corrected for high-pass filtering losses. Application of these functions to correct fluxes is however limited as the low-frequency part of cospectra cannot be measured and is thus is not well defined (Kaimal and Finnigan 1994). Techniques allowing the evaluation of high-pass filtering errors are proposed in Sect. 4.1.3.3. Benefits and disadvantages of the different high-pass filtering methods are discussed in terms of flux uncertainty assessment in Sect. 7.3.3.1.

3.2.3.2 Time Lag Determination

Application of Eq. 3.9a requires that the instantaneous quantities χ_{sj} and u_{j} are measured at the same place and the same time. This is, however, generally impossible. Consequently, before applying Eq. 3.9a, the recorded time series must be lagged by a certain time against each other.

The delay between the two time series is mainly caused by differences in electronic signal treatment, spatial separation between wind and scalar sensors, and air travel through the tubes in closed-path eddy covariance systems. Time delays caused by signal electronic treatment (signal conversion and computation) are generally relatively small, constant, and known and can thus be considered directly. Delay due to sensor separation is more important. The air parcel needs some time to pass both of the instruments, which depends on wind speed, wind direction, and distance between the sensors. New sensor development aims at combining chemical and velocity measurements in one sampling volume. Larger lag times as common for closed-path systems comprise the time needed for the air to travel from the intake to the measurement cell in the analyzer. This delay depends on the inner volume of the air conducting parts of the eddy covariance system (filters, tubes, valves, and detection cell), on the mass flow through the system (and thus may vary with pump aging and filter contamination), and on the considered gas. Indeed, larger time delays may be observed if gases interact with the tube walls, which is notably the case for water vapor (Ibrom et al. 2007a, b; Massman and Ibrom 2008).

Two procedures are generally used to estimate the lag time. In the case of closed-path eddy covariance systems, where the most important cause of delay is due to air travel through the tubes, a mass flow controller can be installed in the pumping systems so that the time lag can be considered as constant. In these conditions, it could be estimated once at the beginning of the measurement period and the time series could be lagged by this constant value during the measurement campaign. It is, however necessary to check this value with empirical methods, because wall interactions are very likely to introduce additional time lags.

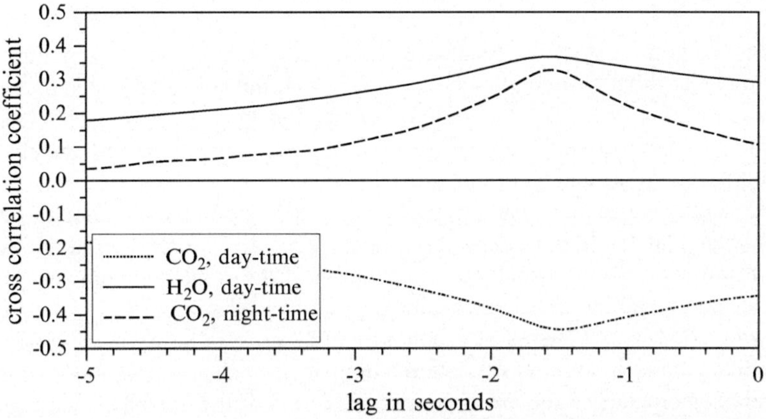

Fig. 3.5 Time lag determination as an example for CO_2 and H_2O compared to the vertical wind component w. *Dashed lines* represent the cross correlation for CO_2 for day- and night-time and the *solid line* represents the cross correlation of H_2O and w for day-time. Data were acquired at Maun, Botswana on DOY 58 in 1999 at 0230 and 1100 hours. Tube length was about 7 m at an inner diameter of one-eighth inch, flow rate was about 7 l min^{-1}

Lag times can be estimated for each averaging interval by performing a cross correlation analysis between the scalar of interest and the vertical wind component. This consists in comparing the correlations between the two signals lagged by different delays (Fig. 3.5). The time lag that is selected is those that produces the highest correlation. However, this procedure could result in ambiguous lag times, especially when the correlation is small. A feasible automatic procedure to determine lag times could thus use a defined search window as determined from mass flow, tube dimensions, and typical wall interactions at times with high enough fluxes (Aubinet et al. 2000; Kristensen et al. 1997; Lee and Black 1994; Moncrieff et al. 1997). In cases where these limits are exceeded, as well as in cases when the change in the lag time is too abrupt, it is recommended to use the value of the preceding averaging interval. Especially for H_2O lag times it can also be useful to determine their dependency on relative humidity and use this dependency for further lag time determinations. Lag times for each of the variables and each averaging interval have then to be included in further postprocessing steps.

3.2.4 *Coordinate Rotation*

3.2.4.1 Requirements for the Choice of the Coordinate Frame and Its Orientation

Each term in the mass balance (Eq. 1.12) is a scalar and so is independent of the coordinate frame. The individual components of the divergence term (all terms but the first in left hand side (LHS) of Eq. 1.13), however, can take different forms in different coordinate systems. As measurements are generally taken from one single point, the coordinate frame must be chosen so that the sole divergence that can be measured (Term IV in Eq. 1.19) approximates the total divergence as closely as possible (Finnigan et al. 2003). This is the basic requirement guiding the choice of the coordinate frame and its orientation.

The setting up of the mass balance (Eq. 1.19) implicitly assumes the choice of a rectangular Cartesian coordinate system with the x direction parallel to the local mean wind vector, usually at the position of the sonic anemometer. The use of other coordinate systems, for example, the physical streamline or the surface-following coordinate systems, can be considered, especially in gentle topography, to facilitate the estimation of extra-terms in the mass balance equation, to combine several anemometers in the estimation of the terms of the mass balance equation, or to incorporate measurements in flow and transport models. These alternative coordinate systems will not be analyzed here. For a thorough discussion on this topic, see Finnigan (2004), Lee et al. (2004), and Sun (2007).

In order to determine the reference frame orientation, a homogeneous boundary layer is assumed where the mean moments of the wind and the scalar field in the surface-normal, cross streamline direction will be much larger than streamwise gradients (i.e.,: $\frac{\partial \overline{u'\chi'_s}}{\partial x}, \frac{\partial \overline{v'\chi'_s}}{\partial y} \ll \frac{\partial \overline{w'\chi'_s}}{\partial z}$). This hypothesis will obviously be met in one-dimensional, horizontally homogeneous mean wind fields above homogeneous sources/sinks but also in two- or three-dimensional flows when the point measurement is not very close to abrupt changes in surface topography or in surface cover. We can consider that this is the case for micrometeorological sites chosen to avoid large inhomogeneities in topography and source distribution, which means most of the long-term flux study sites, even those in complex terrain. In these conditions, the desired orientation of the coordinate system, that is, the one that will yield the best approximation to the divergence using an anemometer at a single point is obtained when the instrument is oriented in the plane spanned by the mean wind vector and the local normal to the surface.

If the vertical axis of the sonic is not aligned with the local normal to the surface, there will be cross-contamination among components of the flux divergence also called shortly "tilt errors." It has been shown that the momentum flux is particularly sensitive to the tilt errors (Wilczak et al. 2001). For a 1° tilt, the error is typically greater than 10% under moderately unstable conditions and can be as large as 100% under free convection conditions. Scalar fluxes are not as sensitive, with a tilt error usually less than 5% for small tilt angles (<2°) but the errors could potentially cause a systematic bias in annually integrated eddy fluxes (Lee et al. (2004) and references therein).

Usually, anemometers are fixed in a permanent position at a tower and it is not possible to align the anemometer coordinate system to the changing flow field. The operator should simply align the sonic in a reasonable, pragmatic way as close as possible to the requested orientation of the z-axis and depending on the technical constraints (usually aligning the z-axis to the gravity field or tilting the sonic according to the expected slope over steep terrains).

As a consequence of the misalignment of the sonic anemometer the mean vertical wind components different from zero may appear (Heinesch et al. 2007). Measurement artifacts which are not discussed within this subsection like electronic problems (Grelle and Lindroth 1994; Wilczak et al. 2001), flow perturbation or insufficient calibration of the anemometer can also contribute to a spurious vertical wind component.

In order to avoid cross-contamination between the flux components due to the above mentioned problems, it is highly recommended to perform a rotation on the data before further corrections are done. The generic way to apply a rotation scheme will be presented in the next section and the definition of the rotation angles will be presented in Sect. 3.2.4.3.

3.2.4.2 Coordinate Transformation Equations

Three degrees of freedom are available, leading to three rotations characterized by the Euler angles α, β, and γ. The first, second and third rotations are performed around the z-axis, new y-axis and new x-axis, respectively, resulting in the angles α, β and γ. If the coordinate system is right-handed and if a positive rotation angle is defined as being a counter-clockwise rotation looking down the axis of rotation, these rotations can be expressed mathematically in matrix form by

$$R_{01} = \begin{pmatrix} \cos\alpha & \sin\alpha & 0 \\ -\sin\alpha & \cos\alpha & 0 \\ 0 & 0 & 1 \end{pmatrix}, R_{12} = \begin{pmatrix} \cos\beta & 0 & \sin\beta \\ 0 & 1 & 0 \\ -\sin\beta & 0 & \cos\beta \end{pmatrix}, R_{23} = \begin{pmatrix} 1 & 0 & 0 \\ 0 & \cos\gamma & \sin\gamma \\ 0 & -\sin\gamma & \cos\gamma \end{pmatrix} \tag{3.14}$$

These rotations are applied successively starting from the wind vector in the sonic anemometer coordinates and ending in the wind vector in the desired coordinate system:

$$\begin{pmatrix} \bar{u}_3 \\ \bar{v}_3 \\ \bar{w}_3 \end{pmatrix} = R_{03}(\alpha, \beta, \gamma) . \begin{pmatrix} \bar{u}_0 \\ \bar{v}_0 \\ \bar{w}_0 \end{pmatrix} \tag{3.15}$$

where: $R_{03}(\alpha, \beta, \gamma) = R_{23}(\gamma).R_{12}(\beta).R_{01}(\alpha)$ is the matrix product of the three sequential rotation matrices.

For the scalar covariance matrix, it gives:

$$\begin{pmatrix} \overline{u'_3 \chi'_s} \\ \overline{v'_3 \chi'_s} \\ \overline{w'_3 \chi'_s} \end{pmatrix} = R_{03}(\alpha, \beta, \gamma) \begin{pmatrix} \overline{u'_0 \chi'_s} \\ \overline{v'_0 \chi'_s} \\ \overline{w'_0 \chi'_s} \end{pmatrix} \tag{3.16}$$

and, for the wind components (co)variance matrix, it gives:

$$\begin{pmatrix} \overline{u_3'u_3'} & \overline{u_3'v_3'} & \overline{u_3'w_3'} \\ \overline{v_3'u_3'} & \overline{v_3'v_3'} & \overline{v_3'w_3'} \\ \overline{w_3'u_3'} & \overline{w_3'v_3'} & \overline{w_3'w_3'} \end{pmatrix} = R_{03}(\alpha,\beta,\gamma)\cdot\begin{pmatrix} \overline{u_0'u_0'} & \overline{u_0'v_0'} & \overline{u_0'w_0'} \\ \overline{v_0'u_0'} & \overline{v_0'v_0'} & \overline{v_0'w_0'} \\ \overline{w_0'u_0'} & \overline{w_0'v_0'} & \overline{w_0'w_0'} \end{pmatrix}\cdot R_{03}^{\tau}(\alpha,\beta,\gamma) \quad (3.17)$$

where R_{03}^{τ} is the transposed R_{03}.

This procedure will always be applied to each flux-averaging interval (typically 30 min averages).

Two methods are available to define these three rotation angles. The so-called double rotation (DR), and the planar-fit (PF) method. The DR has been used since the early years of eddy covariance measurements and is the most common and easiest to use method. The planar-fit method has been shown to have advantages over the DR method in complex terrain. They will both be presented in the next subsections.

3.2.4.3 Determination of Rotation Angles

Double Rotation

In this vector basis orientation, the z-axis is normal to and points away from the (30 min) mean local streamline and the x-axis is parallel to the (30 min) mean flow with x increasing in the direction of the flow.

In order to obtain the desired vector basis, the first rotation has to be performed to align $\overline{u}$ into the mean wind direction, forcing $\bar{v}$ to 0, resulting in the yaw angle α:

$$\alpha_{DR} = \tan^{-1}\left(\frac{\overline{v_0}}{\overline{u_0}}\right) \quad (3.18)$$

The second rotation has to be performed to nullify $\bar{w}$, resulting in the pitch angle β:

$$\beta_{DR} = \tan^{-1}\left(\frac{\overline{w_1}}{\overline{u_1}}\right) \quad (3.19)$$

This DR scheme ends in what is termed a "natural wind system," firstly introduced by Tanner and Thurtell (1969) and further described by McMillen (1988) and Kaimal and Finnigan (1994) among others.

After these two rotations, no further information can be extracted from the velocity vector but there is still infinity of orientations for the vector basis due to the last degree of freedom around the x-axis. A third rotation, introduced by McMillen (1988), was originally intended to minimize the $\overline{v'w'}$ momentum flux. In practice, it was often found that this rotation results in unphysical orientation of the vector basis and is thus not recommended anymore (Finnigan 2004). Instead, the anemometer vertical axis should be aligned as closely perpendicular as possible to the underlying surface and then just the first two rotations have to be applied. In these conditions, Eqs. 3.14–3.17 are still valid, provided that R_{23} is the identity matrix.

The DR is an efficient way to level the anemometer to the surface in an idealized homogeneous flow and has the advantage to be usable online, even when the orientation of the anemometer is modified. However, drawbacks of the DR rotation procedure became apparent when eddy covariance measurements were performed above non-flat terrain and on long-term basis. Limitations are the risk of over-rotation (if there is an electronic offset in the measurement of w, it will be interpreted erroneously as a tilt), the loss of information (information on possible non zero $\overline{w}$ are missed), degradation of data quality (unrealistically large pitch angles in low wind speed conditions) and high-pass filtering of the data (it produces the undesirable effect of having turbulent time series that are discontinuous, see Lee et al. 2004).

Planar-Fit Method

More often and especially above tall vegetation or complex terrain, a non zero mean (30 min) vertical wind velocity may exist and has to be taken into account as pointed out first by Lee (1998) and Paw U et al. (2000). An alternative rotation procedure, the so-called "planar-fit method" was therefore proposed by Wilczak et al. (2001) based on the assumption that the vertical wind component is only equal to zero over longer averaging periods, usually weeks or longer, representing different typical flow features of a site under investigation.

To define this reference system, a mean streamline plane is first defined on the basis of measurements made on periods long enough to encompass all wind directions and a sample size that allows robust averaging. The z-axis is then fixed as perpendicular to this plane, the x-axis as the normal projection of the (30 min) mean wind velocity on this plane, and the y-axis as the normal to the two other axes. To obtain the mean streamline plane, a multiple linear regression is performed on the (30 min) wind components following:

$$\bar{w}_0 = b_0 + b_1\bar{u}_0 + b_2\bar{v}_0 \tag{3.20}$$

from which the regression coefficients b_0, b_1, and b_2 are deduced. b_0 gives the instrumental offset in the vertical velocity component that must be subtracted from w_0 in further calculations. b_1 and b_2 are used to determine the pitch (β_{PF}) and roll angles (γ_{PF}). Combining Eqs. 42 and 44 of Wilczak et al. (2001), these angles can be obtained as

$$\sin\beta_{PF} = \frac{-b_1}{\sqrt{b_1^2 + b_2^2 + 1}}, \quad \cos\beta_{PF} = \frac{\sqrt{b_2^2 + 1}}{\sqrt{b_1^2 + b_2^2 + 1}} \tag{3.21}$$

$$\sin \gamma_{PF} = \frac{b_2}{\sqrt{1+b_2^2}}, \cos \gamma_{PF} = \frac{1}{\sqrt{1+b_2^2}}$$

The pitch and roll rotations are applied with these fixed angles to each (30 min) individual periods that were used for the determination of the coefficients. The z-axis of the reference coordinate system is perpendicular to the long-term local mean streamline plane. Finally, the yaw rotation is applied for each individual period with a varying angle:

$$\alpha_{PF} = \tan^{-1}\left(\frac{\overline{v_2}}{\overline{u_2}}\right) \quad (3.22)$$

Because rotations are not commutative and because the regression coefficients have been computed from the wind components in the sonic anemometer frame, the pitch and roll rotations must be applied before the yaw rotation so that the definition of the R_{03} matrix given above should be modified accordingly.

It is recommended to reject low wind speed conditions (generally below 1 m s^{-1}) for the computation of the regression coefficients, thereby removing the problem of unrealistically large pitch angles. Depending on the complexity of the topography the data set can be split into different wind sectors, to determine different planes for different wind sectors, but it has to be ensured that an appropriate number of data sets contribute to the calculation for each sector. This method is called the "sector-wise planar-fit method".

By relying on an ensemble of observations, the coordinate system is stable through time and the x-y plane is more or less parallel to the local surface. Sites where a systematic vertical motion exists are exceptions to this (forest edges or abrupt changes in the topography) but these sites are very rare and will face a lot of other methodological problems for flux computation. Using the planar-fit method, the drawbacks of the DR are overcome. Indeed, the risk of over rotation is minimized because the z-axis is now independent of the wind direction. Information on the two-dimensional or three-dimensional nature of the flow is now available, including the non zero (30 min) mean vertical velocity which allows the investigation of non-turbulent advective fluxes (see Sect. 5.4.2). Again, these advantages over the DR method will be substantial mainly over nonideal sites and/or under bad weather conditions. The planar-fit method on a site enables valuable insights in the complexity of the flow, especially over forests (Lee et al. 2004).

The planar-fit method has also some drawbacks. Regression coefficients depend on the anemometer orientation, the possible instrumental offset in vertical velocity, and the canopy structure. Therefore, they have to be computed every time one of these parameters is modified. Together with the requirement of a long data set available for the estimation of these coefficients, this can be a limitation of the applicability of the method at particular sites (see Sect. 12.3). Finally, influences of atmospheric stability and strong winds, for example, remain to be investigated and thorough intercomparisons of rotation procedures are still sparse (Su et al. 2008).

3.3 Flux Determination

As described in Sects. 1.3 and 1.4, the vertical turbulent flux of any scalar χ_s can be deduced from the covariance of vertically rotated wind speed (w) and the mixing ratio of this scalar that have been treated as described above.

The general shape of the flux is given by:

$$F_s^{EC} = \overline{\rho_d} \cdot \overline{w'\chi_s'} = \frac{\overline{p_d} \cdot m_d}{R \cdot \bar{\theta}} \cdot \overline{w'\chi_s'} \tag{3.23}$$

Specific shapes for each flux will be detailed below. If, in Eq. 3.23 and in the following equations, the average sonic temperature $\bar{\theta}_s$ was used instead of true air temperature $\bar{\theta}$, Eq. 3.3 should be applied in order to account for the difference between these two variables (Liu et al. 2001; Schotanus et al. 1983). When the concentration is expressed in terms of density or molar concentration, further corrections are needed to take high-frequency dry air density fluctuations into account (Sect. 4.1.4.1).

3.3.1 Momentum Flux

After application of the rotation, the *momentum flux* τ (kg m^{-2} s^{-1}) may be determined from the fluctuations of vertical (w) and horizontal (u) wind components:

$$\tau = \overline{\rho_d} \cdot \overline{w'u'} = \frac{\overline{p_d} \cdot m_d}{R \cdot \bar{\theta}} \cdot \overline{w'u'} \tag{3.24}$$

Friction velocity, u_* (ms^{-1}), may be directly deduced from the covariance of vertical and horizontal wind components, as

$$u_* = \sqrt{-\overline{u'w'}} \tag{3.25}$$

3.3.2 Buoyancy Flux and Sensible Heat Flux

The *buoyancy flux* can be determined from fluctuations of vertical wind component and sonic temperature θ_s fluctuations:

$$H_s = \overline{\rho_d} c_p \overline{w'\theta'_s} = \frac{\overline{p_d} \cdot m_d}{R \cdot \bar{\theta}} \cdot c_p \overline{w'\theta'_s} \tag{3.26}$$

while the sensible heat flux writes:

$$H_s = \overline{\rho_d} c_p \overline{w'\theta'_s} = \frac{\overline{p_d} \cdot m_d}{R \cdot \bar{\theta}} \cdot c_p \overline{w'\theta'_s} \tag{3.27}$$

Fluctuations of true air temperature in Eq. 3.27 may be deduced from Eq. 3.3 only if high-frequency measurements of water vapor pressure and air pressure are available. If not, the conversion of buoyancy flux into sensible heat flux must be done on the basis of the averaged data. This correction is described in detail in Sect. 4.1.4.2.

3.3.3 Latent Heat Flux and Other Trace Gas Fluxes

For all other tracers, the conversion of velocity - mass mixing ratio covariances into mass flux can be performed by using Eq. 3.23. Alternatively, the equations computing fluxes in mass or molar units from scalar concentrations expressed in molar mixing ratio, mass mixing ratios, molar concentration, or density are presented in Table 3.1.

The turbulent mass flux of water vapor, F_v^{EC}, may be deduced from (3.23) where the scalar is water vapor mixing ratio. Often the water vapor flux is expressed as latent heat flux (W m^{-2}) which is then determined as

$$\lambda E = F_v^{EC} \cdot \lambda \tag{3.28}$$

where $\lambda = 3147.5 - 2.372\ \theta$ (θ in K) is the latent heat of vaporization for water (J kg^{-1}).

3.3.4 Derivation of Additional Parameters

One of the most important parameters in micrometeorological applications that describes the atmospheric stratification is the stability parameter ζ, defined as

$$\zeta = \frac{h_m - d}{L} \tag{3.29}$$

where h_{m} is the measuring height, d the zero plane displacement height, and L the Obukhov-length, which relates dynamic, thermal, and buoyant processes. This length is defined as

$$L = -\frac{u_*^3}{\kappa \cdot \left(g/\bar{\theta}_S\right) \cdot \overline{w'\theta'_s}} \tag{3.30}$$

Another parameter of interest is the Bowen ratio that relates sensible and latent heat flux:

$$Bo = \frac{H}{\lambda_v F_v^{EC}} \tag{3.31}$$

which is a helpful measure if energy partitioning is investigated.

Table 3.1 Formulas computing mass or molar fluxes from scalar concentrations expressed in molar mixing ratios, mass mixing ratios, molar concentrations or density. In the two latter cases, Webb-Pearman-Leuning (WPL) corrections should be applied (see Sect. 4.1.3)

	Mass flux $F_s^{EC} =$	Molar flux $F_{s,mol}^{EC} =$
Molar mixing Ratio, χ_s	$\frac{m_s\, p_d}{R\,\overline{\theta}}\overline{w'\chi'_s}$	$\frac{p_d}{R\,\overline{\theta}}\overline{w'\chi'_s}$
Mass mixing Ratio, χ_{sm}	$\frac{m_d\, p_d}{R\,\overline{\theta}}\overline{w'\chi'_{sm}}$	$\frac{m_d\, p_d}{m_s R\,\overline{\theta}}\overline{w'\chi'_{sm}}$
Molar concentration, c_s (incl. WPL correction)	$m_s\overline{w'c'_s} + m_s\frac{\overline{c_s}}{\overline{c_d}}\left[(\overline{c_d}+\overline{c_v})\frac{\overline{w'\theta'}}{\overline{\theta}} + \overline{w'c'_v}\right]$	$\overline{w'c'_s} + \frac{\overline{c_s}}{\overline{c_d}}\left[(\overline{c_d}+\overline{c_v})\frac{\overline{w'\theta'}}{\overline{\theta}} + \overline{w'c'_v}\right]$
Density, ρ_s (incl. WPL correction)	$\overline{w'\rho'_s} + \frac{\overline{\rho_s}}{\overline{\rho_d}}\left[\left(\overline{\rho_d}+\frac{m_d}{m_v}\overline{\rho_v}\right)\frac{\overline{w'\theta'}}{\overline{\theta}} + \frac{m_d}{m_v}\overline{w'\rho'_v}\right]$	$\frac{1}{m_s}\left[\overline{w'\rho'_s} + \frac{\overline{\rho_s}}{\overline{\rho_d}}\left[\left(\overline{\rho_d}+\frac{m_d}{m_v}\overline{\rho_v}\right)\frac{\overline{w'\theta'}}{\overline{\theta}} + \frac{m_d}{m_v}\overline{w'\rho'_v}\right]\right]$

Acknowledgements MA and BH acknowledge financial support by the European Union (FP 5, 6 and 7), the Belgian Fonds de la recherche Scientifique (FNRS-FRS), the Belgian Federal Science Policy Office (BELSPO), and the Communauté française de Belgique (Action de Recherche Concertée). AI was funded by the EU FP6 project IMECC. RQ was funded by the TUD, OK and CR were funded by the MPI for Biogeochemistry, Jena, Germany and Helmholtz Centre for Environmental Research.

References

Aubinet M, Grelle A, Ibrom A, Rannik Ü, Moncrieff J, Foken T, Kowalski A, Martin P, Berbigier P, Bernhofer C, Clement R, Elbers J, Granier A, Grünwald T, Morgenstern K, Pilegaard K, Rebmann C, Snijders W, Valentini R, Vesala T (2000) Estimates of the annual net carbon and water exchange of forests: the EUROFLUX methodology. Adv Ecol Res 30:113–175

Clement RJ (2004) Mass and energy exchange of a plantation forest in Scotland using micrometeorological methods. PhD, University of Edinburgh, Edinburgh, 597 pp

Cuerva A, Sanz-Andres A, Navarro J (2003) On multiple-path sonic anemometer measurement theory. Exp Fluid 34(3):345–357

Culf AD (2000) Examples of the effects of different averaging methods on carbon dioxide fluxes calculated using the eddy correlation method. Hydrol Earth Syst Sci 4(1):193–198

Draper NR, Smith HNYW (1998) Applied regression analysis. Wiley, New York, 736 pp

Eugster W, Pluss P (2010) A fault-tolerant eddy covariance system for measuring CH_4 fluxes. Agric For Meteorol 150(6):841–851

Finnigan JJ (2004) A re-evaluation of long-term flux measurement techniques – Part II: coordinate systems. Bound Layer Meteorol 113(1):1–41

Finnigan JJ, Clement R, Malhi Y, Leuning R, Cleugh HA (2003) A re-evaluation of long-term flux measurement techniques – Part I: averaging and coordinate rotation. Bound Layer Meteorol 107(1):1–48

Foken T (2008) Micrometeorology. Springer, Berlin

Foken T, Göckede M, Mauder M, Mahrt L, Amiro BD, Munger JW (2004) Post-field data quality control. In: Lee X, Massman W, Law B (eds) Handbook of micrometeorology: a guide for surface flux measurements. Kluwer, Dordrecht, pp 81–108

Grelle A, Lindroth A (1994) Flow distortion by a Solent sonic anemometer: wind tunnel calibration and its assessment for flux measurements over forest and field. J Atmos Ocean Technol 11(6):1529–1542

Heinesch B, Yernaux M, Aubinet M (2007) Some methodological questions concerning advection measurements: a case study. Bound Layer Meteorol 122(2):457–478

Hojstrup J (1993) A statistical-data screening-procedure. Meas Sci Technol 4(2):153–157

Ibrom A, Dellwik E, Flyvbjerg H, Jensen NO, Pilegaard K (2007a) Strong low-pass filtering effects on water vapour flux measurements with closed-path eddy correlation systems. Agric For Meteorol 147:140–156

Ibrom A, Dellwik E, Larsen SE, Pilegaard K (2007b) On the use of the Webb-Pearman-Leuning theory for closed-path eddy correlation measurements. Tellus Ser B Chem Phys Meteorol 59B:937–946

Kaimal JC, Businger JA (1963) A continuous wave sonic anemometer-thermometer. J Appl Meteorol 2:156–164

Kaimal JC, Finnigan JJ (1994) Atmospheric boundary layer flows: their structure and measurement. Oxford University Press, New York, 289 pp

Kaimal JC, Gaynor JE (1991) Another look at sonic thermometry. Bound Layer Meteorol 56: 401–410

Kristensen L (1998) Time series analysis: dealing with imperfect data. Risø National Laboratory, Roskilde

Kristensen L, Mann J, Oncley SP, Wyngaard JC (1997) How close is close enough when measuring scalar fluxes with displaced sensors? J Atmos Ocean Technol 14:814–821

Lee X (1998) On micrometeorological observations of surface-air exchange over tall vegetation. Agric For Meteorol 91:39–49

Lee X, Black TA (1994) Relating eddy correlation sensible heat flux to horizontal sensor separation in the unstable atmospheric surface layer. J Geophys Res Atmos 99(D8):18545–18553

Lee X, Finnigan J, Paw UKT (2004) Coordinate systems and flux bias error. In: Lee X, Massman W, Law B (eds) Handbook of micrometeorology. Kluwer, Dordrecht, pp 33–66

Lenschow DH, Mann J, Kristensen L (1994) How long is long enough when measuring fluxes and other turbulence statistics. J Atmos Ocean Technol 11(3):661–673

Liu HP, Peters G, Foken T (2001) New equations for sonic temperature variance and buoyancy heat flux with an omnidirectional sonic anemometer. Bound Layer Meteorol 100(3):459–468

Massman WJ, Ibrom A (2008) Attenuation of concentration fluctuations of water vapor and other trace gases in turbulent tube flow. Atmos Chem Phys 8(20):6245–6259

McMillen RT (1988) An eddy correlation technique with extended applicability to non-simple terrain. Bound Layer Meteorol 43:231–245

Moncrieff JB, Massheder JM, de Bruin H, Elbers J, Friborg T, Heusinkveld B, Kabat P, Scott S, Soegaard H, Verhoef A (1997) A system to measure surface fluxes of momentum, sensible heat, water vapour and carbon dioxide. J Hydrol 188–189:589–611

Moncrieff J, Clement R, Finnigan J, Meyers T (2004) Averaging, detrending, and filtering of eddy covariance time series. In: Lee X, Massman W, Law B (eds) Handbook of micrometeorology. Kluwer, Dordrecht, pp 7–31

Paw UKT, Baldocchi DD, Meyers TP, Wilson KB (2000) Correction of eddy-covariance measurements incorporating both advective effects and density fluxes. Bound Layer Meteorol 97:487–511

Rannik Ü, Vesala T (1999) Autoregressive filtering versus linear detrending in estimation of fluxes by the eddy covariance method. Bound Layer Meteorol 91(2):259–280

Schmid HP, Grimmond CSB, Cropley F, Offerle B, Su H-B (2000) Measurements of CO_2 and energy fluxes over a mixed hardwood forest in the mid-western United States. Agric For Meteorol 103:357–374

Schotanus P, Nieuwstadt FTM, De Bruin HAR (1983) Temperature measurements with a sonic anemometer and its application to heat and moisture fluxes. Bound Layer Meteorol 26:81–93

Su HB, Schmid HP, Vogel CS, Curtis PS (2008) Effects of canopy morphology and thermal stability on mean flow and turbulence statistics observed inside a mixed hardwood forest. Agric For Meteorol 148(6–7):862–882

Sun JL (2007) Tilt corrections over complex terrain and their implication for CO_2 transport. Bound Layer Meteorol 124(2):143–159

Tanner CB, Thurtell GW (1969) Anemoclinometer measurements of Reynolds stress and heat transport in the atmospheric surface layer. US Army Electronics Command, Department of Soil Science, University of Wisconsin, Madison

Vickers D, Mahrt L (1997) Quality control and flux sampling problems for tower and aircraft data. J Atmos Ocean Technol 14:512–526

Vickers D, Thomas C, Law BE (2009) Random and systematic CO_2 flux sampling errors for tower measurements over forests in the convective boundary layer. Agric For Meteorol 149(1):73–83

Vogt R (1995) Theorie, Technik und Analyse der experimentellen Flußbestimmung – ein Beitrag zu den Energiebilanzuntersuchungen über Wäldern beim REKLIP. Geographisches Institut der Universität Basel, University of Basel, Basel, 101 pp

Wilczak JM, Oncley SP, Stage SA (2001) Sonic anemometer tilt correction algorithms. Bound Layer Meteorol 99(1):127–150

Chapter 4
Corrections and Data Quality Control

Thomas Foken, Ray Leuning, Steven R. Oncley, Matthias Mauder, and Marc Aubinet

This chapter describes corrections that must be applied to measurements because practical instrumentation cannot fully meet the requirements of the underlying micrometeorological theory. Typically, measurements are made in a finite sampling volume rather than at a single point, and the maximum frequency response of the sensors is less than the highest frequencies of the turbulent eddies responsible for the heat and mass transport. Both of these cause a loss of the high-frequency component of the covariances used to calculate fluxes. Errors also arise in calculating fluxes of trace gas quantities using open-path analyzers because of spurious density fluctuations arising from the fluxes of heat and water vapor. This chapter gives the reader an overview of how these sources of error can be eliminated or reduced using some model assumptions and additional measurements. Corrections needed for some specific instruments are presented (Sect. 4.1), followed by a discussion

---

T. Foken (✉)
Department of Micrometeorology, University of Bayreuth, 95440 Bayreuth, Germany
e-mail: thomas.foken@uni-bayreuth.de

R. Leuning
Marine and Atmospheric Research, CSIRO, PO Box 3023, Canberra, ACT 2601, Australia
e-mail: ray.leuning@csiro.au

S.R. Oncley
Earth Observing Laboratory, NCAR, P.O. Box 3000, Boulder, CO, 80307-3000, USA
e-mail: oncley@ucar.edu

M. Mauder
Institute for Meteorology and Climate Research, Atmospheric Environmental Research, Karlsruhe Institute of Technology, Kreuzeckbahnstr. 19, 82467 Garmisch-Partenkirchen, Germany
e-mail: matthias.mauder@kit.edu

M. Aubinet
Unit of Biosystem Physics, Gembloux Agro-Bio Tech, University of Liege, 5030 Gembloux, Belgium
e-mail: Marc.Aubinet@ulg.ac.be

M. Aubinet et al. (eds.), *Eddy Covariance: A Practical Guide to Measurement and Data Analysis*, Springer Atmospheric Sciences, DOI 10.1007/978-94-007-2351-1_4,

of the generally observed lack of closure of the energy balance using the sum of latent and sensible heat fluxes (Sect. 4.2). The chapter closes with a discussion of measures needed to determine the quality of the final calculated fluxes (Sect. 4.3).

4.1 Flux Data Correction

4.1.1 Corrections Already Included into the Raw Data Analysis (Chap. 3)

In this chapter, we assume that several steps of preprocessing of the raw velocity and scalar time series have been completed (Sect. 3.2.2). These include despiking of the raw data (Højstrup 1993; Vickers and Mahrt 1997) and cross-correlation of the time series data in order to shift all signals to the same time base. The most important issues are the delay time of closed-path sensors (Sect. 3.2.3.2) and digitization delays in the sensors. The acoustic temperature measured by sonic anemometers must be corrected for crosswind influences (Schotanus et al. 1983; Liu et al. 2001). Current sonic anemometers include this correction in their firmware, but this is not the case for some older anemometers (Sect. 2.3.2) It is also assumed that the coordinate system has been rotated to ensure zero vertical wind velocity over a certain averaging time (Sect. 3.2.4). This rotation is called a tilt correction (Tanner and Thurtell 1969; Hyson et al. 1977) or coordinate rotation (Kaimal and Finnigan 1994) and can be applied to two or three axes, where the three-axis rotation is not recommended (Finnigan et al. 2003). Currently the planar-fit method (Wilczak et al. 2001) is often used, which overcomes some deficiencies of the double rotation for short averaging times. For more details see Chap. 3.

4.1.2 Conversion of Buoyancy Flux to Sensible Heat Flux (SND-correction)

Conversion of buoyancy flux to sensible heat flux is known as SND-correction after the three authors of the paper Schotanus et al. (1983), formerly also called the Schotanus-correction. It is based on the transformation of sonic or acoustic temperature (θ_s) to actual air temperature using (see also Chap. 3, Eq. 3.3)

$$\theta_s = \theta\left(1 + 0.32\frac{e}{p}\right) \tag{4.1}$$

where p is atmospheric pressure and e is the partial pressure of water vapor. Extra measurements of humidity are thus needed to calculate the sensible heat flux ($H = \rho c_p \overline{w'\theta'}$) while $\rho c_p \overline{w'\theta_s'}$ is the buoyancy flux (Kaimal and Gaynor 1991).

Application of Reynolds decomposition to Eq. 4.1 and computation of the temperature variance and covariances lead to the following relations (Schotanus et al. 1983)

$$\sigma_{\theta}^{2} = \sigma_{\theta s}^{2} - 1.02\,\bar{\theta}\,\overline{\chi'_{v}\theta'} - 0.51^{2}\,\overline{\chi'^{2}_{v}}\,\bar{\theta}^{2} \tag{4.2}$$

$$\overline{w'\theta'} = \overline{w'\theta'_{s}} - 0.51\,\bar{\theta}\,\overline{w'\chi'_{v}}, \tag{4.3}$$

where the factor 0.51 results from the multiplication of factor 0.32 in Eq. 4.1 by the ratio of dry air and water molar masses. This correction is straightforward if the latent heat flux is also measured. If such measurements are not available, an estimate of the Bowen ratio $Bo = H/\lambda E$, for example, from the temperature and moisture difference between two levels, can be used (Arya 2001; Hatfield and Baker 2005; Foken 2008b; Monteith and Unsworth 2008). In this case, the sensible heat flux is calculated using (λ: heat of evaporation):

$$H = \rho c_{p}(\overline{w'\theta'}) = \rho c_{p}\frac{(\overline{w'\theta'_{s}})}{1 + \frac{0.51 c_{p}\bar{\theta}}{\lambda Bo}} \tag{4.4}$$

Because the Bowen-ratio involves the sensible and latent heat flux, both must be known to apply this method. Thus, the solution of this equation often is made iteratively to update the sensible heat flux in the Bowen-ratio. However, Oncley et al. (2007) show that the equations can be solved simultaneously using the two Eqs. 4.3 and 4.4 and the calculation of both corrected fluxes of sensible and latent heat.

Note that the buoyancy flux required, for example, in the computation of the Obukhov length (Foken, 2006). The buoyancy flux is $\overline{w'\theta_v'}$ which is close to, but not exactly the same as, $\overline{w'\theta_s'}$ (see Sect. 3.2.1.1). When Bo is large, $\overline{w'\theta'_s} \cong \overline{w'\theta'}$ and no correction to $\overline{w'\theta'}$is needed. However, in the general case, it is necessary to use Eq. 4.3 and then compute $\overline{w'\theta'_v}$ from $\overline{w'\theta'} + 0.51\,\overline{\theta}\overline{w'\chi'_v}$.

4.1.3 Spectral Corrections

4.1.3.1 Introduction

Eddy covariance systems, like all measuring instruments, act as filters, removing both high- and low-frequency components of the signal. High-frequency losses are mainly due to inadequate sensor frequency response, line averaging, sensor separation and, in closed-path systems, air transport through the tubes. The impact of high-frequency losses on cospectral density are illustrated schematically in Fig. 4.1 (see also Sect. 1.5). Correction procedures for high-frequency losses are

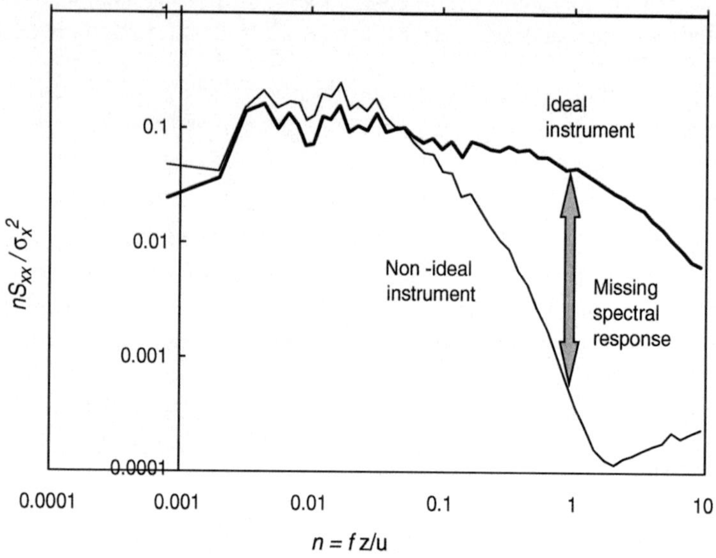

Fig. 4.1 Normalized turbulence spectra for an ideal instrument, which measures the unaffected turbulence spectra, and a nonideal instrument. The missing energy between both response curves must be corrected (n normalized frequency; f frequency, z height, u wind velocity, S_{xx}: energy density of the parameter x; σ_x^2: dispersion of the parameter x)

described in Sect. 4.1.3.2. Low-frequency losses result from the finite sampling duration, with the averaging period not always being sufficiently long to include all relevant low frequencies. The use of detrending or recursive filtering, by attenuating fluctuations at periods larger than the filter time constant, may enhance this effect. They are therefore not recommended in general. Their use could however be necessary, when a sensor calibration drift should be removed. In this case, all information relative to fluctuations at period larger than recursive filter time constant would be lost.

4.1.3.2 High-Frequency Loss Corrections

The relative error on the flux introduced by high-cut (often referred to as low-pass) filtering was described in Chap. 1, Eq. 1.31 by the formula:

$$\delta_s / F_s^{EC} = 1 - \frac{\int_0^\infty C_{ws}(f) T_{ws}(f) df}{\int_0^\infty C_{ws}(f) df}, \tag{4.5}$$

where C_{ws} represents the nonfiltered or 'ideal' cospectral density and T_{ws} represents the transfer function of the system, involving high-cut and, possibly, low-cut (high-pass) filtering effects. This section is focused on the computation of the correction

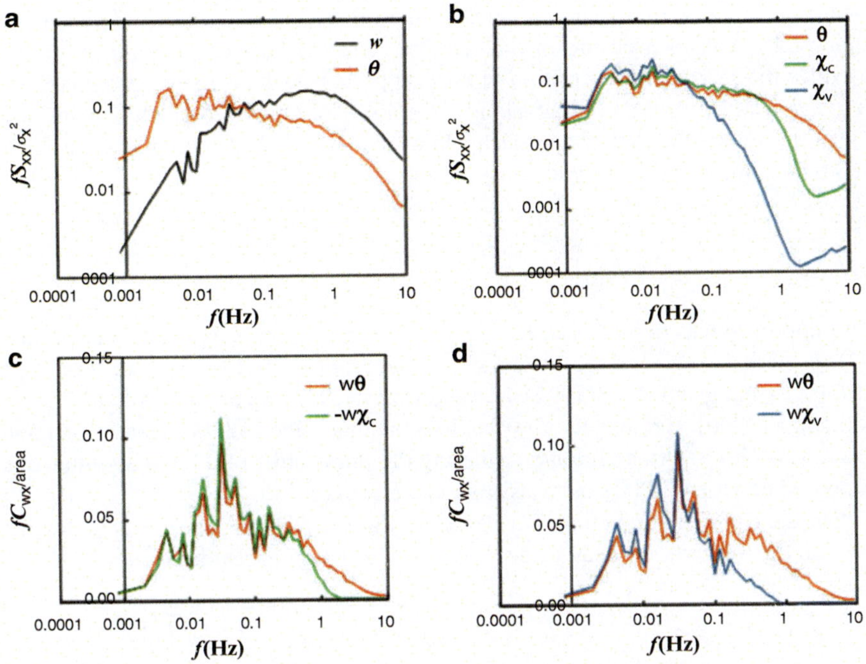

Fig. 4.2 Normalized turbulence power spectra S_{xx} for (a) w and θ and (b) θ, χ_c and χ_v. Normalized cospectra C_{xy} for (c) $w\theta$ and $-w\chi_c$, and (d) for $w\theta$ and $w\chi_v$ (Leuning, unpublished results). Symbols: n: normalized frequency; f frequency, z height, u wind velocity. Power spectra are normalized by σ_x^2: the variance of x; cospectra are normalized by the area under the respective curves

for high-cut filtering effects. This could be evaluated if both the system transfer function $T_{ws}(f)$ and the nonfiltered cospectrum $C_{ws}(f)$ are known.

The effects of high-frequency losses on spectral and cospectral density are illustrated in Fig. 4.2. Poor high-frequency response of the (closed-path) measurement system causes the CO_2 cospectrum (C_{wc}) to roll-off more quickly than for temperature ($C_{w\theta}$), and the roll-off is even more rapid for the water vapor cospectrum (C_{wv}) due to its adsorption/desorption on tubing walls. Reduced high-frequency response causes the normalized C_{wc} cospectrum to be lower than that for the normalized $C_{w\theta}$ cospectrum, and the effect is even greater for the C_{wv} cospectrum. This causes the fluxes of CO_2 and water vapor to be underestimated. The following section describes two approaches commonly used to correct for imperfect instrument high-frequency responses.

In the theoretical approach, the transfer function T_{ws} is deduced from prior knowledge of the measuring system and the cospectral function C_{ws}. In the experimental approach, the fractional error is computed using the normalized ratio of two cospectral density functions measured simultaneously at the same site: one referring to the filtered scalar of interest, the other considered as a reference and supposedly real cospectrum. In practice, the sensible heat cospectrum is usually used for the second. A description of these approaches, their implementation, their respective advantages and disadvantages are discussed below. Many applicants use the more general theoretical approach while the experimental approaches need site- and sensor-specific investigations.

The Theoretical Approach

This approach was first devised for eddy covariance systems by Moore (1986) and later extended to CO_2 closed-path systems, in particular by Leuning and Moncrieff (1990), Leuning and King (1992), Lee and Black (1994), Leuning and Judd (1996), Massman (2000), Ibrom et al. (2007a), Massman and Ibrom (2008), and Horst and Lenschow (2009). An addition considering the phase shift of a low-pass filter was proposed by Horst (2000) and Massman and Ibrom (2008).

The total transfer function T_{ws} describing an eddy covariance system for the vertical flux of a tracer s may be described as a function of real frequency f:

$$T_{ws}(f) = G_w(f) \cdot G_s(f) \cdot T_{ss}(f) \cdot \sqrt{T_{pw}(f)} \cdot \sqrt{T_{ps}(f)} \cdot \sqrt{T_{ta}(f)} \qquad (4.6)$$

where:

- $G_{w(s)}(f)$ describe the high frequency loss by the sensors and are defined as

$$G_{w(s)} = \left[1 + \left(2\pi f \tau_{w(s)}\right)^2\right]^{-1/2} \qquad (4.7)$$

  in which $\tau_{w(s)}$ are time constants, specific to the sensors (Moore 1986), see also Horst (1997)
- $T_{ss}(f)$ describes the high-cut filtering due to lateral separation between two instruments (in this case the sonic anemometer and the open-path analyzer or the inlet of the closed-path analyzer). Moore (1986) gives a simple expression for this displacement, based on empirical fits to an isotropic turbulence model:

$$T_{ss}(n) = e^{-9.9 n_{ss}^{1.5}} \qquad (4.8)$$

  where:

$$n_{ss} = \frac{f\, d_{ss}}{\bar{u}} \qquad (4.9)$$

  Here, $\bar{u}$ is the average wind speed and $d_{ss} = d_{sa}\,|\sin(\beta_d)|$ is the effective lateral separation distance between the two sensors. d_{sa} is the actual separation distance and β_d is the angle between the line joining the sensors and the wind direction. However, only nonisotropic turbulence can produce a flux. For this reason, Horst and Lenschow (2009) created a more realistic nonisotropic turbulence model that depends on stability from a unique set of field observations and used it to obtain somewhat more involved expressions for T_{ss} (n_{ss}).

- T_{pw} represents the transfer function for the wind vector component line averaging. It can be approximated by (Kaimal et al. 1968; Horst 1973):

$$T_{pw}(n) = \frac{2}{\pi n_w}\left(1 + \frac{e^{-2\pi n_w}}{2} - 3\frac{1 - e^{-2\pi n_w}}{4\pi n_w}\right) \tag{4.10}$$

where

$$n_w = \frac{f\,d_{pl}}{\bar{u}} \tag{4.11}$$

and d_{pl} represents the sonic anemometer path length.

- T_{ps} represents the transfer function for the scalar line averaging (Moore 1986):

$$T_{ps}(n) = \frac{1}{2\pi n_s}\left(3 + e^{-2\pi n_s} - 4\frac{1 - e^{-2\pi n_s}}{2\pi n_s}\right) \tag{4.12}$$

where:

$$n_s = \frac{f\,d_s}{\overline{u_{pl}}} \tag{4.13}$$

u_{pl} is the average air speed inside the sensor path length and d_s represents the infrared gas analyzer (IRGA) path length or the path length of the sonic anemometer for temperature measurements.

- T_{ta} describes the fluctuation attenuation due to air transport in the tubes of closed path systems. When flow in the tube is laminar, Lenschow and Raupach (1991) and Leuning and King (1992) proposed the following formulation for passive scalars such as CO_2:

$$T_{ta} = \exp\left\{-\frac{\pi^3 r_t^4 f^2 L_t}{6 D_s Q}\right\} \tag{4.14}$$

or when the flow in the tube is turbulent:

$$T_{ta} = exp\left\{-160 \cdot Re^{-1/8} \cdot \frac{\pi^2 r_t^5 f^2 L_t}{Q^2}\right\} \tag{4.15}$$

Here, r_t and L_t are the tube radius and length, Q is the volumetric flow rate in the tube, and D_s is the molecular diffusivity of scalar s. The Reynolds number is defined as $Re = \frac{2Q}{\pi r_t v}$, in which υ is the kinematic viscosity. Turbulent flow in tubing occurs when Re $\gtrsim$ 2300.

Massman and Ibrom (2008) recently reexamined these formulations and found that Eq. 4.15 tends to underestimate the attenuation of fluctuations. For passive scalars, they proposed the alternative expression

$$T_{ta} = \exp\left\{-\left(160 \cdot Re^{-1/8} + 2666 \cdot Re^{-29/40}\right)\frac{\pi^2 r_t^5 f^2 L_t}{Q^2}\right\} \quad (4.16)$$

while for scalars adsorbed/desorbed on the walls of the tubing they proposed

$$T_{ta} = \exp\left\{-\left(160 \cdot Re^{-1/8} + 2666 \cdot Re^{-29/40} + 8000 \cdot Sc^{-1/2} \times \left[10^9 \cdot Re^{-2} \cdot r_h \cdot e^{l_* r_h}\right]\right)\frac{\pi^2 r_t^5 f^2 L_t}{Q^2}\right\} \quad (4.17)$$

In (4.17), $Sc = \nu / D_s$ is the Schmidt number, r_{h} the relative humidity, and l_* an empirically determined coefficient set to 8.26 by Massman and Ibrom (2008).

Some additional remarks about the use of the preceding equation set:

All transfer functions must be expressed in terms of the real frequency f before being used with Eq. 4.6.

Equation 4.8 can only be used in the unstable case, and if the sensor separation is less than 10% of the aerodynamic measuring height (height above zero-plane displacement). Under stable stratification, the distance between the sensors should not be greater than 0.7% of the Obukhov length (Moore 1986).

The relation describing line averaging of vector quantities is more complex because of the influence of the sensor geometry (Kaimal et al. 1968; Horst 1973). Equation 4.10 is an approximation that is accurate to about 2% for the vertical wind component.

Equation 4.12 is an approximation when the angle between the line-averaging path of the sensor and the wind field is 90° (Gurvitch 1962). For all angles, the equation is given by Silverman (1968).

A correction of the longitudinal sensor separation is only necessary if the covariance was not maximized by cross correlation analysis (see Sect. 3.2.3.2, Mauder and Foken 2004). The transfer function for lateral separation can also be used for the correction of longitudinal separation, (Moore 1986) since in both cases the 3 dB-point (damping of the signal by $1/\sqrt{2}$) is the same in both transfer functions.

Implementation of the theoretical approach needs a spectral model. For unstable conditions, we recommend using spectra presented by Højstrup (1981) for wind components, those of Kaimal et al. (1972) for scalar spectra, and those of Kristensen et al. (1997) for scalar cospectra. Other models for spectra and cospectra are given in textbooks (Kaimal and Finnigan 1994; Foken 2008b).

A careful application of Eqs. 4.5–4.17 shows that for open-path sensors, the high-frequency spectral correction is not directly dependent on windspeed; δ_s / F_s^{EC} depends only on the ratios of the various length scales, d_{ss}, d_{pl} etc. divided by

h_m–d. The correction factor is weakly dependent on stability under unstable and neutral conditions because there is then no significant variation in the shape of the normalized cospectrum (Kaimal et al. 1972). However, the normalized cospectrum does vary significantly under stable conditions and this causes δ_s / F_s^{EC} to vary strongly with stability.

There is a wind speed dependence of δ_s / F_s^{EC} for closed-path systems when the flow in the tube is laminar. In this case the half-power frequency of filtering by the tubing $f_{0,t}$, is less than that for the high-frequency roll-off of atmospheric turbulence: $f_{o,s} = (2 \sim 5)\ \overline{u}/(h_m - d)$. The attenuation increases with wind speed in this case. If flow in the pipe is sufficiently turbulent, $f_{o,s}$ will be greater than f_c and there will be no dependence of the attenuation on wind speed. The transfer function for flow of passive scalars through tubing is a function of volumetric flow rate, tube length, and radius, while adsorption isotherms need to be considered for sorbing scalars.

The theoretical approach has the advantage of relying on a fundamental description of the system and allows a comprehensive description of filtering processes. However, it also has some shortcomings. First, the theoretical cospectral densities that are proposed may not correspond to the real cospectral densities observed at the sites. Amiro (1990) showed notably that at forested sites, the measured sensible heat cospectral density diverged from the Kaimal et al. (Kaimal et al. 1972) cospectra in the inertial range. This was confirmed experimentally by de Ligne et al. (2010). It is also well known that the low-frequency part of the cospectrum is not universally defined in unstable conditions and could depend on mesoscale movements specific to a given site. Secondly, all processes cannot be thoroughly described by the transfer function, especially for closed-path systems. Indeed, in this case, problems may come from uncertainties in volume flow, if the systems do not employ mass-flow controllers; from uncertainties in the flow regime, the Reynolds number value being not always a sufficient criterion for turbulent flow in the tubing; or from the impact of particle filters on the transfer function (Aubinet et al. 2000; Aubinet et al. 2001). In addition, the transfer functions for tube attenuation (4.14–4.15) are exact only in the case of straight horizontal tubes, which is rarely the case in eddy covariance systems.

In practice, if the theoretical approach is well suited to open-path systems, its application to closed-path systems appears more problematic. In any case, it is recommended to compare the theoretical transfer functions obtained with the set of Eqs. 4.5–4.17 with experimental transfer functions obtained with the procedure described in the next section.

The Experimental Approach

The experimental approach assumes that we can measure without significant error the cospectrum of one quantity (e.g. $\overline{w'\theta'}$) and that this can be used to rescale the cospectrum of another quantity that is subject to high-cut filtering. The computation process consists of: (1) Selecting long time periods (3 h at least in order to reduce the uncertainties on the low-frequency part of the cospectra) with sunny, stationary conditions, with different wind velocities (and different air humidity, as far as sorbing scalars are concerned); (2) for each period, calculating the cospectra for heat and for the scalar of interest and, finally, (3) calculating the transfer function as the ratio of the normalized cospectral densities as

$$T_{ws}^{\exp(f)} = \frac{N_\theta C_{ws}(f)}{N_s C_{w\theta}(f)} \tag{4.18}$$

where N_θ and N_s are normalization factors. Similarity would require that $\frac{N_\theta}{N_s} = \frac{\overline{\omega'\theta'}}{\overline{\omega'\chi_s{}'}}$ but, as covariances are affected by high-frequency attenuation, they cannot be calculated exactly. Aubinet et al. (2000) proposed thus to compute them as

$$\frac{N_\theta}{N_s} = \frac{\int_0^{f'} C_{w\theta}^{\exp}(f)df}{\int_0^{f'} C_{ws}^{\exp}(f)df} \tag{4.19}$$

where the limit frequency, f', is high enough to allow computation of the normalization factor with enough precision, and low enough, to not be affected by high-frequency attenuation (Aubinet et al. 2000).

For each period, a sigmoidal function may be fitted to $T_{ws}^{\exp}(f)$, from which a half-power frequency $f_{o,s}$in Hz may be computed. The half-power frequency (or its relation with wind velocity and, if necessary, air humidity in the case of closed paths with laminar flows) can then be deduced and further used to compute the transfer function response to wind velocity.

Different sigmoidal functions are used to fit the experimental transfer functions. The most commonly used equations are the Gaussian (Aubinet et al. 2001):

$$T_{ws}^{fit}(f) = \exp\left[-\ln(2)\left(\frac{f}{f_{o,s}}\right)^2\right] \tag{4.20}$$

and the Lorentzian (Eugster and Senn 1995):

$$T_{ws}^{fit}(f) = \frac{1}{1+\left(\frac{f}{f_{o,s}}\right)^2} \tag{4.21}$$

Eq. (4.21) can also be considered as characteristic of a first-order filter performed by a resistor-inductor (RL) circuit of unitary resistance and inductance equal to L_{self} in Hz^{-1}. Under these conditions, the filter inductance is related to the half-power frequency by:

$$f_{o,s} = \frac{1}{2\pi\, L_{self}} \tag{4.22}$$

This equation is widely used, for example, by Horst (1997), Su et al. (2004), Ibrom et al. (2007a), Hiller et al. (2008), and Mammarella et al. (2009). However, de Ligne et al. (2010) found that the shape of Eqs. 4.20 and 4.21 was not well suited to water vapor transfer functions, their decrease with frequency being too sharp. They proposed an alternative relation:

$$T_{ws}^{fit}(f) = \exp\left[-\ln(2)\left(\frac{f}{f_{o,s}}\right)^n\right] \quad (4.23)$$

where the parameter n was lower than two and varied (like $f_{o,s}$) with air saturation deficit.Other procedures suited to water vapor transfer functions were proposed by Ibrom et al. (2007a) and Mammarella et al. (2009).

Similar to the theoretical method, implementation of the experimental approach also needs a spectral model. Højstrup (1981), Kaimal et al. (1972) or Kristensen et al. (1997) models could be used as above. Alternatively, the use of experimental, site specific, cospectral models could be relevant above forests (de Ligne et al. 2010). The introduction of the experimental transfer function and of the cospectral model (e.g., Eq. 1.30b) in Eq. 4.5 will lead to a correction function which has the same properties as those obtained with the theoretical approach, that is, it is a single function of the wind speed and measurement height in unstable conditions and of the wind speed, measurement height, and stability in stable conditions. It is stable as long as the set up remains unchanged and could be applied to every individual flux estimate.

The experimental approach relies on different hypotheses: First, it assumes that the processes of atmospheric turbulent transport of sensible heat and other tracers are similar and therefore the cospectral densities should be proportional to each other. This hypothesis was proposed notably by Wyngaard and Coté (1971), Panofsky and Dutton (1984), and Othaki (1985), and it has been widely used in spectral correction schemes. It was tested above forests by various authors (Anderson et al. 1986; Monji et al. 1994; Ruppert et al. 2006) who found high scalar similarity at the midday period.

Secondly, it assumes that the high-frequency attenuation of sensible heat cospectral density is negligible compared with those affecting other tracers (i.e., that fluctuation attenuation due to electronic response time or path averaging takes place at much higher frequencies than attenuation due to mixing in the tube and from sorption/desorption). If this hypothesis clearly makes sense in closed-path systems where the high-frequency losses are mainly due to tube attenuation and sensor separation, it is less relevant in the case of open-path systems. As a result, the experimental approach could lead to an underestimation of the high-frequency correction in open-path systems and would not be recommended in this case (Aubinet et al. 2000; Aubinet et al. 2001).

4.1.3.3 Low-Cut Frequency

The low-cut frequency correction is necessary if very large eddies occur that are not completely sampled over the averaging period. Most researchers use an averaging period of 30 min to calculate eddy fluxes, but this may not be long enough to capture all of the low-frequency contributions to the fluxes (Finnigan et al. 2003). It is therefore best to test if the flux has its maximum value within the adopted averaging time. This is done using the so-called *ogive* test (Desjardins et al. 1989; Oncley et al. 1990; Foken et al. 1995). The ogive (Og_{ws}) is calculated using the cumulative integral of the cospectrum of the turbulent flux beginning with the highest frequencies (Fig. 4.3):

$$Og_{ws}(f_0) = \int_{\infty}^{f_0} C_{ws}(f)\, df \tag{4.24}$$

The averaging period is satisfactory if the value of the integral approaches a constant value (the flux) at low frequencies.

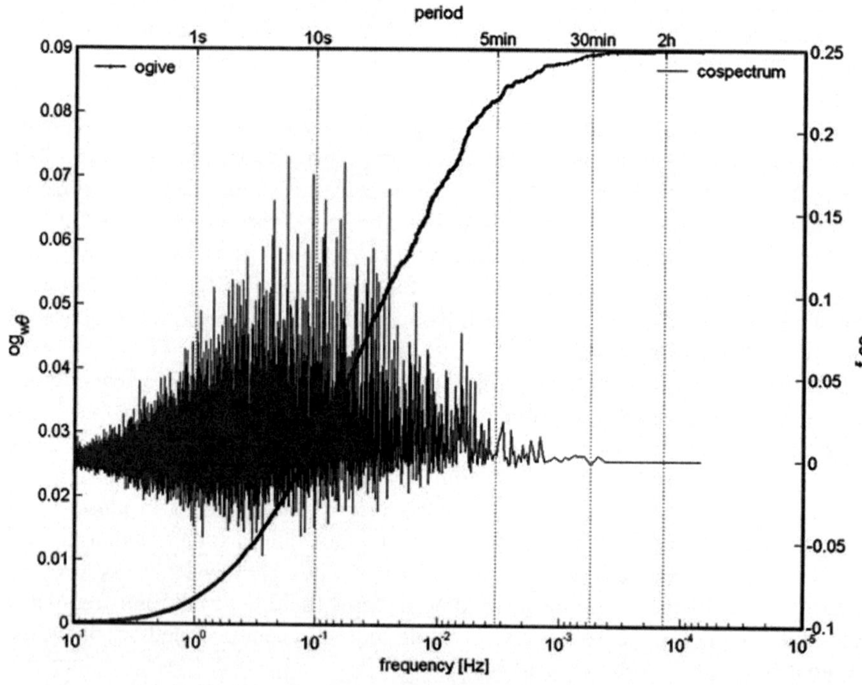

Fig. 4.3 Converging ogive ($Og_{w\theta}$) and cospectrum (f $CO_{w\theta}$) of the sensible heat flux during the LITFASS-2003 experiment (17.06.2003, 12:30–16:30 UTC, Foken et al. 2006)

Foken et al. (2006) have shown for the LITFASS-2003 experiment (Mengelkamp et al. 2006) that in about 80% of all cases, the ogive converged within a period of 30 min. In the remaining cases, mainly in the transition periods of the day, the ogives did not converge or reached a maximum value before the integration time of 30 min, and then decreased in magnitude. In these cases, it would be best to recalculate the fluxes using a different averaging time (longer for the first case and shorter for the second). Since this change of averaging time is difficult to implement in operational data streams, the ogive method is typically used on data from research campaigns. For practical application it may be helpful to check selected time series for unstable, stable, and transition periods and to apply the findings for the data set.

4.1.4 WPL Corrections

4.1.4.1 Introduction

This correction was formerly called the Webb-correction after the first author of a conference paper to correct water vapor fluxes, but now is called WPL-correction after the three authors (Webb, Pearman and Leuning 1980, WPL) who extended its use to measurement of eddy fluxes of trace gases. After the first publication by Webb et al. (1980), several authors discussed this problem, summarized by Fuehrer and Friehe (2002), and Lee and Massman (2011) with different approaches (Bernhardt and Piazena 1988; see Liebethal and Foken 2003; Liebethal and Foken 2004) who found equal results but also controversial solutions (Liu 2005). A clarification of the problem was recently given by Leuning (2004, 2007). The correction is necessary because fluctuations in temperature and humidity cause fluctuations in trace gas concentrations that are not associated with the flux of the trace gas we wish to measure. The correction to the measured flux can be large, for example, the additive correction significantly reduces the CO_2 flux calculated using the covariance of vertical velocity and density (Fig. 4.4). A very careful application of this correction is essential for all trace gases as discussed below.

4.1.4.2 Open-Path Systems

Webb et al. (1980) derived the following expression for the *eddy flux* of a trace gas c to account for the effects of density fluctuations due to temperature and humidity fluctuations when measurements are made using open-path instruments:

$$\overline{F}_c(h_m) = \overline{w'\rho_c'} + \mu\,(\overline{\rho_c}\,/\overline{\rho_d})\,\overline{w'\rho_v'} + (1+\mu\sigma)\left(\overline{\rho_c}/\overline{\theta}\right)\overline{w'\theta'} \tag{4.25}$$

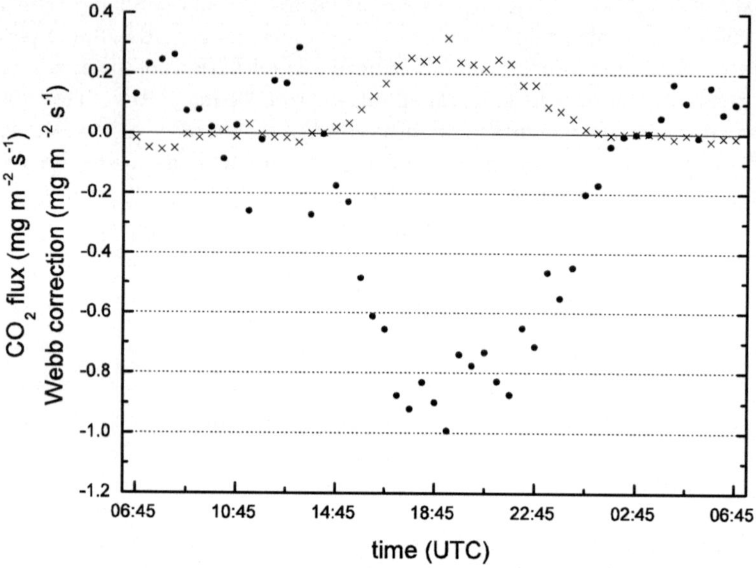

Fig. 4.4 Uncorrected CO_2flux (points) and WPL-correction (crosses) according to Liebethal and Foken (2003). The corrected flux is the sum of both time series

where $\mu = m_d/m_v$ is the ratio of molar masses of dry air and water vapor, and $\sigma = \overline{\rho}_v/\overline{\rho}_d$, the ratio of the densities of water vapor and dry air. The other quantities have been defined earlier.

To derive the last two terms on the right side of this fundamental equation, WPL assumed that there is no flux of dry air through a plane at the measurement height h_m. This assumption is correct for horizontally homogeneous, steady-state conditions, i.e. when there is no change in mass storage within the air layer below h_m. However, for non steady-state flows there *is* a net flux of all components of the air through the plane at h_m, thus violating the assumptions of the original paper and throwing doubt on the WPL equations.

This issue was resolved in Chap. 1 where it was shown that the one-dimensional conservation equation for the trace quantity c under nonsteady-state conditions is given by Eq. 1.23:

$$\underbrace{\overline{F_c} = \overline{F}_c(0)}_{I} + \underbrace{\int_0^{h_m} \overline{S_c} dz}_{II} = \underbrace{\overline{\rho_d}\,\overline{w'\chi'_c}|_{h_m}}_{III} + \underbrace{\int_0^{h_m} \overline{\rho}_d \frac{\overline{\partial \chi_c}}{\partial t} dz}_{IV} + \underbrace{\int_0^{h_m} [\overline{\chi_c}(z) - \overline{\chi_c}(h)] \frac{\overline{\partial \rho_d}}{\partial t} dz}_{V} \tag{4.26}$$

Our objective is to estimate the sum of the flux of c at the ground (term I) plus the integrated contribution of all sources and sinks of c below the measurement height h_m (term II) by measuring the net turbulent flux at h_m (term III), the change in storage (term IV), and the mean flux of c caused by the change in density of dry air (term V). The eddy flux term III in Eq. 4.26 is identical to that derived by WPL for steady-state conditions and it is thus only necessary to add terms IV and V equation for nonsteady-state conditions.

Densities, rather than mixing ratios are typically measured using open-path eddy covariance instruments and following WPL, the eddy flux for a trace gas c is calculated using (Eq. 4.25).

The corresponding eddy flux of water vapor is

$$\overline{F}_v(h_m) = (1 + \mu\sigma)\left[\overline{w'\rho_v'} + \left(\overline{\rho_v}\,/\,\overline{\theta}\right)\overline{w'\theta'}\right] \quad (4.27)$$

4.1.4.3 WPL and Imperfect Instrumentation

The above theory assumes that the eddy fluxes are measured using perfect instrumentation. This requires the cospectral frequency response of the instrument array used to measure $\overline{w'\rho_v'}$ and $\overline{w'\theta'}$ to be identical to that used to measure $\overline{w'\rho_c'}$. As discussed in Sect. 4.1.3, it is necessary to correct for any differences in frequency response of the instrument array *before* applying the WPL corrections. An extreme example is shown in Fig. 4.5 by measurements made by Kondo and Tsukamoto (2008) over an asphalt parking lot where the CO_2 and water vapor fluxes were zero. In this case, the $\overline{w'\theta'}$ correction term should be of equal magnitude but of opposite sign to the raw $\overline{w'\rho_c'}$in each frequency band. The error term arises because of a mismatch in the high-frequency components of the $\overline{w'\theta'}$ and $\overline{w'\rho_c'}$covariances. The true CO_2 flux is obtained by first adjusting the high-frequency component of the raw $\overline{w'\rho_c'}$ covariance to the red line (Sect. 4.1.3) and then applying the WPL corrections.

4.1.4.4 Closed-Path Systems

There is often considerable loss of eddy flux data when open-path gas analyzers are used at sites where rain, mist, and snow impair measurements of trace gas concentrations. Closed-path gas analyzers provide an attractive alternative because of lower rates of data loss but such measurement systems require significantly different corrections for time delays, high-frequency filtering, and density effects compared to those needed for open-path systems (see Sects. 2.4, 4.1.4.2).

Temperature fluctuations in tubing with high thermal conductivity are reduced to 1% of their initial value when the tubing length to radius $L_t/r_t > 1200$ for laminar flow in the tube and when $L_t/r_t > 500$ for turbulent flow (Leuning and Judd 1996; Sect. 4.1.4.2). When temperature fluctuations at all frequencies are eliminated by the air sampling system there is no need for the $\overline{w'\theta'}$correction term in Eq. 4.27 and the fluxes for a trace gas and water vapor are then calculated using

$$\overline{F_c}(h_m) = \left(\bar{p}\bar{\theta}_i\,/\,\bar{p}_i\bar{\theta}\right)\left[\overline{w'\rho'_{c,i}} + \mu\left(\overline{\rho_{c,i}}\,/\,\overline{\rho_{d,i}}\right)\overline{w'\rho'_{v,i}}\right] \quad (4.28)$$

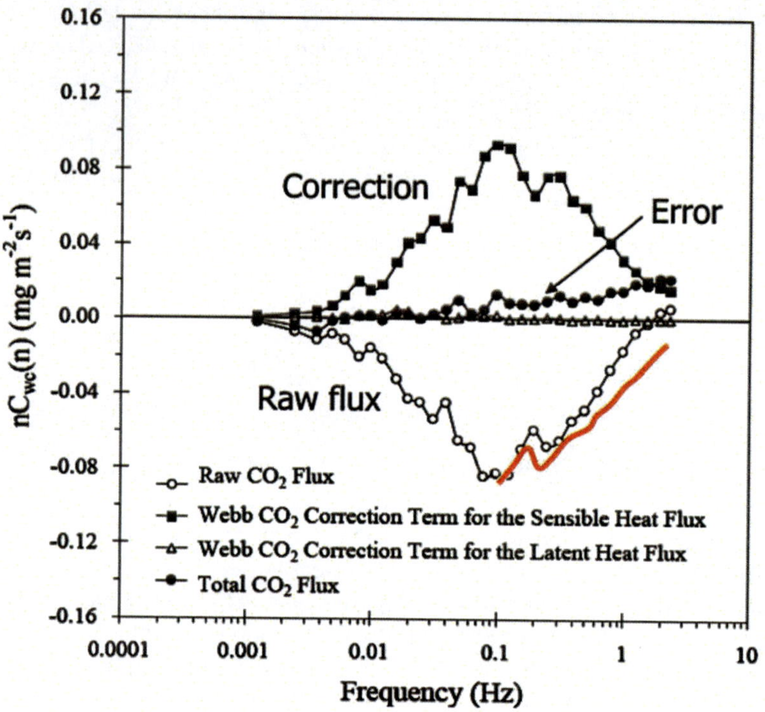

Fig. 4.5 Cospectra for vertical velocity w and CO_2 concentration c, measured using a sonic anemometer and an open-path sensor located 0.25 m apart at a height of 1.62 m over an asphalt parking lot (adapted from Kondo and Tsukamoto 2008). Also shown are the WPL correction terms to the raw CO_2 flux due to sensible and latent heat

and

$$\overline{F_v}(h_m) = \left(\bar{p}\bar{\theta}_i / \bar{p}_i\bar{\theta}\right)\left[(1 + \mu\ \overline{\rho_{v,i}} / \overline{\rho_{d,i}})\ \overline{w'\rho'_{v,i}}\right] \tag{4.29}$$

where $\overline{p}$ and $\overline{\theta}$ are the mean pressure and absolute temperature in the ambient air and $\overline{p_i}$ and $\overline{\theta}_i$ are the corresponding quantities measured within the gas analyzer.

Complete elimination of temperature fluctuations may not be achievable for practical lengths of tubing with walls of low thermal conductivity, in which case some unknown fraction of the $\overline{w'\theta'}$ density correction must be applied. The solution is to measure the temperature and pressure fluctuations within the gas analyzer at the 10 or 20 Hz normally used for the rest of the eddy covariance system. Eddy fluxes of the trace gas and water vapor are then calculated using

$$\overline{F_c}(h_m) = \overline{\rho}_d \overline{w'\chi'_c}, \quad \overline{F_v}(h_m) = \overline{\rho}_d \overline{w'\chi'_v} \tag{4.30}$$

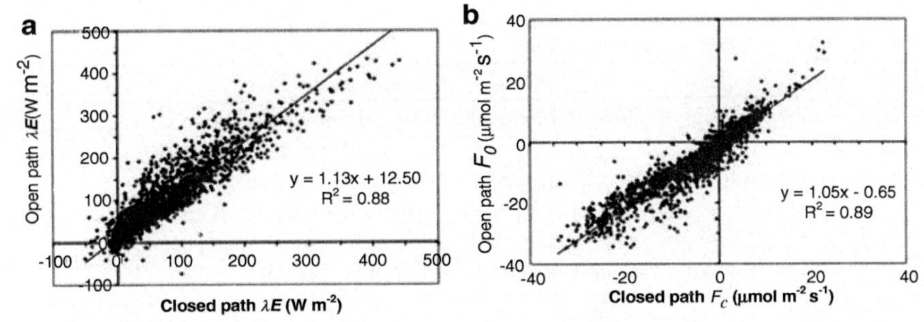

Fig. 4.6 Comparison of hourly fluxes of (**a**) latent heat and (**b**) CO_2, measured using a common sonic anemometer combined with open- and closed-path gas analyzers (Leuning, unpublished results)

where the instantaneous mixing ratios χ'_c and χ'_v of c and v relative to dry air are given by

$$\chi'_c \frac{\rho'_c/m_d}{p'_i/(R\theta'_i) - \rho'_v/m_v}, \quad \chi'_v \frac{\rho'_v/m_d}{p'_i/(R\theta'_i) - \rho'_v/m_v} \tag{4.31}$$

Normalized CO_2 and water vapor cospectra shown in Fig. 4.2, d were calculated using these formulas. We note that some closed-path instruments measure the wall temperature of the gas analysis cell rather than the desired air temperature and use of these measurements thus introduces some high-cut filtering into the θ'signal.

The rapid attenuation of fluctuations in CO_2 and water vapor at high frequencies seen in Fig. 4.2b causes a corresponding loss of high-frequency covariance. The resultant underestimate of the fluxes is proportional to the difference in area beneath the normalized $w\theta$ cospectra and the corresponding CO_2 and water vapor cospectra in Fig. 4.2. Thus, while using a closed-path gas analyzer eliminates the need for the WPL density corrections, alternative corrections are needed for loss of $w\chi_c$ and $w\chi_v$ covariances at high frequencies. As an example, Fig. 4.6 shows that λE and F_c^{EC} from the open-path system (after the appropriate high-cut frequency and WPL density corrections) were 13% and 5% higher than the corresponding fluxes calculated using the closed-path analyser. This is even after application of the theoretical transfer functions for high-frequency losses due to air flow in tubes as described in Sect. 4.1.3.1.

4.1.5 Sensor-Specific Corrections

4.1.5.1 Flow Distortion Correction of Sonic Anemometers

Flow distortion has been a well-known problem since the beginning of sonic anemometry (Dyer 1981). The reasons are the installations of the sensors and the

Table 4.1 Classification of sonic anemometers (Based on a classification by Foken and Oncley 1995; Mauder et al. 2006)

Anemometer class		Sensor type
A	Basic research for flux measurements	Kaijo-Denki A-Probe
		Campbell CSAT3, Solent HS
B	General use for flux measurements	Kaijo-Denki B-Probe
		Solent Wind Master, R2, R3
		METEK USA-1, Young 81000
C	General use for wind measurements	sensors of class B
		2D-anemometer of different producers

size of the transmitters/receivers. For new sensors, a large ratio of the path length, d_{pl}, to the transmitter/receiver diameter, a, of up to $d_{pl}/a = 20$ is required to minimize the influence of flow distortion. Furthermore, the angle between the wind vector and the transmitter-receiver path should be large (Kaimal and Finnigan 1994). From these requirements follows a classification of sonic anemometers (Table 4.1) into those having low flow distortion but a limited open angle, which suggested for research and in omni-directional sonic anemometers for routine applications.

One approach is for manufacturers to use wind tunnel measurements to guide the design of the anemometer array in a way that reduces the amount of flow distortion and then report winds with no correction for a specified acceptance angle. This is the approach taken by Campbell Sci. with the CSAT3. Other sonic anemometer manufacturers include a flow distortion correction for horizontal flow in the firmware, unfortunately, often not described in the manual. For the sonic anemometer USA-1 and Solent HS, the correction may be turned off. These corrections not only compensate for wind tunnel measurements but also take into account experience from in situ comparisons. Here, the user has no possibility of manipulating this correction. A still open problem is whether due to this correction a self correlation between the wind components is generated which influences the flux measurements. Because of the smaller eddies close to the ground, the influence of flow distortion decreases with height. Therefore, omni-directional sonic anemometers and other sensors that do not have a large ratio between the path length and the transmitter/receiver diameter should not be used close to the ground but have no problems well above the surface.

Recently, it was proposed that this correction should also include the angle between the wind vector and the horizontal level. This correction is called angle of attack correction (van der Molen et al. 2004; Nakai et al. 2006). Anemometer-specific data are available for Solent R3 (van der Molen et al. 2004; Nakai et al. 2006); the H4 head correction for USA-1 is comparable to this correction. The application of this correction increases the fluxes significantly (Cava et al. 2008). Because all correction functions were determined in the wind tunnel, the fluxes will be overestimated and the correction should not be used (Wyngaard 1981; Högström and Smedman 2004) or used only with care.

A final approach is to use a single-path model to a three-dimensional array, incorporating the path geometry. This approach can be successful if the geometry is relatively simple, as in the ATI-K probe or the CSAT3 (van Dijk 2002). Clearly, the caveats mentioned above apply – the single-path correction must be measured in a flow with turbulence levels similar to the atmosphere.

4.1.5.2 Correction Due to Sensor Head Heating of the Open-Path Gas Analyzer LiCor 7500

The sensor head of the LiCor 7500 open-path gas analyzer is heated which can generate convection within the sampling volume and therefore (some modifications were made in LiCor 7500A) has an influence on the application of the WPL correction. Potential corrections for this effect are discussed by Burba et al. (2008), Järvi et al. (2009), and Burba and Anderson (2010). A correction using additional fine-wire thermometers within the measurement volume is described by Grelle and Burba (2007). No general consensus has been reached yet as to which method is most effective and efficient. All corrections depend on wind speed and the inclination of the sensor. Therefore, any correction should be applied with care. Overall, a correction would be larger in cold weather than in warm weather, because it is generally a function of the temperature differences between the instrument surface and the ambient air, where the instrument surface temperature is a function of thermal control, set at about 30°C, and radiation load.

Specifically, during warm weather conditions (e.g. +30°C), the difference between air and instrument surface temperatures is mostly affected by solar load, which is usually less than 1–2°C at noon. For colder ambient temperatures, the instrument surface temperature is increasingly affected by the electronics. This means that, even in summer, there is some effect due to sensor heating. However, a potential summer correction is usually smaller in absolute values, due to less need for electronics heating, and it is much smaller in relative contribution, in relation to large summer ecosystem fluxes, as compared to winter correction, which is enhanced in absolute values by strong electronics heating and in relative contribution because of very small ecosystem fluxes. A simple solution for this problem is to deploy the sensor head upside down (though still a bit inclined to allow rain to run off). In this way, the heat generated at the base of the head rises away from the sensor path. Another solution for applications where low power consumption is required could be the deployment of an enclosure around the measurement path of the LI-7500 with short tubing (Clement et al. 2009). This idea resulted in the development of the LiCor 7200 sensor, which is supposed to combine the benefits of open- and closed-path systems.

4.1.5.3 Corrections to the Krypton Hygrometer KH20

Krypton hygrometers are used to measure the water vapor content of the air by absorption of H_2O molecules in the ultraviolet spectrum. Due to the wave length used, there is a cross sensitivity to O_2 molecules, which has to be corrected for as recommended by Tanner et al. (1993) and van Dijk et al. (2003).

$$\overline{w'\chi'_v} = \overline{w'\chi'_{vKH20}} + C_{ko}\left(\frac{\overline{\rho_v}}{\overline{\theta}}\right)\overline{w'\theta'} \tag{4.32}$$

where

$$C_{ko} = \frac{C_o m_o}{m_d} \cdot \frac{k_o}{k_v} = 0.23\frac{k_o}{k_v} \tag{4.33}$$

k_o and k_w are the KH20 extinction coefficients for oxygen and water vapor, $Co = 0.21$ is the molar fraction of oxygen in the atmosphere, and m_o is the molar mass of oxygen. The coefficients k_w and k_o are specific for each instrument. The extinction coefficient for water k_w is given in the calibration certificate by the manufacturer. The extinction coefficient for oxygen k_o can be determined experimentally. Tanner et al. (1993) recommend using a value of $k_o = -0.0045$, if the instrument-specific coefficient is not known.

4.1.5.4 Corrections for CH_4 and N_2O Analyzers

In recent years, fast-response sensors for trace gases other than CO_2 have become commercially available. Particularly, analyzers for the greenhouse gases CH_4 and N_2O are gaining more and more popularity in climate change research. As for the CO_2/H_2O analyzers, these sensors are also based on light absorption of a specific kind of molecule. However, since atmospheric concentrations of these gas species are much smaller than for CO_2 a more defined light source is necessary and that is why these sensors generally use lasers instead of noncoherent light. In this context, two basic measurement principles can be distinguished (Werle et al. 2008):

Tunable Diode Laser (TDL) Spectroscopy, where the output wavelength of the laser is tunable over a certain spectral range, so that the laser can scan across a specific absorption band of a molecule. For many of those analyzers, the actual measurement is carried out in an optical cell at a very low pressure. This has the disadvantage of requiring high-power pumps but leads to a wanted narrowing of the absorption bands and therefore better separation between different gas species. Examples for this type of analyzer are the Campbell TGA-100/200 or the Aerodyne QCL. The Licor LI-7700 also uses a tunable laser source but in an open-path measurement setup, similar to the LI-7500 but with a longer path length, and at ambient pressure.

Cavity Ring-Down Spectroscopy (CRDS) or Integrated Cavity Output Spectroscopy (ICOS), which generally rely upon vacuum pumps to draw down pressure in the optical cell. The intensity decay rate of light trapped in an optical cavity is measured, which is a function of the concentration of a gas species that absorbs light at that specific wavelength. The CRDS principle is applied, for example, by the PICARRO analyzers, whereas ICOS is used by the Los Gatos Fast Greenhouse Gas Analyzer.

The corrections required for CH_4 and N_2O analyzers are the same as for the commonly used H_2O or CO_2 analyzers, depending whether they have an open or closed measurement path. Sometimes a scrubber is deployed between the tube inlet and the measurement cell to remove water vapor from the sampling air. However, only if all temperature and pressure fluctuations are eliminated and any humidity is completely removed, the application of the WPL-correction is unnecessary, (Sect. 4.1.5.4). As mentioned before, for closed-path instruments in general, a correct determination of the time delay between the analyzer signal and the sonic signal is crucial to obtain accurate flux estimates.

4.1.6 Nonrecommended Corrections

As already shown above, not all corrections are recommended for general use. The reason is mainly that these corrections are not adequately tested or have significant limitations. During the past 40 years several corrections have been proposed and many are now applied, sometimes in modified versions. However, some should not be used and these are addressed below.

Due to the presentation by Stull (1988) of the correction of humidity-dependent fluctuations of the specific heat proposed by Brook (1978), which is some percentage of the flux, this correction is often used. However, shortly after the publication of this correction several authors (Leuning and Legg 1982; Nicholls and Smith 1982; Webb 1982) showed that this correction is based on incorrect conditions, and should never be used.

Liu et al. (2006) proposed to apply the effect of the energy balance closure by using the WPL-correction. This is a correct assumption in the case that the energy balance closure is based on an incorrect determination of the fluxes by the eddy covariance method. According to the results given in Sect. 4.2, this is not the reason for the missing energy. Therefore, this correction should not be used.

An energy balance (EB) closure adjustment of the sensible and latent heat flux according to the Bowen and also of the CO_2 flux (Desjardins 1985; Twine et al. 2000) is not recommended. The nonclosure of the EB indicates that there is a problem, but this solution would probably be too simple, since it is unknown whether scalar similarity can be assumed for the processes that cause an underestimation of the eddy covariance (EC) flux. Further studies about the causes of non-closure are warranted (see Sect. 4.2).

Furthermore, the published aliasing correction (Moore 1986) should not be applied. Aliasing is the transformation of the energy of higher frequencies to lower ones if the measuring system has no low-pass filters.

4.1.7 Overall Data Corrections

As shown in Sect. 4.1.2–4.1.5 most of the corrections are stability dependent or need the turbulent fluxes as input parameters. Therefore, the corrections often are performed iteratively. Though the math is tedious, Oncley et al. (2007) showed that the above corrections can be solved without iteration as a set of simultaneous equations. In Fig. 4.7, the schematic shows how to organize the system of raw data correction (Sect. 3.2.2), covariance correction (Sect. 4.1) and data quality tests as discussed in Sect. 4.3. This iteration has an effect of about 1% on the fluxes.

Furthermore, in Fig. 4.8 the effect of all correction for an about 6 week data set of the LITFASS-2003 experiment (Mengelkamp et al. 2006) is presented by Mauder and Foken (2006). Most relevant are the spectral correction (here only the shortwave part) on all fluxes and the transformation of the buoyancy flux into the sensible heat flux with a significant flux reduction. The figure shows also that the WPL correction needs a very careful application. While the latent heat flux is only slightly changed, the effect on the CO_2 flux and many other trace gas fluxes is significant (Box 4.1).

Box 4.1: Recommendations for Flux Correction

- Applying of flux correction after all raw data modifications presented in Sect. 3.2.2.
- First, all fluxes must be corrected for spectral losses (Sect. 4.1.3)

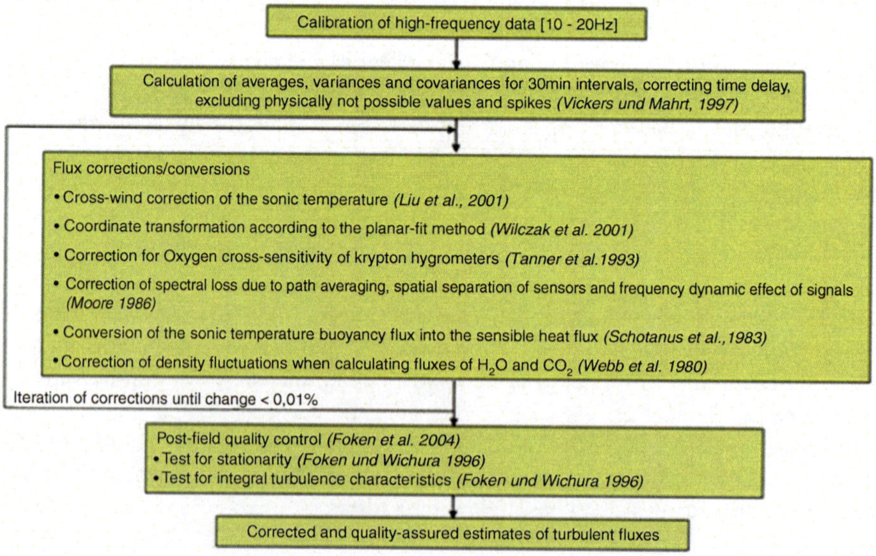

Fig. 4.7 Schematic of the postfield data processing (Mauder and Foken 2006)

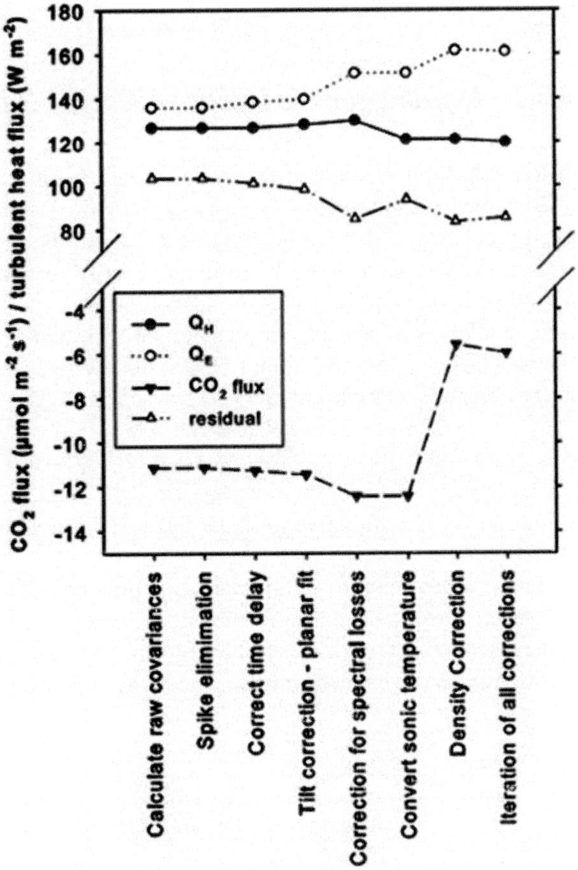

Fig. 4.8 Impact of single postfield data processing steps on the sensible heat flux Q_H, the latent heat flux Q_E, the CO_2 flux, and the surface energy balance residual (see Sect. 4.2), based on 30-min averaging time of the afternoon hours only (1200–1300 UTC) after Mauder and Foken (2006). Data set from a selected maize site of LITFASS-2003

- The buoyancy flux must be transferred into the sensible heat flux, which is used in further corrections and most of the application (Sect. 4.1.2)
- Water vapor and trace gas fluxes must be corrected for density fluctuations, which is different for open and closed path sensors (Sect. 4.1.4)
- Several sensors need specific corrections, which may be still in development for recent sensors (Sect. 4.1.5)
- The corrections should be calculated with an iterative system or a combined system of all equations (Sect. 4.1.7)
- The calculation of the atmosphere-ecosystem flux needs further nonsensor-specific corrections like the storage and night time flux correction (Sect. 5.4)

4.2 Effect of the Unclosed Energy Balance

4.2.1 Reasons for the Unclosed Energy Balance

After wide application of the eddy covariance method including all corrections and after the availability of highly accurate net radiometers it became obvious that the energy balance at the Earth's surface could not be closed with experimental data (Foken and Oncley 1995). The available energy, that is, the sum of the net radiation and the ground heat flux, was found in most cases to be larger than the sum of the turbulent fluxes of sensible and latent heat. For many field experiments and also for the CO_2flux networks (Aubinet et al. 2000; Wilson et al. 2002), a closure of the energy balance of approximately 80% was found. The residual is

$$\text{Res} = R_n - G - H - \lambda E, \tag{4.34}$$

With: R_n: net radiation, G: soil heat flux, H: sensible heat flux, and λE: latent heat flux (Fig. 4.9).

The problem cannot be described as only an effect of statistically distributed measuring errors because of the clear underestimation of turbulent fluxes or overestimation of the available energy. In the literature, several reasons for this incongruity have been discussed, most recently in an overview paper by Foken (2008a).

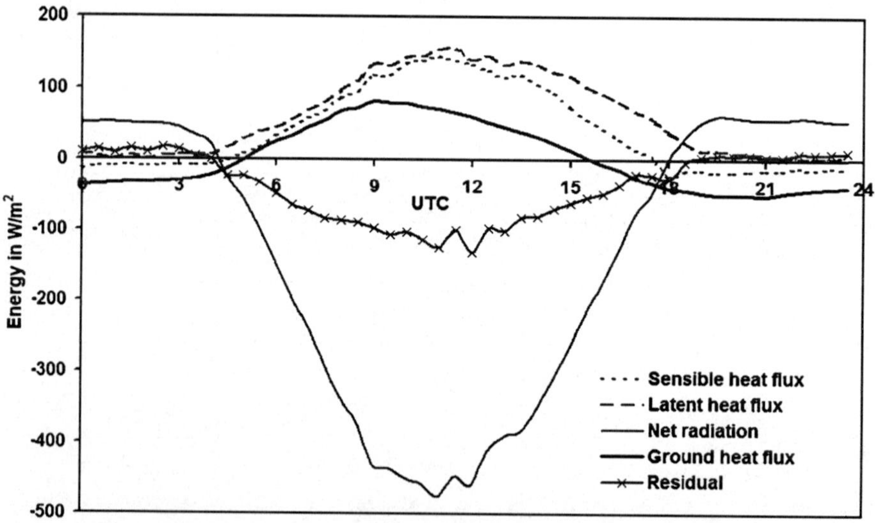

Fig. 4.9 Mean diurnal cycle of all energy balance components for the maize site during the LITFASS-2003 (Mengelkamp et al. 2006) period after Liebethal (2006)

In recent papers, it was found that time-averaged fluxes (Finnigan et al. 2003) or spatially averaged fluxes including turbulent-organized structures (Kanda et al. 2004) can close the energy balance. Therefore, it must be assumed that the phenomena of the unclosed energy balance at the earth's surface is not related to errors in the eddy covariance technique but related to atmospheric phenomena which cannot be measured with this technique. Thus, a simple correction is impossible, and how to handle this phenomena for energy fluxes and probably also for trace gas fluxes (carbon dioxide) is still an open question Combining all findings about the problem it can be concluded that (Foken 2008a; Foken et al. 2010):

In the past, the most common point of discussion with respect to the energy balance closure problem was measurement errors, especially those of the eddy covariance technique which were assumed to cause a systematic underestimation of the turbulent fluxes. Improvements in the sensors as well as in the correction methods, and the application of a more stringent determination of the data quality have made this method much more accurate over the past 10 years (Foken et al. 2004; Moncrieff 2004; Mauder and Foken 2006; Mauder et al. 2007b). Also the analysis of the data quality of eddy covariance measurements (Mauder et al. 2006) had no remarkable effect. As shown in Sect. 4.1.7, even a careful application of all corrections of the turbulent fluxes can reduce the residual only slightly (Mauder and Foken 2006).

Different reference levels and different sampling scales of the measuring methods for net radiation, turbulent fluxes, and soil heat flux were often seen as another possible reason for the lack of energy-balance closure. Moreover, the role of energy storage in the canopy and in the soil was discussed by several authors. Most of these energy storages appear to be not significant to the problem for low vegetation canopies (Oncley et al. 2007) with the exception of the heat storage in the soil (see e.g. Culf et al. 2004; Heusinkveld et al. 2004; Meyers and Hollinger 2004; Foken 2008a).

The nonclosure of the energy balance has also been explained by the heterogeneity of the land surface (Panin et al. 1998). These authors assumed that the heterogeneity in the vicinity of a flux-measurement site generates eddies at larger time scales, but such turbulent structures generated by heterogeneities close to the measuring tower can be measured with the eddy covariance method (Thomas and Foken 2007; Zhang et al. 2007). Therefore, the low-frequency part of the spectra (Foken et al. 2006) up to about 2 h has no significant influence on the closure problem.

This problem is also closely connected with advection and fluxes associated with longer wavelengths. Some recent studies have found that fluxes averaged over long time periods of several hours (Sakai et al. 2001; Finnigan et al. 2003; Mauder and Foken 2006) or spatially averaged fluxes (Kanda et al. 2004; Inagaki et al. 2006; Steinfeld et al. 2007) could close the energy balance. During the EBEX-2000 experiment, it was found that advection can play a significant role (Oncley et al. 2007). Due to the reduction of the residual by accounting for advection, the energy balance closure problem for EBEX-2000 is smaller than similar experiments.

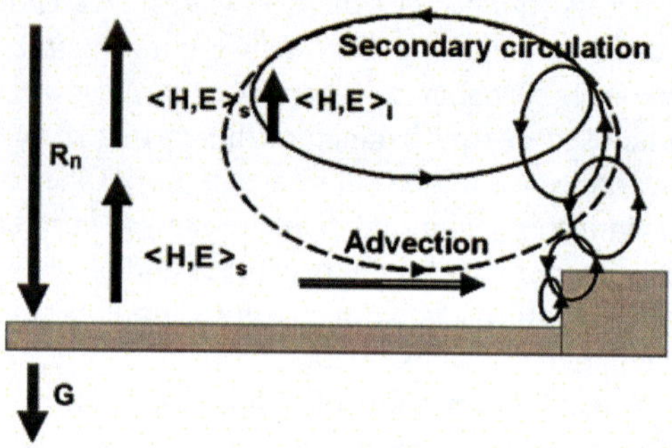

Fig. 4.10 Schematic figure of the generation of secondary circulations and the hypothesis of turbulent fluxes in different scales based on small eddies (s) and large eddies (l) according to Foken (2008a) where $\langle H, E\rangle_s$ is the sensible or latent heat flux by averaging over small eddies and $\langle H, E\rangle_l$ by averaging over large eddies. R_n is the net radiation and G is the ground heat flux

Over homogeneous surfaces like deserts (Heusinkveld et al. 2004) or bush land (Mauder et al. 2007a) the surface energy balance can be closed. Therefore, heterogeneities on a scale larger than 100 m and up to more than 10 km currently are the focus of possible explanations.

To verify these results, area-averaged flux measurements were used during the LITFASS-2003 experiment (Beyrich and Mengelkamp 2006; Mengelkamp et al. 2006), with large aperture scintillometers (Meijninger et al. 2006), aircraft measurements, and Large-Eddy Simulations. With these area-averaging techniques a much better energy balance could be reached (Foken et al. 2010).

Combining these previous investigations, it is obvious that the correction of the unclosed energy balance cannot be a part of the eddy covariance method and its correction procedures. A conceptual picture considering the land surface–atmosphere interaction at different scales was shown by Foken (2008a). This is also based on numerical studies which have shown that at steps of heterogeneities, the fluxes are significantly larger than over more homogeneous areas (e.g. Schmid and Bünzli 1995a; Schmid and Bünzli 1995b). This was underlined by the experiments by Klaassen et al. (2002). If the size of the heterogeneities or the difference of the characteristic heterogeneities (e.g. roughness, heat fluxes) is too small, this effect disappears (Friedrich et al. 2000). In Fig. 4.10 is shown that near the surface the smaller eddies are measured with micrometeorological methods such as the eddy covariance technique and the long-wave part is not available (Steinfeld et al. 2007). The transfer of the energy from the surface to the larger eddies happens mainly at significant heterogeneities and is not uniformly distributed over the area. In the sum of the fluxes by the smaller and larger eddies, the energy can be closed as measured with area-averaging techniques and also with long-term integration. Such long-term integration shows (Fig. 4.11) that the sensible heat flux increases and closes the residual while the latent heat flux is not affected. This underlines a nonsimilarity of both fluxes which may be dependent on the transport at the heterogeneities and should be different for different sites.

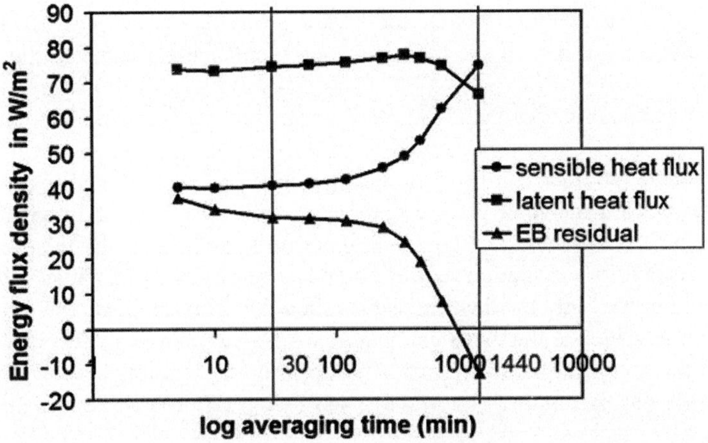

Fig. 4.11 Influence of averaging time on the sensible and latent heat flux and the residual of the energy balance closure (all in W m^{-2}) for the maize site of the whole LITFASS-2003 period (Mauder and Foken 2006)

4.2.2 Correction of the Unclosed Energy Balance

As shown above, the correction of the residual of the energy balance is not an issue of the correction of the eddy covariance method because the missing energy is not a missing flux at the measuring point and can be, at most, measured as advection. But such experimental setups are nearly impossible, as the EBEX-Experiment has shown (Oncley et al. 2007). If heterogeneities in the vicinity of the measuring stations are responsible, the analysis of the footprints and the footprint quality (Göckede et al. 2008) of tower sites (Chap. 8) should give a hint on this subject. But the quality of the footprint, that is, the percentages of the target area in the footprint, is not correlated with the residual. Instead, the existence of heterogeneities with a spatial scale larger than 500 m had a significant influence on the residual (Falge and Foken, 2007, personal communication). Because area-averaging flux measurements and Large Eddy Simulations are usually not available, only long-term integration (Sakai et al. 2001; Finnigan et al. 2003; Mauder and Foken 2006) can be used for the correction.

Therefore, eddy covariance measurements near the surface should not be corrected and the problem should be discussed for the lower part of the atmospheric boundary layer at spatial scales larger than the flux footprint.

As a first step, the energy exchange between the atmosphere and the underlying surface on scales larger than 1 km can be corrected with the Bowen ratio (Twine et al. 2000; Foken 2008a). According to the Bowen ratio, the residual will be distributed to the sensible and latent heat flux. This method is only valid if a similarity of both fluxes is given, which is obviously not always realized. There are some papers available (Mauder and Foken 2006; Ingwersen et al. 2011) which propose that a large part of the unclosed energy balance is related to the sensible heat flux.

As a more realistic approach, long-term integration is proposed. About 3–5 days with similar synoptic situations should be used for each station. However, the results may have to be modified for different weather situations, wind directions, and times of the year. From the final picture comparable with Fig. 4.8, the change of the sensible and latent heat flux should be determined and used for the correction. Because secondary circulation as the possible reason for the unclosure does not occur at nighttime, only the daytime values should be corrected. On the other hand, the nighttime values of the turbulent heat fluxes are very small and each correction is within the possible statistical error. But up to now not published investigations have shown that the findings by Mauder and Foken (2006) can be site- and time-dependent.

The problem will be more complicated for trace gas fluxes like the CO_2 flux, because this flux cannot be corrected according to the degree of the energy balance closure as earlier proposed (Twine et al. 2000). At nighttime, no correction is necessary. For the daytime values long-term integration may be possible. But such methods are dependent on many open questions and are still in progress. A panel discussion about this subject in October 2009 could show some possible paths of research but no solution for correction (Foken et al. 2011; Box 4.2).

Box 4.2: Recommendations Energy Balance Closure

The phenomena of the "unclosed" energy balance in the surface layer is not a technical problem of the eddy-covariance method itself. It is related to the heterogeneous terrain and its influence on the turbulent exchange. The sensible and latent heat flux can be as a first guess corrected with the Bowen-ration under the assumption that the scalar similarity is fulfilled. All trace gas fluxes should not be corrected.

4.3 Data Quality Analysis

A quality assurance (QA) and quality control (QC) process are essential for all meteorological measurements. For eddy covariance measurements, it is particularly recommended because of the very complex calculation procedure. This issue was extensively presented by Foken et al. (2004). Therefore, this chapter gives only an overview of an already published summary paper with some additional remarks.

In contrast to standard meteorological measurements (Essenwanger 1969; Smith et al. 1996; DeGaetano 1997), there are only a few papers available that discuss QC of eddy covariance measurements (Foken and Wichura 1996; Vickers and Mahrt 1997). QC of eddy covariances should include not only tests for instrument errors and problems with the sensors but also evaluations of how closely the conditions fulfil the theoretical assumptions underlying the method. Because the latter depends on meteorological conditions, eddy covariance QC tools must be a combination of a typical test for high-resolution time series and examination of the turbulent conditions. A second problem is connected with the representativeness of the measurements depending on the footprint of the measurement. The fraction of the footprint that is in the area of interest must be calculated (Chap. 8).

Quality assurance is one of the most important issues for creation and management of a measuring program. Issues of QA are widely known for routine meteorological measuring programs (Shearman 1992). This whole book is an update of already available QA programs (e.g. Moncrieff et al. 1997; Aubinet et al. 2000; Foken et al. 2004).

The most important part of QA is QC. Several tests are discussed in this paper. QC must be done in real-time or shortly after the measurements to minimize data loss by reducing the time to detect and fix instrument problems.

4.3.1 Quality Control of Eddy Covariance Measurements

The QC for meteorological data follows a scheme which is similar for most of the data and is illustrated in Fig. 4.12. The first steps are automatic tests that the signal is in the typical range of the sensor. This is, in most cases, already done in the sensor software. For the raw data, several tests are necessary. The first is a check if the data are in a meteorologically possible range and the second is a set of statistical tests.

The following test is a comparison with other meteorological measurements. It is important to compare the averaged temperature, moisture, and trace gas concentrations with additional measurements. For the wind velocity, this is often not necessary.

A uniform scheme does not exist for QC of eddy covariance measurements. There is only a discussion of several aspects in the literature. In the following lines, an overview of different QC steps is given:

- The first steps of data analysis are basic tests of the raw data (Vickers and Mahrt 1997) such as automatic tests of the amplitude, the resolution of the signal, a check of the electrical and meteorological range of the data, and spikes (Højstrup 1993), which are discussed in Sect. 3.2.2.
- Statistical and uncertainty tests must be applied to sampling errors of the time series (Haugen 1978; Vickers and Mahrt 1997; Finkelstein and Sims 2001; Richardson et al. 2006) and are discussed in Chap. 7. Also steps in the time series, or reasons for nonstationarity must be identified (Mahrt 1991; Vickers and Mahrt 1997).

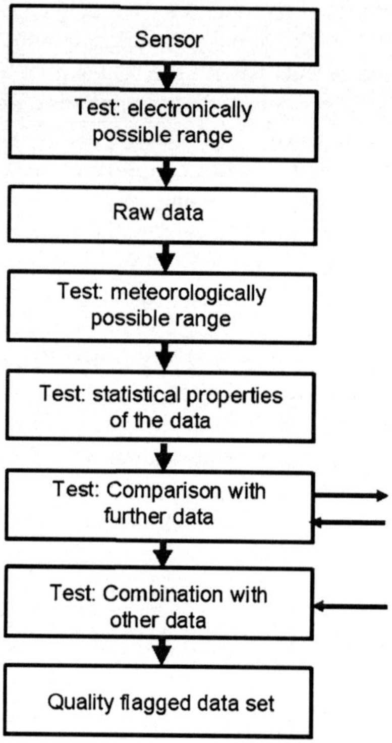

Fig. 4.12 Schema of the quality control of meteorological data (VDI 2012)

- A main issue for QC are tests on fulfilment of the requirements for eddy covariance measurements. Steady state conditions and a developed turbulent regime are influenced not from the sensor configuration but from the meteorological conditions (Foken and Wichura 1996). The fulfilment of these conditions is given in Sect. 4.3.2.
- A system of general quality flagging of the data is discussed in Sect. 4.3.3.
- A site-dependent QC based on footprint analysis is presented in Sect. 8.5.

4.3.2 *Tests on Fulfilment of Theoretical Requirements*

Foken and Wichura (1996) applied criteria for fast-response turbulence data to test for non-stationarity and substantial deviations from flux-variance similarity theory, whether due to instrumental or physical causes. The following presentation is based on Foken et al. (2004):

4.3.2.1 Steady State Tests

Steady state conditions means that all statistical parameters do not vary in time (e. g. Panofsky and Dutton 1984). Typical nonstationarity is driven by the change of meteorological variables with the time of the day, changes of weather patterns, significant mesoscale variability, or changes of the measuring point relative to the measuring events such as the phase of a gravity wave. The latter may occur because of changing footprint areas, changing internal boundary layers (especially internal thermal boundary layers in the afternoon), or by the presence of gravity waves. Presently, there are two main tests used to identify nonsteady state conditions. The first is based on the trend of a meteorological parameter over the averaging interval of the time series (Vickers and Mahrt 1997) and the second method indicates nonsteady state conditions within the averaging interval (Foken and Wichura 1996).

Vickers and Mahrt (1997) regressed the meteorological element χ_s over the averaging interval of a time series and determined the difference of χ_s between the beginning and the end of the time series according to this regression, $\delta\chi_s$. With this calculation they determined the parameter of relative nonstationarity, mainly for wind components, as

$$RN_{\chi} = \frac{\delta\chi_s}{\overline{\chi_s}} \tag{4.35}$$

Measurements made over the ocean exceeded the threshold ($RN_{\chi} > 0.50$) 15% of the time and measurements over forest exceeded the threshold 55% of the time. A more rigorous measure of stationarity can be found in Mahrt (1998).

The steady state test used by Foken and Wichura (1996) is based on developments of Russian scientists (Gurjanov et al. 1984). It compares the statistical parameters determined for the averaging period and for short intervals within this period. For instance, the time series for the determination of the covariance of the measured signals w (vertical wind) and χ_s (horizontal wind component or scalar) of about 30 min duration will be divided into $M = 6$ intervals of about 5 min. N is the number of measuring points of the short interval ($N = 6{,}000$ for 20 Hz scanning frequency and a 5 min interval):

$$\begin{aligned}
\left(\overline{w'\chi_s'}\right)_i &= \frac{1}{N-1}\left[\sum_j w_j \cdot \chi_{sj} - \frac{1}{N}\left(\sum_j w_j \cdot \sum_j \chi_{sj}\right)\right] \\
\overline{w'\chi_s'}\big|_{SI} &= \frac{1}{M}\sum_i \left(\overline{w'\chi_s'}\right)_i
\end{aligned} \tag{4.36}$$

This value will be compared with the covariance determined for the whole interval:

$$\left|\overline{w'\chi_s'}\right|_{WI} = \tfrac{1}{M\cdot N-1}\left[\sum_i\left(\sum_j w_j \cdot \chi_{sj}\right)_i - \tfrac{1}{M\cdot N}\sum_i\left(\sum_j w_j \cdot \sum_j \chi_{sj}\right)_i\right] \tag{4.37}$$

The authors proposed that the time series is steady state if the difference between both covariances

$$RN_{Cov} = \left| \frac{\overline{(w'\chi'_s)}_{SI} - \overline{(w'\chi'_s)}_{WI}}{\overline{(w'\chi'_s)}_{WI}} \right| \tag{4.38}$$

is less than 30%. This value has been found by long experience but is in good agreement with other test parameters including those of other authors (Foken and Wichura 1996).

4.3.2.2 Test on Developed Turbulent Conditions

Flux-variance similarity is a good measure to test the development of turbulent conditions. This similarity means that the ratio of the standard deviation of a turbulent parameter and its turbulent flux is nearly constant or a function of stability. These so-called integral turbulence characteristics are basic similarity characteristics of the atmospheric turbulence (Obukhov 1960; Wyngaard et al. 1971) and are routinely discussed in boundary layer and micrometeorology textbooks (Stull 1988; Kaimal and Finnigan 1994; Arya 2001; Foken 2008b). Foken and Wichura (1996) used functions determined by Foken et al. (1991). These functions depend on stability and have the general form for standard deviations of wind components:

$$\frac{\sigma_{u,v,w}}{u_*} = c_1 \left(\frac{h_m - \mathrm{d}}{L} \right)^{c_2} \tag{4.39}$$

where u is the horizontal or longitudinal wind component, v the lateral wind component, u_* the friction velocity, and L the Obukhov length. For scalar fluxes, the standard deviations are normalized by their dynamical parameters:

$$\frac{\sigma_{\chi_s}}{\chi_{s*}} = c_1 \left(\frac{h_m - \mathrm{d}}{L} \right)^{c_2} \tag{4.40}$$

The constant values in Eqs. 4.39 and 4.40 are given in Table 4.2. For the neutral range, the external forcing assumed by Johansson et al. (2001) and analyzed for the integral turbulence characteristics by Thomas and Foken (2002) was considered in Table 4.3 with the latitude (Coriolis parameter f). The parameters given for the temperature can be assumed for most of the scalar fluxes. It must be mentioned that under nearly neutral conditions, the integral turbulence characteristics of the scalars have extremely high values (Table 4.2) and the test fails.

The test can be done for the integral turbulence characteristics of both parameters used to determine the covariance. The measured and the modeled parameters according to Eqs. 4.39 or 4.40 will be compared according to (χ: u, v, w, or χ_s)

Table 4.2 Coefficients of the integral turbulence characteristics (Foken et al. 1991, 1997; Thomas and Foken 2002)

Parameter	$(h_m\text{-}d)/L$	c_1	c_2
σ_w/u_*	$0 > (h_m\text{-}d)/L > -0.032$	1.3	0
	$-0.032 > (h_m\text{-}d)/L$	2.0	1/8
σ_u/u_*	$0 > (h_m\text{-}d)/L > -0.032$	2.7	0
	$-0.032 > (h_m\text{-}d)/L$	4.15	1/8
σ_θ/θ_*	$0.02 < (h_m\text{-}d)/L < 1$	1.4	−1/4
	$0.02 > (h_m\text{-}d)/L > -0.062$	0.5	−1/2
	$-0.062 > (h_m\text{-}d)/L > -1$	1.0	−1/4
	$-1 > (h_m\text{-}d)/L$	1.0	−1/3

Table 4.3 Coefficients of the integral turbulence characteristics for wind components under neutral conditions (Thomas and Foken 2002)

Parameter	$-0{,}2 < (h_m\text{-}d)/L < 0{,}4$
$\frac{\sigma_w}{u_*}$	$0,21\ \ln\left(\frac{z_+ \cdot f}{u_*}\right) + 3.1;\ z_+ = 1\ m$
$\frac{\sigma_u}{u_*}$	$0,44\ \ln\left(\frac{z_+ \cdot f}{u_*}\right) + 6.3;\ z_+ = 1\ m$

$$ITC_\sigma = \left| \frac{\left(\frac{\sigma\chi}{\chi^*}\right)_{model} - \left(\frac{\sigma\chi}{\chi^*}\right)_{measurement}}{\left(\frac{\sigma\chi}{\chi^*}\right)_{model}} \right| \tag{4.41}$$

If the test parameter, integral turbulence characteristic (ITC_σ) is <30%, a well-developed turbulence can be assumed.

4.3.3 Overall Quality Flag System

This section also is based on the overview paper by Foken et al. (2004). The quality tests given above open the possibility to also flag the quality of a single measurement. Foken and Wichura (1996) proposed to classify the tests according to Eqs. 4.38 and 4.41 into different steps and to combine different tests. An important parameter which must be included in the classification scheme is the orientation of the sonic anemometer, if the anemometer is not an omnidirectional probe and the measuring site does not have an unlimited fetch in all directions. For these three tests, the definition of the flags is given in Table 4.4. Further tests, such as an acceptable range of the mean vertical wind velocity, can be included into this scheme.

The most important part of a flagging system is the combination of all flags into a general flag for easy use. This is done in Table 4.5 for the flags given in Table 4.4. The user of such a scheme must know the appropriate use of the flagged data. The presented scheme was classified by micrometeorological experiences so

Table 4.4 Classification of the data quality by the steady state test according to Eq. 4.38, the integral turbulence characteristics according to Eq. 4.41, and the horizontal orientation of a sonic anemometer of the type CSAT3 (Foken et al. 2004)

Steady state test according to Eq. 4.38		Integral turbulence characteristics according to Eq. 4.41		Horizontal orientation of the sonic anemometer	
Class	Range	Class	Range	Class	Range
1	0–15%	1	0–15%	1	±0–30°
2	16–30%	2	16–30%	2	±31–60°
3	31–50%	3	31–50%	3	±60–100°
4	51–75%	4	51–75%	4	±101–150°
5	76–100%	5	76–100%	5	±101–150°
6	101–250%	6	101–250%	6	±151–170°
7	251–500%	7	251–500%	7	±151–170°
8	501–1,000%	8	501–1,000%	8	±151–170°
9	>1,000%	9	>1,000%	9	> ± 171°

Remark: The classes 1–5 for the horizontal orientation of the sonic anemometer have the same influence on the overall flagging system (Table 4.5)

Table 4.5 Proposal for the combination of the single quality flags into a flag of the general data quality (Foken et al. 2004)

Flag of the general data quality	Steady state test according to Eq. 4.38	Integral turbulence characteristics according to Eq. 4.41	Horizontal orientation of the sonic anemometer
1	1	1–2	1–5
2	2	1–2	1–5
3	1–2	3–4	1–5
4	3–4	1–2	1–5
5	1–4	3–5	1–5
6	5	≤5	1–5
7	≤6	≤6	≤8
8	≤8	≤8	≤8
	≤8	6–8	≤8
9	one flag equal to 9		

that classes 1–3 can be used for fundamental research, such as the development of parameterizations. Classes 4–6 are available for general use such as for continuously running systems of the FLUXNET program. Classes 7 and 8 are only for orientation. Sometimes it is better to use such data instead of a gap-filling procedure, but then these data should not differ significantly from the data located before and after these data in the time series. Data of class 9 should be excluded under all circumstances. Such a scheme gives the user a good opportunity to use eddy covariance data. Finally, the data can be presented together with the quality flag as in Fig. 4.13. Most of the unusual values can be explained by the data quality flag. At night, other factors can influence the measurements. For analysis of integrated fluxes rejected

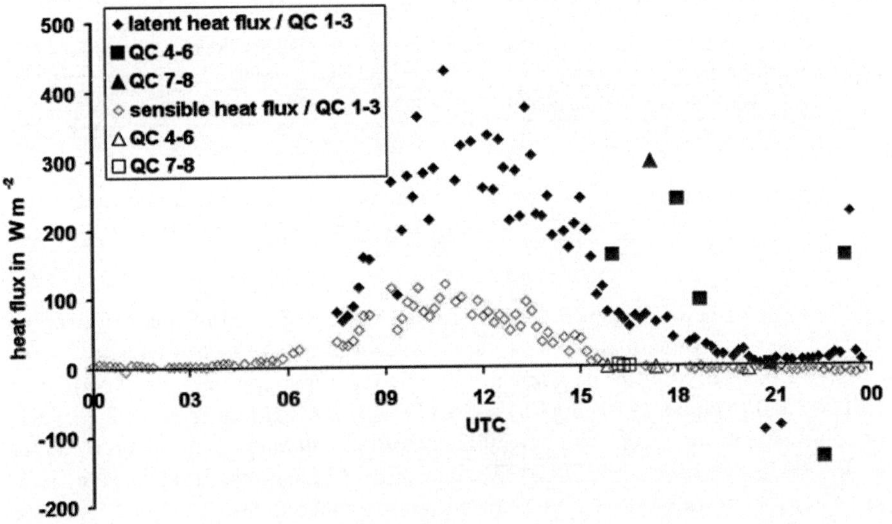

Fig. 4.13 Daily cycle of the sensible and latent heat flux with quality classes measured by the University of Bayreuth during the LITFASS-1998 experiment (Beyrich et al. 2002) on June 02, 1998 in Lindenberg/Germany over grassland (Foken et al. 2004)

data will need to be filled in. Obviously, investigations to infer process relationships should exclude both flagged data and the gap-filled values (Box 4.3).

Box 4.3: Recommendations or Data Quality Analysis

The use of a data quality analysis is essential for the application of the eddy covariance technique. A physical and meteorological control of the range of the input data is insufficient. Steady state conditions and a fully developed turbulent regime are important requirements for the application of the eddy covariance technique and special tests are necessary. If the tests fail the data should be replaced with a gap-filling procedure. The application of an adequate flagging system is important for the users of the data.

4.4 Accuracy of Turbulent Fluxes After Correction and Quality Control

The very complex algorithm of the eddy covariance method does not allow the determination of the errors according to the error propagation law. But statistical analyses are possible to determine the uncertainties of the method (Richardson et al.

Table 4.6 Evaluation of the accuracy of the eddy covariance method on the basis of the experimental results (Mauder et al. 2006), the data quality (Sect. 4.3), and the type of the sonic anemometer (Table 4.1, Foken and Oncley 1995)

Sonic anemometer	Data quality class	Sensible heat flux	Latent heat flux
type A, *e.g.*. CSAT3	1–3	5% or 10 Wm^{-2}	10% or 20 Wm^{-2}
	4–6	10% or 20 Wm^{-2}	15% or 30 Wm^{-2}
type B, *e.g.*. USA-1	1–3	10% or 20 Wm^{-2}	15% or 30 Wm^{-2}
	4–6	15% or 30 Wm^{-2}	20% or 40 Wm^{-2}

2006), as described in Chap. 7. In this chapter, more empirical results are given so that the user has some hints for assessing the accuracy of the measured data.

On the basis of long-term experience in sensor comparisons and the experiments EBEX-2000 (Mauder et al. 2007b) and LITFASS-2003 (Mauder et al. 2006) and other investigations (Loescher et al. 2005) as well as software comparisons (Mauder et al. 2007b; Mauder et al. 2008), Mauder et al. (2006) have tried to give some numbers for the possible accuracy of eddy covariance measurements if they are obtained according to the present state of knowledge. A significant dependence was found on the type of sonic anemometer (Table 4.1) and on the data quality (Sect. 4.3). The results are summarized in Table 4.6. To transfer these data to the CO_2 flux, the results for the latent heat flux should be used with a threshold of about 0.2 mg m^{-2} s^{-1}.

Aside from these errors, the problem of energy balance closure (Sect. 4.2) and the influence of the surrounding surface must be taken into account. The latter is discussed in relation to the footprint (Sect. 8.5). For this topic, Göckede et al. (2008) gave a classification of the measuring sites depending on the flux in the target area for different wind directions and stratifications.

Furthermore, it is important to discuss the effects of an internal boundary layer. Such a layer can exist due to a sudden change of the surface roughness or thermal conditions. The fetch from this change should be long enough that the height of the new equilibrium layer is larger than the measuring height (Stull 1988; Garratt 1990). A simple equation can be used to calculate the height of the new equilibrium layer in dependence on the fetch x_f (Raabe 1991):

$$h_e = 0.3\sqrt{x_f} \tag{4.42}$$

A simple version to characterize a measuring point related to footprint and internal boundary layers can be obtained by combination of the percentage of the target area in the footprint with the height of the new equilibrium layer (Mauder et al. 2006). Fluxes should be accepted as good in quality if the percentage of the target area is higher than 80% (Göckede et al. 2008) and the new equilibrium layer is higher than the measuring height (Table 4.7).

Table 4.7 Fetch x, height of the new equilibrium layer h_e, and flux contribution from the target land use type dependent on the wind direction and stability for a maize site during the LITFASS-2003 experiment (Mauder et al. 2006)

	30°	60°	90°	120°	150°	180°	210°	240°	270°	300°	330°	360°
x_f in m	29	41	125	360	265	203	211	159	122	81	36	28
h_e in m	1.6	1.9	3.4	5.7	4.9	4.3	4.4	3.8	3.3	2.7	1.8	1.6
Flux contribution from target land use type in %												
Stable	26	37	76	97	93	84	86	81	76	61	37	26
Neutral	56	67	100	100	100	100	100	100	100	88	67	56
Unstable	76	87	100	100	100	100	100	100	100	98	87	76

Table 4.8 Overview about software packages and included corrections and quality checks (Mauder et al. 2008, updated)

Software	TK3	Alteddy	ECPack	EddySoft	EdiRE	eth-flux	TUDD	S + packages	ECO_2S[a]
	Univ. of Bayreuth	Alterra	University of Wageningen	Max-Planck-Institute Jena	University of Edinburgh	Technical University Zürich	Technical University Dresden	NCAR/EOL	IMECC-EU Univ. of Tuscia
Data sampling	CSAT3, USA-1,HS, R2,R3,ATI-K,NUW, Young; 6262,7000, 7500,KH20, ADC OP-2	R2, R3, WMPro; CSAT3, USA-1, 6262, 7500, 7000, KH20, TGA100A, Los Gatos DLT100	R2, R3, CSAT3 KDTR90/TR61 7500, KH20, Lyman-α	R2,R3,Young; 6262, 7000, 7500, ADC OP-2	Any	CSAT3, R2, R3, HS; 6262, 7000, 7500, FM-100, MonitorLabs, Scintrex LMA3, Los Gatos FMA and FGGA, Aerodyne QCL	R2, R3, HS, USA-1; 6262, 7000	CSAT3, KH20,7500 standard, other sensors possible	R2, R3, WMPro; CSAT3, USA-1, 6262, 7500, 7000, 7550 (7200/7700)
Data preparation	Test plausibility, spikes; Block average; Time lag const/auto	Test plausibility, spikes; Block average; optional detrending (filter); Time lag const/auto	Test plausibility, spikes; optional detrending (linear); Time lag const.	Test plausibility, spikes; Block average, optional detrending; Time lag const/auto	Test plausibility, spikes; optional detrending (linear/filter); Block average; Time lag const/auto	Test plausibility, spikes; Block average, optional detrending; Time lag const/auto	Test plausibility; Block average; Time lag for closed path sensor	Test/replace spikes; Block average; Time lag; (all in NIDAS software)	Test plausibility, spikes; Block average; optional detrending and running mean; Time lag const/auto/ RH-dependent
Coordinate rotation	Planar fit/2D rotation; Head-correction	2D rotation; Nakai et al. (2006)	planar fit/2D/3D rotation	planar fit/2D/3D rotation	planar fit/2D/3D rotation	2D/3D rotation	3D rotation	planar fit	planar fit/2D/3D rotation; head correction for Gill sonics (Nakai et al. 2006)

(continued)

Table 4.8 (continued)

Software	TK3	Alteddy	ECPack	EddySoft	EdiRE	eth-flux	TUDD	S + packages	ECO_2S
	Univ. of Bayreuth	Alterra	University of Wageningen	Max-Planck-Institute Jena	University of Edinburgh	Technical University Zürich	Technical University Dresden	NCAR/EOL	IMECC-EU Univ. of Tuscia
Buoyancy flux → sensible heat flux	Schotanus et al. (1983)/Liu et al. (2001)	Schotanus et al. (1983)	Schotanus et al. (1983)	Schotanus et al. (1983)/Liu et al. (2001)	Schotanus et al. (1983)/Liu et al. (2001)	–	Schotanus et al. (1983)/Liu et al. (2001)	Schotanus et al. (1983)	Van Dijk et al. (2004)/Liu et al. (2001)
Oxygen correction hygrometer	Tanner et al. (1993)	Tanner et al. (1993)/van Dijk et al. (2003)	Tanner et al. (1993)/van Dijk et al. (2003)	–	–	–	–	van Dijk et al. (2003)	–
High-frequency loss	Moore (1986)	Moore (1986)/Eugster and Senn (1995)	Moore (1986)	Eugster and Senn (1995)	Moore (1986)/Eugster and Senn (1995)	Eugster and Senn (1995)	Moore (1986)/ Eugster and Senn (1995)	Horst and Lenschow (2009)	Moncrieff (1997), Horst (1997), Ibrom et al. (2007a); implement-ting Horst and Lenshow (2009)
WPL correction	Webb et al. (1980)	Webb et al. (1980), Burba et al. (2008)	Webb et al. (1980)	Webb et al. (1980)	Webb et al. (1980)	Webb et al. (1980)	Webb et al. (1980)	Webb et al. (1980)	Webb et al. (1980), Burba et al. (2008), Ibrom et al. (2007b) for closed-path; LI-COR-compliant for 7200

(continued)

Table 4.8 (continued)

Software	TK3	Alteddy	ECPack	EddySoft	EdiRE	eth-flux	TUDD	S + packages	ECO_2S
	Univ. of Bayreuth	Alterra	University of Wageningen	Max-Planck-Institute Jena	University of Edinburgh	Technical University Zürich	Technical University Dresden	NCAR/EOL	IMECC-EU Univ. of Tuscia
Iteration of all corrections	yes	–	yes	–	yes	–	–	equations solved simultaneously, Oncley et al. (2007)	partial
Calculation	$\lambda(\theta)$; $c_p(c_{p,\text{dry}},q)$.; $\rho(\theta,p)$	$\lambda(\theta)$; $c_p = \text{const.}$; $\rho(\theta,p)$	$\lambda(\theta)$; $c_p = \text{const.}$; $\rho(\theta,p)$	$\lambda(\theta)$; $c_p = \text{const.}$; $\rho(\theta,p)$	$\lambda(\theta)$; $c_p(c_{p,\text{dry}},q)$.; $\rho(\theta,p)$	$\lambda(\theta)$; $c_p = \text{const.}$; $\rho(\theta,p)$	$\lambda(\theta)$; $c_p = \text{const.}$; $\rho(I)$	$\lambda(\theta)$; $c_{p,\text{dry}}$ and $c_{p,\text{water}}$; $\rho(\theta,p)$	$\lambda(\theta)$; $c_p(c_{p,\text{dry}},p)$; $\rho(\theta,p)$
Quality control	Test steady state, integral turbulence characteristics: Foken and Wichura (1996); Kormann and Meixner (2001) Ogive, Footprint	uncertainty factor	statistical error: van Dijk et al. (2004)	Test steady state, integral turbulence characteristics: Foken and Wichura (1996); Footprint	Test steady state, integral turbulence characteristics: Foken and Wichura (1996), Vickers and Mahrt (1997); Footprint	Test steady state, integral turbulence characteristics: Foken and Wichura (1996)	nighttime critical u_*	manual; use redundant sensors	Test steady state, integral turbulence characteristics: Foken and Wichura (1996), Vickers and Mahrt (1997); Footprint (Schuepp et al. 1990, Kljun et al. 2004)

[a] Similar to EddyPro by LiCor

4.5 Overview of Available Correction Software

The application of the different corrections is not really uniform in the applicable program packages. Nevertheless, differences between the programs are not present in basic questions but in some specific details. The comparison of the different software packages has shown (Mauder et al. 2008) that the results differ much less than the accuracy of the method. The selection of the software by the user depends much on whether the user needs an online or offline software, or software with a very fixed procedure or allowing many possibilities. But it is very important that the selection and application of the software needs micrometeorological experience. Furthermore, all constructive details which are necessary for the successful application of the software must be carefully documented during the installation of the software package. Table 4.8 gives an overview about the use of data correction and data quality testing in different applicable or widely used software packages.

Acknowledgments MA acknowledges financial support by the European Union (FP 5, 6, and 7), the Belgian Fonds de la recherche Scientifique (FNRS-FRS), the Belgian Federal Science Policy Office (BELSPO), and the Communauté française de Belgique (Action de Recherche Concertée). TF and MM acknowledge financial support by the European Union (FP 5, 6) an the Federal Ministry of Education and Research of Germany (project DEKLIM).

References

Amiro BD (1990) Comparison of turbulence statistics within three boreal forest canopies. Bound Layer Meteorol 51:99–121

Anderson DE, Verma SB, Clement RJ, Baldocchi DD, Matt DR (1986) Turbulence spectra of CO_2, water vapour, temperature and velocity over a deciduous forest. Agric For Meteorol 38:81–99

Arya SP (2001) Introduction to micrometeorology. Academic, San Diego, 415 pp

Aubinet M et al (2000) Estimates of the annual net carbon and water exchange of forests: the EUROFLUX methodology. Adv Ecol Res 30:113–175

Aubinet M, Chermanne B, Vandenhaute M, Longdoz B, Yernaux M, Laitat E (2001) Long term carbon dioxide exchange above a mixed forest in the Belgian Ardennes. Agric For Meteorol 108:293–315

Bernhardt K, Piazena H (1988) Zum Einfluß turbulenzbedingter Dichteschwankungen auf die Bestimmung turbulenter Austauschströme in der Bodenschicht. Z Meteorol 38:234–245

Beyrich F, Mengelkamp H-T (2006) Evaporation over a heterogeneous land surface: EVA_GRIPS and the LITFASS-2003 experiment – an overview. Bound Layer Meteorol 121:5–32

Beyrich F, Herzog H-J, Neisser J (2002) The LITFASS project of DWD and the LITFASS-98 experiment: the project strategy and the experimental setup. Theor Appl Climatol 73:3–18

Brook RR (1978) The influence of water vapor fluctuations on turbulent fluxes. Bound Layer Meteorol 15:481–487

Burba G, Anderson D (2010) A brief practical guide to eddy covariance flux measurements. Li-COR Inc., Lincoln

Burba G, McDermitt DK, Grelle A, Anderson DJ, Xu L (2008) Addressing the influence of instrument surface heat exchange on the measurements of CO2 flux from open-path gas analyzers. Glob Chang Biol 14:1854–1876

Cava D, Contini D, Donateo A, Martano P (2008) Analysis of short-term closure of the surface energy balance above short vegetation. Agric For Meteorol 148:82–93

Clement RJ, Burba GG, Grelle A, Anderson DJ, Moncrieff JB (2009) Improved trace gas flux estimation through IRGA sampling optimization. Agric For Meteorol 149:623–638

Culf AD, Foken T, Gash JHC (2004) The energy balance closure problem. In: Kabat P et al (eds) Vegetation, water, humans and the climate. A new perspective on an interactive system. Springer, Berlin/Heidelberg, pp 159–166

de Ligne A, Heinesch B, Aubinet M (2010) New transfer functions for correcting turbulent water vapour fluxes. Bound Layer Meteorol 137(2):205–221

DeGaetano AT (1997) A quality-control routine for hourly wind observations. J Atmos Ocean Technol 14:308–317

Desjardins RL (1985) Carbon dioxide budget of maize. Agric For Meteorol 36:29–41

Desjardins RL, MacPherson JI, Schuepp PH, Karanja F (1989) An evaluation of aircraft flux measurements of CO_2, water vapor and sensible heat. Bound Layer Meteorol 47:55–69

Dyer AJ (1981) Flow distortion by supporting structures. Bound Layer Meteorol 20:363–372

Essenwanger OM (1969) Analytical procedures for the quality control of meteorological data. In: Proceedings of the American meteorological society symposium on meteorological observations and instrumentation. Meteorol Monogr 11(33):141–147

Eugster W, Senn W (1995) A cospectral correction for measurement of turbulent NO_2 flux. Bound Layer Meteorol 74:321–340

Finkelstein PL, Sims PF (2001) Sampling error in eddy correlation flux measurements. J Geophys Res D106:3503–3509

Finnigan JJ, Clement R, Malhi Y, Leuning R, Cleugh HA (2003) A re-evaluation of long-term flux measurement techniques, part I: averaging and coordinate rotation. Bound Layer Meteorol 107:1–48

Foken T (2006) 50 years of the Monin-Obukhov similarity theory. Bound Layer Meteorol 119: 431–447

Foken T (2008a) The energy balance closure problem - an overview. Ecol Appl 18:1351–1367

Foken T (2008b) Micrometeorology. Springer, Berlin/Heidelberg, 308 pp

Foken T, Oncley SP (1995) Results of the workshop ‘Instrumental and methodical problems of land surface flux measurements’. Bull Am Meteorol Soc 76:1191–1193

Foken T, Wichura B (1996) Tools for quality assessment of surface-based flux measurements. Agric For Meteorol 78:83–105

Foken T, Skeib G, Richter SH (1991) Dependence of the integral turbulence characteristics on the stability of stratification and their use for Doppler-Sodar measurements. Z Meteorol 41:311–315

Foken T, Dlugi R, Kramm G (1995) On the determination of dry deposition and emission of gaseous compounds at the biosphere-atmosphere interface. Meteorol Z 4:91–118

Foken T et al (1997) Results of the LINEX-96/2 experiment, vol 48, Dt Wetterdienst, Forsch. Entwicklung, Arbeitsergebnisse. Dt Wetterdienst, Geschäftsbereich Forschung und Entwicklung, Offenbach am Main, 75 pp

Foken T, Göckede M, Mauder M, Mahrt L, Amiro BD, Munger JW (2004) Post-field data quality control. In: Lee X et al (eds) Handbook of micrometeorology: a guide for surface flux measurement and analysis. Kluwer, Dordrecht, pp 181–208

Foken T, Wimmer F, Mauder M, Thomas C, Liebethal C (2006) Some aspects of the energy balance closure problem. Atmos Chem Phys 6:4395–4402

Foken T et al (2010) Energy balance closure for the LITFASS-2003 experiment. Theor Appl Climatol 101:149–160

Foken T, Aubinet M, Finnigan J, Leclerc MY, Mauder M, PawU KT (2011) Results of a panel discussion about the energy balance closure correction for trace gases. Bull Am Meteorol Soc 92:ES13–ES18

Friedrich K, Mölders N, Tetzlaff G (2000) On the influence of surface heterogeneity on the Bowen-ratio: a theoretical case study. Theor Appl Climatol 65:181–196

Fuehrer PL, Friehe CA (2002) Flux correction revised. Bound Layer Meteorol 102:415–457

Garratt JR (1990) The internal boundary layer - a review. Bound Layer Meteorol 50:171–203
Göckede M et al (2008) Quality control of CarboEurope flux data – part 1: coupling footprint analyses with flux data quality assessment to evaluate sites in forest ecosystems. Biogeosciences 5:433–450
Grelle A, Burba G (2007) Fine-wire thermometer to correct CO_2 fluxes by open-path analyzers for artificial density fluctuations. Agric For Meteorol 147:48–57
Gurjanov AE, Zubkovskij SL, Fedorov MM (1984) Mnogokanalnaja avtomatizirovannaja sistema obrabotki signalov na baze EVM (Automatic multi-channel system for signal analysis with electronic data processing). Geod Geophys Veröff, R II 26:17–20
Gurvitch AS (1962) Spectry pulsacii vertikalnoj komponenty skorosti vetra i ich svjazi s mikrometeorologitcheskimi uslovijach (Spectra of the fluctuations of the vertical wind component and the connection to micrometeorological conditions). Atmos Turbulent – Trudy inst fiziki atmos AN SSSR 4:101–136
Hatfield JL, Baker JM (eds) (2005) Micrometeorology in agricultural systems. American Society of Agronomy, Madison, 584 pp
Haugen DA (1978) Effects of sampling rates and averaging periods on meteorological measurements. In: Fourth symposium meteorological observations and Instrumentation, Am Meteorol Soc, Boston, pp 15–18
Heusinkveld BG, Jacobs AFG, Holtslag AAM, Berkowicz SM (2004) Surface energy balance closure in an arid region: role of soil heat flux. Agric For Meteorol 122:21–37
Hiller R, Zeeman MJ, Eugster W (2008) Eddy-covariance flux measurements in the complex terrain of an Alpine valley in Switzerland. Bound Layer Meteorol 127:449–467
Högström U, Smedman A (2004) Accuracy of sonic anemometers: laminar wind-tunnel calibrations compared to atmospheric in situ calibrations against a reference instrument. Bound Layer Meteorol 111:33–54
Højstrup J (1981) A simple model for the adjustment of velocity spectra in unstable conditions downstream of an abrupt change in roughness and heat flux. Bound Layer Meteorol 21:341–356
Højstrup J (1993) A statistical data screening procedure. Meas Sci Technol 4:153–157
Horst TW (1973) Spectral transfer functions for a three component sonic-anemometer. J Appl Meteorol 12:1072–1075
Horst TW (1997) A simple formula for attenuation of eddy fluxes measured with first-order-response scalar sensors. Bound Layer Meteorol 82:219–233
Horst TW (2000) On frequency response corrections for eddy covariance flux measurements. Bound Layer Meteorol 94:517–520
Horst TW, Lenschow DH (2009) Attenuation of scalar fluxes measured with spatially-displaced sensors. Bound Layer Meteorol 130:275–300
Hyson P, Garratt JR, Francey RJ (1977) Algebraic und elektronic corrections of measured uw covariance in the lower atmosphere. Bound Layer Meteorol 16:43–47
Ibrom A, Dellwik E, Flyvbjerg H, Jensen NO, Pilegaard K (2007a) Strong low-pass filtering effects on water vapour flux measurements with closed-path eddy correlation systems. Agric For Meteorol 147:140–156
Ibrom A, Dellwik E, Larsen SE, Pilegaard K (2007b) On the use of the Webb–Pearman–Leuning theory for closed-path eddy correlation measurements. Tellus B 59:937–946
Inagaki A, Letzel MO, Raasch S, Kanda M (2006) Impact of surface heterogeneity on energy balance: a study using LES. J Meteorol Soc Jpn 84:187–198
Ingwersen J et al (2011) Comparison of Noah simulations with eddy covariance and soil water measurements at a winter wheat stand. Agric For Meteorol 151:345–355
Järvi L, Mammarella I, Eugster W, Ibrom A, Siivola E, Dellwik E, Keronen P, Burba G, Vesala T (2009) Comparison of net CO_2 fluxes measured with open- and closed-path infrared gas analyzers in urban complex environment. Boreal Environ Res 14:499–514
Johansson C, Smedman A, Högström U, Brasseur JG, Khanna S (2001) Critical test of Monin-Obukhov similarity during convective conditions. J Atmos Sci 58:1549–1566

Kaimal JC, Finnigan JJ (1994) Atmospheric boundary layer flows: their structure and measurement. Oxford University Press, New York, 289 pp

Kaimal JC, Gaynor JE (1991) Another look to sonic thermometry. Bound Layer Meteorol 56:401–410

Kaimal JC, Wyngaard JC, Haugen DH (1968) Deriving power spectra from a three component sonic anemometer. J Appl Meteorol 7:827–834

Kaimal JC, Wyngaard JC, Izumi Y, Coté OR (1972) Spectral characteristics of surface layer turbulence. Q J R Meteorol Soc 98:563–589

Kanda M, Inagaki A, Letzel MO, Raasch S, Watanabe T (2004) LES study of the energy imbalance problem with eddy covariance fluxes. Bound Layer Meteorol 110:381–404

Klaassen W, van Breugel PB, Moors EJ, Nieveen JP (2002) Increased heat fluxes near a forest edge. Theor Appl Climatol 72:231–243

Kljun N, Calanca P, Rotach M, Schmid HP (2004) A simple parameterization for flux footprint predictions. Bound Layer Meteorol 112:503–523

Kondo F, Tsukamoto O (2008) Evaluation of Webb correction on CO_2 flux by eddy covariance technique using open-path gas analyzer over asphalt. J Agric Meteorol 64:1–8

Kormann R, Meixner FX (2001) An analytical footprint model for non-neutral stratification. Bound Layer Meteorol 99:207–224

Kristensen L, Mann J, Oncley SP, Wyngaard JC (1997) How close is close enough when measuring scalar fluxes with displaced sensors. J Atmos Ocean Technol 14:814–821

Lee X, Black TA (1994) Relating eddy correlation sensible heat flux to horizontal sensor separation in the unstable atmospheric surface layer. J Geophys Res 99(D9):18545–18553

Lee X, Massman W (2011) A perspective on thirty years of the Webb, Pearman and Leuning density corrections. Bound Layer Meteorol 139:37–59

Lenschow DH, Raupach MR (1991) The attenuation of fluctuations in scalar concentrations through sampling tubes. J Geophys Res 96:5259–5268

Leuning R (2004) Measurements of trace gas fluxes in the atmosphere using eddy covariance: WPL corrections revisited. In: Lee X et al (eds) Handbook of micrometeorology: a guide for surface flux measurements and analysis. Kluwer, Dordrecht, pp 119–132

Leuning R (2007) The correct form of the Webb, Pearman and Leuning equation for eddy fluxes of trace gases in steady and non-steady state, horizontally homogeneous flows. Bound Layer Meteorol 123:263–267

Leuning R, Judd MJ (1996) The relative merits of open- and closed path analysers for measurements of eddy fluxes. Glob Chang Biol 2:241–254

Leuning R, King KM (1992) Comparison of eddy-covariance measurements of CO_2 fluxes by open- and closed-path CO_2 analysers. Bound Layer Meteorol 59:297–311

Leuning R, Legg BJ (1982) Comments on 'The influence of water vapor fluctuations on turbulent fluxes' by Brook. Bound Layer Meteorol 23:255–258

Leuning RL, Moncrieff JB (1990) Eddy covariance CO_2 flux measurements using open and closed path CO_2 analysers: correction for analyser water vapour sensitivity and damping of fluctuations in air sampling tubes. Bound Layer Meteorol 53:63–76

Liebethal C (2006) On the determination of the ground heat flux in micrometeorology and its influence on the energy balance closure. PhD thesis, University of Bayreuth

Liebethal C, Foken T (2003) On the significance of the Webb correction to fluxes. Bound Layer Meteorol 109:99–106

Liebethal C, Foken T (2004) On the significance of the Webb correction to fluxes, Corrigendum. Bound Layer Meteorol 113:301

Liu H (2005) An alternative approach for CO_2 flux correction caused by heat and water vapour transfer. Bound Layer Meteorol 115:151–168

Liu H, Peters G, Foken T (2001) New equations for sonic temperature variance and buoyancy heat flux with an omnidirectional sonic anemometer. Bound Layer Meteorol 100:459–468

Liu H, Randerson JT, Lindfors J, Massman WJ, Foken T (2006) Consequences of incomplete surface energy balance closure for CO_2 fluxes from open-path CO_2/H_2O infrared gas analyzers. Bound Layer Meteorol 120:65–85

Loescher HW et al (2005) Comparison of temperature and wind statistics in contrasting environments among different sonic anemometer–thermometers. Agric For Meteorol 133: 119–139

Mahrt L (1991) Eddy asymmetry in the sheared heated boundary layer. J Atmos Sci 48: 472–492

Mahrt L (1998) Flux sampling errors for aircraft and towers. J Atmos Ocean Technol 15: 416–429

Mammarella I, Launiainen S, Grönholm T, Keronen P, Pumpanen J, Rannik Ü, Vesala T (2009) Relative humidity effect on the high frequency attenuation of water vapour flux measured by a closed-path eddy covariance system. J Atmos Ocean Technol A26:1856–1866

Massman WJ (2000) A simple method for estimating frequency response corrections for eddy covariance systems. Agric For Meteorol 104:185–198

Massman WJ, Ibrom A (2008) Attenuation of concentration fluctuations of water vapor and other trace gases in turbulent tube flow. Atmos Chem Phys 8:6245–6259

Mauder M, Foken T (2004) Documentation and instruction manual of the eddy covariance software package TK2, vol 26, Arbeitsergebnisse, Universität Bayreuth, Abteilung Mikrometeorologie. Universität Bayreuth, Abteilung Mikrometeorologie, Bayreuth, 42 pp. ISSN 1614–8916

Mauder M, Foken T (2006) Impact of post-field data processing on eddy covariance flux estimates and energy balance closure. Meteorol Z 15:597–609

Mauder M, Liebethal C, Göckede M, Leps J-P, Beyrich F, Foken T (2006) Processing and quality control of flux data during LITFASS-2003. Bound Layer Meteorol 121:67–88

Mauder M, Jegede OO, Okogbue EC, Wimmer F, Foken T (2007a) Surface energy flux measurements at a tropical site in West-Africa during the transition from dry to wet season. Theor Appl Climatol 89:171–183

Mauder M et al (2007b) The energy balance experiment EBEX-2000. Part II: Intercomparison of eddy covariance sensors and post-field data processing methods. Bound Layer Meteorol 123:29–54

Mauder M, Foken T, Clement R, Elbers J, Eugster W, Grünwald T, Heusinkveld B, Kolle O (2008) Quality control of CarboEurope flux data - part 2: inter-comparison of eddy-covariance software. Biogeosciences 5:451–462

Meijninger WML, Lüdi A, Beyrich F, Kohsiek W, DeBruin HAR (2006) Scintillometer-based turbulent surface fluxes of sensible and latent heat over heterogeneous a land surface - a contribution to LITFASS-2003. Bound Layer Meteorol 121:89–110

Mengelkamp H-T et al (2006) Evaporation over a heterogeneous land surface: the EVA_GRIPS project. Bull Am Meteorol Soc 87:775–786

Meyers TP, Hollinger SE (2004) An assessment of storage terms in the surface energy of maize and soybean. Agric For Meteorol 125:105–115

Moncrieff J (2004) Surface turbulent fluxes. In: Kabat P et al (eds) Vegetation, water, humans and the climate. A new perspective on an interactive system. Springer, Berlin/Heidelberg, pp 173–182

Moncrieff JB et al (1997) A system to measure surface fluxes of momentum, sensible heat, water vapor and carbon dioxide. J Hydrol 188–189:589–611

Monji N, Inoue M, Hamotani K (1994) Comparison of eddy heat fluxes between inside and above a coniferous forest. J Agric Meteorol 50:23–31

Monteith JL, Unsworth MH (2008) Principles of environmental physics, 3rd edn. Elsevier/Academic Press, Amsterdam/Boston, 418 pp

Moore CJ (1986) Frequency response corrections for eddy correlation systems. Bound Layer Meteorol 37:17–35

Nakai T, van der Molen MK, Gash JHC, Kodama Y (2006) Correction of sonic anemometer angle of attack errors. Agric For Meteorol 136:19–30

Nicholls S, Smith FB (1982) On the definition of the flux of sensible heat. Bound Layer Meteorol 24:121–127

Obukhov AM (1960) O strukture temperaturnogo polja i polja skorostej v uslovijach konvekcii (Structure of the temperature and velocity fields under conditions of free convection). Izv AN SSSR, ser Geofiz 1392–1396

Oncley SP, Businger JA, Itsweire EC, Friehe CA, LaRue JC, Chang SS (1990) Surface layer profiles and turbulence measurements over uniform land under near-neutral conditions. In: 9th symposium on boundary layer and turbulence, Roskilde, Denmark, April 30–May 3, 1990, Am Meteorol Soc City pp 237–240

Oncley SP et al (2007) The energy balance experiment EBEX-2000, part I: overview and energy balance. Bound Layer Meteorol 123:1–28

Othaki E (1985) On the similarity in atmospheric fluctuations of atmospheric carbon dioxide, water vapour and temperature over vegetated fields. Bound Layer Meteorol 32:25–37

Panin GN, Tetzlaff G, Raabe A (1998) Inhomogeneity of the land surface and problems in the parameterization of surface fluxes in natural conditions. Theor Appl Climatol 60:163–178

Panofsky HA, Dutton JA (1984) Atmospheric turbulence – models and methods for engineering applications. Wiley, New York, 397 pp

Raabe A (1991) Die Höhe der internen Grenzschicht. Z Meteorol 41:251–261

Richardson AD et al (2006) A multi-site analysis of random error in tower-based measurements of carbon and energy fluxes. Agric For Meteorol 136:1–18

Ruppert J, Thomas C, Foken T (2006) Scalar similarity for relaxed eddy accumulation methods. Bound Layer Meteorol 120:39–63

Sakai R, Fitzjarrald D, Moore KE (2001) Importance of low-frequency contributions to eddy fluxes observed over rough surfaces. J Appl Meteorol 40:2178–2192

Schmid HP, Bünzli D (1995a) The influence of the surface texture on the effective roughness length. Q J R Meteorol Soc 121:1–21

Schmid HP, Bünzli D (1995b) Reply to comments by E. M. Blyth on 'The influence of surface texture on the effective roughness length'. Q J R Meteorol Soc 121:1173–1176

Schotanus P, Nieuwstadt FTM, DeBruin HAR (1983) Temperature measurement with a sonic anemometer and its application to heat and moisture fluctuations. Bound Layer Meteorol 26:81–93

Schuepp PH, Leclerc MY, MacPherson JI, Desjardins RL (1990) Footprint prediction of scalar fluxes from analytical solutions of the diffusion equation. Bound Layer Meteorol 50:355–373

Shearman RJ (1992) Quality assurance in the observation area of the Meteorological Office. Meteorol Mag 121:212–216

Silverman BA (1968) The effect of the spectral averaging on spectral estimation. J Appl Meteorol 7:168–172

Smith SR, Camp JP, Legler DM (1996) Handbook of quality control, procedures and methods for surface meteorology data. Center for Ocean Atmospheric Prediction Studies,TOGA/COARE, Technical Report. 96–3:60 pp. [Available from Florida State University, Tallahassee, FL, 32306–33041]

Steinfeld G, Letzel MO, Raasch S, Kanda M, Inagaki A (2007) Spatial representativeness of single tower measurements and the imbalance problem with eddy-covariance fluxes: results of a large-eddy simulation study. Bound Layer Meteorol 123:77–98

Stull RB (1988) An introduction to boundary layer meteorology. Kluwer Acad. Publ, Dordrecht/Boston/London, 666 pp

Su HB, Schmid HP, Grimmond CSB, Vogel CS, Oliphant AJ (2004) Spectral characteristics and correction of long-term eddy-covariance measurements over two mixed hardwood forests in non-flat terrain. Bound Layer Meteorol 110:213–253

Tanner CB, Thurtell GW (1969) Anemoclinometer measurements of Reynolds stress and heat transport in the atmospheric surface layer. ECOM, United States Army Electronics Command, Research and Development, ECOM-66-G22-F, Fort Huachuca, AZ, 82 pp

Tanner BD, Swiatek E, Greene JP (1993) Density fluctuations and use of the krypton hygrometer in surface flux measurements. In: Allen RG (ed) Management of irrigation and drainage systems: integrated perspectives. American Society of Civil Engineers, New York, pp 945–952

Thomas C, Foken T (2002) Re-evaluation of integral turbulence characteristics and their parameterisations. In: 15th conference on turbulence and boundary layers, Wageningen, NL, 15–19 July 2002, Am Meteorol Soc, City, pp 129–132

Thomas C, Foken T (2007) Flux contribution of coherent structures and its implications for the exchange of energy and matter in a tall spruce canopy. Bound Layer Meteorol 123:317–337

Twine TE, Kustas WP, Norman JM, Cook DR, Houser PR, Meyers TP, Prueger JH, Starks PJ, Wesely ML (2000) Correcting eddy-covariance flux underestimates over a grassland. Agric For Meteorol 103:279–300

van der Molen MK, Gash JHC, Elbers JA (2004) Sonic anemometer (co)sine response and flux measurement: II the effect of introducing an angle of attack dependent calibration. Agric For Meteorol 122:95–109

van Dijk A (2002) Extension to 3D of "The effect of line averaging on scalar flux measurements with a sonic anemometer near the surface" by Kristensen and Fitzjarrald. J Atmos Ocean Technol 19:80–82

van Dijk A, Kohsiek W, DeBruin HAR (2003) Oxygen sensitivity of krypton and Lyman-alpha hygrometers. J Atmos Ocean Technol 20:143–151

van Dijk A, Kohsiek W, DeBruin HAR (2004) The principles of surface flux physics: theory, practice and description of the ECPACK library. University of Wageningen, Wageningen

VDI (2012) Umweltmeteorologie – Meteorologische Messungen - Grundlagen (Environmental meteorology – Meteorological measurements - Basics). Beuth-Verlag, Berlin, VDI 3786, Blatt 1, in print

Vickers D, Mahrt L (1997) Quality control and flux sampling problems for tower and aircraft data. J Atmos Ocean Technol 14:512–526

Webb EK (1982) On the correction of flux measurements for effects of heat and water vapour transfer. Bound Layer Meteorol 23:251–254

Webb EK, Pearman GI, Leuning R (1980) Correction of the flux measurements for density effects due to heat and water vapour transfer. Q J R Meteorol Soc 106:85–100

Werle P, D'Amato F, Viciani S (2008) Tunable diode-laser spectroscopy: principles, performance, perspectives. In: Lackner M (ed) Lasers in chemistry – probing matter. Wiley-VCH, Weinheim, pp 255–275

Wilczak JM, Oncley SP, Stage SA (2001) Sonic anemometer tilt correction algorithms. Bound Layer Meteorol 99:127–150

Wilson KB et al (2002) Energy balance closure at FLUXNET sites. Agric For Meteorol 113:223–234

Wyngaard JC (1981) The effects of probe-induced flow distortion on atmospheric turbulence measurements. J Appl Meteorol 20:784–794

Wyngaard JC, Coté OR (1971) The budgets of turbulent kinetic energy and temperature variance in the atmospheric surface layer. J Atmos Sci 28:190–201

Wyngaard JC, Coté OR, Izumi Y (1971) Local free convection, similarity and the budgets of shear stress and heat flux. J Atmos Sci 28:1171–1182

Zhang G, Thomas C, Leclerc MY, Karipot A, Gholz HL, Foken T (2007) On the effect of clearcuts on turbulence structure above a forest canopy. Theor Appl Climatol 88:133–137

Chapter 5
Nighttime Flux Correction

Marc Aubinet, Christian Feigenwinter, Bernard Heinesch, Quentin Laffineur, Dario Papale, Markus Reichstein, Janne Rinne, and Eva Van Gorsel

5.1 Introduction

5.1.1 History

Since the early tests performed with eddy covariance systems (Ohtaki 1984; Anderson et al. 1984), and the paper of Goulden et al. (1996), it became clear that the eddy covariance method underestimates the CO_2 flux in stable conditions. This underestimation acts as a selective systematic error (Moncrieff et al. 1996) and could lead to a strong overestimation of net ecosystem exchange (NEE) at an annual scale.

M. Aubinet (✉) • B. Heinesch • Q. Laffineur
Unit of Biosystem Physics, Gembloux Agro-Bio Tech, University of Liege,
5030 Gembloux, Belgium
e-mail: Marc.Aubinet@ulg.ac.be

C. Feigenwinter
Institute of Meteorology, Climatology and Remote Sensing, University of Basel, Basel, Switzerland
e-mail: feigenwinter@metinform.ch

D. Papale
DIBAF, University of Tuscia, Viterbo, Italy
e-mail: darpap@unitus.it

M. Reichstein
Max Planck Institute für Biogeochemistry, Jena, Germany
e-mail: mreichstein@bgc-jena.mpg.de

J. Rinne
Department of Physics, University of Helsinki, FI-00014 Helsinki, Finland
e-mail: Janne.Rinne@helsinki.fi

E. Van Gorsel
CSIRO, Canberra, Australia

M. Aubinet et al. (eds.), *Eddy Covariance: A Practical Guide to Measurement and Data Analysis*, Springer Atmospheric Sciences, DOI 10.1007/978-94-007-2351-1_5,

The problem has now been confirmed by many researchers working on very different sites: In tropical (Grace et al. 1996; Loescher et al. 2006; Hutyra et al. 2008), boreal (Jarvis et al. 1997; Pattey et al. 1997), temperate mixed (Aubinet et al. 2001; Teklemariam et al. 2009), broadleaved (Pilegaard et al. 2001; Cook et al. 2004), or coniferous (Berbigier et al. 2001; Carrara et al. 2003; Turnipseed et al. 2003) forests as grasslands (Wohlfahrt et al. 2005) or crops (Moureaux et al. 2006). First intersite evaluations of this error were proposed by Aubinet et al. (2000) (ten forested sites) and Gu et al. (2005) (five forest and two grassland sites). They confirmed that practically all the sites were affected significantly by a night flux error which necessitates an adequate correction.

5.1.2 *Signs Substantiating the Night Flux Error*

Like all systematic errors, the night flux error is not easy to distinguish as its detection would require a comparison of eddy fluxes with independent evaluations of ecosystem respiration at the same spatial and temporal scale. As such measurements are not available, the sole possibility is to refer to indirect proofs. Goulden et al. (1996) put two symptoms forward: First, total ecosystem respiration estimates are generally lower when estimated by eddy covariance than when obtained by a bottom up approach. Second, at night, the turbulent flux is sensitive to the friction velocity (u_*) while there is no evident reason for the biotic flux being sensitive to this variable. These two indices are discussed in the paragraphs below.

5.1.2.1 Comparison with Bottom Up Approaches

A comparison of eddy flux measurements with alternative flux estimates is always delicate as it is not possible to find measurements that are performed at the same spatial and temporal scale. Generally, such estimates are extrapolated from soil and plant respiration measurements obtained with soil chamber and branch bags. These estimates are themselves subject to instrumental errors and to a large uncertainty due to spatial variability. In addition, the comparison requires both spatial and temporal extrapolation as chamber measurements are performed at smaller scale and, most often, at weekly or monthly scale. Several studies based on this approach (Goulden et al. 1996; Lavigne et al. 1997; Lindroth et al. 1998; Kutsch et al. 2008; Hutyra et al. 2008) confirmed the underestimation of night fluxes by the eddy covariance approach. In addition, these studies provide a procedure to evaluate the importance of this error and to correct it.

5.1.2.2 Sensitivity of Flux to Friction Velocity

The second symptom supporting the existence of a night flux error is the sensitivity of eddy flux to friction velocity in stable conditions (Fig. 5.1). Indeed, as the mechanisms controlling night fluxes are linked to plant and soil respiration, they

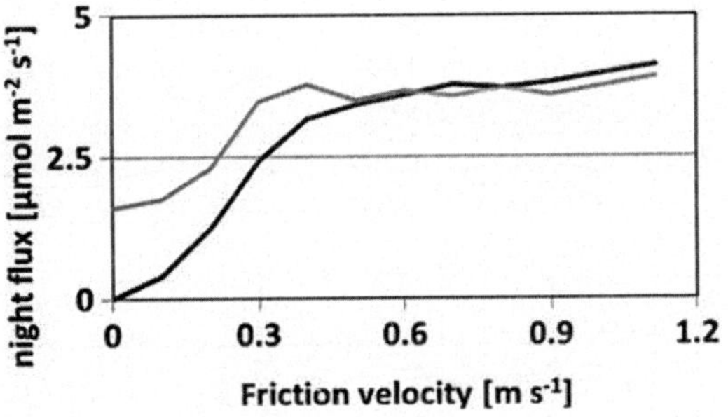

Fig. 5.1 Typical evolution of night flux with friction velocity. Average on three successive vegetation periods (May to September) at the Vielsalm site. *Black curve*: eddy flux only, *Gray curve*: eddy flux and storage change

are expected to be independent of u_*. Therefore, any dependence on u_* should come from an artifact. This assertion calls, however, for some comments: First, it could be true only if friction velocity does not covary with respiration driving variables, as temperature and soil humidity. In order to avoid any confounding effect of these variables with friction velocity it is thus recommended to normalize respiration by these variables before to establish the relation with u_* (Aubinet et al. 2000).

Second, the independence of soil respiration to friction velocity is questioned by several authors who mentioned the possibility of a pressure-pumping mechanism. Gu et al. (2005) suggested that, as the CO_2 mixing ratio difference between air and the first soil layers is large, air movement into and out of the soil induced by pressure fluctuations may introduce a significant physical component to the soil efflux that adds to the biological component. Such component could be related to turbulence inducing a relation between night flux and friction velocity. However, this component is mainly significant at sites where the soil exhibits a large porosity (Takle et al. 2004), especially above snow (Massman et al. 1997; Massman and Lee 2002) or on volcanic soils (Rogie et al. 2001). In addition, as such fluctuations could explain an increase of night flux at large u_*, it could not explain the turbulent flux decrease that is observed in very low turbulent conditions.

5.1.3 *The Causes of the Problem*

Massman and Lee (2002) listed and discussed in detail the possible instrumental errors affecting turbulent flux measurements (see also Chaps. 4 and 7). However, as instrument problems contribute to the flux loss, they suggested that it was mainly meteorological in nature. Meteorological problems are generally identified as follows:

1. Sublayers develop between the measurement system and the surface so that the system is decoupled from the surface and the eddy flux and change in storage terms are no longer representative of the local flux.
2. Even in the absence of a sublayer, the flux may be not representative of the surface because the extent of the flux footprint.

3. In low turbulence, the advection terms gain importance and are no longer negligible (Lee 1998; Aubinet et al. 2003, 2005; Feigenwinter et al. 2004; Marcolla et al. 2005).
4. Strong concentration or velocity changes could appear so that conditions become nonstationary, which invalidates the hypotheses underlying the eddy covariance method.
5. Similarity conditions are not always fulfilled in the stable boundary layer (Mahrt 1999), which makes quality tests, corrections, and footprint evaluation to some extent impossible.

From these different problems, the third appears as the most important that explains a systematic underestimation of the flux. To better understand this problem we will refer to the CO_2 mass conservation (Eq. 1.25).

$$\underbrace{\int_0^{h_m} \overline{\rho_d}\frac{\partial \overline{\chi_s}}{\partial t}dz}_{I} + \underbrace{\int_0^{h_m} \left[\overline{\rho_d w}\frac{\partial \overline{\chi_s}}{\partial z}\right]dz}_{IIa} + \underbrace{\int_0^{h_m} \left[\overline{\rho_d u}\frac{\Delta\overline{\chi}_{s,x}}{\Delta x} + \overline{\rho_d v}\frac{\Delta\overline{\chi}_{s,y}}{\Delta y}\right]dz}_{IIb} + \underbrace{\overline{\rho_d w'\chi'_s}|_{h_m}}_{IV} = \underbrace{F_s}_{V} \tag{1.25}$$

In the generalized eddy covariance method, it is assumed that the stationarity and homogeneity criteria are met so that advection terms (II) can be considered as negligible compared with the change of storage (I) and eddy covariance (IV) terms. These conditions are probably not met in night conditions, which leads either to an incorrect evaluation of terms I and IV, or to increased terms II that can no more be neglected compared to the two former.

5.2 Is This Problem Really Important?

Box 5.1

- The night CO_2 flux error appears at all sites during low turbulent nights. In most cases, it leads to an underestimation of the scalar source/sink intensity.
- When a complete data set is not necessary (which is the case when establishing functional relationships, for instance), it is recommended to discard data collected during low turbulence using a filtering procedure.
- When these data are necessary (for long-term budgets), they should be corrected.
- Storage is most often not enough to correct the fluxes but it has to be considered when a filtering/parameterization procedure is applied.

5.2.1 In Which Case Should the Night Flux Error Be Corrected?

There is now experimental evidence that night flux underestimation affects practically all the sites (Schimel et al. 2008). As the night flux error acts as a systematic error, it seems clear that a data treatment is necessary in order to offset it.

This treatment cannot be simply the addition of storage to the turbulent flux, as will be shown in Sect. 5.2.2. It could be different according to the data purpose: if the aim of the data analysis is to infer functional relationships, a data filtering could be sufficient. On the other hand, if long-term flux budgets are required, all data affected by the error should be corrected.

The way filtering procedures should be implemented is presented in Sect. 5.3, while correction procedures are described and evaluated in Sect. 5.4. In the following parts of this section, we will discuss the role of the storage (Sect. 5.2.2), present some assessments of the night flux error on cumulated sequestration (Sect. 5.2.3) and on functional relationships (Sect. 5.2.4), and, finally, evaluate its impact on other tracer fluxes (Sect. 5.2.5).

5.2.2 What Is the Role of Storage in This Error?

This section tries to answer two questions: (1) Can the night flux error be corrected by only adding the storage term to the turbulent flux? (2) How to introduce the storage in filtering and correction procedures?

From Sect. 5.1, it arises that the main cause of the night flux error is that storage flux and advection become important compared to the turbulent flux in low turbulent conditions. However, the problem is not the same if the term that competes with the turbulent flux is the storage or the advection (Fig. 5.2).

In the first case, it means that the CO_2 that is respired by the ecosystem accumulates in the air below the measurement system and would be released as soon as turbulence would onset (Fig. 5.2b). In these conditions, the flux capture by the measurement system would simply be delayed. This would be without impact on long-term budget but would however induce a bias on half hourly flux estimates and, consequently on the relationships between these fluxes and climate variables. Grace et al. (1996), Berbigier et al. (2001), and Dolman et al. (2002) considered in particular that the night flux underestimation at their site resulted only from storage and, consequently, did not apply any further night-data filtering to their data when computing annual sums. However, we think that these cases remain the exception rather than the rule.

In the second case, the respired CO_2 is removed from the ecosystem by advection and is definitively lost by the measurement system (Fig. 5.2c). In this case, a treatment is necessary not only for half hourly estimates but also for long-term budgets.

In most cases, both these processes take place simultaneously (Fig. 5.2d). As a consequence, a data filtering or correction is necessary, but there is a risk that it

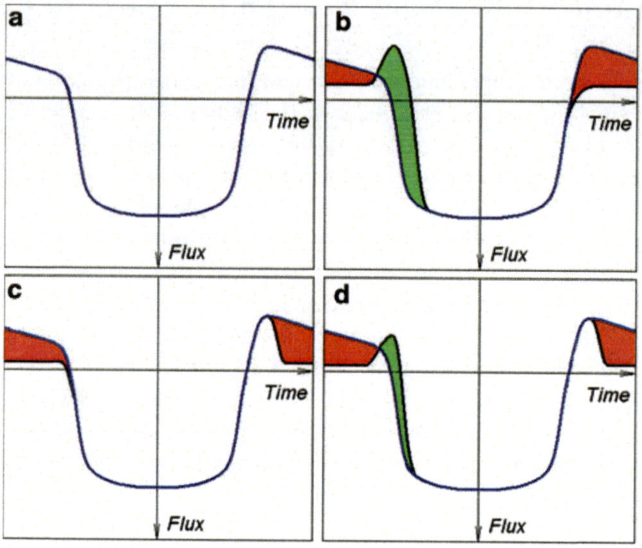

Fig. 5.2 Idealized dial evolution of CO_2 flux exchanged by an ecosystem. (**a**) (and *blue curve* in all other figures): Expected evolution of the biotic flux (the flux decrease at night mimics a response to temperature), (**b**) (*black curve*): Expected measured flux if the night flux underestimation is only due to storage change (the *red* and *green* surfaces compensate), (**c**) (*black curve*): Expected measured flux if the night flux underestimation is only due to a nonturbulent evacuation of CO_2 respired at night. See Sect. 5.4.1 for the peak explanation in early evening, (**d**) (*black curve*): Expected measured flux when both storage change and nonturbulent transport are responsible for the night flux underestimation (the *red* and *green* surfaces do not compensate)

leads to an overstated correction. This point should be considered with care when applying the filtering or the correction procedures, and will be discussed in detail in Sects. 5.3 and 5.4.

5.2.3 *What Is the Impact of Night Flux Error on Long-Term Carbon Sequestration Estimates?*

The night flux error acts as a selective systematic error (Moncrieff et al. 1996), that is, it affects much more night flux measurements, when the ecosystem behaves as a source, than day flux measurements, when the ecosystem behaves as a sink. As a result, it always leads to a carbon sequestration overestimation. The importance of the error varies from site to site and depends at the same time on local average meteorological conditions (frequency of occurrence of nonturbulent periods), on site topography, on land cover heterogeneity, on soil and plant biology (importance of respiration), and on canopy architecture (vegetation height, canopy density).

An assessment of this error could be obtained by comparing u_* corrected and noncorrected NEE estimates. Such estimates have been extensively presented in the literature. These results are gathered in Table 5.1. Tropical forests appear to be the most sensitive to the error that could reach 200 to more than 400 $g\,C\,m^{-2}\,year^{-1}$. This is because these forests are high, dense, and generally subjected to intense respiration fluxes. It could reach more than 100 $g\,C\,m^{-2}\,year^{-1}$ in Mediterranean forests, 50–90 $g\,C\,m^{-2}\,year^{-1}$ in temperate forests, and generally lesser than 50 $g\,C\,m^{-2}\,year^{-1}$ in crops and grasslands.

5.2.4 What Is the Impact of the Night Flux Error on Functional Relationships?

Night flux underestimation may also affect flux–climate relationships. Most usual flux relationships relative to CO_2 fluxes are the photosynthetically active photon flux density (PPFD) response of day flux and the temperature response of night flux. Night flux error induces both random and systematic error in the night flux response to temperature as it increases data spread and leads to an underestimation of the relationship parameters, that is, respiration at 10°C and temperature sensitivity. The response to PPFD of day flux may also be affected as the left end of the curve corresponds to low PPFD, generally associated to the beginning or the end of the night. Contrasting results may be observed: at sunrise, stable conditions often reduce turbulence while soil cooling is not large enough to generate advection. In these conditions, the CO_2 accumulation is especially important and turbulent fluxes underestimate the source/sink term. At sunset, following turbulence onset, the CO_2 accumulated at night is evacuated which may lead, on the contrary, to turbulent fluxes that overestimate the source/sink term. Conjunction of sunrise and sunset data in the flux to PPFD relationships may thus lead to both over- and underestimation of the flux. This generates an important data spread and, consequently, large uncertainties on the intercept (dark respiration) and the initial slope (quantum yield) of the light response. One could think that the problem could be solved by adding storage change to the turbulent flux. Unfortunately, half hourly storage estimates are themselves subject to a large spread so that this rarely improves the problem.

5.2.5 What Is the Impact of the Night Flux Error on Other Fluxes?

As the night flux problem results mainly from atmospheric processes that hinder the turbulent transport of tracers, it should affect any passive tracer that, similarly than for CO_2, could be exchanged by the surfaces at night and whose flux would be mainly controlled by production/absorption mechanisms that carry out independent of the presence or absence of turbulent transport.

Table 5.1 Impact of night flux correction on annual carbon sequestration at different sites

Site	Author	noncorrected	corrected	Δ	Threshold	% missing
Forest temperate conifer 15						
Tharandt 2001	Papale et al. (2006)	−623	−574	49	0.19	
Tharandt 2002	Papale et al. (2006)	−674	−623	51	0.18	
Hainich 2001	Papale et al. (2006)	−591	−559	32	0.35	
Hainich 2002	Papale et al. (2006)	−593	−530	63	0.35	
Braschaat 1997–2001	Carrara et al. (2003)	−171	−110	61	0.2	56
Loobos 1997	Dolman et al. (2002)	−338	−338	0	/	0
Kiryu 2003–2004	Ohkubo et al. (2007)	−798	−589	209	0.5	
Yamashiro 2000–2002 (mixed)	Kominami et al. (2008)	−312	−127		0.4	80
Forest boreal 2						
Hyytiala 2001	Papale et al. (2006)	−221	−178	43	0.2	
Hyytiala 2002	Papale et al. (2006)	−299	−215	84	0.25	
Forest temperate deciduous 9						
Vielsalm 2001	Papale et al. (2006)	−600	−538	62	0.3	
Vielsalm 2002	Heinesch et al. (2007)	−680	−545	75	0.5	62
Hesse 2001	Papale et al. (2006)	−575	−592	−18	0.1	
Hesse 2002	Papale et al. (2006)	−582	−608	−26	0.1	
Soroe 1996–1997	Pilegaard et al. (2001)	−240	−183	57	0.25	
Takayama 1999–2001	Saigusa et al. (2005)	−323	−266	57	0.2	
Forest mediterranean 3						
Puechabon	Papale et al. (2006)	−445	−302	143	0.18	
Yatir	Papale et al. (2006)	−240	−174	66	0.21	
Roccarespampani	Papale et al. (2006)	−151	−12	139	0.13	

(continued)

Table 5.1 (continued)

Site	Author	non corrected	corrected	Δ	Threshold	% missing
Forest tropical 6						
Santarem (Brasil)	Saleska et al. (2003)	−390	+40	430		
Tapajos Nat. For.	Hutyra et al. (2008)	−340	+95	435	0.22	
Pasoh (Malaysia)	Kosugi et al. (2008)	−850	−580	270	0.2	
Grasslands 2						
Schidler	Falge et al. (2001)	−431	−355, −383[a]	48, 76		
Sierra Nevada	Xu and Baldocchi (2004)	−90	−51	39		
Crops 6						
Lonzée SB 2004	Moureaux et al. (2006)	−620	−590	30	0.2	27
Lonzée, WW 2005	Moureaux et al. (2008)	−670	−620	50	0.3	
Gebesee, WW	Anthoni et al. (2004)	−320	−215	105	0.3	
Bondville, Corn	Falge et al. (2001)	−547	−526, −563[a]	−16, +21		
Bondville, Soybean	Falge et al. (2001)	129	125, 165[a]	−36, +4		
Ponca City, WW 1997	Falge et al. (2001)	−249	−147, −174[a]	75, 102		

Nc noncorrected sums, *c* sum corrected using u^* filtering, Δ difference, *Threshold* u^* threshold selected for u^* filtering, *% missing* percentage of data removed by the filtering

[a]Two different values are given, depending on the data gap filling method

First, tracers whose fluxes are negligible at night, such as water vapor and isoprene, could be considered as not concerned by such problem. For other tracers, like sensible heat, methane, monoterpenes, methanol, nitrous oxide, ozone, or NOx the situation is more complex. In these cases, a careful and specific analysis is needed for each tracer to determine if the flux decrease under low turbulence (if any) is the result of a measurement artifact or of a real flux slowing down. When the flux is not controlled by production/absorption processes at the surface but rather result from a diffusive exchange between a reservoir and the atmosphere, as is the case in deposition processes for example, the dependence of the flux on turbulence could be real. In these conditions, the night flux correction is not recommended for long-term budgets as it could lead to a large flux overestimation.

In addition, night flux effect could be very different if the gas is passive or reactive. In the first case, a behavior similar to those of CO_2 would be expected while the second situation would be more complex. Indeed, the turbulence limitation, by hindering atmospheric transport, would limit not only the tracer flux but also the reactive transport and, by this, the reactive encounters and their mutual destruction. In these conditions, the residence time of reactive components could therefore be prolonged under low turbulence.

The effect of chemical destruction of an emitted compound on its above canopy flux depends on the chemical lifetime of the compound, and the effectiveness of turbulent transport. The ratio of the turbulent mixing time scale to the chemical life time, called Damköhler number (Damköhler 1940), can be used to assess the importance of chemistry on fluxes. The Damköhler number can be written as

$$Da = \frac{\tau_*}{\tau_c}, \tag{5.1}$$

where mixing time scale can be estimated as $\tau_* = (h_m - d)/u_*$. The chemical lifetime, τ_c is the time constant characterizing the degradation of the compound characterized by its mixing ratio χ_R. The differential equation describing this degradation may write:

$$\frac{d\chi_R}{dt} = -\sum_{i=1}^{N} k_i \chi_i \chi_R - k_{photolysis} \chi_R \tag{5.2}$$

From which τ_c can be deduced:

$$\tau_c = \left(\sum_{i=1}^{N} k_i \chi_i + k_{photolysis} \right)^{-1} \tag{5.3}$$

where χ_i refers to different oxidant concentrations, k_i is the rate constant for the reaction between the oxidant and the compound, and $k_{photolysis}$ is the photolysis rate.

Using stochastic Lagrangian transport model, Rinne et al. (2007b) estimated that the above canopy flux is significantly reduced already at Damköhler number values well below 0.1. As the friction velocity is typically lower during night, the mixing time scale tends to be longer. Also the chemical lifetime of a compound can be different during day and night. For example, hydrocarbon compounds (e.g. isoprene and monoterpenes) react in the surface layer with ozone, hydroxyl radical, and nitrate radical, all of which have their different diurnal cycles. Thus one needs to calculate the chemical lifetime for different conditions (day, night) to assess the possible importance of the chemistry on fluxes.

Dependence of sensible heat on u_* has been supported indirectly by analyzing energy balance closure in night conditions. Indeed, at night, the numerator of the closure fraction (CF), defined as: $CF = \frac{H+\lambda E}{Rn+G}$, depends only on turbulent fluxes (i.e., mainly on sensible heat as latent heat is negligible at night), so that the evolution of CF with the friction velocity is an indication of the sensible heat underestimation at night.

Decreases of the CF at low friction velocities were pointed out in particular by Aubinet et al. (2000), Turnipseed et al. (2002), Wilson et al. (2002), Barr et al. (2006), and Tanaka et al. (2008). In addition, Barr et al. (2006) highlighted the similarity between CF and normalized NEE evolutions with u_* at night, showing in particular that the u_* threshold were similar for the two tracers.

Evidence for a night flux dependence on u_* were found notably for ozone (Fig. 5.3a) by Rannik et al. (2009) and for monoterpenes (Fig. 5.3b) by Laffineur (comm. Pers.). However, in none of these cases there is an evidence for mechanism that should produce or absorb these gases independently of turbulence. It is thus possible that these responses reflect a real flux dependency on turbulence.

Many authors systematically sort their data by the mean of a u_* filter before analyzing them. This is especially the case of Rinne et al. (2007a) for methane or Davison et al. (2009) for methanol, acetaldehyde, acetone, and monoterpenes. Here again, a careful analysis of the mechanisms underlying the exchange is necessary in order to determine if the flux dependency on turbulence is the result of a measurement artifact or of a real production/absorption slowing down. The application of a night flux correction for long-term budgets would be relevant only in the first case.

5.3 How to Implement the Filtering Procedure?

5.3.1 General Principle

Filtering methods consist in discarding eddy flux measurements taken during conditions where the eddy covariance measurement is considered as nonrepresentative of the biotic flux. When necessary (for computing sums, e.g.) the gaps created

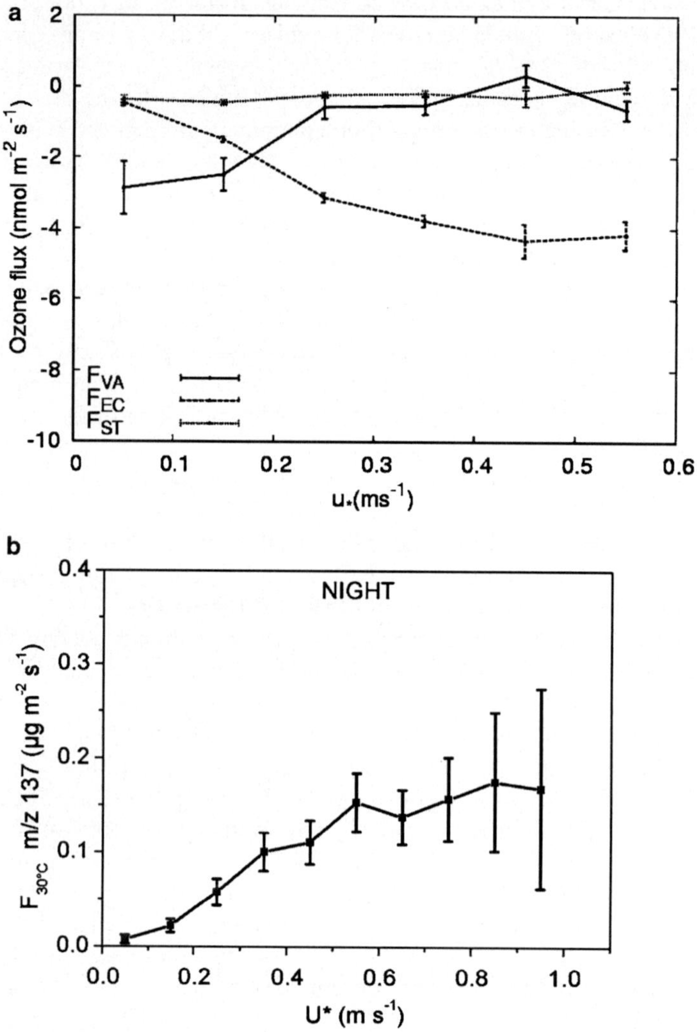

Fig. 5.3 *u** response of other tracer fluxes (**a**) Ozone fluxes at Hyytialla. F_{EC} eddy covariance measurements, F_{ST} storage change, F_{VA} advection (Courtesy of Rannik et al. 2009), (**b**) Monoterpene fluxes at Vielsalm (Laffineur, comm. pers.)

by the filtering could be filled. These aspects are discussed in Sect. 5.4. Here we concentrate on the filtering itself, the main questions relevant to this procedure being the determination of the most adapted criterion to discard periods affected by the night flux error and the implementation of the filtering procedure.

5.3.2 *Choice of the Selection Criterion*

Considering that the night flux problem arises essentially when turbulence is insufficient, Goulden et al. (1996) proposed to use a criterion based on the friction velocity, that is, data measured when u_* is below a given threshold, $u_{*\text{crit}}$, being discarded.

This threshold is identified by looking at flux vs u_* relations: As the biotic flux is expected to not depend on friction velocity, $u_{*\text{crit}}$ could be identified as the threshold below which the flux decreases with decreasing u_*. Alternative criteria have been proposed. However, if some of them could appear relevant, we will concentrate in the following sections on u_* filtering, as this is the procedure mostly used at present. We will show in Sect. 5.3.3 how the method may be implemented and discuss some difficulties that could appear during this implementation. Finally, in Sect. 5.3.4, we will discuss the pros and cons of the approach and introduce some of these alternative filtering criteria.

5.3.3 *Filtering Implementation*

The most critical question is to choose correctly the friction velocity threshold $u_{*\text{crit}}$, that is, to determine the u_* range in which eddy fluxes can be considered as reliable. This range depends on local topography, surface roughness and heterogeneity, source distribution and intensity, so it varies from site to site and, at a given site, may vary from season to season. The use at a given site of a "standard" threshold derived from literature may indeed lead either to an excessive selection of the data (if too large) or, worse, to a bias in the correction (if too small). It is therefore recommended to make a specific evaluation of the threshold at each site.

Gu et al. (2005) suggested that the data selection should be operated not only below a lower threshold but also above a higher threshold, in order to take account of turbulent flux contamination by pressure pumping under high turbulence. The relevance of such an upper threshold is still a matter of discussion and is not confirmed at all sites.

The lower threshold is site specific and, even, could vary at one given site according to the period. It is especially the case on crops (Moureaux et al. 2008; Béziat et al. 2009). It needs thus to be evaluated individually. This evaluation results from a compromise: on the one hand, the threshold should be as small as possible in order to minimize the number of data that are discarded and, therefore, the random uncertainty on night flux data; on the other hand it should be large enough in order not to introduce any systematic bias on the cumulated NEE value. One can define the lower threshold as the lowest value above which NEE becomes insensitive to the threshold changes. This threshold can be identified by sorting nighttime NEE data by u_* classes and performing statistical comparison between each class-averaged NEE. The threshold is then defined by the lowest u_* value for which the difference

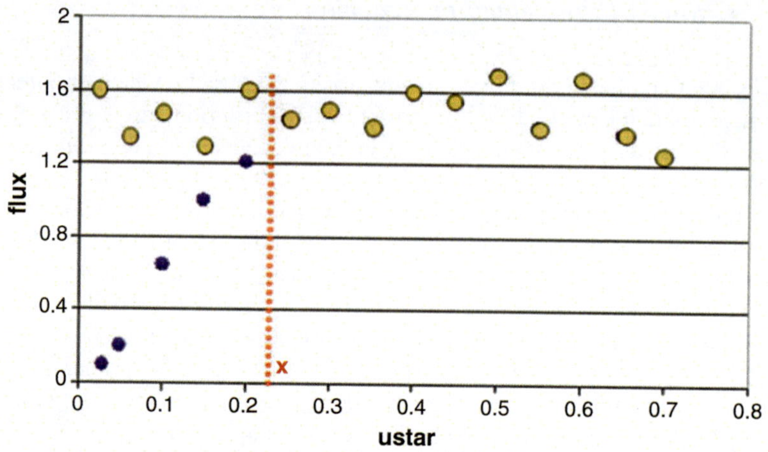

Fig. 5.4 u^* threshold selection (theoretical and optimal situation): *blue dots* are normalized nighttime fluxes (or nighttime fluxes acquired in a narrow temperature range) storage corrected, *yellow dots* are the theoretical pattern of the fluxes (independent by u^*), *red x* is the u^* threshold

between the corresponding averaged NEE is not significantly different from the plateau-averaged NEE (Fig. 5.4). Gu et al. (2005) and Reichstein et al. (2005) proposed algorithms implementing this procedure. In practice, the lowest threshold varies typically from 0.1 to 0.5 m s^{-1} according to the sites.

The preceding approach is valid only if there is a guarantee that friction velocity does not covary with other climatic respiration-driving variables. Indeed, if such covariation exists, it could induce some night flux response to u_* that would not design necessarily a night flux error. In order to avoid such covariation, it is recommended, before sorting NEE according to u_* classes, to plot u_* against the main driving variables and, if any relation is detected, to first perform a normalization by using functions describing NEE response to these driving factors. In temperate regions, normalization by a temperature function is generally used (Aubinet et al. 2000) (Fig. 5.4).

The filtering approach may lead to a flux overestimation in presence of storage. If a part of the flux underestimation during calm periods resulted from CO_2 accumulation below the measurement point, this flux would be restored as soon as turbulence onsets. In these conditions, the u_* filtering, by removing the calm period where the flux is underestimated and keeping the turbulence onset period where it is overestimated, would lead to a global respiration overestimation. If, in addition, the calm period data were replaced by any parameterization (filtering – data gap-filling approach, see Sect. 5.4.2), the part of emitted flux that would have been stored for a short while would have been counted twice. To avoid such bias, NEE estimate must take storage change into account. The introduction of storage leads thus generally to a reduction of the correction brought by u_* filtering (Fig. 5.2d; Box 5.2).

Box 5.2: Steps Recommended in Order to Apply the u^* Filtering Procedure

1. Compute the storage change. It is given by $\int_o^{h_m} \rho d \overline{\frac{\partial \chi_c}{\partial t}} dz$. This term is computed by approximating the spatial integral by a weighted sum of different concentrations measured along a vertical profile. In forest sites, the vertical profiles should include as many sampling points as possible (at least four), distributed following a logarithmic pattern along the vertical. In grasslands and croplands, where the measuring height is lower, the profile could be approximated by a single point. The time derivative is approximated by a finite difference between instantaneous concentrations during consecutive half hours.
2. Compute night flux data as the sum of turbulent flux and storage change.
3. Sort night flux data by increasing u_*.
4. Evaluate if there is covariation between u_* and other respiration-driving variables (most often the temperature). If yes, normalize the data in order to get rid of covariation of respiration with this variable.
5. Set a number of u_* classes (normally between 20 and 30) and calculate the mean NEE for each class.
6. Determine the threshold by comparison between NEE in each u_* class and the average of the mean NEE values measured at higher u_*. The new threshold is reached when the NEE of a given u_* class become significantly different from the mean NEE at higher u_*.
7. Remove data situated below the under threshold.
8. If an upper threshold is relevant, same scheme should be followed.

5.3.4 Evaluation

An absolute evaluation of the approach is, however, difficult in the absence of independent flux measurement methods as it aims at correcting a selective systematic error which is unknown. As a result, uncertainties may remain after u^* filtering that are due to an incomplete (or an overstated) correction.

The first criticism is that the way the selection is operated is empirical. First, the choice of the criterion is questionable, second there is no guarantee that the data filtering removes all bad data and neither that it removes only bad data.

Acevedo et al. (2009) recalled that u_* represents a flux and could also be contaminated by mesoscale movements. They proposed to use the standard deviation of vertical velocity component, σ_w, as an alternative criterion to u_* as this latter variable did not suffer from this flaw. By applying filtering procedures based either on σ_w or on u_* on three Amazonian sites, they showed that the first procedure represented a significant improvement, with two main consequences: easier determination of the threshold and larger respiration rates of the series classified as turbulent.

Another flaw of the approach is that u_* is generally evaluated from turbulence measurements made at the canopy top. However, in the case of tall vegetation, decoupling may appear between the wind field above and below the canopy so that the value of u_* above the canopy may be not representative of the turbulence and wind field in the canopy.

Some experiences highlight events during which the u_* filtering failed in keeping wrong measurements: abnormally high turbulent fluxes have been observed at two forest sites (Cook et al. 2004; Rebmann et al. 2010) under well-mixed periods (not removed by the u_* filter). These fluxes were supposed to result from CO_2 advection from pools where it had accumulated.

Wohlfahrt et al. (2005), measuring in an alpine pasture, showed that the direct application of the u_* criterion led to an overestimation of the selected flux data, compared to chamber estimates. By adding a stationarity screening to the u_* filtering, he got more defensible flux estimates. A hypothesis could be that the site is subjected to intermittent turbulence so that turbulent events correspond to releases of CO_2 accumulated during the preceding calm periods.

Finally, the method remains questionable in the cases of sites where no discernable plateau can be observed in the flux/u_* relationship.

Despite these flaws, the u_* filtering has been successfully used in many cases: in particular, u_*-filtered data have often been used to find ecologically relevant functional relationships (see, in particular: Janssens et al. 2001; Suyker et al. 2005; Moureaux et al. 2006; Zhao et al. 2006). This method also has the advantage of simplicity as the selection criterion is based on a variable that is immediately available from eddy flux measurements. In addition, when appropriate data gap-filling algorithms are used, it does not need any modeling, which could pose a problem when the data are used later for model calibration or validation.

5.4 Correction Procedures

As specified above, correction procedures are necessary at least to establish long-term budgets. In these conditions indeed, a full cover of the measurement period is needed so that underestimated fluxes should be corrected. Two correction approaches are discussed here: the u_* filtering + data gap-filling method and the advection corrected mass balance (ACMB) approach.

5.4.1 Filtering + Gap Filling

The approach consists simply in combining the filtering procedure described in Section 5.3 and a data gap-filling procedure as presented in Chap. 6. Most often the friction velocity is used as a filtering criterion. For gap filling, different methods were used, like parameterization (Goulden et al. 1996; Aubinet et al. 2000), lookup tables (Falge et al. 2001), neural networks (Papale and Valentini 2003), constrained source optimization (Juang et al. 2006) or modeling (Lavigne et al. 1997; Lindroth et al. 1998; Kutsch et al. 2008; Hutyra et al. 2008).

As based on the filtering procedure, the approach of course suffers from the same defects, as described above. Despite these flaws, this method remains the most often applied approach because of its simplicity and its relative robustness in many cases. However, some researchers are seeking for alternative filtering criteria. In particular, van Gorsel et al. (2007) proposed a filtering based on the peak sum of the turbulent flux and the change in storage. Aubinet et al. (2005) showed that this peak occurs at the beginning of the night at most sites. van Gorsel et al. (2007, 2008) argue that this is the result of the following sequence of events: After sunset, when the boundary layer becomes sufficiently stably stratified through radiative cooling of the canopy, much of the respired CO_2 is stored in the canopy, and CO_2 mixing ratios close to the ground start to increase. The cool layer within the canopy modifies buoyancy and hence the hydrostatic pressure gradient. Gravity flows start once the air close to the surface has cooled to the extent that the hydrostatic pressure gradient exceeds the sum of hydrodynamic pressure gradient and foliage drag (Finnigan 2007). Entrainment of air with a lower CO_2 mixing ratio at the hill crests leads to the development of horizontal CO_2 gradients. Once these gradients have developed advection starts to drain CO_2 out of the control volume, which results in a decrease in the sum of eddy flux and change in storage term. They hypothesize that there is a hiatus between sunset and the onset of advection during which the sum of eddy flux and storage of CO_2 may be considered as a reliable estimate of the biotic flux. They suggest, thus, keeping these measurements only and filling data gaps with one of the above-cited procedure. By applying this method to 25 tower flux sites covering a wide range of vegetation, climate, and topography, they found higher nocturnal respiration rates than estimated with u_*-threshold filter, and – where available – excellent agreement with independent estimates such as ones derived from upscaled chamber measurements (van Gorsel et al. 2009). A disadvantage of the method is that the procedure keeps very little data so that functional relationships based on these data sets are subject to large random uncertainties. Another restriction of the method is that there is no guarantee that the event sequence which is at its base takes place everywhere in all conditions. The method could thus be not applicable at some sites.

5.4.2 *The ACMB Procedure*

5.4.2.1 History

The ACMB (Aubinet et al. 2010) approach consists in estimating the NEE by completing eddy covariance and storage estimates by direct measurements of horizontal and vertical advection. A first attempt to estimate vertical advection was made by Lee (1998). By assuming a linear increase of the vertical velocity with height, he proposed an expression of the vertical advection based on the vertical c_c profile and on one vertical velocity measurement made at the control volume top. The advantage of this method is that it is based on a single point measurement and does not require any additional measurement. It was notably used by Baldocchi et al. (2000) and Schmid et al. (2000) to revise NEE estimations. In his reply to the Lee paper, Finnigan (1999) suggested that horizontal advection should not be neglected as it was of the same order of magnitude as the vertical advection. Following this recommendation, direct horizontal advection measurements were performed using simple single level 2D (Aubinet et al. 2003), multilevel 2D (Marcolla et al. 2005; Heinesch et al. 2007, 2008; Tóta et al. 2008), single level 3D (Staebler and Fitzjarrald 2004, 2005), and multilevel 3D (Feigenwinter et al. 2004; Sun et al. 2007; Leuning et al. 2008; Yi et al. 2008) set ups. The most advanced set up was probably those installed at three European sites in the frame of the ADVEX experiment (Feigenwinter et al. 2008). A system made up of four towers equipped each with four-point temperature, velocity, and χ_c profiles was installed at sites already equipped with eddy covariance systems. Continuous measurements were performed during 2–4 months of campaigns (Feigenwinter et al. 2010a,b). An alternative sampling system, based on continuous sampling using perforated tubing arranged parallel to the ground, was used by Leuning et al. (2008).

5.4.2.2 Procedure

The ACMB approach requires estimates of horizontal and vertical advection. Lee (1998) proposed to compute vertical advection as

$$F_{VA} = \overline{w}\left(\overline{\chi_c}\,|_{h_m} - \langle \chi_c \rangle\right) \tag{5.4}$$

where, w and $\overline{\chi_c}\,|_{h_m}$ represent the vertical component of velocity and CO_2 mixing ratio at control volume top and $\langle \chi_c \rangle$ a CO_2 mixing ratio averaged between this height and the soil. In practice, the vertical component of velocity is deduced from 3D velocity measurements performed with a sonic anemometer. In order to obtain this component, a planar-fit approach or a sectorwise planar fit is necessary, as classical 2D and 3D approaches systematically nullify w. Different methods have been proposed (Lee 1998; Paw et al. 2000; Wilczack et al. 2001), (see also Sect. 3.2.4).

Horizontal advection should require at the same time the estimation of the horizontal velocity and of the χ_c gradient in the same direction. This constitutes a strong limitation of the approach as, in sites where horizontal velocity changes often, it should require high spatial resolution χ_c samplings. In addition, these measurements should be integrated on all the control volume height, requiring in practice a multiplication of towers. In sloping sites where a sloping wind regime takes place, some authors (Aubinet et al. 2003; Marcolla et al. 2005; Heinesch et al. 2007, 2008) postulated that the wind regime was mainly 2D so that a simpler set up, based on two profiles aligned along the slope, could be used. In addition, as CO_2 build up is expected to be larger close to the soil, concentration gradients were supposed to be more important at this place and a simpler system that sampled χ_c only in the lowest layers was sometimes used. However, these hypotheses require careful verification, because also nearly uniform nighttime profiles in the canopy with largest vertical gradients near the top of the canopy were reported for a number of sites (e.g. Reiners and Anderson 1968; Goulden et al. 2006; de Araújo et al. 2008; Tóta et al. 2008; Feigenwinter et al. 2010b).

5.4.2.3 Evaluation

Unfortunately, the ACMB approach was found to give deceptive results as affected both by random and systematic uncertainties and giving nonrobust NEE estimates. Aubinet et al. (2010) showed indeed that ACMB estimates obtained at the three ADVEX sites were often one order of magnitude larger than the expected biotic fluxes and that they were not stable according to u_* changes. They found in addition that they vary with wind direction, while biotic fluxes should not vary with this variable at homogeneous sites.

Uncertainties on horizontal advection result mainly from uncertainties on horizontal χ_c gradients. Firstly, in many cases, these gradients are small and need good resolution set ups to be correctly measured. In addition, sampling point positioning is critical: as vertical gradients are generally one order of magnitude larger than the horizontal gradients, a bad vertical positioning of the sensor can lead to important systematic errors. Moreover, large horizontal gradient heterogeneities may appear in the control volume, due to source heterogeneities or to air circulation in the control volume. As a consequence, large uncertainties may also result from an insufficient spatial resolution of the χ_c sampling. Finally, in presence of large horizontal gradients that are almost perpendicular to the average wind velocity, a small error on the angle between the concentration gradient and the wind velocity could lead to erroneously large horizontal advection estimates. On the other hand, large horizontal velocities together with small horizontal gradients can also cause unrealistic high advective fluxes.

Uncertainties on vertical advection estimates are mainly due to the measurement errors that affect the vertical component of the velocity. Uncertainties relate as well to its value at the control volume top as to its vertical profile shape. Large uncertainties result notably from the computation method: none of them can be considered as better. A comparison between these methods, performed by Vickers and Mahrt (2006), pointed out significant differences between these approaches. Facing such inconsistencies, an alternative approach based on the mass continuity equation has also been proposed by Vickers and Mahrt (2006) and Heinesch et al. (2007). However, as based on an estimate of horizontal velocity divergence, it suffers from a large uncertainty, though it may be theoretically the most justified approach (Box 5.3).

Box 5.3: Recommended and Nonrecommended Correction Procedures

1. ACMB is not recommended for correcting eddy covariance measurements because
   (a) It is difficult to implement, requiring heavy set up and many workforce.
   (b) Advection measurements are affected by large random errors introducing a relative uncertainty often larger than 100% on half hourly estimates.
   (c) In most cases, huge systematic errors affect advection measurements so that ACMB lead to non realistic results even after averaging on long time periods.
2. At present, despite their different shortcomings, the filtering – gap filling approach remains the recommended correction procedure.
3. u_* is at present the most often used parameter for data selection. Criteria based on vertical velocity variance and night flux chronology are promising alternatives.

Acknowledgments Authors acknowledge financial support by the EU (FP 5, 6, and 7), the Belgian Fonds de la recherche Scientifique (FNRS-FRS), the Belgian Federal Science Policy Office (BELSPO) and the Communauté française de Belgique (Action de Recherche Concertée).

References

Acevedo OC, Moraes OLL, Degrazia GA, Fitzjarrald DR, Manz AO, Campos JG (2009) Is friction velocity the most appropriate scale for correcting nocturnal carbon dioxide fluxes? Agric For Meteorol 149:1–10

Anderson DE, Verma SB, Rosenberg NJ (1984) Eddy correlation measurements of CO2: latent heat and sensible heat fluxes over a crop surface. Agric For Meteorol 29:263–272

Anthoni PM, Freibauer A, Kolle O, Schulze ED (2004) Winter wheat carbon exchange in Thuringia, Germany. Agric For Meteorol 121(1–2):55–67

Aubinet M, Grelle A, Ibrom A, Rannik Ü, Moncrieff J, Foken T, Kowalski AS, Martin PH, Berbigier P, Bernhofer C, Clement R, Elbers J, Granier A, Grünwald T, Morgenstern K, Pilegaard K, Rebmann C, Snijders W, Valentini R, Vesala T (2000) Estimates of the annual net carbon and water exchange of forests: the EUROFLUX methodology. Adv Ecol Res 30: 113–175

Aubinet M, Chermanne B, Vandenhaute M, Longdoz B, Yernaux M, Laitat E (2001) Long term carbon dioxide exchange above a mixed forest in the Belgian Ardennes. Agric For Meteorol 108:293–315

Aubinet M, Heinesch B, Yernaux M (2003) Horizontal and vertical CO2 advection in a sloping forest. Bound Layer Meteorol 108:397–417

Aubinet M, Berbigier P, Bernhofer Ch, Cescatti A, Feigenwinter C, Granier A, Grünwald Th, Havrankova K, Heinesch B, Longdoz B, Marcolla B, Montagnani L, Sedlak P (2005) Comparing CO2 storage and advection conditions at night at different CARBOEUROFLUX sites. Bound Layer Meteorol 116:63–93

Aubinet M, Feigenwinter C, Bernhofer C, Canepa E, Heinesch B, Lindroth A, Montagnani L, Rebmann C, Sedlak P, Van Gorsel E (2010) Advection is not the solution to the nighttime CO2 closure problem—evidence from three inherently different forests. Agric For Meteorol 150(5):655–664

Baldocchi D, Finnigan J, Wilson K, Paw UKT, Falge E (2000) On measuring net ecosystem carbon exchange over tall vegetation on complex terrain. Bound Layer Meteorol 96:257–291

Barr AG, Morgenstern K, Black TA, McCaughey JH, Nesic Z (2006) Surface energy balance closure by the eddy-covariance method above three boreal forest stands and implications for the measurement of the CO2 flux. Agric For Meteorol 140:322–337

Berbigier P, Bonnefond J-M, Mellmann P (2001) CO2 and water vapour fluxes for 2 years above Euroflux forest site. Agric For Meteorol 108(3):183–197

Béziat P, Ceschia E, Dedieu G (2009) Carbon balance of a three crop succession over two cropland sites in South West France. Agric For Meteorol 149:1628–1645

Carrara A, Kowalski AS, Neirynck J, Janssens IA, Curiel Yuste J, Ceulemans R (2003) Net ecosystem CO2 exchange of mixed forest in Belgium over 5 years. Agric For Meteorol 119:209–227

Cook B, Davis KJ, Wang W, Desai A, Berger BW, Teclaw RM, Martin JG, Bolstad PV, Bakwin PS, Yi C, Heilman W (2004) Carbon exchange and venting anomalies in an upland deciduous forest in northern Wisconsin, USA. Agric For Meteorol 126:271–295

Damköhler G (1940) Der Einfluss der Turbulenz auf die Flammengeschwindigkeit in Gasgemischen. Z Elektrochem Angew Phys Chem 46:601–626

Davison B, Taipale R, Langford B, Misztal P, Fares S, Matteucci G, Loreto F, Cape JN, Rinne J, Hewitt CN (2009) Concentrations and fluxes of biogenic volatile organic compounds above a Mediterranean macchia ecosystem in western Italy. Biogeosciences 6:1655–1670

de Araújo A, Kruijt B, Nobre AD, Dolman AJ, Waterloo MJ, Moors EJ, de Souza JS (2008) Nocturnal accumulation of CO_2 underneath a tropical forest canopy along a topographical gradient. Ecol Appl 18(6):1406–1419

Dolman AJ, Moors EJ, Elbers JA (2002) The carbon uptake of a mid latitude pine forest growing on sandy soil. Agric For Meteorol 111:157–170

Falge E, Baldocchi D, Olson R, Anthoni P, Aubinet M, Bernhofer C, Burba G, Ceulemans R, Clement R, Dolman H, Granier A, Gross P, Grünwald T, Hollinger D, Jensen N-O, Katul G, Keronen P, Kowalski A, Ta Lai C, Law B, Meyers T, Moncrieff J, Moors EJ, Munger JW, Pilegaard K, Rannik Ü, Rebmann C, Suyker A, Tenhunen J, Tu K, Verma S, Vesala T, Wilson K, Wofsy S (2001) Gap filling strategies for defensible annual sums of net ecosystem exchange. Agric For Meteorol 107:43–69

Feigenwinter C, Bernhofer C, Vogt R (2004) The influence of advection on short-term CO2 budget in and above a forest canopy. Bound Layer Meteorol 113:201–224

Feigenwinter C, Bernhofer C, Eichelmann U, Heinesch B, Hertel M, Janous D, Kolle O, Lagergren F, Lindroth A, Minerbi S, Moderow U, Mölder M, Montagnani L, Queck R, Rebmann C, Vestin P, Yernaux M, Zeri M, Ziegler W, Aubinet M (2008) Comparison of horizontal and vertical advective CO_2 fluxes at three forest sites. Agric For Meteorol 148:12–24

Feigenwinter C, Montagnani L, Aubinet M (2010a) Plot-scale vertical and horizontal transport of CO2 modified by a persistent slope wind system in and above an alpine forest. Agric For Meteorol 150(5):665–673

Feigenwinter C, Mölder M, Lindroth A, Aubinet M (2010b) Spatiotemporal evolution of CO2 concentration, temperature, and wind field during stable nights at the Norunda forest site. Agric For Meteorol 150(5):692–701

Finnigan J (1999) A comment on the paper by Lee (1998): on micrometeorological observation of surface-air exchange over tall vegetation. Agric For Meteorol 97:55–64

Finnigan JJ (2007) Turbulent flow in canopies on complex topography and the effects of stable stratification. In: Flow and transport processes with complex obstructions, vol 236, NATO science series II: mathematics, physics and chemistry. Springer, Dordrecht, 414 pp

Goulden ML, Munger JW, Fan S-M, Daube BC, Wofsy SC (1996) Measurements of carbon sequestration by long-term eddy covariance: methods and a critical evaluation of accuracy. Glob Change Biol 2:169–182

Goulden ML, Miller SD, da Rocha HR (2006) Nocturnal cold air drainage and pooling in a tropical forest. J Geophys Res 111:D08S04. doi:10.1029/2005JD006037

Grace J, Malhi Y, Lloyd J, McIntyre J, Miranda AC, Meir P, Miranda HS (1996) The use of eddy covariance to infer the net carbon dioxide uptake of Brazilian rain forest. Glob Change Biol 2:209–218

Gu L, Falge E, Boden T, Baldocchi DD, Black TA, Saleska SR, Suni T, Vesala T, Wofsy S, Xu L (2005) Observing threshold determination for nighttime eddy flux filtering. Agric For Meteorol 128:179–197

Heinesch B, Yernaux M, Aubinet M (2007) Some methodological questions concerning advection measurements: a case study. Bound Layer Meteorol 122:457–478

Heinesch B, Yernaux Y, Aubinet M (2008) Dependence of CO2 advection patterns on wind direction on a gentle forested slope. Biogeosciences 5:657–668

Hutyra LR, Munger JW, Hammond-Pyle E, Saleska SR, Restrepo-Coupe N, Daube BC, de Camargo PB, Wofsy SC (2008) Resolving systematic errors in estimates of net ecosystem exchange of CO2 and ecosystem respiration in a tropical forest biome. Agric For Meteorol 148:1266–1279

Janssens IA, Lankreijer H, Matteucci G, Kowalski AS, Buchmann N, Epron D, Pilegaard K, Kutsch W, Longdoz B, Grunwald T, Dore S, Montagnani L, Rebmann C, Moors EJ, Grelle A, Rannik U, Morgenstern K, Oltchev S, Clement R, Gudmundsson J, Minerbi S, Berbigier P, Ibrom A, Moncrieff J, Aubinet M, Bernhofer C, Jensen NO, Vesala T, Granier A, Schulze ED, Lindroth A, Dolman AJ, Jarvis PG, Ceulemans R, Valentini R (2001) Productivity and disturbance overshadow temperature in determining soil and ecosystem respiration across European forests. Glob Change Biol 7(3):269–278

Jarvis PG, Massheder J, Hale SE, Moncrieff J, Rayment M, Scott SL (1997) Seasonal variation of carbon dioxide, water vapor and energy exchanges of boreal black spruce forest. J Geophys Res 102:28953–28966

Juang JY, Katul GG, Siqueira MB, Stoy PC, Palmroth S, McCarthy HR, Kim HS, Oren R (2006) Modeling nighttime ecosystem respiration from measured CO2 concentration and air temperature profiles using inverse methods. J Geophys Res 111(D8):D08S05

Kominami Y, Jomura M, Dannoura M, Goto Y, Tamai K, Miyama T, Kanazawa Y, Kaneko S, Okumura M, Misawa N, Hamada S, Sasaki T, Kimura H, Ohtani Y (2008) Biometric and eddy-covariance-based estimates of carbon balance for a warm-temperate mixed forest in Japan. Agric For Meteorol 148:723–737

Kosugi Y, Takanashi S, Ohkubo S, Matsuo N, Tani M, Mitani T, Tsutsumi D, Nik AR (2008) CO2 exchange of a tropical rainforest at Pasoh in Peninsular Malaysia. Agric For Meteorol 148:439–452

Kutsch K, Kolle O, Rebmann C, Knohl A, Ziegler W, Schulze ED (2008) Advection and resulting CO_2 exchange uncertainty in a tall forest in central Germany. Ecol Appl 18(6):1391–1405

Lavigne MB, Ryan MG, Anderson DE, Baldocchi DD, Crill PM, Fitzjarrald DR, Goulden ML, Gower ST, Massheder JM, McCaughey JH, Rayment M, Striegl RG (1997) Comparing nocturnal eddy covariance measurements to estimates of ecosystem respiration made by scaling chamber measurements at six coniferous boreal sites. J Geophys Res Atmos 102:28977–28985

Lee X (1998) On micrometeorological observations of surface air exchange over tall vegetation. Agric For Meteorol 91:39–49

Leuning R, Zegelin SJ, Jones K, Keith H, Hughes D (2008) Measurement of horizontal and vertical advection of CO2 within a forest canopy. Agric For Meteorol 148:1777–1797

Lindroth A, Grelle A, Morén A-S (1998) Long-term measurements of boreal forest carbon balance reveals large temperature sensitivity. Glob Change Biol 4:443–450

Loescher HW, Law BE, Mahrt L, Hollinger DY, Campbell J, Wofsy SC (2006) Uncertainties in, and interpretation of, carbon flux estimates using the eddy covariance technique. J Geophys Res Atmos 111:D21S90. doi:10.1029/2005JD006932

Mahrt L (1999) Stratified atmospheric boundary layers. Bound Layer Meteorol 90:375–396

Marcolla B, Cescatti A, Montagnani L, Manca G, Kerschbaumer G, Minerbi S (2005) Role of advective fluxes in the carbon balance of an alpine coniferous forest. Agric For Meteorol 130:193–206

Massman WJ, Lee X (2002) Eddy covariance flux corrections and uncertainties in long term studies of carbon and energy exchanges. Agric For Meteorol 113:121–144

Massman WJ, Sommerfeld RA, Mosier AR, Zeller KF, Hehn TJ, Rochelle SG (1997) A model investigation of turbulence driven pressure-pumping effects on the rate of diffusion of CO2, N2O, and CH4 through layered snowpacks. J Geophys Res 102:18851–18863

Moncrieff JB, Malhi Y, Leuning R (1996) The propagation of errors in long-term measurements of land atmosphere fluxes of carbon and water. Glob Change Biol 2:231–240

Moureaux C, Debacq A, Bodson B, Heinesch B, Aubinet M (2006) Carbon sequestration by a sugar beet crop. Agric For Meteorol 139:25–39

Moureaux C, Debacq A, Hoyaux J, Suleau M, Tourneur D, Vancutsem F, Bodson B, Aubinet M (2008) Carbon balance assessment of a Belgian winter wheat crop (Triticum aestivum L.). Glob Change Biol 14(6):1353–1366

Ohkubo S, Kosugia Y, Takanashia S, Mitania T, Tani M (2007) Comparison of the eddy covariance and automated closed chamber methods for evaluating nocturnal CO2 exchange in a Japanese cypress forest. Agric For Meteorol 142(1):50–65

Ohtaki E (1984) Application of an infrared carbon dioxide and humidity instrument to studies of turbulent transport. Bound Layer Meteorol 29:85–107

Papale D, Valentini R (2003) A new assessment of European forests carbon exchanges by eddy fluxes and artificial neural network spatialization. Glob Change Biol 9:525–535

Papale D, Reichstein M, Canfora E, Aubinet M, Bernhofer C, Longdoz B, Kutsch W, Rambal S, Valentini R, Vesala T, Yakir D (2006) Towards a standardized processing of Net Ecosystem Exchange measured with eddy covariance technique: algorithms and uncertainty estimation. Biogeosciences 3:571–583

Pattey E, Desjardin RL, St-Amour G (1997) Mass and energy exchange over a black spruce forest during key periods of BOREAS 1994. J Geophys Res 102:28967–28975

Paw U KT, Baldocchi DD, Meyers TP, Wilson KB (2000) Correction of eddy covariance measurements incorporating both advective effects 5 and density fluxes. Bound Layer Meteorol 97:487–511

Pilegaard K, Hummelshoj P, Jensen NO, Chen Z (2001) Two years of continuous CO2 eddy flux measurements over a Danish beech forest. Agric For Meteorol 107:29–41

Rannik Ü, Mammarella I, Keronen P, Vesala T (2009) Vertical advection and nocturnal deposition of ozone over a boreal Alpine forest. Atmos Chem Phys 9:2089–2095

Rebmann C, Zeri M, Lasslop G, Mund M, Kolle O, Schulze ED, Feigenwinter C (2010) Influence of meso-scale transport processes on CO_2-exchange at a complex forest site in Thuringia – Germany. Agric For Meteorol 150(5):684–691

Reichstein M, Falge E, Baldocchi D, Papale D, Aubinet M, Berbigier P, Bernhofer C, Buchmann N, Gilmanov T, Granier A, Grünwald T, Havrankova K, Ilvesniemi H, Janous D, Knohl A, Laurila T, Lohila A, Loustau D, Matteucci G, Meyers T, Miglietta F, Ourcival J-M, Pumpanen J, Rambal S, Rotenberg E, Sanz M, Tenhunen J, Seufert G, Vaccari F, Vesala T, Yakir D, Valentini R (2005) On the separation of net ecosystem exchange into assimilation and ecosystem respiration: review and improved algorithm. Glob Change Biol 11:1424–1439

Reiners WA, Anderson RO (1968) CO2 concentrations in forests along a topographic gradient. Am Midl Nat 80(1):111–117

Rinne J, Riutta T, Pihlatie M, Aurela M, Haapanala S, Tuovinen J-P, Tuittila E-S, Vesala T (2007a) Annual cycle of methane emission from a boreal fen measured by the eddy covariance technique. Tellus 59B:449–457

Rinne J, Taipale R, Markkanen T, Ruuskanen TM, Hellén H, Kajos MK, Vesala T, Kulmala M (2007b) Hydrocarbon fluxes above a Scots pine forest canopy: measurements and modeling. Atmos Chem Phys 7:3361–3372

Rogie JD, Kerrick DM, Sorey ML, Chiodini G, Galloway DL (2001) Dynamics of carbon dioxide emission at Mammoth Mountain California. Earth Planet Sci Lett 188:535–541

Saigusa N, Yamamoto S, Murayama S, Kondo H (2005) Inter-annual variability of carbon budget components in an AsiaFlux forest site estimated by long-term flux measurements. Agric For Meteorol 134:4–16

Saleska SR, Miller SD, Matross DM, Goulden ML, Wofsy SC, da Rocha HR, de Camargo PB, Crill P, Daube BC, de Freitas HC, Hutyra L, Keller M, Kirchhoff V, Menton M, Munger JW, Pyle EH, Rice AH, Silva H (2003) Carbon in Amazon forests: unexpected seasonal fluxes and disturbance-induced losses. Science 302:1554–1557

Schimel D, Aubinet M, Finnigan J (2008) Eddy flux measurements in difficult conditions. Ecol Appl 18(6):1338–1339

Schmid HP, Grimmond S, Cropley F, Offerle B, Su HB (2000) Measurements of CO2 and energy fluxes over a mixed hardwood forest in the mid-western United States. Agric For Meteorol 103:357–374

Staebler RM, Fitzjarrald DR (2004) Observing subcanopy CO2 advection. Agric For Meteorol 122:139–156

Staebler RM, Fitzjarrald DR (2005) Measuring canopy structure and the kinematics of subcanopy flows in two forests. J Appl Meteorol 44:1161–1179

Sun J, Burns SP, Delany AC, Oncley SP, Turnipseed AA, Stephens BB, Lenschow DH, LeMone MA, Monson RK, Anderson DE (2007) CO2 transport over complex terrain. Agric For Meteorol 145:1–21

Suyker AE, Verma SB, Burba GG, Arkebauer TJ (2005) Gross primary production and ecosystem respiration of irrigated maize and irrigated soybean during a growing season. Agric For Meteorol 131(3–4):180–190

Takle ES, Massman WJ, Brandle JR, Schmidt RA, Zhou XH, Litvina IV, Garcia R, Doyle G, Ric CW (2004) Influence of high-frequency ambient pressure pumping on carbon dioxide efflux from soil. Agric For Meteorol 124:193–206

Tanaka H, Hiyama T, Kobayashi N, Yabuki H, Ishii Y, Desyatkin RV, Maximov TV, Ohta T (2008) Energy balance and its closure over a young larch forest in eastern Siberia. Agric For Meteorol 148:1954–1967

Teklemariam T, Staebler RM, Barr AG (2009) Eight years of carbon dioxide exchange above a mixed forest at Borden, Ontario. Agric For Meteorol 149:2040–2053

Tóta J, Fitzjarrald DR, Staebler RM, Sakai RK, Moraes OMM, Acevedo OC, Wofsy SC, Manzi AO (2008) Amazon rain forest subcanopy flow and the carbon budget: Santarém LBA-ECO site. J Geophys Res 113. doi:10.1029/2007JG00597 G00B02

Turnipseed AA, Blanken PD, Anderson DE, Monson RK (2002) Surface energy balance above a high-elevation subalpine forest. Agric For Meteorol 110:177–201

Turnipseed AA, Anderson DE, Blanken PD, Baugh W, Monson RK (2003) Air flows and turbulent flux measurements in mountainous terrain. Part 1: canopy and local effects. Agric For Meteorol 119:1–21

van Gorsel E, Leuning R, Cleugh HA, Keith H, Suni T (2007) Carbon efflux: reconciliation of eddy covariance and chamber measurements using an alternative to the u* threshold filtering technique. Tellus 59B:397–403

van Gorsel E, Leuning R, Cleugh HA, Keith H, Kirschbaum MU, Suni T (2008) Application of an alternative method to derive reliable estimates of nighttime respiration from eddy covariance measurements in moderately complex topography. Agric For Meteorol 148:1174–1180

van Gorsel E, Delpierre N, Leuning R, Black A, Munger JW, Wofsy S, Aubinet M, Feigenwinter C, Beringer J, Bonal D, Chen B, Chen J, Clement RR, Davis KJ, Desai AR, Dragoni D, Etzold S, Grünwald T, Gu L, Heinesch B, Hutyra LR, Jans WW, Kutsch W, Law BE, Leclerc MY,

Mammarella I, Montagnani L, Noormets A, Rebmann C, Wharton S (2009) Estimating nocturnal ecosystem respiration from the vertical turbulent flux and change in storage of CO2. Agric For Meteorol 149:1919–1930

Vickers DA, Mahrt L (2006) Contrasting mean vertical motion from tilt correction methods and mass continuity. Agric For Meteorol 138(1–4):93–103

Wilczack J, Oncley SP, Stage SA (2001) Sonic anemometer tilt correction algorithms. Bound Layer Meteorol 99:127–150

Wilson K, Goldstein A, Falge E, Aubinet M, Baldocchi D, Berbigier P, Bernhofer C, Ceulemans R, Dolman H, Field C, Grelle A, Ibrom A, Law B, Kowalski A, Meyers T, Moncrieff J, Monson R, Oechel W, Tenhunen J, Valentini R, Verma S (2002) Energy balance closure at FLUXNET sites. Agric For Meteorol 113:223–243

Wohlfahrt G, Anfang C, Bahn M, Haslwanter A, Newesely C, Schmitt M, Drösler M, Pfadenhauer J, Cernusca A (2005) Quantifying nighttime ecosystem respiration of a meadow using eddy covariance, chambers and modelling. Agric For Meteorol 128:141–162

Xu LK, Baldocchi DD (2004) Seasonal variation in carbon dioxide exchange over a Mediterranean annual grassland in California. Agric For Meteorol 1232:79–96

Yi C, Anderson D, Turnipseed A, Burns S, Sparks J, Stannard D, Monson R (2008) The contribution of advective fluxes to net ecosystem exchange in a high-elevation, subalpine forest. Ecol Appl 18(6):1379–1390

Zhao LA, Li YN, Xu SX, Zhou HK, Gu S, Yu GR, Zhao XQ (2006) Diurnal, seasonal and annual variation in net ecosystem CO2 exchange of an alpine shrubland on Qinghai-Tibetan plateau. Glob Change Biol 12(10):1940–1953

Chapter 6
Data Gap Filling

Dario Papale

6.1 Introduction

The eddy covariance (EC) technique provides data at high temporal resolution, continuously, day and night and potentially for multiple years. Despite the recent developments in the EC technique and the availability of instruments with low power consumption, system failures are unavoidable and create gaps in the measurements. Common problems in the data acquisition are power breaks, in particular when the power system is based on solar panels; damages to instruments, for example, due to animals or lightning; incorrect system calibrations; maintenances; and also human actions like vandalism or robbery. In addition to these events related to the data acquisition phase, there are also gaps introduced by the data quality filtering, where measurements are discarded if acquired under non ideal conditions. Examples of these filters are the raw data tests described in Sects. 3.2.2, 4.3.3 and the nighttime filtering depicted in Sect. 5.3. Falge et al. (2001) found on average 35% of data missing due to system failures and data rejections across 19 EC sites while Papale et al. (2006) estimated that 20–60% of the data was rejected by the different quality filters applied.

Are these gaps a problem in our analyses? When should we fill these gaps in the measured fluxes and which are the methods available? In this chapter, the flux measurement gap filling will be discussed, focusing in particular on the differences between the methods available and providing indications about the best way to fill gaps in the data set on the basis of the data use and ecosystem characteristics.

D. Papale (✉)
DIBAF, University of Tuscia, Viterbo, Italy
e-mail: darpap@unitus.it

M. Aubinet et al. (eds.), *Eddy Covariance: A Practical Guide to Measurement and Data Analysis*, Springer Atmospheric Sciences, DOI 10.1007/978-94-007-2351-1_6,

6.2 Gap Filling: Why and When Is It Needed?

Do we need to fill the gaps in an EC time-series? It depends on the use of the data and the analysis that we plan to do. Thanks to its high temporal resolution, the EC technique provides a large amount of data that are often acquired under similar situation in terms of vegetation status and meteorological conditions. This "redundancy" of data is fundamental in the gap-filling methods; it is also sufficient to perform specific analyses, when no gap-free data sets are needed. Examples are the analysis of functional relationships between fluxes and drivers or models validation and parameterization when the model time resolution is the same of the EC measurements. In these cases, it is not needed to fill the gaps present in the time-series and only the measured and not-rejected data can be used.

Instead, whenever it is needed to calculate aggregated values, for example, sums to estimate annual budgets or daily averages needed in model evaluations, the completeness of the data set is required. If missing and rejected values in the half-hourly data set would be perfectly random distributed, the calculation of an integrated value could be easily performed by taking the average of all available data. Unfortunately, data gaps do not occur randomly. For example, $u*$ filtering removes mainly nighttime data or power failures occur principally in winter and night when the solar panels are used. This nonrandomness of the gaps in the data set leads to the need to apply more sophisticated gap-filling methods to reconstruct the missing periods.

6.3 Gap-Filling Methods

There are different gap-filling methods, in particular for carbon fluxes, that have been proposed in scientific literature. These can be classified according to different characteristics:

- *Principles*: All the gap-filling methods make use of the valid data to reconstruct the missing period. This reconstruction however can be based on completely empirical techniques or on the use of "functional models." In the first case, there are no assumptions imposed in the shape of the relations between drivers and fluxes and the data are used to find this relation and parameterize it. In the "functional models," the knowledge about the process under study is used to prescribe the way how drivers and fluxes are linked and the data are used only to parameterize these functions. In general, functional models are not recommended when the data are used in models evaluation activities because the same knowledge about the processes involved could be used in both the gap-filling method and the model to validate, leading to spurious correlations and circularities. However, if the empirical methods are in general good in the interpolation, they have led to high uncertainty in extrapolation, where the empirical relation found using the available data could be not valid (e.g., filling winter time data with a relation found and parameterized using summer time data). In these cases the functional models are more suitable since the knowledge about the dynamic of the system and the role of the different drivers in the different periods of the year is included in the method.

- *Drivers*: The drivers are variables, which can explain at least partially the variability of the flux measured with the EC technique that needs to be filled. Generally, the meteorological variables are used as drivers in most of the gap-filling methods since they influence the ecosystem responses in terms of carbon, water, energy, and other greenhouse gas fluxes. Incoming shortwave radiation, air and soil temperature, vapor pressure deficit, and soil water content are in general the most used drivers; however, other variables like precipitation, diffuse and reflected radiation, and wind speed can also be important in specific sites or conditions. The gap-filling method flexibility in the requested or accepted drivers could be an important criterion to select the most appropriate technique. Methods that have a fixed list of drivers are clearly less flexible and cannot be applied if one of the drivers is also missing or if a variable that is supposed to be relevant in the flux reconstruction is not included in the model. In general, empirical methods are fully flexible in this respect and, for this reason, preferable in these conditions. There are, however, conditions when all the meteorological data are also missing. In these situations, if it is not possible to reconstruct at least some of the drivers the only method that can be applied is the *Mean Diurnal Variation*.
- *Variables simulated*: The variables that need to be gap-filled could be different. In addition to the fluxes (CO_2, H_2O, Energy, CH_4, N_2O, volatile organic compounds (VOCs) and all the other species that can be measured by the EC technique), the meteorological variables can also be filled, to construct a complete driver data set that can be used as input in the fluxes gap-filling. The ability of the methods to simulate different variables and be available for this reason as a gap-filling tool for different fluxes and meteorological data set should be taken into consideration.
- *Noise conservation*: Fluxes measured with the EC technique are affected by random errors that introduce noise in the data. Most of the gap-filling methods are based on interpolations and for this reason tend to remove the noise signal from the data. There are few methods that conserve the noise in the data, for example, the *Kalman filters* (Gove and Hollinger 2006) and the *Multiple Imputation* (Hui et al. 2004) approaches.
- *Implementation*: The computer computational power available today is more than sufficient to run all the existing gap-filling techniques. However, the implementation of some of the existing methods could be complicated and would need a good knowledge of programming languages. In these cases, the centralized services offered by databases and portals could play an important role, implementing these methods and giving a complete and robust gap-filling tool available to the users (see Sect. 17.3).

6.3.1 Meteorological Data Gap Filling

The gap-filling techniques presented in this chapter have been proposed mainly for CO_2 data; however, part of them, in particular the empirical methods, can be easily adapted to be used for other fluxes. All of them (except *Mean Diurnal Variation*) however require as input meteorological variables that for this reason should be available as continuous and gap-free data set. Although data quality filtering applied to meteorological measurements has minor impact in terms of data points removed, gaps can occur, in particular, due to sensors malfunctioning or power breaks. In these cases, it is needed to first fill the gaps in the drivers and then use gap-filled meteorological data in the fluxes gap-filling. This is clearly a delicate step since errors and uncertainties introduced in the drivers will be reflected also in the fluxes. In addition, it is important to underline that gap-filled meteorological data should not be used in the flux gap-filling model parameterization.

The best way to fill gaps in meteorological data is to have a back-up meteorological station with main variables measured (incoming radiation, air temperature, relative humidity, precipitation, wind speed) close to the main EC tower but independent regarding the power supply system. When this back-up system is not available, the empirical methods described later in this chapter can be used when only part of the meteorological variables are missing using as driver the variables present and additional inputs like top of atmosphere incoming radiation or indicators of date and time.

In the unfortunate but also quite common cases where all the meteorological data are missing and no meteorological stations are available in the area, linear interpolation of variables with slow changes (like temperature) or the use of *Mean Diurnal Variation* method are simplest solutions to implement. Other more complex possibilities that, however, would give more trustable results involve the use of remote sensing data or meteorological reanalysis data. New generation meteorological satellites like the European Meteosat MSG (http://www.esa.int/SPECIALS/MSG/) provide high temporal resolution images (15–30 min) that can be used to derive variables like incoming radiation, surface temperature, or albedo (http://landsaf.meteo.pt/). Finding the regression between the site-level measurements and the data produced for the pixel where the tower is located it is then possible to apply such regression to rescale the remote sensing products at site level when the tower measurements are missing. The same approach can be applied using meteorological reanalysis data instead of remote sensing products. These data are also gridded data set produced integrating observations and models, generally with daily temporal resolution (see as example the ERA-Interim data set produced by ECMWF: http://www.ecmwf.int/research/era/do/get/era-interim) that can be downscaled at local level using site-specific relations parameterized using periods where the variables of interest are present.

6.3.2 General Rules and Strategies (Long Gaps)

Gap-filling method setting and parameterization are crucial steps and they are directly linked to the quality of the results. The relative abundance of data due to the high time resolution of the EC technique and the number of meteorological variables measured should not lead to an underestimation of the importance of this phase that must be carefully implemented. The drivers, for example, should be selected carefully finding the right compromise between the known biological importance of a specific meteorological variable to explain the flux to be reproduced, its presence and quality through the year, and its possible correlation with other drivers used that for some of the methods could lead to an over-parameterization.

In general, the dataset used in the model parameterization should be as much as possible representative of the different conditions with an even distribution of samples measured in the diverse situations. This means, for example, that there should be equilibrium between data acquired during daytime and nighttime or in the different seasons. In addition, also the length of the time-windows used to parameterize and apply the model has an important role. In practice, a model could be parameterized using data from the whole year and then applied to all the present gaps. This, however, must imply that the model is able to distinguish the different "ecosystem states," for example, phenological phases or different agricultural periods (see next section). In fact, fluxes acquired under similar meteorological conditions but during completely different "ecosystem state" could be completely different. As an alternative, the model could be parameterized and applied on the basis of shorter time-windows, in the order of weeks or months, where it is assumed that certain conditions (e.g., phenology or biomass) are stable and the fluxes are explained only by the meteorological conditions. These time windows could be fixed in terms of length and position (e.g., a different model parameterization for each month) or, more sophisticatedly and correctly, be centered on each single gap to fill and have an increasing length, up to the minimum window size that provides a sufficient number of data points to parameterize the model (see as example Reichstein et al. 2005).

One problem, in particular when short time-windows are used, is the presence of long gaps where no data are available for the parameterization. In these cases, when multiple years are available and the ecosystem state did not change in the period, the model can be parameterized using data acquired in the same period of a year (e.g., season) but in different years. The basic assumption that justifies this approach is that the fluxes, in the same period but different years, are mainly function of the meteorological conditions. This is probably true, for example, in mature or old forests that did not experience substantial disturbances or management events or during the growing season of crops when the species and season are the same.

6.3.2.1 Sites with Management and Disturbances

Managed or disturbed ecosystems experience rapid changes of their conditions that drastically change the fluxes and their relations with the meteorological drivers due to changes of the ecosystem dynamics between the period just before and just after the perturbation. This is typical in cropland and managed grassland, in particular, after tillage or after harvesting and grazing (Hammerle et al. 2008; Wohlfahrt et al. 2008) when the green biomass is removed in a few days and rapid vegetation regrowth could start.

These are conditions that should be considered when the gap-filling method is selected and implemented. In theory, the method should to be able to recognize that the system status changed. This could be possible using as driver a variable related to the ecosystem characteristic affected by the disturbance or management practices. In case of harvesting, for example, a spectral reflectance measurement in the spectral bands linked to the green vegetation (e.g., the normalized difference vegetation index (NDVI) bands) or a below-canopy radiation sensor could help identify the abrupt change of ecosystem status, but the gap-filling method must be flexible enough to take this information as input.

An alternative method to take into consideration management and disturbances during gap filling is to change the parameterization strategy. Parameters of the gap-filling method are set using valid data measured in a time window that could have different sizes from multiple years to few days (see Sect. 6.3.2), and clearly in ecosystems where management and disturbances occurs, it is important to keep this window as short as possible to avoid the use of data acquired in periods with similar drivers values (e.g., meteorological conditions) but completely different fluxes due to the change of status in the same parameterization step. However, even a small window could include data before and after the disturbance event, in particular, when, in croplands or grassland, the EC system needs to be removed during harvesting and the gap is long and centered around the critical period.

The best way to solve this problem is probably the use of disturbance or management indicators (DI) to split the data set in subsets that do not include abrupt ecosystem status changes. In practice, registering the date and time of management practices or disturbances events that are supposed to have a direct and immediate effect on the fluxes it is possible to identify periods where the fluxes are function only of time (e.g., regrowth) and meteorological conditions. The gap-filling method can be then parameterized using only data acquired during the homogeneous subperiod (Fig. 6.1). In addition, in case of similar management across the years, for example, in grassland where generally the 3–4 cutting events per year happen in the same periods or in cropland if the same species is cultivated for different years, the gap-filling model can be parameterized using data from the same subperiod of previous years (if the others states conditions remained stable). In this way, the number of data points available increase making the parameterization more robust.

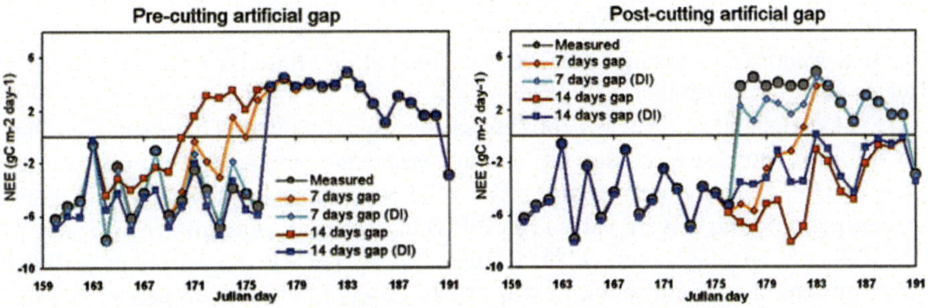

Fig. 6.1 Example of the Disturbance Indicator (DI) uses in gap filling NEE data measured in a managed grassland. Two artificial gaps of 7 days and 14 days have been added before (*left*) and after (*right*) the cutting dates that create a discontinuity. The artificial gaps have been filled using the MDS method (Sect. 6.3.3.2). It is possible to see how the gap-filling method performances improve when the DI is used (These data have been gently processed and provided by Arnaud Carrara)

6.3.3 Methods Description

6.3.3.1 Mean Diurnal Variation

The Mean Diurnal Variation method (MDV) is an interpolation technique that is based on the temporal auto-correlation of the fluxes (Falge et al. 2001). Missing observation is replaced by the mean of valid values measured on adjacent days at the same time (the same half-hour or with a buffer of +/− 1 h). The length and definition of the averaging period (window) can vary between different method implementations. In general, a window length not larger than 2 weeks is recommended since for longer periods nonlinear dependence on environmental variables could introduce large uncertainty and errors (Falge et al. 2001). Also the position of the window could be fixed or variable where in the first case the windows are predecided and fixed and all gaps occurring in each of the windows are replaced applying the MDV in the same period, while in the second case the windows are defined around each single gap. Clearly, the second method is preferable because the gap will always be centered in the window.

The MDV method does not require drivers and it is the only method applicable when all the meteorological data are missing; it is an empirical method and can be in theory applied to fill all the variables when temporal auto-correlation is expected. The method implementation is easy but the accuracy and performances are lower with respect to the others methods (see Sect. 6.4).

6.3.3.2 Look-Up Tables

The Look-up table (LUT) is an empirical method, easy to implement, where the missing values are replaced with the average of valid measurements occurring under similar meteorological conditions. In practice, a multidimensional table is created where the missing value can be "looked up" based on the values of the meteorological drivers.

For example, in Falge et al. (2001) four tables were created in the year (according to the different seasons) and the drivers used were photosynthetic photon flux density (PPFD) and air temperature. The valid NEE data were binned and averaged according with the drivers values (23 PPFD classes of 100 μmol m^{-2} s^{-1} and 35 Air Temperature classes of 2°C for a total of $35 \times 23 = 805$ classes per period) and each missing data point replaced with the NEE value in the table in same drivers combination class occurring during the gap. Gaps in the table, where no valid NEE data were present for a given combination of the two drivers, were filled with linear interpolation.

The drivers used in the table preparation should be selected according to the site characteristics, taking into consideration the environmental variables that are more important in the processes of interest (i.e., the flux to be gap-filled) without selecting too many variables that would lead to the impossibility of finding a sufficient number of valid data to calculate robust averages for each of the driver class combinations. In general, 3–4 variables selected among incoming and diffuse radiation, air and soil temperature, soil water content, and vapor pressure deficit are sufficient.

Also, the number of tables created in each year is an important aspect to consider. Monthly or biweekly LUTs are possible if the amount of valid data is sufficient, and in these cases the number of drivers classes can be smaller. In addition, the drivers considered could change according to the period of the phenological cycle and the daily course, for example, not using incoming radiation as a driver during nighttime.

Reichstein et al. (2005) proposed a method (Marginal Distribution Sampling – MDS) where they consider both the covariation of fluxes with meteorological variables and the temporal auto-correlation of the fluxes. In their approach, similar meteorological conditions are sampled in the temporal vicinity of the gap to be filled looking in a window around the gap as small as possible to include a sufficient number of valid data with similar meteorological conditions to calculate the average flux. In their method, the drivers used to evaluate the similarity in the meteorological conditions vary in order to find a compromise between number of drivers and window length. Incoming radiation, air temperature, and vapor pressure deficit are first considered; then, if the window needed exceeds predefined maximum length, only incoming radiation is considered, and finally the MDV method (described in Sect. 6.3.3.1). For a given gap, in the impossibility to find a sufficient number of valid data to calculate the average for certain window length and drivers set, the next step could be to increase the size of the window or to reduce the numbers of drivers considered; the strategy to decide which of the two options to follow in the different conditions is well explained in their paper and can be used as example. The MDS method has been implemented in the European database as one of the standard gap-filling methods available in the central processing (see Chap. 17).

6.3.3.3 Artificial Neural Networks

The Artificial Neural Networks (ANNs) are purely empirical nonlinear regression models with a medium level of implementation difficulties. The ANN consists in a set of nodes, often organized in layers and connected by weights that are equivalent to the regression parameters (Bishop 1995; Rojas 1996). The first step to use an ANN is the network parameterization process called "training." The ANN is trained by presenting it with sets of input data (drivers) and associated output data that, in the case of a gap-filling application, are valid fluxes. Once the ANN is trained, the underlying dependencies of the output on the driver variables are mapped onto the weights and the ANN can be then used to predict the missing values.

There are different algorithms to train the ANNs and one of the most used is the back-propagation algorithm, where the training of the ANN is performed by propagating the input data through the nodes via the weighted connections and then back-propagating the error calculated as difference between the predicted and real output and adjusting the weights to minimize this error (Papale and Valentini 2003; Braswell et al. 2005).

Similar to the LUT method, also in the ANN, it is important to select as input the appropriate and relevant environmental variables that drive the flux variability. These could be a large set (e.g., all the meteorological variables measured at the site) or just a preselected subset. In the first case, the ANN has the possibility to use (i.e., assign high weights) variables commonly not considered as drivers that would be probably excluded in the second case, but it is also important to keep in mind that increasing the number of input variables leads to an increase of the degree of freedom (number of weights) and requires the use of a larger training data set to avoid model over-fitting and consequent loss of generalization ability.

The quality and the representativeness of the training data set play also an important role. The ANNs, as all the purely empirical models, can only map and extract information present in the data set used in the parameterization; for this reason the data set must be accurate and cover as much and as homogenously as possible the different ecosystem conditions (e.g., seasons, phenological phases, daily courses). Presampling of the training data set to ensure an equal coverage of the different conditions and the use of fuzzy values to represent additional information such as time have been tested and used showing good results (Papale and Valentini 2003; Moffat et al. 2007). Also, the training of different ANNs for daytime and nighttime (using different drivers) or the training of different ANNs for different periods and using data from adjacent years, as explained in Sect. 6.3.2.1, can improve the method performances.

The performances of the ANN method in carbon flux gap-filling are good (see Sect. 6.4) and for this reason this technique is used as standard in the European database and in FLUXNET, together with the MDS method explained above. The ANNs require a gap-free driver dataset and for this reason it is needed to first gap-fill the meteorological variables (Sect. 6.3.1) or, when this is impossible to implement, a second method (e.g., the MDV, Sect. 6.3.3.1) to be used when one or more drivers are missing.

6.3.3.4 Nonlinear Regressions

The nonlinear regressions method is based on parameterized non-linear equations which express semi-empirical relationships between the flux and environmental variables, often temperature and light for CO_2 fluxes. There are different versions and implementations that have been proposed (Falge et al. 2001; Hollinger et al. 2004; Barr et al. 2004; Desai et al. 2005; Richardson et al. 2006; Noormets et al. 2007) but in general two different equations are used, one for nighttime data often estimated as function of temperature and one for daytime data using a light response function.

The response of fluxes to the photosynthetic photon flux density PPFD is commonly modeled using the rectangular hyperbola function like the *Michaelis and Menten* equation (Eq. 9.6, Sect. 9.3.3.4) or an exponential function like *Mitscherlich* equation (Eq. 9.8, Sect. 9.3.3.4) (Falge et al. 2001). For nighttime data the most used functions are the *Lloyd and Taylor* and the *Arrhenius* (Eq. 9.5, Sect. 9.3.2.2) (Lloyd and Taylor 1994; Falge et al. 2001; Moffat et al. 2007). Both the equations have temperature as driver and can use either air or soil temperatures.

The parameters estimation for all these functions is done using measured valid data. Also in this case it is important to carefully check the data before and use only accurate measurements. In addition, the regression parameters can be kept constant only for a certain period of time to accommodate the variation over the year of all the other drivers not considered in the equations (i.e., season, water availability etc.).

The method is semi-empirical because, although the parameters are estimated using the measurements, the shape of functions between drivers and fluxes are imposed. This is an important aspect to consider when the gap-filled data are needed in modeling activities because the model to validate or parameterize could have the same function, linking for example, temperature and respiration and leading to the risk of spurious correlations (Sect. 6.2). In addition, it is applicable only when the functions linking meteorological variables and fluxes are well known and consolidated.

6.3.3.5 Process Models

In the process models, we can include all the models that have been developed to estimate and predict fluxes, simulating all the processes occurring and using generally as input not only the meteorological variables but also state variables like soil and vegetation characteristics and others quantities like leaf area index (LAI) and biomass. These models are generally not developed specifically to be used as gap-filling techniques.

In these models, which make full use of our knowledge of the processes involved in the ecosystem functioning, the data are used to constrain some of the model parameters. The advantage is that, assuming that the processes are well represented in the model, it is possible to apply it to reconstruct also long periods of gaps or even the fluxes under different climate, for example, in different years with respect to the one with EC measurements. The disadvantages are related to the uncertainty in the reproduction processes in the model and the risk that some important process could be completely missing or not correctly reproduced.

The implementation of this method is quite complex and require knowledge of the model and parameters optimization techniques. The results can be used in the site-level analysis but not for model validation and parameterization if the two models have similar routines or functions.

6.4 Uncertainty and Quality Flags

Uncertainty estimation is an important information that should be always included when data are gap filled. There are two main different uncertainty sources in the gap-filled values: One is represented by the diverse estimations that different gap-filling methods give for the same missing data point and the other is due to the uncertainty in the selected gap-filling model parameterization that is, for example, larger when the gaps are longer.

Moffat et al. (2007) showed in a comparison of gap-filling techniques for carbon fluxes that most of the methods implemented give good results, often with errors with magnitudes similar to the noise component in the data (see also Sect. 7.3.3.3) but with a slightly higher performances for empirical methods like ANN and MDS. Based on these results, it is possible to conclude that the uncertainty related to the selection of the gap-filling method is relatively small when one of the high-performance methods is used if the gap length is not too long and if the data set available to set the gap-filling model parameters is sufficiently large and with of good quality.

The best way to assess the uncertainty due to the parameterization is linked to the method selected. The quality of the parameterization is a function of the number of data points, the data quality, and the number of the variables used to constrain the model. In general, long gaps, during which the general ecosystem conditions can change (e.g., growing season phase, ground water table, nutrients availability), are more difficult to fill and the uncertainty associated with the gap-filled values will be in general higher with respect to short gaps, with highest uncertainty values in the middle of the gap due to the distance (in time and for this reason also in terms of ecological conditions) from the measured data used to estimate the parameters values.

It is important to assess an uncertainty or a "confidence level" to associate with each gap-filled value; this information is essential for a correct data analysis and interpretation. For some of the methods introduced in this chapter, an estimate of the uncertainty level is relatively simple. In the LUT method (Sect. 6.3.3.2) for example, the standard deviation of the flux values in the same drivers class gives an indication of the variability inside a group of data that the method assumes to be similar. The same is valid for the MDV method (Sect. 6.3.3.1) where the standard deviation or percentile distribution of the measurements at the same time in adjacent days gives information about the uncertainty in the gap-filled values. In others cases, like when ANN or NRL are used (Sects. 6.3.3.3 and 6.3.3.4), the uncertainty can be estimated using subsets of the available data to parameterize different versions of the same model that then can be all applied obtaining different values for the same gap.

Independently of the estimation of an uncertainty value to associate to each gap-filled value, it is important to create additional information about the method applied that can be included in the data set. This information can include the distance of each single half-hourly missing from the first valid value, indication of the drivers used to fill the gap, length of the window needed to find sufficient data to parameterize the model, and number of the data points used. In addition, quality flags to summarize the expected quality of each gap-filled value can be defined and added; an example of these quality flags is presented in the appendix of the Reichstein et al. (2005) paper.

6.5 Final Remarks

Gap-filling is a process that is sometimes unavoidable, in particular when daily to annual integrals are needed, and different methods exist. The compherensive analysis conducted by Moffat et al. (2007) showed that all the gap-filling techniques give on average good results when the gaps are shorter than 10 days and the relevant meteorological drivers available. In addition it has been also shown that including information about discontinuity (6.3.2.1) can improve the results in case of sites with management.

The decision about which method to select should be then based on different considerations. First, the availability of drivers: if no meteorological data are available, the MDV method is often the only one available and the uncertainty associated to the simulated values will be large. Another important aspect to consider is the possibility of spurious or circular correlation between data and model results when the gap-filled measurements are used in process model validation. In these cases, it is important to use a purely empirical method.

Also the difficulties in the implementation could preclude the use of some of the methods. In these cases, however, the use of centralized gap-filling services often provided by the databases could help to use the best methods without the need to implement them locally.

Finally, due to the strong link between the gap-filling quality and uncertainty (always important to estimate), the availability of meteorological data, and information about management and disturbance events, it is fundamental to carefully register all the ancillary data about the site and to install a back-up meteorological station close to the EC tower and independent in terms of energy.

References

Barr AG, Black TA, Hogg EH, Kljun N, Morgenstern K, Nesic Z (2004) Inter-annual variability in the leaf area index of a boreal Aspen-Hazelnut forest in relation to net ecosystem production. Agric For Meteorol 126:237–255

Bishop CM (1995) Neural networks for pattern recognition. Oxford University Press, Oxford

Braswell BH, Sacks B, Linder E, Schimel DS (2005) Estimating ecosystem process parameters by assimilation of eddy flux observations of NEE. Glob Change Biol 11:335–355

Desai AR, Bolstad P, Cook BD, Davis KJ, Carey EV (2005) Comparing net ecosystem exchange of carbon dioxide between an old-growth and mature forest in the upper Midwest, USA. Agric For Meteorol 128(1–2):33–55

Falge E, Baldocchi D, Olson RJ, Anthoni P, Aubinet M, Bernhofer C, Burba G, Ceulemans R, Clement R, Dolman H, Granier A, Gross P, Grünwald T, Hollinger D, Jensen N-O, Katul G, Keronen P, Kowalski A, Ta Lai C, Law BE, Meyers T, Moncrieff J, Moors E, Munger JW, Pilegaard K, Rannik U, Rebmann C, Suyker A, Tenhunen J, Tu K, Verma S, Vesala T, Wilson K, Wofsy S (2001) Gap filling strategies for defensible annual sums of net ecosystem exchange. J Agric For Meteorol 107:43–69

Gove JH, Hollinger DY (2006) Application of a dual unscented Kalman filter for simultaneous state and parameter estimation in problems of surface-atmosphere exchange. J Geophys Res 111:D08S07. doi:10.1029/2005JD006021

Hammerle A, Haslwanter A, Tappeiner U et al (2008) Leaf area controls on energy partitioning of a temperate mountain grassland. Biogeosciences 5(421):431

Hollinger DY, Aber J, Dail B, Davidson EA, Goltz SM, Hughes H, Leclerc M, Lee JT, Richardson AD, Rodrigues C, Scott NA, Varier D, Walsh J (2004) Spatial and temporal variability in forest-atmosphere CO2 exchange. Glob Change Biol 10:1689–1706

Hui D, Wan S, Su B, Katul G, Monson R, Luo Y (2004) Gap-filling missing data in eddy covariance measurements using multiple imputation (MI) for annual estimations. Agric For Meteorol 121:93–111

Lloyd J, Taylor JA (1994) On the temperature dependence of soil respiration. Funct Ecol 8: 315–323

Moffat AM, Papale D, Reichstein M, Hollinger DY, Richardson AD, Barr AG, Beckstein C, Braswell BH, Churkina G, Desai AR, Falge E, Gove JH, Heimann M, Hui D, Jarvis AJ, Kattge J, Noormets A, Stauch VJ (2007) Comprehensive comparison of gap-filling techniques for eddy covariance net carbon fluxes. Agric For Meteorol 147:209–232. doi:10.1016/j.agrformet.2007.08.011, ISSN: 0168–1923

Noormets A, Chen J, Crow TR (2007) Age-dependent changes in ecosystem carbon fluxes in managed forests in northern Wisconsin, USA. Ecosystems 10:187–203

Papale D, Valentini R (2003) A new assessment of European forests carbon exchanges by eddy fluxes and artificial neural network spatialization. Glob Change Biol 9:525–535

Papale D, Reichstein M, Aubinet M, Canfora E, Bernhofer C, Longdoz B, Kutsch W, Rambal S, Valentini R, Vesala T, Yakir D (2006) Towards a standardized processing of Net Ecosystem Exchange measured with eddy covariance technique: algorithms and uncertainty estimation. Biogeosciences 3:571–583

Reichstein M, Falge E, Baldocchi D, Papale D, Aubinet M, Berbigier P, Bernhofer C, Buchmann N, Gilmanov T, Granier A, Grünwald T, Havrankova K, Ilvesniemi H, Janous D, Knohl A, Laurila T, Lohila A, Loustau D, Matteucci G, Meyers T, Miglietta F, Ourcival JM, Pumpanen J, Rambal S, Rotenberg E, Sanz M, Tenhunen J, Seufert G, Vaccari F, Vesala T, Yakir D, Valentini R (2005) On the separation of net ecosystem exchange into assimilation and ecosystem respiration: review and improved algorithm. Glob Change Biol 11:1424–1439

Richardson AD, Braswell BH, Hollinger DY, Burman P, Davidson EA, Evans RS, Flanagan LB, Munger JW, Savage K, Urbanski SP, Wofsy SC (2006) Comparing simple respiration models for eddy flux and dynamic chamber data. Agric For Meteorol 141:219–234
Rojas R (1996) Neural networks. Springer, Berlin
Wohlfahrt G, Hammerle A, Haslwanter A et al (2008) Seasonal and inter-annual variability of the net ecosystem CO2 exchange of a temperate mountain grassland: effects of weather and management. J Geophys Res 113:D08110. doi:10.1029/2007JD009286

Chapter 7
Uncertainty Quantification

Andrew D. Richardson, Marc Aubinet, Alan G. Barr, David Y. Hollinger, Andreas Ibrom, Gitta Lasslop, and Markus Reichstein

7.1 Introduction

> There are known knowns. These are things we know that we know. There are known unknowns. That is to say, there are things that we know we don't know. But there are also unknown unknowns. These are things we don't know we don't know. (Donald Rumsfeld, February 12, 2002)

Despite our best efforts, measurements are never perfect, and thus all measurements are subject to errors or uncertainties (Taylor 1991). Sources of uncertainty include operator errors (insufficient vigilance, blunders), population sampling errors

A.D. Richardson (✉)
Department of Organismic and Evolutionary Biology, Harvard University Herbaria, 22 Divinity Avenue, Cambridge, MA, 02138 USA
e-mail: arichardson@oeb.harvard.edu

M. Aubinet
Unit of Biosystem Physics, Gembloux Agro-Bio Tech., University of Liege, 5030 Gembloux, Belgium
e-mail: Marc.Aubinet@ulg.ac.be

A.G. Barr
Environment Canada, 11 Innovation Blvd, Saskatoon, SK S7N 3H5 Canada
e-mail: Alan.Barr@ec.gc.ca

D.Y. Hollinger
USDA Forest Service, Northern Research Station, 271 Mast Road, Durham, NH, 03824 USA
e-mail: dhollinger@fs.fed.us

A. Ibrom
Risø National Laboratory for Sustainable Energy, Technical University of Denmark (DTU), Frederiksborgvej 399, 4000 Roskilde, Denmark
e-mail: anib@risoe.dtu.dk

G. Lasslop • M. Reichstein
Max-Planck Institute for Biogeochemistry, 07745 Jena, Germany
e-mail: gitta.lasslop@zmaw.de; mreichstein@bgc-jena.mpg.de

M. Aubinet et al. (eds.), *Eddy Covariance: A Practical Guide to Measurement and Data Analysis*, Springer Atmospheric Sciences, DOI 10.1007/978-94-007-2351-1_7,

(poor sampling design), instrument errors (glitches or bugs), calibration errors (zero and span), instrument limitations (limited resolution or an inappropriate application), and measurement conditions that are in conflict with the underlying theory. While errors are unavoidable and inevitable, to some degree they can always be reduced, as for example through improvements in design and greater attention to calibration.

Identifying sources and quantifying the nature and magnitude of error is essential for two reasons. First, the largest sources of error can be targeted for efforts at error reduction; second, the uncertainties can be taken into consideration during data analysis and interpretation. For example, is a measurement 10.0 ± 0.1, 10 ± 1, or 10 ± 10 g? – the size of the uncertainty may influence how we perceive the data, or the questions to which the data are applied, as larger uncertainties (or in other words, limited information content) reduce the usefulness of the data.

There is a long history in physics and engineering (e.g., Kline and McClintock 1953) of conducting and reporting detailed error analyses. In environmental and earth sciences, it is only now being recognized that greater attention should be paid to quantifying uncertainties, especially given potential applications of these data to management strategies and policy decision-making (Ascough et al. 2008). Examples of policy-relevant issues where this is essential include carbon accounting and climate change mitigation efforts, and quantification of water balances under climate change or land-use change.

With respect to eddy covariance measurements of surface-atmosphere fluxes, particularly of CO_2, there are specific applications where uncertainty information is needed. Three examples are as follows:

1. Uncertainty estimates are needed to make statistically valid comparisons between two sets of measurements (comparing "site A" and "site B"), or between measurements and models (model "validation" or "evaluation"; Hollinger and Richardson 2005; Medlyn et al. 2005; Ibrom et al. 2006). Only if the data uncertainties are known can confidence limits, at a particular level of statistical significance, be generated for individual observations, or can statistics (e.g., X^2) be calculated for a set of observation. Even in a less formal sense, knowledge of uncertainty can also guide our interpretation of the data; we should have more confidence in measurements with smaller uncertainties, and less confidence in measurements with larger uncertainties.
2. Although scaling of data in space (regional-to-continental extrapolation) or time (calculating flux integrals at annual or decadal time scales) does not strictly require uncertainty estimates, this information is critical if the resulting data products are to be used to set policy or for risk analysis. As an example, the question "what are realistic confidence intervals on the estimated regional C sink strength?" cannot be answered without a full accounting of uncertainty, and propagation of this forward in the scaling analysis.
3. Flux data are commonly being used in "data-model fusion," which refers to the systematic and rigorously quantitative means by which observational data, including flux and stock measurements, can be used to constrain process models

(Raupach et al. 2005; Williams et al. 2009; Wang et al. 2009). To conduct such an analysis in a statistically defensible manner, information about uncertainties in all data streams must be incorporated into the objective function (or "cost function") specified as the basis for optimization of data-model agreement. Thus, what is known or assumed about the data uncertainties directly influences the posterior distributions of parameter estimates and model predictions, as demonstrated in the recent OptIC (Trudinger et al. 2007) and REFLEX (Fox et al. 2009) experiments. On this basis, Raupach et al. (2005) suggested that "data uncertainties are as important as the data values themselves."

7.1.1 Definitions

The Cooperation on International Traceability in Analytical Chemistry (CITAC) initiative maintains an Internet-based guide (http://www.measurementuncertainty.org/) to quantifying uncertainty in analytical measurements, where a distinction is drawn between "error" and "uncertainty." Here, we follow these definitions: *Error* is a single value indicating the difference between an individual measurement and the actual or true quantity being measured, whereas *uncertainty* is a range of values characterizing the limits within which the quantity being measured could be expected to fall. If the error is known, a correction for this error can be applied. On the other hand, the uncertainty estimate cannot be used as the basis for such a correction, because uncertainty is a range and not a single number.

7.1.2 Types of Errors

Measurement errors have traditionally been classified into two groups with fundamentally different intrinsic properties: Random errors and systematic (or bias) errors. In this approach, these errors propagate in different ways when measurements are combined or aggregated. A direct consequence of this is that random and systematic errors have very different effects on our interpretation of data.

The International Organization for Standardization (ISO) takes a different approach (ISO/IEC 2008), classifying uncertainty into errors that can be determined by statistical measures (type "A") and those that are evaluated by other means (type "B"), but then treating (propagating) them together in a similar fashion. Because systematic errors in flux measurements may not be constant, we prefer to follow the traditional approach and propagate them separately. As an example, consider our measurement (x) of a particular quantity ($\hat{x}$); note that $x \neq \hat{x}$, because measured x incorporates both random (ε) and systematic (δ) errors, that is, we actually observe $x = \hat{x} + \varepsilon + \delta$. The random error, ε, is stochastic and thus unpredictable, and is characterized by a probability distribution function (pdf), commonly assumed to be

Gaussian (normal) with a standard deviation of σ. Random errors cause "noise" or "scatter" in the data, and reduce the precision of measurements; because they are random, it is impossible to correct for them. Repeated measurements can be used to characterize the pdf of the total random error (e.g., what is the standard deviation of 10 measurements of the diameter of a particular tree?). In addition, averaging over n measurements improves the precision by a factor of $1/\sqrt{n}$, resulting in the so-called standard error of the mean.

On the other hand, the systematic error, δ, is a bias that is considered to remain constant but is unknown (Abernethy et al. 1985). It thus must be estimated based on judgment and experience (often the direction of the error is known, but there is uncertainty about its magnitude), theoretical considerations, or with complementary measurements (e.g., comparing tower-based and inventory estimates of ecosystem C storage). Unlike random errors, systematic errors cannot be identified through statistical analysis of the measurements themselves, nor can they be reduced through averaging. Systematic errors are an important consideration in flux measurement because they may differ between day and night (Moncrieff et al. 1996) and thus often have a significant impact on the annual net flux estimate.

Comments above about the impact of averaging on random and systematic errors imply that these errors accumulate, or propagate, in different ways, for example, when arithmetic operations are carried out on multiple measurements. Random errors accumulate "in quadrature": if we measure x_1 and x_2 ($x_i = \hat{x} + \varepsilon_i$), and assume that the random errors (ε_1 and ε_2, where ε_i is a random variable with mean 0 and standard deviation σ_i) on these measurements are independent of one another (zero covariance between ε_1 and ε_2), then the expected error on the sum ($x_1 + x_2$) is given by $\sqrt{\sigma_1^2 + \sigma_2^2}$, which is always less than$(\sigma_1 + \sigma_2)$. Thus it is often said that random errors "average out." This is, however, somewhat misleading as the random error never truly "disappears" (except in the limit of an infinite sample size), although by definition the expected value, $E[\varepsilon_i]$, equals 0. By comparison, systematic errors accumulate linearly: In this case, if we measure x_1 and x_2 ($x_i = \hat{x} + \delta_i$), then the expected error on the sum ($x_1 + x_2$) is simply ($\delta_1 + \delta_2$). More thorough treatments of formal error propagation are given elsewhere (e.g., Taylor 1991).

In the context of data-model fusion, as described above, an important distinction should be made between random and systematic errors (Lasslop et al. 2008; Williams et al. 2009). Random errors place an upper limit on the agreement between data and models. Because random errors are stochastic, they cannot be modeled (Grant et al. 2005; Richardson and Hollinger 2005; Ibrom et al. 2006). Random errors also lead to greater uncertainty in model parameterization and process attribution (essentially a problem of "equifinality," sensu Franks et al. 1997: with random errors or noise in the data, the set of model parameters that provide similarly good model fits becomes larger as the data uncertainties become larger). By comparison, uncorrected systematic errors can potentially bias data-model fusion analyses but do not necessarily increase parameter or model prediction uncertainties (Lasslop et al. 2008). Even in the absence of model error, uncorrected systematic errors may also lead to inconsistencies between model predictions and data constraints that cannot be reconciled given what is known or assumed about the random errors.

7.1.3 Characterizing Uncertainty

For random errors, we would like to describe the associated uncertainty in terms of the full pdf of the error distribution: Is it normal, lognormal, uniform, or double-exponential? What are its moments? In addition to standard deviation, we may also be interested in higher order moments, for example, skewness and kurtosis. Is the error variance constant (homoscedastic), or is it in some way time varying or otherwise correlated with one or more independent variables (heteroscedastic)? Are errors in successive measurements in time fully independent, or are they positively (or negatively) autocorrelated? These questions need to be answered in order for the appropriate statistical or analytical methods to be chosen.

For systematic errors, we are particularly interested in knowing whether the bias influences all measurements to the same degree ("fully systematic"), or only measurements made under certain conditions ("selectively systematic") (Moncrieff et al. 1996). Systematic errors may also result in a fixed bias, or the bias may be relative and scale with the magnitude of what is being measured, or it may change over time. In terms of CO_2 concentration measurements, a zero offset would result in a fixed bias, whereas calibration against a mislabeled standard, that is, causing sensitivity or span bias, would lead to a relative bias.

7.1.4 Objectives

In this chapter, we focus on describing and quantifying the random and systematic errors affecting eddy covariance flux measurements. Our emphasis is on some of the more recent work that was not synthesized in previous reviews (e.g., Goulden et al. 1996; Moncrieff et al. 1996; Aubinet et al. 2000; Baldocchi 2003; Kruijt et al. 2004; Loescher et al. 2006).

Random errors tend to be quite large at the half-hourly time scale and cannot be ignored even in the context of annual flux integrals, especially as they propagate through to gap-filled and partitioned net ecosystem exchange (NEE) time series. A number of methods have been developed to quantify the random errors; these are summarized here and the general patterns presented.

Some of the systematic errors in flux measurements are well characterized, and corrections (sometimes drawing from improvements in our theoretical understanding and treatment) have been developed for these biases (see Sects. 3.2.2, 4.1, 5.4). However, in many cases, the corrections for these errors are imperfect, and thus some uncertainty remains even after the correction is applied. For some systematic errors, particularly advection, current practices (e.g., u_* filtering) allow us to reduce, but not completely eliminate, the associated uncertainties; here we aim to quantify the uncertainty that still remains. As an aside, we note that while in principle the distinction between random and systematic errors is clear, in practice this can be more difficult, as many errors have both a random and a systematic component and operate at varying time scales. This idea is discussed more fully by Moncrieff et al. (1996), as well as by Kruijt et al. (2004) and Richardson et al. (2008).

We do not address measurements in other types of flux measurements, such as cuvette or chamber measurements of photosynthesis or respiration, or other ecological measurements that are made at many sites, as these are beyond the scope of this book and are discussed elsewhere. For example, Smith and Hollinger (1991) discussed and quantified uncertainty in chamber measurements, soil respiration measurement uncertainty is described and quantified by Davidson et al. (2002) and Savage et al. (2008), and an approach to estimate ecosystem biomass and nutrient budget uncertainty is presented by Yanai et al. (2010). An evaluation of uncertainties in disjunct eddy covariance measurements (DEC) is presented in Sect. 10.5.

7.2 Random Errors in Flux Measurements

Random error in flux measurements arise from a variety of sources. These include:

1. The stochastic nature of turbulence (Wesely and Hart 1985) and, associated sampling errors, including incomplete sampling of large eddies, and uncertainty in the calculated covariance between the vertical wind velocity (w) and the scalar of interest (c);
2. Errors due to the instrument system, including random errors in measurements of both w and c; and
3. Uncertainty attributable to changes in wind direction and velocity which influence the footprint over which the measurements integrate, and thus the degree to which any individual 30-min measurement is representative of the point in space where the measurement system is located, or, more generally, the surrounding ecosystem (Aubinet et al. 2000).

While it could be argued that (3) is distinctly different in nature from (1) and (2), we included it here as a source of uncertainty because footprint variability is typically not taken into account, neither when 30-min measurements are aggregated to annual ecosystem carbon budgets, nor when the 30-min measurements are analyzed statistically or used in a more sophisticated data-model fusion scheme.

We will discuss each of these sources of uncertainty in greater detail below, but note that the methods developed to date to quantify random uncertainty for the most part focus on the total uncertainty – this being needed for most applications where uncertainty information is used – rather than attempt to parse this aggregate value to the three components listed above.

7.2.1 Turbulence Sampling Error

Finkelstein and Sims (2001) provide an overview of the uncertainties associated with turbulence sampling errors. They note that these errors occur because large eddies, which are responsible for much of the total flux, cannot be adequately sampled during a 30-min integration period. They also improve on previous methods to estimate the variance of the calculated covariance through incorporation of necessary lag and cross-correlation terms. A conceptual framework is provided by the equation, developed by Lenschow et al. (1994) and Mann and Lenschow (1994) from the basic equations of turbulence, to estimate for the relative error in an aircraft flux measurements. Hollinger and Richardson (2005) and Richardson et al. (2006a) adapted this approach to provide an approximation of uncertainty in tower-based flux measurements. This framework separates out (1) an estimate of the uncertainty in the variance of the covariance from (2) uncertainty associated with the organization of turbulence into large eddies and a finite integration period (full details are given in Richardson et al. 2006a).

While micrometeorological approaches such as this are appealing, they require an estimate of the integral timescale (a measure of how long turbulence remains correlated with itself, signifying the scale of the most energetic eddies and corresponding to the peak of the spectral density; Finnigan 2000), as well as knowledge of the turbulence statistics, which means not only that the measurement and the error estimate are based on the same flux variances and covariances, but also that the necessary information should be made available in standard 30-min data files.

7.2.2 Instrument Errors

Random errors resulting from the measurement system have been quantified using a number of different approaches. Similar to the paired measurement approaches described below, Eugster et al. (1997) used simultaneous measurements from two collocated towers in the Alaskan tundra to quantify instrument uncertainties; these were estimated to be 7% for H, 9% for λE, and 15% for F_c. Using essentially the same approach, Dragoni et al. (2007) estimated that instrument uncertainty was about 13% for F_c at the 30-min time step, and calculated that at the annual time step, this accumulated to an uncertainty of ± 10 g C m^{-2} year^{-1}, or 3% of annual NEE at a temperate deciduous site, Morgan Monroe. By comparison, Oren et al. (2006) used the variability in nocturnal λE as an indicator of measurement system uncertainty and, assuming analogous errors in F_c, estimated that at the annual time step, this accumulated to an uncertainty of ± 8–28 g C m^{-2} year^{-1} for the Duke pine plantation.

All these comparisons are built on assumptions that are difficult to test. Such comparisons always risk confusing instrument and noninstrument errors. The only unequivocal solution is to adopt the conventional engineering approach (e.g.,

Coleman and Steele 2009) and investigate instrument uncertainty from the bottom up, that is, from the component uncertainties of the eddy flux instrumentation.

7.2.3 *Footprint Variability*

Flux measurements integrate across a time-varying, and usually somewhat heterogeneous, footprint. Oren et al. (2006) reanalyzed data from an experiment described by Katul et al. (1999), in which simultaneous eddy covariance measurements were made at six towers within the Duke pine plantation, to distinguish the relative contribution of (1) spatial variability (i.e., differences in "ecosystem activity") and (2) turbulent sampling errors to the measurement uncertainty. This study found that at the 30-min time step, spatial variability ($\approx$10% of the measured flux, during the day) accounted for 50% of the measurement uncertainty, even in a relatively homogeneous forest. At the annual time step, the spatial variability accumulated to an uncertainty of $\pm$25–65 g C m^{-2} year^{-1}, or in some years as much as 50% of total (including gap-filling) annual NEE uncertainty ($\pm$79–127 g C m^{-2} year^{-1}). Related to this, the observation by Schmid et al. (2003) that annual NEE integrals for the University of Michigan Biological Station (UMBS) deciduous forest could differ by up to 80 g C m^{-2} year^{-1}, depending on whether data measured at a height of 34 or 46 m were used, presumably also partially reflects footprint differences.

7.2.4 *Quantifying the Total Random Uncertainty*

If each of the sources of random error could be independently quantified, then the total random flux measurement uncertainty could be estimated by adding the individual uncertainties together in quadrature. A more straightforward approach is to conduct statistical analyses that directly yield estimates of the total random uncertainty. Three methods have been developed; these are referred to as the "paired tower," "24 h differencing," and "model residual" approaches.

As proposed by Finkelstein and Sims (2001), the paired tower approach is based on the premise that repeated, independent measurements of a quantity can be used to estimate the statistical properties of the random error (ε) in those measurements. Hollinger et al. (2004) and Hollinger and Richardson (2005) used simultaneous measurements ($x_{1,t}$ and $x_{2,t}$) from two towers separated by $\approx$ 800 m at the Howland Forest AmeriFlux site to estimate the moments of ε, assuming that the measurement errors ($\varepsilon_{1,t}$ and $\varepsilon_{2,t}$) at the two towers were independent and identically distributed. For this assumption to hold, the footprints must be nonoverlapping, so that the turbulence sampling errors at tower 1 and tower 2 are uncorrelated (cf. Rannik et al. 2006, who estimated uncertainties using data from two towers which, because they were separated by only 30 m, had overlapping footprints and thus correlated sampling errors, and Dragoni et al. 2007, who used simultaneous flux measurements from two instrument systems separated by approximately 1 m to quantify random instrument errors). Then, the standard deviation of the measurement error can be estimated as in Eq. 7.1, using multiple realizations (i.e., repeated over time) of $x_{1,t}$ and $x_{2,t}$ to obtain more precise estimates of the statistics of ε.

$$\sigma(\varepsilon_t) = \frac{\sigma(x_{1,t} - x_{2,t})}{\sqrt{2}} \tag{7.1}$$

For this method to work, it is critical that (1) in a given half-hour, the environmental conditions in the footprint of tower 1 are nearly identical to those in the footprint of tower 2; and (2) the vegetation, soils, etc. are extremely similar between the footprints of tower 1 and tower 2, so that the biological response to the abiotic forcing is the same. Together, these ensure $x_{1,t}$ and $x_{2,t}$ are essentially measurements of the same quantity, and thus that the difference between the measurement pair is due only to measurement error (including random variation of the sampled footprint) and not to differences in biotic or abiotic factors.

Recognizing that there are few eddy covariance sites around the world where two towers would satisfy the "similar but independent" criteria required for the paired tower approach, the 24-h differencing approach, which trades time for space, was developed by Hollinger and Richardson (2005) and subsequently implemented at a range of AmeriFlux and CarboEurope sites by Richardson et al. (2006a, 2008). With this method, two flux measurements ($x_{1,t}$, $x_{1,t+24}$) made at a single tower, exactly 24 h apart (to minimize diurnal effects) and under similar environmental conditions, are considered analogs of the simultaneous two-tower paired measurements described above. The similar environmental conditions criterion is included so that the difference between $x_{1,t}$ and $x_{1,t+24}$ can largely be attributed to random error rather than environmental forcing; for this filtering, PPFD within 75 μmol m^{-2} s^{-1}, air temperature within 3°C, wind speed within 1 m s^{-1}, and vapor pressure deficit within 0.2 kPa has been found to yield an acceptable balance between the requirement that environmental conditions be "similar" and the desire for a sufficiently large sample size of measurement pairs so that the statistics of ε could be adequately estimated (Richardson et al. 2006a, 2008). More stringent filtering (e.g., excluding measurement pairs if the mean half-hourly wind directions differed by more than ±15°) was reported to only result in a modest (≈10%) reduction in estimated uncertainty, and a large reduction in the number of accepted measurement pairs.

The third, or model residual approach, uses the difference between a highly tuned empirical model and the measured fluxes as an estimate of ε (Richardson and Hollinger 2005; Richardson et al. 2008; Stauch et al. 2008; Lasslop et al. 2008). In principle, it is assumed that model error is negligible and that the model residual can be attributed almost entirely to random measurement error. This assumption has been largely confirmed in Moffat et al. (2007) and Richardson et al. (2008). An advantage of this method over the 24-h differencing approach is that many more estimates of the inferred error are available for use in estimating statistics of ε.

Hollinger and Richardson (2005) demonstrated not only that the paired tower and 24-h differencing approaches provided roughly comparable estimates of flux measurement uncertainty but also that these were both in reasonable agreement with predictions of the Mann and Lenschow (1994) sampling error model (see Sect. 7.2.1, above). Richardson et al. (2008) showed that uncertainty estimates from the model residual approach were larger (by 20% or more; the actual amount depended on the model used) than those derived by 24-h differencing, presumably because even in the best case, model error could not be completely eliminated. However, overall patterns, particularly with respect to the pdf of ε, and the way in which $\sigma(\varepsilon)$ scales with flux magnitude, have been found to be extremely similar (especially considering that uncertainty estimates are inherently uncertain) regardless of the method. That being said, a key difference among methods is that the two approaches relying on paired observations are unable to estimate odd moments such as skewness, because the differencing implies symmetry in the resulting pdf. While positive skewness has been demonstrated with the model residual approach (Richardson et al. 2008), particularly for near-zero fluxes, this may simply be the result of selective data editing by the investigators, and the preferential elimination of positive or negative outliers.

7.2.5 *Overall Patterns of the Random Uncertainty*

Regardless of the method used to quantify the random flux measurement uncertainty, two characteristics of the uncertainty have been shown to be extremely robust, both with respect to different fluxes (i.e., for H and λE as well as F_c) and across a variety of sites and ecosystem types (Hollinger and Richardson 2005; Richardson et al. 2006a, 2008; Stauch et al. 2008; Lasslop et al. 2008; Liu et al. 2009).

First, the standard deviation of the random measurement uncertainty (in μmol m^{-2} s^{-1}) generally increases with the magnitude of the flux ($|F_s|$) in question, and this relationship can be approximated as in Eq. 7.2 (see Table 7.1 and Fig. 7.1):

$$\sigma(\varepsilon_s) = a + b\,|F_s| \tag{7.2}$$

For F_c, the nonzero y-axis intercept, a, varies among sites, with typical values between 0.9 and 3.5 μmol m^{-2} s^{-1} (Richardson et al. 2008). By comparison, the slope, b, lies in a relatively narrow range across sites, usually between 0.1 and 0.2. A consequence of the nonzero intercept, a, is that there is a baseline of residual uncertainty even when the flux is zero; this implies that relative errors decrease with increasing flux magnitude (cf. the error model based on turbulence statistics, Sect. 7.2.1, for which relative error is assumed to be constant).

Second, the overall distribution of the flux measurement uncertainty is non-Gaussian, most notably because it is strongly leptokurtic – meaning that it is peaky, with heavy tails; the Laplace, or double exponential, distribution is a good

Table 7.1 For H, λE, and F_c, random flux measurement error ($\sigma(\varepsilon)$) scales linearly with the magnitude of the flux (F). Results are summarized below from three previous studies. Standard errors for parameter estimates (where available) are in parentheses. All slope coefficients are significantly different from zero ($P < 0.01$)

(A) Hollinger and Richardson (2005); two towers

Site		Uncertainty
Howland	H	$10 + 0.22\ \lvert H \rvert$
	λE	$10 + 0.32\ \lvert \lambda E \rvert$
	F_c	$2 + 0.1\ F_c\ (F \leq 0)$
		$2 + 0.4\ F_c\ (F \geq 0)$

(B) Richardson et al. (2006a); 24 h differencing

Flux		Uncertainty	
		$F \geq 0$	$F \leq 0$
H	Forested	$19.7\ (3.5) + 0.16\ (0.01)\ H$	$10.0\ (3.8) - 0.44\ (0.07)\ H$
	Grassland	$17.3\ (1.9) + 0.07\ (0.01)\ H$	$13.3\ (2.5) - 0.16\ (0.04)\ H$
λE	Forested	$15.3\ (3.8) + 0.23\ (0.02)\ \lambda E$	$6.2\ (1.0) - 1.42\ (0.03)\ \lambda E$
	Grassland	$8.1\ (1.7) + 0.16\ (0.01)\ \lambda E$	No data
F_c	Forested	$0.62\ (0.73) + 0.63\ (0.09)\ F_c$	$1.42\ (0.31) - 0.19\ (0.02)\ F_c$
	Grassland	$0.38\ (0.25) + 0.30\ (0.07)\ F_c$	$0.47\ (0.18) - 0.12\ (0.02)\ F_c$

(C) Richardson et al. (2008); Forested sites

Method	Uncertainty
Model residuals (neural network)	$1.69(0.20) + 0.16(0.02)\ \lvert F_c \rvert$
Paired observations	$1.47(0.22) + 0.17(0.02)\ \lvert F_c \rvert$

approximation of the pdf. As a result, not only are very large errors more common than if the error distribution was normal, but also very small errors are more common than if the error distribution was normal. It was proposed that the leptokurtic distribution could result from the superposition of Gaussian distributions with nonconstant variances (Hollinger and Richardson 2005; Stauch et al. 2008; Lasslop et al. 2008). Indeed, Lasslop et al. (2008) showed that after normalizing the error (by dividing with the expected standard deviation for each flux observation) the overall distribution generally became approximately Gaussian. However, for some sites, even when flux data are binned into relatively narrow classes, nonnormal random errors are observed for fluxes close to zero (e.g., $-1 < F_c < 1$, as in Fig. 7.2), whereas for large uptake fluxes ($F_c < -10\ \mu\text{mol m}^{-2}\ \text{s}^{-1}$, Fig. 7.2), the errors tend to be much more Gaussian (see also Fig. 3 in Richardson et al. 2008). We conducted an analysis of the whole LaThuile FLUXNET dataset using the "model residual" approach (Fig. 7.3). We find the patterns discussed above, that is, a positive kurtosis for the overall distribution of the model residuals, but this is largely (although not completely) reduced when the nonconstant variances are accounted for by normalization. Skewness is also apparent in the error distribution for some sites, particularly at night (Richardson and Hollinger 2005, Barr et al. unpublished results). Richardson et al. (2008) found trimming the top and bottom 1% of residuals

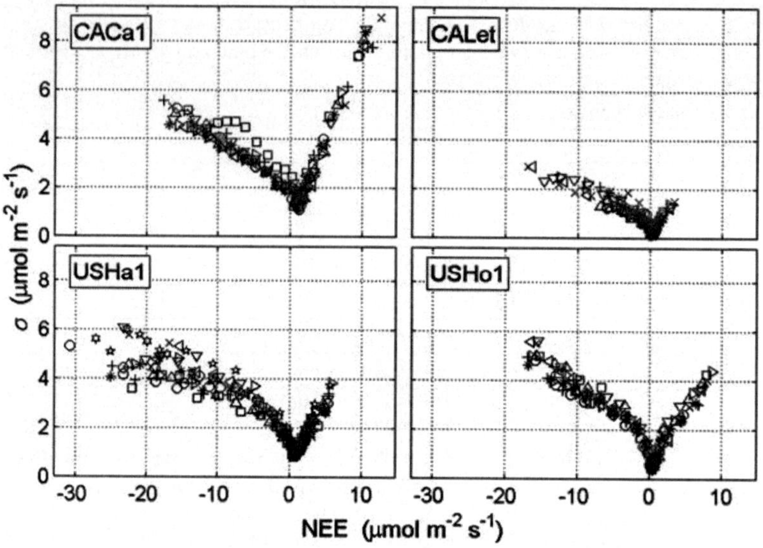

Fig. 7.1 Scaling of random uncertainty (1σ) with flux magnitude (NEE, μmol m^{-2} s^{-1}) for four temperate sites: CaCa1 – Campbell River mature stand, a Douglas-fir-dominated evergreen coniferous site; CaLet – Lethbridge, a Great Plains grassland; USHa1 – Harvard Forest EMS tower, an oak-dominated deciduous broadleaf forest; USHo1 – Howland Forest Main tower, a spruce-dominated evergreen coniferous site. Random uncertainty estimated using the residuals from calibrated Fluxnet-Canada gap-filling algorithm, which was also used to predict NEE (Source: Barr, Hollinger and Richardson, unpublished). *Different symbols* indicate different years of data, showing that uncertainty estimates are estimated consistently over time

typically resulted in a much more symmetric distribution of ε, and also reduced kurtosis (see, e.g., Fig. 7.3). However, blindly filtering outlier points that cause accentuated kurtosis and skewness is not recommended, as, in addition to changing the apparent pdf of the random measurement error, this may have an impact on annual flux estimates.

Thus, although there are some general patterns across sites, differences in site characteristics, as well as differences in the data acceptance practices used by site investigators, may necessitate careful site-specific analyses of the random error following the methods described here (see also Richardson et al. 2006a, 2008; Lasslop et al. 2008). We note that at each site decisions must be made concerning the degree to which valid flux data are contaminated with data from a separate (nonbiological or atmospheric) process. If this is judged to be the case, then approaches can be used to identify and remove such outliers (Barnett and Lewis 1994). However, data-trimming methods are sensitive to the underlying statistical distribution of the data and the appropriate method of identifying outliers should be used based on the error pdf; Barnett and Lewis (1994) present methods for both Gaussian and double exponential distributions.

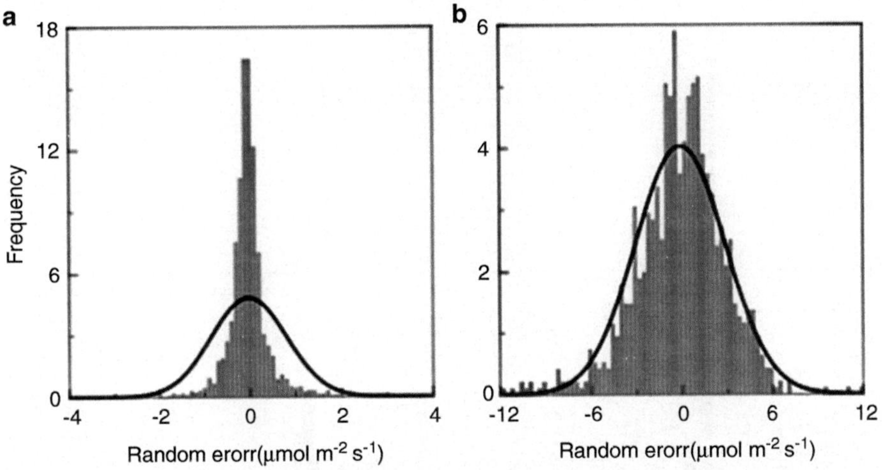

Fig. 7.2 Comparison of probability distributions of inferred random error for (**a**) near-zero fluxes ($-1 \leq F_c \leq 1$ μmol m^{-2} s^{-1}; $n = 2{,}544$, standard deviation = 0.82, kurtosis = 123.92) and (**b**) large uptake fluxes ($F_c \leq -10$ μmol m^{-2} s^{-1}; $n = 949$, standard deviation = 2.97, kurtosis = 1.99). Random errors estimated using paired tower approach ("Main" and "West" towers at Howland Forest AmeriFlux site). In both cases, the normal distribution is shown as a *black line*

The maximum likelihood method is used to determine model parameters (which may range from coefficients of simple regression models to physiological parameters in complex carbon cycle models) that maximize the probability (likelihood) of the sample data. This method takes into account prior knowledge of data uncertainties, using estimators (likelihood functions) that depend upon the error structure of the data. For normally distributed data with constant variance, the maximum likelihood is calculated via ordinary least squares. Minimizing the sum of absolute deviations (rather than squared deviations) is appropriate if the error distribution is deemed to follow the Laplace distribution. If the errors are heteroscedastic, as is typically the case with eddy flux data, then observations should also be appropriately down-weighted, that is, by $1/\sigma(\varepsilon)$ (weighted absolute deviations) or $1/\sigma^2(\varepsilon)$ (weighted least squares). It should be noted that different minimization criteria may result in different best-fit parameter sets, parameter covariances, and uncertainty estimates – not to mention different interpretations of the data (Richardson and Hollinger 2005; Lasslop et al. 2008).

Several additional details about the random measurement error are worth noting:

1. At some sites, the relationship between flux magnitude and uncertainty appears to level off for large negative fluxes (US – Ha1 in Fig. 7.1);
2. At many, but not all (Richardson et al. 2006a, 2008; Barr et al., unpublished results) sites, the slope, *b*, is larger for positive (i.e., nocturnal release) than negative (i.e., daytime uptake) fluxes, which may have to do with outlier removal

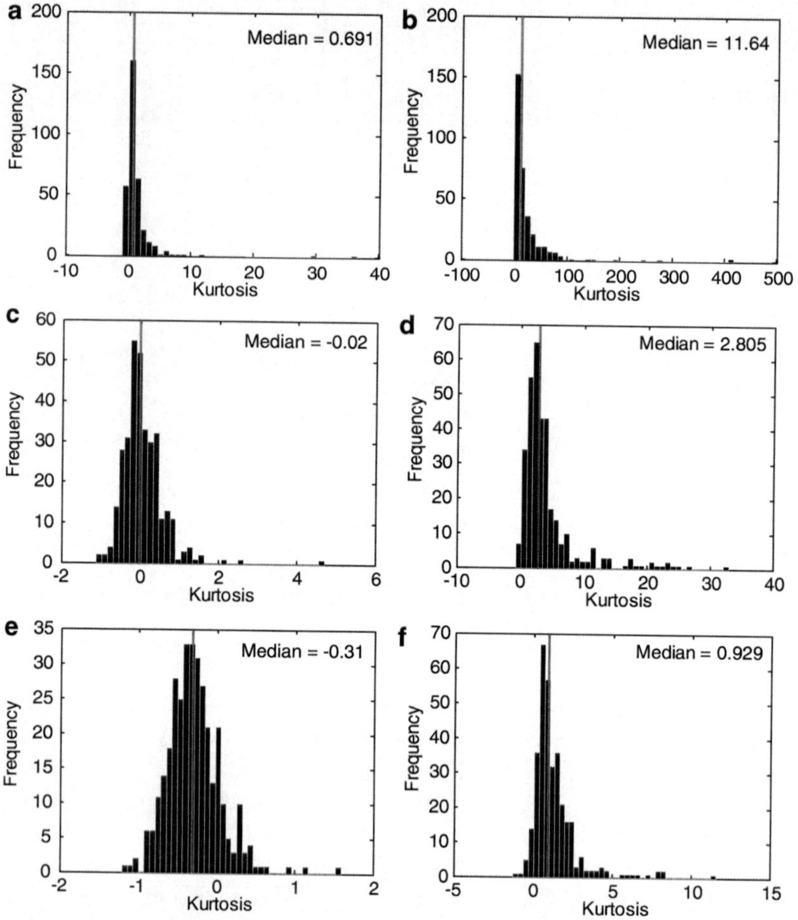

Fig. 7.3 Histograms of the kurtosis of the half hourly random error estimates for 332 FLUXNET site-years. In the *first column*, only error estimates of high-magnitude fluxes (NEE $< -20\ \mu\text{mol m}^{-2}\ \text{s}^{-1}$) are used; in the *second*, only fluxes with $|\text{NEE}| < 1\ \mu\text{mol m}^{-2}\ \text{s}^{-1}$. The *first row* shows the kurtosis of the errors not accounting for the variable standard deviation, the *second row* the kurtosis of errors normalized with their standard deviation, in the *third row* the tails of the error distribution trimmed (1%) and the errors were normalized

and data editing by site investigators, or to differences in the turbulent transport statistics between unstable conditions during the day and stable conditions at night;

3. While Raupach et al. (2005) suggested that errors in measured fluxes would be cross-correlated (i.e., positive correlation between error in F_c and error in λE), Lasslop et al. (2008) reported that this was not the case. This is surprising given that different scalars are carried by the same turbulent eddies, but a possible explanation for this observation is that the exchange sites within the

ecosystem differ among fluxes (as discussed in Hollinger and Richardson 2005). In contrast to the results of Lasslop et al. (2008), data from the two-tower system at Howland (Hollinger unpublished) indicate that between-tower differences (errors) of various fluxes are weakly correlated at night (e.g., for F_c and λE, $r = 0.2$) while during active daytime periods correlations are higher (e.g., during the growing season when PPFD $\geq$ 1,000 μmol m^{-2} s^{-1}, F_c: λE $r = -0.33$, F_c:H $r = -0.46$, H: λE $r = 0.52$). Lasslop et al. (2008) also found that the autocorrelation of flux measurement errors dropped off rapidly, and is typically less than 0.6 for a 30 min lag;

4. Consistent with theory, the CO_2 flux measurement uncertainty decreases with increasing wind speed (Hollinger et al. 2004), although this was not generally observed for H or λE (Richardson et al. 2006a);
5. Differences in random flux measurement error between open- and closed-path systems appear to be more or less negligible (Richardson et al. 2006a; Ocheltree and Loescher 2007; Haslwanter et al. 2009).

7.2.6 Random Uncertainties at Longer Time Scales

Over time (days, months, years), the total random uncertainty on a flux integral increases with the length of the integration period. However, at the same time, the random uncertainty on the *mean* flux becomes smaller. For example, Rannik et al. (2006) reported the random uncertainty (1σ) on half-hourly fluxes at the Hyytiälä site was ±1.1 μmol m^{-2} s^{-1} (±23 mg C m^{-2}), whereas the random uncertainty on the daily *mean* flux was ±0.2 μmol m^{-2} s^{-1} (±4 mg C m^{-2}), which is consistent with the rule that random errors decrease with averaging as $1/\sqrt{n}$(whereas for the integral they increase as $n/\sqrt{n}$). On the daily flux *integral*, however, this translates to ±195 mg C m^{-2}. This emphasizes the importance of distinguishing between uncertainties on means and uncertainties on integrals; the latter is n times larger than the former. And, whereas diurnal and seasonal differences in the sign of the measured flux may cancel each other so the net flux is near zero, this is not the case with uncertainties on the flux integral, which always grow over time. Finally, it should be noted that what seems a trivial error on the mean half-hour flux (e.g., ±0.1 μmol m^{-2} s^{-1}) is certainly not insignificant when considered in terms of daily (±0.1 g C m^{-2} day^{-1}) or yearly (±40 g C m^{-2} year^{-1}) integrals.

Propagation of uncertainties to longer time scales is conveniently done using some sort of Monte Carlo or resampling technique (e.g., Richardson and Hollinger 2005), especially as this permits incorporation of uncertainties due to gap filling (e.g., Moffat et al. 2007; Richardson and Hollinger 2007). Using a bootstrapping approach, Liu et al. (2009) quantified random uncertainties in flux integrals at various time scales (30-min, day, month, quarter, year) for a young conifer plantation; relative uncertainty dropped from ≈ 100% at subdaily timescales to 7–22% (±10–40 g C m^{-2} year^{-1}) at the annual timescale. Other studies have similarly attempted to quantify the random uncertainty for annual NEE integrals; across a range of sites.

Stauch et al. (2008) and Richardson and Hollinger (2007) reported that random uncertainties on integrated NEE accumulated to roughly ±30 g C m^{-2} $year^{-1}$ (95% confidence); these estimates are consistent with the observation by Hollinger et al. (2004) that, over a 3-year period, annual NEE integrals from the Howland "main" and "west" towers never differed by more than 25 g C m^{-2} $year^{-1}$, which was substantially less than the observed interannual variability.

7.3 Systematic Errors in Flux Measurements

We now address the sources of systematic error, or bias, in flux measurements. These can be grouped into three categories. The first two categories have to do with measurement issues, due to the underlying assumptions of the eddy covariance technique not being satisfied (Sect. 7.3.1), or resulting from instrument calibration and design errors (Sect. 7.3.2). The third category relates to processing issues, for example, how both the raw high-frequency measurements and also the 30-min covariances are treated in preparation of a "final" quality-controlled, corrected, and gap-filled data set (e.g., Kruijt et al. 2004) (Sect. 7.3.3).

As noted above, systematic errors, unlike random errors, can and should be corrected; if the correction has been applied correctly, this error disappears completely. However, uncertainties appear because the correction is not complete, or is not sufficiently accurate to entirely eliminate the error. In this section, our focus is on a brief overview (as these are treated in greater detail in separate chapters) of the major systematic errors and the method(s) used to correct them, and we attempt to quantify any uncertainty that remains after having applied the correction.

7.3.1 Systematic Errors Resulting from Unmet Assumptions and Methodological Challenges

Calculation of the eddy flux from the conservation equation requires several simplifying assumptions (Baldocchi et al. 1988, 1996; Dabberdt et al. 1993; Foken and Wichura 1996; Massman and Lee 2002), most important of which are that the surrounding terrain is homogeneous and flat, that the transport processes are stationary in time, that there is adequate turbulence to drive transport, and that the vertical turbulent flux is the only significant transport mechanism. Violation of these assumptions will induce errors and uncertainties in the measured flux; we note that Foken and Wichura (1996) have proposed quality tests with which suspect data, violating the underlying assumptions, can be flagged and filtered (see Sect. 4.3). We now discuss in greater detail some of these uncertainties, as well as a related methodological challenge: the problem of nocturnal measurements, which Massman and Lee (2002) described as "a co-occurrence of all eddy covariance limitations."

Surface heterogeneity is thought to be a key factor contributing both to advection (Sects. 5.1.3 and 5.4.2) and to energy balance nonclosure (Sect. 4.2) errors. For example, Finnigan (2008) notes that even in flat terrain, advection can occur if the canopy source-sink strength is not spatially homogeneous. It is increasingly recognized that without accounting for advection, annual estimates of CO_2 sink strength are likely biased upward, because advection tends to be a selectively systematic error and usually results in underestimation of nocturnal CO_2 efflux (Staebler and Fitzjarrald 2004). Quantifying the advection bias is challenging (Finnigan 2008), and the size of the bias likely varies widely among sites (Feigenwinter et al. 2008). However, Aubinet (2008) recently proposed a scheme to classify sites to one of five different advection patterns, suggesting that a general model may be possible.

With respect to energy balance closure, Foken (2008) concluded that this was "a scale problem" resulting from surface heterogeneity and the omission of low-frequency fluxes associated with large eddies generated at edges or changes in land use. Barr et al. (2006) observed an increased energy imbalance at low wind speeds that may be related to the onset of organized mesoscale circulations that produce stationary cells that add horizontal and vertical advection (Kanda et al. 2004). We note that if either or both of the turbulent energy fluxes are systematically underestimated, then this suggests the potential for a corresponding error in the measured CO_2 flux because atmospheric transport processes are similar for all scalars and the calculation of all scalar fluxes rests on the same theoretical assumptions (Twine et al. 2000; Wilson et al. 2002). The CO_2 flux bias and energy imbalance have been shown to respond similarly to u_* and atmospheric stability (Barr et al. 2006). However, using the energy imbalance to "correct" CO_2 fluxes is not widely accepted (Foken et al. 2006). We do not recommend its use at this time (see also Sect. 4.2).

Nonstationarity of the turbulent statistics can result from underlying diurnal cycles or from changes in weather (Foken and Wichura 1996). When nonstationarity occurs, a key consequence is that the surface exchange is not exactly equal to the sum of the measured flux and storage terms (Finnigan 2008). Measurements taken under nonsteady-state conditions may be identified and then filtered by application of the stationarity test described in (Sect. 4.3.2). The resulting uncertainty is mainly random and depends on the gap frequency and gap-filling algorithm. In a comparison of 18 European sites, Rebmann et al. (2005) showed that the test eliminated on average 23% of the data. However, they did not study the impact of this elimination on annual NEE. At Vielsalm (forested) and Lonzée (crop) sites, Heinesch (not published) found a similar percentage of eliminated data in day conditions but, at night, this percentage was larger, reaching 30–40%. However, nonsteady-state nighttime data are often also removed by u_* filtering (see below).

Under stable conditions with poorly developed turbulence, the eddy covariance method is unable to accurately measure the surface exchange because nonturbulent fluxes (storage, advection) may become as important as the turbulent fluxes themselves (see Sect. 5.1.3). The error resulting from assumptions of adequate turbulence not being satisfied is probably the most important in eddy covariance measurement. In addition, as it acts as a selective systematic error (Moncrieff et al. 1996) its impact on annual fluxes is especially critical.

Recent experiments have shown unambiguously that correcting for advection, although attractive from a theoretical point of view, is impractical because direct advection measurements introduce not only large uncertainties (Aubinet et al. 2003; Feigenwinter et al. 2008; Leuning et al. 2008) but also large systematic biases (Aubinet et al. 2010) in flux estimates (see also Sect. 5.4.2.3).

For these reasons, filtering nocturnal measurements during poorly mixed periods remains the best method. A filtering procedure based on a friction velocity threshold was proposed by Goulden et al. (1996). The method, its advantages, and shortcomings are discussed in Sects. 5.3 and 5.4.1 and some alternatives are proposed.

By comparing 12 site-years in certain European forests where the nocturnal flux error is thought to be large, Papale et al. (2006) reported the error associated with not correcting for low turbulence always induced a systematic NEE overestimation, varying by site and year, but generally in the range of 20–130 $\mathrm{g\,C\ m^{-2}\ year^{-1}}$ (based on the difference between annual CO_2 flux integrals calculated with and without u_* filtering). Uncertainties resulting from u_* filtering may have two sources: Uncertainty regarding determination of the specific u_* threshold ($u_{*\mathrm{crit}}$) applied, and uncertainty from the algorithm used to fill the resulting data gaps. Uncertainties linked with data gap-filling algorithms are discussed in Sects. 7.2 and 7.3.3.3. Impact of the uncertainty on $u_{*\mathrm{crit}}$ was analyzed by Papale et al. (2006) (see also Hollinger et al. 2004). They reported confidence intervals on $u_{*\mathrm{crit}}$ of 0.15–0.25 $\mathrm{m\,s^{-1}}$, which lead to 10–70 $\mathrm{g\,C\ m^{-2}\ year^{-1}}$ uncertainties on annual NEE. NEE declined when $u_{*\mathrm{crit}}$ was increased, that is, sites became smaller carbon sinks. Analyzing a winter wheat crop, Moureaux et al. (2008) obtained values in the lower range of these estimates, that is, 10 (1.6%), 50 (5.2%), and 30 (1.9%) $\mathrm{g\,C\,m^{-2}}$ on NEE, R_{eco}, and GEP, respectively.

7.3.2 *Systematic Errors Resulting from Instrument Calibration and Design*

The eddy covariance measurement system itself can also be a source of systematic errors. These include errors related to calibration and drift, as well as errors resulting from the infrared gas analyzers (IRGA) and sonic anemometer instruments themselves. Many of these errors can be minimized by careful attention to system design (see Sects. 2.3 and 2.4). A list of these errors, their order of magnitude, the recommended correction procedure, and the possible uncertainty remaining after the correction is given in Table 7.2.

7.3.2.1 Calibration Uncertainties

For any type of instrument, calibration errors and drift result in biased measurements. These errors are, in principle, systematic, but there is a random component operating at longer timescales (days to weeks) because both the sign and magnitude of the error are often unknown.

Table 7.2 Systematic errors due to instruments

Error Type (sensor concerned)	Error impact on annual C exchange. Default unit: $gC\ m^{-2}\ year^{-1}$	Possible corrections	Uncertainty on annual C exchange after correction Default unit: $gC\ m^{-2}\ year^{-1}$	Remarks Remaining uncertainty	References
Calibration drift (mainly gas analyzers)	5% per week Ocheltree and Loescher (2007)	Frequent calibrations, interpolations between calibrations.	→ 0	Due to non linearity of calibration drift	Section 2.4.2.3, Section 7.3.2.1
Instrument spikes	<10 Papale et al. (2006)	Peak detection algorithms, removal and interpolation	→ 0	Due to uncertainty on peak limit detection.	Section 3.2.2
Sheltering/Distortion (sonic anemometers)	3–13% Nakai et al. (2006)	Calibration by manufacturer. User calibration (requires wind tunnel).	→ 0		
High frequency losses (sonics + open-paths)	3% Järvi et al. (2009)	Spectral corrections based on theoretical co-spectra and transfer functions.	2% Järvi et al. (2009)	Due to uncertainties on transfer function and cospectra shapes.	Section 1.5, Section 4.1.3
High frequency losses (sonics + closed-paths)	11% Järvi et al. (2009) <5% (day), <12% (night) Berger et al. (2001) 4% Ibrom et al. (2007a)	Spectral corrections based: on theoretical co-spectra and transfer functions; on experimental co-spectra and transfer functions.	3% Järvi et al. (2009) 10 per 0.1 impedance Anthoni et al. (2004)	Due to uncertainties on transfer function and cospectra shapes. Due to uncertainties in impedance	Section 1.5 Section 4.1.3

(continued)

Table 7.2 (continued)

Error Type (sensor concerned)	Error impact on annual C exchange. Default unit: $gC\ m^{-2}\ year^{-1}$	Possible corrections	Uncertainty on annual C exchange after correction Default unit: $gC\ m^{-2}\ year^{-1}$	Remarks Remaining uncertainty	References
Density fluctuation (Open-path and potentially closed-path)	CP: 0–160, 2.90% Ibrom et al. (2007b)	WPL corrections: CP: water vapor (Calculate covariances with molar mixing ratios relative to dry air to avoid needing WPL water vapor correction)	CP: Potentially 0.09 per MJ error in cumulated latent heat (see Table 7.3)	CP: Using the original WPL equation (24) overcorrects the effects (cf. Ibrom et al. 2007b).	Section 4.1.4
	OP: 190–920 See Table 7.3	OP: sensible heat and water vapor (needs additional sensible heat measurement)	OP: see Table 7.3	OP: Due to uncertainties in sensible and latent heat estimates. Additional error due to sensor surface heating.	
Sensor surface heating (Open-path IRGA)	100 Burba et al. (2008) 140 Järvi et al. (2009) forest 330 Järvi et al. (2009) urban	Burba corrections. Different algorithms to estimate sensible heat excess.	5–13 Burba et al. (2008) 40 Järvi et al. (2009) forest 170 Järvi et al. (2009) urban	Due to differences between correction algorithms.	Section 4.1.5.2

CP: closed-path gas analyzer; *OP*: open-path gas analyzer

Calibration uncertainties result either from uncertainties in the concentration of calibration standards or from calibration drift. The relative error on the eddy covariance flux resulting from uncertainties in the standard gases is equal to the relative error on the gas concentration. This error is often as high as 2.5%, although 0.5% accuracy is easily achieved.

Calibration drift error is due to instrument instability and affects mainly gas analyzers. For the AmeriFlux Portable Eddy Covariance System, Ocheltree and Loescher (2007) found that over a week-long period, calibration drift between two different measurement systems resulted in a 5% difference in the measured fluxes. Regular (daily to weekly) calibrations are thus required to minimize this source of uncertainty. The set up of an automatic calibration procedure facilitates its regular application. Uncertainty resulting from the calibration drift largely depends on the time interval between two successive calibrations and on the procedure that is used to account for drift. Three different procedures could be followed: centered, averaged, and linearly interpolated calibration. In order to estimate the uncertainty in each case, we assume that at each calibration the relation between the quantity being measured (x) and the electronic signal (V) is given by $x_j = f_j(V)$ and that calibration drift is monotonic. In the case of centered calibration, each intercalibration period (between j and $j + 1$) is divided in two parts, $f_j(V)$ being used in the first half and $f_{j+1}(V)$ in the second. In these conditions, an upper limit to calibration error is given by:

$$\delta_{Cal} = \left| f_j(V) - f_{j+1}(V) \right| \quad (7.3)$$

In the case of averaged calibration, during the intercalibration period, the signal is computed as the average between $f_j(V)$ and $f_{j+1}(V)$. An upper limit to calibration error is then given by:

$$\delta_{Cal} = \frac{\left| f_j(V) - f_{j+1}(V) \right|}{2} \quad (7.4)$$

For interpolated calibration, the calibration function $f_t(V)$ is computed at each moment of the intercalibration period as

$$f_t(V) = f_j(V) + \frac{t}{T}\left(f_{j+1}(V) - f_j(V) \right) \quad (7.5)$$

where T is the period duration between the two calibration and t is the time since the last calibration, j. In case of linear drift with time, this procedure reduces the error due to calibration drift to zero. However, in case of nonlinear drift, an uncertainty may remain whose upper limit is given by Eq. 7.4.

7.3.2.2 Spikes

Spikes in high-frequency raw data can be caused by instrumental problems (electronic spikes) or by any perturbation of the measurement volume (bird droppings, cobwebs, precipitation, etc.). Algorithms that detect spikes but also abnormally large variances, skewnesses, kurtosis, and discontinuities are currently available and correction procedures are discussed in Sect. 3.2.2. In the case of short peaks, the algorithm removes the spike and fills the resulting gap, in other cases the measurement may be flagged, leaving to the user the possibility to remove it from the data set or not. Papale et al. (2006) showed that spikes generally have a small impact on annual NEE (usually $<10\ \mathrm{g\,C\,m^{-2}\,year^{-1}}$ and only occasionally $>20\ \mathrm{g\,C\,m^{-2}\,year^{-1}}$). The uncertainty remaining after elimination of flagged data depends mainly on the quantity of flagged data and on the data gap-filling algorithm (see Sect. 7.3.3.3).

7.3.2.3 Sonic Anemometer Errors

Systematic errors associated with sonic anemometers can be due to its misalignment or to the limitations of a particular instrument design. Dyer et al. (1982) pointed out that, after adequate coordinate rotation (Sect. 3.2.4) the error on scalar fluxes due to sensor misalignment was about 3% per degree tilt. In addition, because of their design, which results in self-sheltering by transducers and flow distortion by the anemometer frame, sonic anemometers have an imperfect cosine response. This results in what are known as "angle of attack" errors (Sect. 4.1.5.1, see also Sect. 2.3.2). Corrections for these have been published and are typically applied to the raw u, v, and w measurements, often by the instrument internal software. An improved correction was found to increase measured F_c, H, and λE fluxes by 3–13% (Nakai et al. 2006). In addition, because sonic anemometers differ in design, the measured turbulent statistics (means and variances) and air temperature tend to vary somewhat depending on manufacturer and model. For short averaging periods in particular, this may result in substantial uncertainty in measured scalar fluxes (Loescher et al. 2005). Distortion due to tower and infrastructure may also affect turbulence. This point is discussed in detail in Sect. 2.2.

7.3.2.4 Infrared Gas Analyzer Errors

Open- and closed-path IRGAs are subject to different errors and biases (Sects. 2.4 and 4.1). However, these can be practically eliminated by careful system design and an adequate correction, so that remaining uncertainty is small. Indeed, Ocheltree and Loescher (2007) compared open- and closed-path IRGA measurements of F_c made with the AmeriFlux Portable Eddy Covariance System and reported good agreement ($R^2 = 0.96$) between the two fluxes, once the appropriate corrections had been made (see also Haslwanter et al. 2009). The significant errors attributable to the gas analyzer are reviewed below.

7.3.2.5 High-Frequency Losses

All sensors (we focus here on IRGAs, but similar problems affect other gas analyzers and sonic anemometers as well) are affected by high-frequency damping due to several reasons including instrument time response, sensor separation, volume averaging, etc. (Sect. 4.1.3). Closed-path systems (IRGA, tunable diode laser (TDL), Proton Transfer Reaction Mass Spectrometry (PTR-MS)) are in addition affected by a damping due to fluctuation attenuation in the sampling tube, so that spectral corrections are generally larger for closed-path analyzers than for open-path analyzers (Sects. 2.4.2 and 4.1.3). The negative effects of damping can be minimized by the use of short, clean tubes and flow rates that are high enough to produce fully turbulent flow. A comparison between open- and closed-path IRGAs in an urban environment showed that these high-frequency losses for CO_2 were about $11 \pm 3\%$ (SD) for a closed-path analyzer, and $3 \pm 2\%$ for an open-path analyzer (Järvi et al. 2009).

Spectral corrections (Sect. 4.1.3) are used to adjust the measured flux for high-frequency losses. The appropriate correction can be estimated both theoretically and empirically (Massman 2000); the theoretical approach yields spectral correction factors for F_c ranging from 4% to 25%, and for λE between 6 and 35% (Aubinet et al. 2000). The high-frequency losses are larger for λE than for F_c because of adsorption and desorption of water in the sampling tube that increases attenuation by the system dramatically at high relative humidity (Ibrom et al. 2007a; De Ligne et al. 2010); high-frequency losses for λE generally increase with the age of sampling tubes (Su et al. 2004; Mammarella et al. 2009). In practice, this means that the spectral transfer function of the eddy covariance system that is used for spectral correction needs to be sensitive to weather conditions (relative humidity), tube aging, and changes in the mass flow through the system. These corrections are described more fully in Sect. 4.1.3 and elsewhere (Aubinet et al. 2000; Massman 2000; Massman and Lee 2002; Ibrom et al. 2007a; Massman and Ibrom 2008).

7.3.2.6 Density Fluctuations

The need to apply the WPL (Webb et al. 1980) correction for density fluctuations in sampled air is well established (Sect. 4.1.4). Its application is required for open-path analyzers and may be needed in part for closed-path IRGAs if the CO_2 concentration is not reported relative to dry air. The correction has been described in Sect. 4.1.4 and consists in two terms (Eq. 4.25), one taking account of density fluctuations related to sensible heat transport, the second, of density fluctuations due to water vapor flux. In the case of an open-path system, both terms must be introduced in

Table 7.3 Expected order of magnitude of density corrections on annual CO_2 flux

	Annual average energy fluxes		Density correction on annual CO_2 flux	
Climate	Sensible heat ($GJ\ m^{-2}\ year^{-1}$)	Latent heat ($GJ\ m^{-2}\ year^{-1}$)	Due to temperature fluctuations ($gC\ m^{-2}\ year^{-1}$)	Due to water vapor fluctuations ($gC\ m^{-2}\ year^{-1}$)
Boreal	0.3	0.6	138	53
Temperate	0.9	0.9	413	80
Tropical	1.8	0.9	826	80
Equatorial	0.9	1.8	413	160

Derived from Webb et al. (1980) and from climatological data from Bonan (2008)

NB: No WPL correction is necessary with closed-path analyzers if the CO_2 concentration is expressed relative to dry air and the flux equation is adapted accordingly (see eq. 4 and Appendix Ibrom et al. (2007b))

NB2: In cases of closed-path systems, where CO_2 is expressed relative to moist air, the WPL vapor correction presented in this table may overcorrect because water vapor concentration variations may lag CO_2 variations (see text)

the correction while in the case of a closed-path system, only the water vapor flux correction is potentially needed as temperature-driven density fluctuations caused by a cooccurring sensible heat flux are attenuated by passage of the air sample through the intake tube (Rannik et al. 1997). If the closed-path analyzer reports dry mole fraction (corrects for water vapor fluctuations internally), then this correction does not need to be made by the experimenter. The impact of these corrections on annual sums can be substantial, varying strongly according to the site and the meteorological conditions. An evaluation of their order of magnitude showing the potential importance of these corrections as derived from Webb et al. (1980) and average climatological data (Bonan 2008) is presented in Table 7.3.

In closed-path sensors where the CO_2 concentration is not reported relative to dry air, the dilution effect of water vapor on CO_2 concentrations is different from what it is in the atmosphere or open-path sensors. As water vapor fluctuations are dampened and phase shifted in the tubes of the closed-path system, using the original formulation, that is, the true latent heat flux in the atmosphere, to correct the dilution of CO_2 concentrations by water vapor fluctuations will overcorrect the CO_2 flux. Ibrom et al. (2007b) found the magnitude of the overcorrection to be about 30 $g\,C\ m^{-2}\ year^{-1}$, a 21% underestimation of the annual carbon budget at the Danish beech forest, Sorø, although this effect will depend upon details of the closed-path system (tube length, flow rate, age of tubes). It is thus recommended that instead of applying the WPL water vapor correction to calculated fluxes from closed-path instruments, researchers instead apply the dilution correction by transforming densities into dry mixing ratios before computing the (co)variances. Many IRGAs measure both water vapor and CO_2 and some of them (LiCor 6262 or 7200) but not all (LiCor 7000) have the option available in the instrument software of correcting the CO2 output for water vapor density fluctuations.

Uncertainties remaining after this correction are relatively small, and can in the case of open-path sensors, be attributed to uncertainties in measured energy fluxes (Liu et al. 2006), and also CO_2 density (Serrano-Ortiz et al. 2008), which propagate through the correction. Liu et al. (2006) determined that minimizing both random and systematic errors in H was essential, as otherwise these have a potentially large negative impact on the accuracy of the "corrected" F_c. Serrano-Ortiz et al. (2008) calculated that underestimation of CO_2 density by just 5% (due to e.g., dirty open-path IRGA optics) resulted in a 13% overestimation (at the monthly time scale) of net C uptake by a semi-arid shrubland in Spain; these biases are most pronounced in ecosystems such as this where H is large at midday (see also Sect. 4.1.4.3).

7.3.2.7 Instrument Surface Heat Exchange

With respect to open-path analyzers, Burba et al. (2008) have demonstrated the influence of instrument surface heat exchange on measured CO_2 fluxes for a widely used instrument (Sect. 4.1.5.2). They showed that the surface of the open-path became warmer than ambient air during daytime, which induced natural convection and a nonzero vertical velocity in the instrument path. This leads to a flux overestimation that appears to be most pronounced in cold climates during the nongrowing season, and leads to a substantial overestimation of ecosystem C uptake. The error on half hourly fluxes varies from 40% to 770% in winter (when the absolute magnitude of fluxes is generally small) but never exceeds 5% in summer conditions. The impact on annual carbon budget was found to be around 90–100 g C m^{-2} $year^{-1}$ (14–16%) for crops (Burba et al. 2008) and 450 g C m^{-2} $year^{-1}$ (17%) for emissions from an urban area (Järvi et al. 2009).

To correct, it is recommended to apply the WPL correction with sensible heat flux measured inside the open-path rather than in the atmosphere (Burba et al. 2008). However, this procedure is seldom workable as this flux is generally not available. A series of empirical corrections were thus proposed by Burba et al. (2008) to overcome this problem. However, they are empirical, instrument-specific (LI-7500), and apply to vertically oriented instruments only.

The residual uncertainty remaining after application of Burba et al. (2008) correction is estimated to be about 5% on annual CO_2 fluxes (Burba et al. 2008); Järvi et al. (2009) estimated (by comparison with closed-path systems) that after correction for self-heating, errors were reduced from 140 to 20 gC m^{-2} in a temperate forest environment and from 330 to 30 g C m^{-2} in an urban environment.

7.3.3 Systematic Errors Associated with Data Processing

Sources of uncertainty associated with processing raw (5–20 Hz) data to obtain 30-min estimates of F_c include detrending, coordinate rotation, and both high- and low-frequency corrections (Kruijt et al. 2004). The uncertainties have been quantified individually and also together in the context of different software packages for data processing. A list of these errors, their order of magnitude, the recommended correction procedure, and the possible uncertainty remaining after the correction is given in Table 7.4.

7.3.3.1 Detrending and High-Pass Filtering

Detrending and high-pass filtering are carried out to reduce random or systematic noise in flux estimates caused by low-frequency bias in turbulent time series. The bias originates either from diurnal or sporadic changes in scalar concentrations, wind speed and direction; or from measurement artifacts such as sudden or transient instrument drifts (Aubinet et al. 2000).

High-pass filtering is unavoidable when calculating covariances from a finite measurement period (low-frequency eddies with periods longer than the averaging period are excluded from the calculated flux) and thus corrections are always required. Detrending of time series (by application of linear detrending or recursive filtering, see (Sect. 3.2.3.1)) is a special case of high-pass filtering, which is more effective than simple averaging, to exclude low-frequency variance. It is up to the investigator to choose the length of the measurement period and whether or not detrending is applied, or in other words, which part of the turbulent signal is deemed to be disturbed and thus needs to be replaced by theory and which not. It has been debated whether detrending is in conflict with common derivations of the flux equation, because only simple block averaging over the measurement period ensures that some flux terms disappear after Reynolds averaging. Despite this debate, detrending is still being widely used when separating the true turbulent flux from the possibly biased measured signal. However, if one interprets the detrended signal as the undisturbed turbulent signal, Reynolds averaging rules are compromised if the measured time series were used. In-depth discussion on this topic is beyond the scope of this overview; again, we aim to provide examples relating to the uncertainty associated with detrending.

Rannik and Vesala (1999) were the first to compare the effects of using three different high-pass filtering approaches (Sect. 3.2.3.1), block averaging (BA), linear detrending (LD) and autoregressive filtering (AF), on flux estimations from measured time series. They calculated theoretical random errors in covariance estimates from finite time series by assuming an exponential covariance function and found random errors of the CO_2 daily averaged fluxes ranging from 0.29 to 0.38 μmol m^{-2} s^{-1}, when using the different detrending methods as compared to 0.32 μmol m^{-2} s^{-1} as the theoretical value. Table 7.5 presents part of a multisite analysis from the European Infrastructure for Measurement of the European Carbon Cycle (IMECC) project where the random error and the systematic error were quantified on measured covariances using the "model residual" approach.

The general effects of using different high-pass filtering methods at this site are relatively small, provided appropriate corrections are made. The larger the filtering effect, the lower the random error. Using the most efficient filter (AF with $\tau = 225$ s) reduced the random error compared to plain averaging by 8%. Simple LD reduces random error by more than 6%.

Table 7.4 Systematic uncertainties due to processing

Error Type	Processing	Uncertainty on annual C exchange after processing	Remarks: Remaining questions	References
High-pass filtering	The use of block averaging (when applicable) allows reducing this error. Apply spectral correction with filtering specific transfer functions and site specific model spectra.	27 gC m^{-2} $year^{-1}$ Table 7.5	Detrending beyond simple block averaging can reduce random error by removing low frequency noise. There is large resulting uncertainty after spectral correction because the co-spectrum is not known in the low frequency range. Problem more critical at high measurement heights.	Section 3.2.3.1 Section 4.1.3.3
Coordinate rotation	Use of planar fit reduces this error.	15 g m^{-2} $year^{-1}$ Anthoni et al. (2004) about 0 Mahrt et al. (2000)	Planar fit method is more difficult to apply above changing surfaces.	Section 3.2.4
Gap filling	Many different algorithms for gap filling.	10–30 g m^{-2} $year^{-1}$ Richardson and Hollinger (2007)		Chapter 6
Flux partitioning	Different algorithms	Less than 10% Desai et al. (2008)		Chapter 9

Table 7.5 Systematic and random errors due to the choice of the detrending algorithm in an annual CO_2 flux data set above Beech forest, Sorø, Denmark (Pilegaard et al. 2003)

			AF	AF	AF
	BA	*LD*	$\tau = 225$ s	$\tau = 450$ s	$\tau = 900$ s
Absolute random error: RMSE of linear regression between F_n and $\hat{F}_n$ (μmol m^{-2} s^{-1})	3.32	3.11	3.05	3.09	3.15
Relative random error (% of the averaged RMSE)	5.5	−0.9	−2.9	−1.7	0.1
Absolute systematic error after correction (difference, in g C m^{-2} year^{-1}, between the annual CO_2 flux estimate to the average of the 5 estimates, −259 g C m^{-2} year^{-1})	−13	−2	14	4	0
Relative systematic error after correction (difference, in %, of the mean slopes of the regressions of $\hat{F}_n$ with $\overline{\hat{F}_n}$ and 1)	0.8	−0.2	−0.8	−0.2	0.0
Systematic error using Horst's peak frequency parameterization (Horst 1997) at this site (difference, in %, of slopes of the regression $\hat{F}_n^H$ with $\hat{F}_n^H$ and 1)	−2.4	−2.2	−1.8	−2.2	−2.5

The raw data were processed using five different high-pass filtering methods, block averaging (BA), linear detrending (LD), and autoregressive filtering (AF) with different time constants (τ) and corrected according to Rannik and Vesala (1999) using either model spectra that have been adapted to the site, yielding storage change corrected net CO_2 fluxes F_n, or using the parameterization of Horst (1997), (F_n^H). Random errors were estimated by the "model residual" approach, i.e., comparing F_n with the expected value $\hat{F}_n$, and systematic errors by comparing $\hat{F}_n$ from different data treatments. Expected net ecosystem exchange values, $\hat{F}_n$, were obtained by using a 2D binned moving averaging (look-up table approach of Falge et al. 2001b). $\overline{\hat{F}_n}$ is the average of $\hat{F}_n$ for the different data treatments

The remaining systematic differences between corrected CO_2 flux estimates from different detrending procedures were $< 1\%$, as shown by the regression slopes of $\hat{F}_n$ (expected flux values computed with one given procedure) with $\overline{\hat{F}_n}$ (average of the expected flux values computed with the different procedures. The intercepts were all smaller than 0.01 μmol m^{-2} s^{-1}) or $\pm$ 16 g m^{-2} year^{-1}, when looking at the annual sums. However, the choice of the model spectra mattered. Net flux estimates were 2–3% higher when using site-adapted cospectral models rather than the often-used Horst parameterization.

Compared to the other systematic errors in the estimation of carbon budgets, the additional systematic error resulting from detrending is small and can be largely removed when corrections from the appropriate cospectral models are applied. Since detrending also has the desirable property of reducing random error, we recommend its general use. The results presented here are from forest sites and similar investigations need to be performed with data from other sites, site conditions, and climates in order to develop a general picture about cospectral models as well as the benefits and disadvantages of detrending in terms of random and systematic flux estimate errors.

7.3.3.2 Coordinate Rotation

Coordinate rotation is intended to eliminate errors resulting from a sonic anemometer that is imperfectly mounted (i.e., not level), and differences between "streamline" and "planar fit" approaches are discussed in Sect. 3.2.4. Anthoni et al. (2004) found only differences of ± 15 g C m^{-2} year^{-1} in annual NEE when different coordinate rotation strategies were applied. Comparing different coordinate rotation methods, Mahrt et al. (2000) found that differences were insignificant. However, Finnigan et al. (2003) noted that coordinate rotation results in high-pass filtering of the scalar covariance, meaning the issues discussed in the previous section (and in Sect. 4.1.3.3) must be addressed. Forcing the mean vertical wind velocity to zero during short (15–30 min) averaging periods resulted in systematic underestimation of H and λE by 10–15%, contributing to the energy balance closure problem at three forest sites (Tumbarumba, Griffin, and Manaus) studied by Finnigan et al. (2003). The proposed solution is to use a longer period (up to 4 h or more) for averaging and coordinate rotation, so that the low-frequency component is not lost. However, Finnigan et al. (2003) did not discuss applying high-pass filtering corrections as an alternative to increasing the averaging time.

7.3.3.3 Gap Filling

There are numerous uncertainties associated with imputation of missing values in eddy flux time series ("gap filling"). For example, Richardson and Hollinger (2007) quantified the way in which random errors in measured fluxes are propagated through gap filling: when measurements are more uncertain (or sparse), there is correspondingly greater uncertainty in the filled values and thus the annual carbon budget. Richardson and Hollinger (2007) showed how this covariance could be quantified using Monte Carlo approaches.

There are also quasirandom uncertainties due to the timing and length of the gaps. Filling long gaps is a particular challenge, especially when these occur during periods when the ecosystem is actively changing (Falge et al. 2001a). This adds additional uncertainty to annual NEE integrals. For example, within deciduous forests, Richardson and Hollinger (2007) found that gaps of 3 weeks during the winter dormant season could be filled with reasonable accuracy, whereas a one-week gap during the spring green-up period was associated with an uncertainty of ± 30 g C m^{-2} year^{-1} at 95% confidence. Although the uncertainty associated with gaps of more than a day in length will depend on the specific site and data-year in question, Richardson and Hollinger (2007) reported values that were typically in the range of ± 10–30 g C m^{-2} year^{-1} when integrated across the entire year; this range is comparable in magnitude to the aggregate uncertainty due to random errors in measurements and as propagated through gap filling.

Finally, there are systematic uncertainties associated with choosing any particular algorithm for gap filling (Falge et al. 2001a; Moffat et al. 2007). The recent gap-filling comparison by Moffat et al. (2007) found that in most cases, the algorithms being used were approaching the noise limit (uncertainty) of the measurements. However, highly empirical approaches, including artificial neural networks and marginal distribution sampling, consistently performed the best (better than non-linear regression models, for example) across a range of forested European sites. At the annual time step, differences among algorithms were generally modest, as most produced annual NEE integrals that were within ± 25 g C m^{-2} year^{-1} of the mean.

By comparison, relatively little effort has been directed at developing and testing algorithms for gap-filling H and λE time series; the early analysis by Falge et al. (2001b) reported that H could vary by up to 140 MJ m^{-2} year^{-1} (19%), and λE by up to 205 MJ m^{-2} year^{-1} (39%) depending on the method used. As eddy flux data are increasingly being used to evaluate and improve ecosystem and land surface models, more emphasis will have to be placed on quantifying these uncertainties for water and energy fluxes.

7.3.3.4 Flux Partitioning

To obtain better insights into the process-level controls over NEE, there is considerable interest in partitioning the measured net flux of CO_2 to two component fluxes, gross ecosystem productivity (GEP) and total ecosystem respiration (R_{eco}) (see Chap. 9 for a review of methods). At night, the partitioning is simple, as $R_{eco} = $ NEE. During the day, the partitioning is dependent on the model used. Therefore there are substantial uncertainties associated with the resulting estimates of GEP and R_{eco} (Hagen et al. 2006; Richardson et al. 2006b). For example, daytime respiration can be estimated by extrapolation of nighttime measurements using some sort of temperature response function, but this approach does not account for daytime inhibition of foliar respiration, which is estimated to be 11–17% of GEP according to a modeling analysis by Wohlfahrt et al. (2005). An alternative method estimates daytime respiration from the *y*-axis intercept of a light response curve. These approaches are compared systematically by Lasslop et al. (2010). Desai et al. (2008) conducted a broad survey of partitioning algorithms; results indicated that most methods differed by less than10% in terms of annual integrals, although there was more variability among methods when additional gaps were added to the data. Patterns across sites tended to be consistent when a single algorithm was applied to all data sets, indicating that choice of partitioning algorithm mostly results in systematic bias of unknown magnitude, since the "true" GEP is not known. At shorter time scales (e.g., with respect to diurnal cycles), there was more variability among algorithms, particularly with respect to R_{eco} (see also Lasslop et al. 2010).

7.4 Closing Ecosystem Carbon Budgets

The above discussion of random errors and systematic biases in eddy covariance measurements of surface-atmosphere exchange raises questions about whether ecosystem C budgets derived from these measurements are in any way consistent with budgets estimated using other types of data, such as inventory-based approaches. Taking data uncertainties into account is critical for these kinds of comparisons. Schelhaas et al. (2004) reported that although the "best" estimates of C uptake by the Loobos pine forest differed by roughly 40% (eddy flux: 295 g C m^{-2} $year^{-1}$; inventory: 202 g C m^{-2} $year^{-1}$), confidence intervals were sufficiently wide that the two estimates were not inconsistent with each other. In an earlier study, Curtis et al. (2002) found that tower-based estimates of forest C uptake from four temperate deciduous forests were in "reasonable" agreement with estimates derived from changes in wood and soil C pools. At a fifth site (Walker Branch), where annual NEE integrals are suspect because of likely advection issues, the agreement was, not surprisingly, poor (eddy flux: 575 g C m^{-2} $year^{-1}$; inventory: 250 g C m^{-2} $year^{-1}$). Gough et al. (2008) emphasized the importance of making such comparisons over several years; there was poor agreement when annual tower- and inventory-based estimates of carbon storage were compared, but surprisingly close agreement (within 1%) with respect to 5-year averages.

Rather than comparing estimates of total C sequestration, Luyssaert et al. (2009) developed a two-stage "consistency cross-check" to compare C balance components based on flux tower and inventory methods. For 13 of the 16 sites examined, the data were judged to pass the test. While this does not necessarily imply that the absolute fluxes are accurate (consistency tests were based on estimating C balance closure terms, and examining ratios of different C balance components), it does give increased confidence in our use of eddy covariance fluxes for model evaluation and hypothesis testing, in spite of the substantial uncertainties described in this chapter.

7.5 Conclusion

Numerous previous studies, including Goulden et al. (1996), Lee et al. (1999), Anthoni et al. (1999, 2004), and Flanagan and Johnson (2005) have quantified various sources of flux measurement uncertainty and have attempted to attach confidence intervals to published annual sums of NEE; Baldocchi (2003) estimated that on ideal sites, the uncertainty in annual NEE was less than ± 50 g C m^{-2} $year^{-1}$, which is about the range that has been estimated in other studies. In this chapter, we have attempted to conduct a comprehensive evaluation of both random and systematic errors, with an emphasis on how these affect our use and interpretation of both 30-min and annual CO_2 fluxes. In our review, we have presented methods for quantifying the random errors, and have discussed the major sources of systematic error, and the degree to which these can be corrected. Of these, biases due to advection appear to represent the most significant "known unknown," and while we do not recommend that attempts be made to use measurements of the advective fluxes directly as a correction, ongoing efforts to quantify advective losses (and to strive to find sites where advection is less likely to be an issue) are clearly justified.

We conclude by noting that, given the challenges and research questions to which eddy covariance measurements of carbon, water, and energy fluxes are now being applied – particularly with respect to regional-to-continental scaling, C accounting and policy decision making, and data-model fusion – it is more important than ever that flux measurement uncertainties be quantified and reported. In one of the earliest reviews of flux measurement uncertainty, Moncrieff et al. (1996) remarked that in some fields it is common to separately report estimates of random (ε) and systematic (δ) uncertainties on measured quantities, for example, $x \pm \varepsilon + \delta$; while this approach has not been widely adopted within the eddy covariance community, it certainly has much to recommend (Aubinet et al. 2000).

Acknowledgments ADR and DYH acknowledge support from the Office of Science (BER), U.S. Department of Energy, through the Terrestrial Carbon Program under Interagency Agreement No. DE-AI02–07ER64355 and through the Northeastern Regional Center of the National Institute for Climatic Change Research.We also acknowledge funding by the European infrastructure project IMECC (http://imecc.ipsl.jussieu.fr/).

References

Abernethy RB, Benedict RP, Dowdell RB (1985) ASME measurement uncertainty. J Fluid Eng 107:161–164

Anthoni PM, Law BE, Unsworth MH (1999) Carbon and water vapor exchange of an open-canopied ponderosa pine ecosystem. Agric For Meteorol 95:151–168

Anthoni PM, Freibauer A, Kolle O, Schulze ED (2004) Winter wheat carbon exchange in Thuringia, Germany. Agric For Meteorol 121:55–67

Ascough JC, Maier HR, Ravalico JK, Strudley MW (2008) Future research challenges for incorporation of uncertainty in environmental and ecological decision-making. Ecol Model 219:383–399

Aubinet M (2008) Eddy covariance CO_2 flux measurements in nocturnal conditions: an analysis of the problem. Ecol Appl 18:1368–1378

Aubinet M, Grelle A, Ibrom A, Rannik Ü, Moncrieff J, Foken T, Kowalski AS, Martin PH, Berbigier P, Bernhofer C, Clement R, Elbers J, Granier A, Grünwald T, Morgenstern K, Pilegaard K, Rebmann C, Snijders W, Valentini R, Vesala T (2000) Estimates of the annual net carbon and water exchange of forests: the EUROFLUX methodology. Adv Ecol Res 30(30):113–175

Aubinet M, Heinesch B, Yernaux M (2003) Horizontal and vertical CO_2 advection in a sloping forest. Bound Layer Meteorol 108:397–417

Aubinet M, Feigenwinter C, Bernhofer C, Canepa E, Heinesch B, Lindroth A, Montagnani L, Rebmann C, Sedlak P, Van Gorsel E (2010) Advection is not the solution to the nighttime CO_2 closure problem – evidence from three different forests. Agric For Meteorol 150:655–664

Baldocchi DD (2003) Assessing the eddy covariance technique for evaluating carbon dioxide exchange rates of ecosystems: past, present and future. Glob Chang Biol 9:479–492

Baldocchi DD, Hicks BB, Meyers TP (1988) Measuring biosphere-atmosphere exchanges of biologically related gases with micrometeorological methods. Ecology 69:1331–1340
Baldocchi D, Valentini R, Running S, Oechel W, Dahlman R (1996) Strategies for measuring and modelling carbon dioxide and water vapour fluxes over terrestrial ecosystems. Glob Chang Biol 2:159–168
Barnett V, Lewis T (1994) Outliers in statistical data. Wiley, Chichester/New York, 604 pp
Barr AG, Morgenstern K, Black TA, McCaughey JH, Nesic Z (2006) Surface energy balance closure by the eddy-covariance method above three boreal forest stands and implications for the measurement of the CO_2 flux. Agric For Meteorol 140:322–337
Berger BW, Davis KJ, Yi CX, Bakwin PS, Zhao CL (2001) Long-term carbon dioxide fluxes from a very tall tower in a northern forest: flux measurement methodology. J Atmos Ocean Technol 18(4):529–542
Bonan G (2008) Ecological climatology, concepts and applications. Cambridge University Press, Cambridge, 550 p
Burba GG, McDermitt DK, Grelle A, Anderson DJ, Xu LK (2008) Addressing the influence of instrument surface heat exchange on the measurements of CO_2 flux from open-path gas analyzers. Glob Chang Biol 14:1854–1876
Coleman HW, Steele WG (2009) Experimentation, validation, and uncertainty analysis for engineers. Wiley, Hoboken, 317 p
Curtis PS et al (2002) Biometric and eddy-covariance based estimates of annual carbon storage in five eastern North American deciduous forests. Agric For Meteorol 113:3–19
Dabberdt WF, Lenschow DH, Horst TW, Zimmerman PR, Oncley SP, Delany AC (1993) Atmosphere-surface exchange measurements. Science 260:1472–1481
Davidson EA, Savage K, Verchot LV, Navarro R (2002) Minimizing artifacts and biases in chamber-based measurements of soil respiration. Agric For Meteorol 113:21–37
de Ligne A, Heinesch B, Aubinet M (2010) New transfer functions for correcting turbulent water vapour fluxes. Bound Layer Meteorol 137:205–221
Desai AR et al (2008) Cross-site evaluation of eddy covariance GPP and RE decomposition techniques. Agric For Meteorol 148:821–838
Dragoni D, Schmid HP, Grimmond CSB, Loescher HW (2007) Uncertainty of annual net ecosystem productivity estimated using eddy covariance flux measurements. J Geophys Res Atmos 112: Art. No. D17102
Dyer AJ, Garratt JR, Francey RJ, Mcilroy IC, Bacon NE, Hyson P, Bradley EF, Denmead OT, Tsvang LR, Volkov YA, Koprov BM, Elagina LG, Sahashi K, Monji N, Hanafusa T, Tsukamoto O, Frenzen P, Hicks BB, Wesely M, Miyake M, Shaw W (1982) An International Turbulence Comparison Experiment (ITCE 1976). Bound Layer Meteorol 24:181–209
Eugster W, McFadden JP, Chapin ES (1997) A comparative approach to regional variation in surface fluxes using mobile eddy correlation towers. Bound Layer Meteorol 85:293–307
Falge E et al (2001a) Gap filling strategies for long term energy flux data sets. Agric For Meteorol 107:71–77
Falge E, Baldocchi D, Olson R, Anthoni P, Aubinet M, Bernhofer C, Burba G, Ceulemans R, Clement R, Dolman H, Granier A, Gross P, Grünwald T, Hollinger D, Jensen NO, Katul G, Keronen P, Kowalski A, Ta Lai C, Law B, Meyers T, Moncrieff J, Moors EJ, Munger JW, Pilegaard K, Rannik Ü, Rebmann C, Suyker A, Tenhunen J, Tu K, Verma S, Vesala T, Wilson K, Wofsy S (2001b) Gap filling strategies for defensible annual sums of net ecosystem exchange. Agric For Meteorol 107:43–69
Feigenwinter C, Bernhofer C, Eichelmann U, Heinesch B, Hertel M, Janous D, Kolle O, Lagergren F, Lindroth A, Minerbi S, Moderow U, Mölder M, Montagnani L, Queck R, Rebmann C, Vestin P, Yernaux M, Zeri M, Ziegler W, Aubinet M (2008) Comparison of horizontal and vertical advective CO_2 fluxes at three forest sites. Agric For Meteorol 148:12–24
Finkelstein PL, Sims PF (2001) Sampling error in eddy correlation flux measurements. J Geophys Res Atmos 106:3503–3509
Finnigan J (2000) Turbulence in plant canopies. Ann Rev Fluid Mech 32:519–571

Finnigan J (2008) An introduction to flux measurements in difficult conditions. Ecol Appl 18:1340–1350

Finnigan JJ, Clement R, Malhi Y, Leuning R, Cleugh HA (2003) A re-evaluation of long-term flux measurement techniques – part I: averaging and coordinate rotation. Bound Layer Meteorol 107:1–48

Flanagan LB, Johnson BG (2005) Interacting effects of temperature, soil moisture and plant biomass production on ecosystem respiration in a northern temperate grassland. Agric For Meteorol 130:237–253

Foken T (2008) The energy balance closure problem: an overview. Ecol Appl 18:1351–1367

Foken T, Wichura B (1996) Tools for quality assessment of surface-based flux measurements. Agric For Meteorol 78:83–105

Foken T, Wimmer F, Mauder M, Thomas C, Liebethal C (2006) Some aspects of the energy balance closure problem. Atmos Chem Phys 6:4395–4402

Fox A et al (2009) The REFLEX project: comparing different algorithms and implementations for the inversion of a terrestrial ecosystem model against eddy covariance data. Agric For Meteorol 149:1597–1615

Franks SW, Beven KJ, Quinn PF, Wright IR (1997) On the sensitivity of soil–vegetation–atmosphere transfer (SVAT) schemes: equifinality and the problem of robust calibration. Agric For Meteorol 86:63–75

Gough CM, Vogel CS, Schmid HP, Su HB, Curtis PS (2008) Multi-year convergence of biometric and meteorological estimates of forest carbon storage. Agric For Meteorol 148:158–170

Goulden ML, Munger JW, Fan SM, Daube BC, Wofsy SC (1996) Measurements of carbon sequestration by long-term eddy covariance: methods and a critical evaluation of accuracy. Glob Chang Biol 2:169–182

Grant RF et al (2005) Intercomparison of techniques to model high temperature effects on CO_2 and energy exchange in temperate and boreal coniferous forests. Ecol Model 188:217–252

Hagen SC, Braswell BH, Linder E, Frolking S, Richardson AD, Hollinger DY (2006) Statistical uncertainty of eddy flux-based estimates of gross ecosystem carbon exchange at Howland Forest, Maine. Journal of Geophysical Research-Atmospheres 111: Art. No. D08S03

Haslwanter A, Hammerle A, Wohlfahrt G (2009) Open-path vs. Closed-path eddy covariance measurements of the net ecosystem carbon dioxide and water vapour exchange: a long-term perspective. Agric For Meteorol 149:291–302

Hollinger DY, Richardson AD (2005) Uncertainty in eddy covariance measurements and its application to physiological models. Tree Physiol 25:873–885

Hollinger DY et al (2004) Spatial and temporal variability in forest-atmosphere CO_2 exchange. Glob Chang Biol 10:1689–1706

Horst TW (1997) A simple formula for attenuation of eddy fluxes measured with first-order-response scalar sensors. Bound Layer Meteorol 82:219–233

Ibrom A, Jarvis PG, Clement RB, Morgenstern K, Oltchev A, Medlyn B, Wang YP, Wingate L, Moncrieff J, Gravenhorst G (2006) A comparative analysis of simulated and observed photosynthetic CO_2 uptake in two coniferous forest canopies. Tree Physiol 26:845–864

Ibrom A, Dellwik E, Flyvbjerg H, Jensen NO, Pilegaard K (2007a) Strong low-pass filtering effects on water vapour flux measurements with closed-path eddy correlation systems. Agric For Meteorol 147:140–156

Ibrom A, Dellwik E, Larsen SE, Pilegaard K (2007b) On the use of the Webb – Pearman – Leuning – theory for closed-path eddy correlation measurements. Tellus B 59B:937–946

ISO/IEC (International Organization for Standardization) (2008) ISO/IEC Guide 98-3: 2008 – Guide to the expression of uncertainty in measurement, Geneva, Switzerland

Järvi L, Mammarella I, Eugster W, Ibrom A, Siivola E, Dellwik E, Keronen P, Burba G, Vesala T (2009) Comparison of net CO_2 fluxes measured with open- and closed-path infrared gas analyzers in urban complex environment. Boreal Environ Res 14:499–514

Kanda M, Ianagaki A, Letzel MO, Raasch S, Wataqnabe T (2004) LES study of the energy imbalance problem with eddy covariance fluxes. Bound Layer Meteorol 110:381–404

Katul G et al (1999) Spatial variability of turbulent fluxes in the roughness sublayer of an even-aged pine forest. Bound Layer Meteorol 93:1–28

Kline SJ, McClintock FA (1953) Describing uncertainties in single-sample experiments. Mech Eng 75:3–7

Kruijt B et al (2004) The robustness of eddy correlation fluxes for Amazon rain forest conditions. Ecol Appl 14:S101–S113

Lasslop G, Reichstein M, Kattge J, Papale D (2008) Influences of observation errors in eddy flux data on inverse model parameter estimation. Biogeosciences 5:1311–1324

Lasslop G, Reichstein M, Papale D, Richardson AD, Arneth A, Barr A, Stoy P, Wohlfahrt G (2010) Separation of net ecosystem exchange into assimilation and respiration using a light response curve approach: critical issues and global evaluation. Glob Chang Biol 16:187–208

Lee XH, Fuentes JD, Staebler RM, Neumann HH (1999) Long-term observation of the atmospheric exchange of CO_2 with a temperate deciduous forest in southern Ontario, Canada. J Geophys Res Atmos 104:15975–15984

Lenschow DH, Mann J, Kristensen L (1994) How long is long enough when measuring fluxes and other turbulence statistics. J Atmos Ocean Technol 11:661–673

Leuning R, Zegelin SJ, Jones K, Keith H, Hughes D (2008) Measurement of horizontal and vertical advection of CO2 within a forest canopy. Agric For Meteorol 148:1777–1797

Liu HP, Randerson JT, Lindfors J, Massman WJ, Foken T (2006) Consequences of incomplete surface energy balance closure for CO_2 fluxes from open-path CO_2/H_2O infrared gas analysers. Bound Layer Meteorol 120:65–85

Liu M, He HL, Yu GR, Luo YQ, Sun XM, Wang HM (2009) Uncertainty analysis of CO_2 flux components in subtropical evergreen coniferous plantation. Sci China Ser D Earth Sci 52:257–268

Loescher HW et al (2005) Comparison of temperature and wind statistics in contrasting environments among different sonic anemometer-thermometers. Agric For Meteorol 133:119–139

Loescher HW, Law BE, Mahrt L, Hollinger DY, Campbell J, Wofsy SC (2006) Uncertainties in, and interpretation of, carbon flux estimates using the eddy covariance technique. J Geophys Res Atmos 111: Art. No. D21S90

Luyssaert S et al. (2009) Toward a consistency cross-check of eddy covariance flux-based and biometric estimates of ecosystem carbon balance. Glob Biogeochem Cycles 23: Art. No. GB3009

Mahrt L, Lee X, Black A, Neumann H, Staebler RM (2000) Nocturnal mixing in a forest subcanopy. Agric For Meteorol 101:67–78

Mammarella I, Launiainen S, Gronholm T, Keronen P, Pumpanen J, Rannik U, Vesala T (2009) Relative humidity effect on the high-frequency attenuation of water vapor flux measured by a closed-path eddy covariance system. J Atmos Ocean Technol 26:1852–1866

Mann J, Lenschow DH (1994) Errors in airborne flux measurements. J Geophys Res Atmos 99:14519–14526

Massman WJ (2000) A simple method for estimating frequency response corrections for eddy covariance systems. Agric For Meteorol 104:185–198

Massman W, Ibrom A (2008) Attenuation of trace gas fluctuations associated with turbulent flow in tubes: application to sampling water vapor with closed-path eddy covariance systems. In: European Geosciences Union General Assembly, 2008. EGU, Vienna, EGU2008-A-02259

Massman WJ, Lee X (2002) Eddy covariance flux corrections and uncertainties in long-term studies of carbon and energy exchanges. Agric For Meteorol 113:121–144

Medlyn BE, Robinson AP, Clement R, McMurtrie RE (2005) On the validation of models of forest CO2 exchange using eddy covariance data: some perils and pitfalls. Tree Physiol 25:839–857

Moffat AM et al (2007) Comprehensive comparison of gap-filling techniques for eddy covariance net carbon fluxes. Agric For Meteorol 147:209–232

Moncrieff JB, Malhi Y, Leuning R (1996) The propagation of errors in long-term measurements of land-atmosphere fluxes of carbon and water. Glob Chang Biol 2:231–240

Moureaux C, Debacq A, Hoyaux J, Suleau M, Tourneur D, Vancutsem F, Bodson B, Aubinet M (2008) Carbon balance assessment of a Belgian winter wheat crop (*Triticum aestivum* L.). Glob Chang Biol 14(6):1353–1366

Nakai T, van der Molen MK, Gash JHC, Kodama Y (2006) Correction of sonic anemometer angle of attack errors. Agric For Meteorol 136:19–30

Ocheltree TW, Loescher HW (2007) Design of the AmeriFlux portable eddy covariance system and uncertainty analysis of carbon measurements. J Atmos Ocean Technol 24:1389–1406

Oren R et al (2006) Estimating the uncertainty in annual net ecosystem carbon exchange: spatial variation in turbulent fluxes and sampling errors in eddy-covariance measurements. Glob Chang Biol 12:883–896

Papale D, Reichstein M, Canfora E, Aubinet M, Bernhofer C, Longdoz B, Kutsch W, Rambal S, Valentini R, Vesala T, Yakir D (2006) Towards a standardized processing of Net Ecosystem Exchange measured with eddy covariance technique: algorithms and uncertainty estimation. Biogeosciences 3:571–583

Pilegaard K, Mikkelsen TN, Beier C, Jensen NO, Ambus P, Ro-Poulsen H (2003) Field measurements of atmosphere-biosphere interactions in a Danish beech forest. Boreal Environ Res 8:315–333

Rannik Ü, Vesala T (1999) Autoregressive filtering versus linear detrending in estimation of fluxes by the eddy covariance method. Bound Layer Meteorol 91:259–280

Rannik Ü, Vesala T, Keskinen R (1997) On the damping of temperature fluctuations in a circular tube relevant to the eddy covariance measurement technique. J Geophys Res Atmos 102(11D):12789–12794

Rannik U, Kolari P, Vesala T, Hari P (2006) Uncertainties in measurement and modelling of net ecosystem exchange of a forest. Agric For Meteorol 138:244–257

Raupach MR et al (2005) Model-data synthesis in terrestrial carbon observation: methods, data requirements and data uncertainty specifications. Glob Chang Biol 11:378–397

Rebmann C et al (2005) Quality analysis applied on eddy covariance measurements at complex forest sites using footprint modelling. Theor Appl Climatol 80(2–4):121–141

Richardson AD, Hollinger DY (2005) Statistical modeling of ecosystem respiration using eddy covariance data: maximum likelihood parameter estimation, and Monte Carlo simulation of model and parameter uncertainty, applied to three simple models. Agric For Meteorol 131:191–208

Richardson AD, Hollinger DY (2007) A method to estimate the additional uncertainty in gap-filled NEE resulting from long gaps in the CO_2 flux record. Agric For Meteorol 147:199–208

Richardson AD et al (2006a) Comparing simple respiration models for eddy flux and dynamic chamber data. Agric For Meteorol 141:219–234

Richardson AD, Hollinger DY, Burba GG, Davis KJ, Flanagan LB, Katul GG, Munger JW, Ricciuto DM, Stoy PC, Suyker AE, Verma SB, Wofsy SC (2006b) A multi-site analysis of random error in tower-based measurements of carbon and energy fluxes. Agric For Meteorol 136:1–18

Richardson AD et al (2008) Statistical properties of random CO_2 flux measurement uncertainty inferred from model residuals. Agric For Meteorol 148:38–50

Savage K, Davidson EA, Richardson AD (2008) A conceptual and practical approach to data quality and analysis procedures for high-frequency soil respiration measurements. Funct Ecol 22:1000–1007

Schelhaas MJ, Nabuurs GJ, Jans W, Moors E, Sabate S, Daamen WP (2004) Closing the carbon budget of a Scots pine forest in the Netherlands. Clim Chang 67:309–328

Schmid HP, Su HB, Vogel CS, Curtis PS (2003) Ecosystem-atmosphere exchange of carbon dioxide over a mixed hardwood forest in northern lower Michigan. J Geophys Res Atmos 108(D14): Art. No. 4417

Serrano-Ortiz P, Kowalski AS, Domingo F, Ruiz B, Alados-Arboledas L (2008) Consequences of uncertainties in CO2 density for estimating net ecosystem CO_2 exchange by open-path eddy covariance. Bound Layer Meteorol 126:209–218

Smith WK, Hollinger DY (1991) Stomatal behavior. In: Lassoie JP, Hinckley TM (eds) Techniques and approaches in forest tree ecophysiology. CRC Press, Boca Raton, pp 141–174

Staebler RM, Fitzjarrald DR (2004) Observing subcanopy CO_2 advection. Agric For Meteorol 122:139–156

Stauch VJ, Jarvis AJ, Schulz K (2008) Estimation of net carbon exchange using eddy covariance CO_2 flux observations and a stochastic model. J Geophys Res Atmos 113: Art. No. D03101

Su HB, Schmid HP, Grimmond CSB, Vogel CS, Oliphant AJ (2004) Spectral characteristics and correction of long-term eddy-covariance measurements over two mixed hardwood forests in non-flat terrain. Bound Layer Meteorol 110:213–253

Taylor JR (1991) An introduction to error analysis. University Science, Sausalito

Trudinger CM et al (2007) OptIC project: an intercomparison of optimization techniques for parameter estimation in terrestrial biogeochemical models. J Geophys Res Biogeosci 112: Art. No. G02027

Twine TE, Kustas WP, Norman JM, Cook DR, Houser PR, Meyers TP, Prueger JH, Starks PJ, Wesely ML (2000) Correcting eddy-covariance flux underestimates over a grassland. Agric For Meteorol 103:279–300

Wang YP, Trudinger CM, Enting IG (2009) A review of applications of model-data fusion to studies of terrestrial carbon fluxes at different scales. Agric For Meteorol 149:1829–1842

Webb EK, Pearman GI, Leuning R (1980) Correction of flux measurements for density effects due to heat and water-vapor transfer. Q J R Meteorol Soc 106:85–100

Wesely ML, Hart RL (1985) Variability of short term eddy-correlation estimates of mass exchange. In: Hutchinson BA, Hicks BB (eds) The forest-atmosphere interaction. D. Reidel, Dordrecht, pp 591–612

Williams M et al (2009) Improving land surface models with FLUXNET data. Biogeosciences 6:1341–1359

Wilson KB, Goldstein AH, Falge E, Aubinet M, Baldocchi D, Berbigier P, Bernhofer C, Ceulemans R, Dolman H, Field C, Grelle A, Law B, Meyers T, Moncrieff J, Monson R, Oechel W, Tenhunen J, Valentini R, Verma S (2002) Energy balance closure at FLUXNET sites. Agric For Meteorol 113:223–243

Wohlfahrt G, Bahn M, Haslwanter A, Newesely C, Cernusca A (2005) Estimation of daytime ecosystem respiration to determine gross primary production of a mountain meadow. Agric For Meteorol 130:13–25

Yanai RD, Battles JJ, Richardson AD, Rastetter EB, Wood DM, Blodgett C (2010) Estimating uncertainty in ecosystem budget calculations. Ecosystems 13:239–248

Chapter 8
Footprint Analysis

Üllar Rannik, Andrey Sogachev, Thomas Foken, Mathias Göckede, Natascha Kljun, Monique Y. Leclerc, and Timo Vesala

8.1 Concept of Footprint

Ideally a flux tower should be installed on a homogeneous and flat terrain. The surface should be physically homogeneous (same forest height and thermal properties) as well as be covered by same tree species, or in the case of the mixed forest, the distribution of the different species should be even ("well-mixed"). The fetch, the outreach of the homogeneous surface, should be longer than the extension of source area of the measurement (footprint). However, many sites are not homogeneous enough in all directions from the tower. In the case of an inhomogeneous surface, knowledge of both the source area and strength is needed

Ü. Rannik (✉) • T. Vesala
Department of Physics, University of Helsinki, Helsinki, Finland
e-mail: ullar.rannik@heuristica.ee; timo.vesala@helsinki.fi

A. Sogachev
Risø National Laboratory for Sustainable Energy, Technical University of Denmark, Roskilde, Denmark
e-mail: anso@risoe.dtu.dk

T. Foken
Department of Micrometeorology, University of Bayreuth, Bayreuth, Germany
e-mail: thomas.foken@uni-bayreuth.de

M. Göckede
Department of Forest Ecosystems & Society, Oregon State University, Corvallis, OR, USA
e-mail: mathias.goeckede@oregonstate.edu

N. Kljun
Department of Geography, Swansea University, Swansea, UK
e-mail: n.kljun@swansea.ac.uk

M.Y. Leclerc
Laboratory for Environmental Physics, The University of Georgia, Griffin, GA, USA
e-mail: mleclerc@uga.edu

M. Aubinet et al. (eds.), *Eddy Covariance: A Practical Guide to Measurement and Data Analysis*, Springer Atmospheric Sciences, DOI 10.1007/978-94-007-2351-1_8,

to interpret the measured signal. Note that inhomogeneity modifies the footprint by modifying the turbulent flow field. Thus, strictly speaking, any method not accounting for heterogeneities is useless for source area estimation. Namely, either the footprint model is fundamentally wrong because of the implicit assumption of homogeneity or, in the case of the fully homogeneous case, the outcome is trivial and no estimation is needed. Nevertheless, footprint models based on the assumption of horizontally homogeneous turbulence field serve as first approximation for evaluation of source contribution to measured flux in real observation conditions. An alternative is to take the flow inhomogeneity into account in footprint estimation by models capable of simulating such flow fields (cf. Sect. 8.4.1).

The footprint defines the field of view of the flux/concentration sensor and reflects the influence of the surface on the measured turbulent flux (or concentration). Strictly speaking, a source area is the fraction of the surface (mostly upwind) containing effective sources and sinks contributing to a measurement point (see Kljun et al. 2002). The footprint is then defined as the relative contribution from each element of the surface area source/sink to the measured vertical flux or concentration (see Schuepp et al. 1990; Leclerc and Thurtell 1990). Functions describing the relationship between the spatial distribution of surface sources/sinks and a signal are called the footprint function or the source weight function as shown in (Horst and Weil 1992, 1994; see also Schmid 1994 for details). The fundamental definition of the footprint function ϕ is given by the integral equation of diffusion (Wilson and Swaters 1991; see also Pasquill and Smith 1983):

$$\eta = \int_{\mathbb{R}} \phi\left(\vec{x}, \vec{x}'\right) Q(\vec{x}') d\vec{x}' \tag{8.1}$$

where η is the quantity being measured at location $\vec{x}$ (note that $\vec{x}$ is a vector) and $Q(\vec{x}')$ is the source emission rate/sink strength in the surface-vegetation volume R. η can be the concentration or the vertical eddy flux and ϕ is then concentration or flux footprint function, respectively.

The footprint problem essentially deals with the calculation of the relative contribution to the mean concentration $<c_s>$ or flux $<wc_s>$, with $<>$ denoting ensemble averaging, at a fixed point in the presence of an arbitrary given source of a compound. Concentration footprints tend to be generally longer than flux footprints (cf. Sect. 8.2.4). The source area naturally depends on measurement height and wind direction. The footprint is also sensitive to both atmospheric stability and surface roughness, as first pointed out by Leclerc and Thurtell (1990). The stability dependence of crosswind-integrated flux footprint function for four different stability regimes is illustrated in Fig. 8.1. It can be seen that the peak location is closer to the receptor and less skewed in the upstream direction with increasingly convective conditions. In unstable conditions, the turbulence intensity is high, resulting in the upward transport of any compound and a shorter travel distance/time. Typically, the location of the footprint peak ranges from a few times the measurement height (unstable) to a few dozen times (stable). In the lateral direction, the stability influences footprints in a similar fashion. Note also the small contribution of the

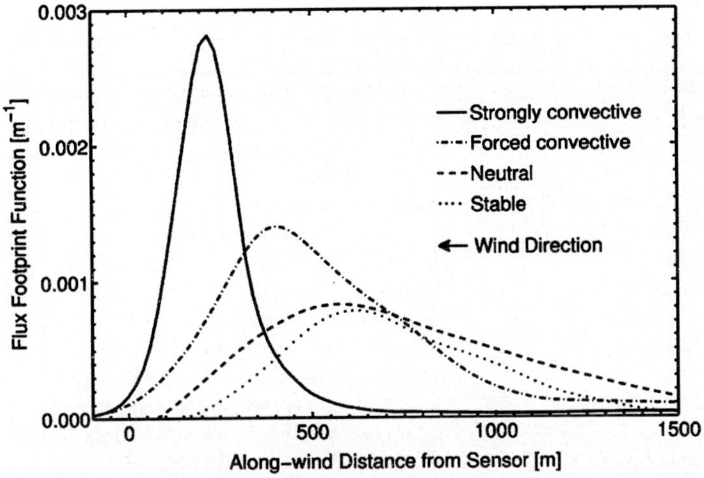

Fig. 8.1 Crosswind-integrated footprint for flux measurements for four different cases of stabilities (strongly convective, forced convective, neutral, and stable conditions; measurement height: 50 m, roughness length: 0.05 m) obtained by Lagrangian simulation according to Kljun et al. (2002)

downwind turbulent diffusion in convective cases. Mathematically, the surface area of influence on the entire flux goes to infinity and thus one must always define the %-level for the source area (see Schmid 1994). Often 50%, 75%, or 90% source areas contributing to a point flux measurement are considered.

The concentration footprint function is always between 0 and 1 whereas the flux footprint function may be even negative for a complex, convergent flow over a hill (Finnigan 2004). In a horizontally homogeneous shear flow, the flux footprint ϕ_f does satisfy $1 > \phi_f > 0$, as it is the case always for the concentration footprint. The vertical distribution of the source/sink can also lead to an anomalous behavior (e.g., Markkanen et al. 2003). Then, the flux footprint represents in fact a combined footprint function that is a source strength-weighted average of the footprints of individual layers. Because of the principle of superposition, the combined function may become negative if one or more of the layers have a source strength that is opposite in sign to the net flux between vegetation and atmosphere (Lee 2003). The combined function is not anymore a footprint function in the sense of Eq. 8.1 and we suggest that it would be called *(normalized) flux contribution function* (see also Markkanen et al. 2003).

The determination of the footprint function ϕ is not a straightforward task and several theoretical approaches have been derived over the previous decades. They can be classified into four categories: (1) analytical models, (2) Lagrangian stochastic particle dispersion models, (3) large-eddy simulations, and (4) ensemble-averaged closure models. Additionally, parameterizations of some of these approaches have been developed, simplifying the original algorithms for use in practical applications (e.g., Horst and Weil 1992, 1994; Schmid 1994; Hsieh et al.

Table 8.1 Overview about the most important footprint models (if no remark: analytical model)

Author	Remarks
Pasquill (1972)	First model description, concept of effective fetch
Gash (1986)	Neutral stratification, concept of cumulative fetch
Schuepp et al. (1990)	Use of source areas, but neutral stratification and averaged wind velocity
Leclerc and Thurtell (1990)	Lagrangian footprint model
Horst and Weil (1992)	One-dimensional footprint model
Schmid (1994, 1997)	Separation of footprints for scalars and fluxes
Leclerc et al. (1997)	LES model for footprints
Baldocchi (1997)	Footprint model within forests
Rannik et al. (2000, 2003)	Lagrangian model for forests
Kormann and Meixner (2001)	Analytical model with exponential wind profile
Kljun et al. (2002)	Three-dimensional Lagrangian model for various turbulence stratifications with backward trajectories
Sogachev and Lloyd (2004)	Boundary-layer model with 1.5 order closure
Sogachev et al. (2004)	Footprint estimates for a non-flat topography
Strong et al. (2004)	Footprint model with reactive chemical compounds
Cai and Leclerc (2007)	Footprints from backward and forward in-time particle simulations driven with LES data
Klaassen and Sogachev (2006)	Footprint estimates for a forest edge
Vesala et al. (2008a)	Footprint estimates for a complex urban surface
Steinfeld et al. (2008)	Footprint model with LES-embedded particles

Adopted from Foken (2008) and Vesala et al. (2010)

2000; Kljun et al. 2004a). The parameterization by Kljun et al. (2004a) is available at http://footprint.kljun.net. The SCADIS closure model (cf. Sect. 8.4.1) was also simplified (two-dimensional domain, neutral stratification, flat topography, etc.) and provided with a user-friendly menu. The operating manual for the set of basic and new created programs, called "Footprint calculator," was presented by Sogachev and Sedletski (2006) and is available freely by request to the authors or from Nordic Centre for Studies of Ecosystem Carbon Exchange (NECC) site (http://www.necc.nu/NECC/home.asp). A thorough overview over the development of the footprint concept is given in Schmid (2002) with Foken and Leclerc (2004), Vesala et al. (2008b), and Vesala et al. (2010) providing more recent information on the subject. Table 8.1 lists the most important studies on footprint modeling.

8.2 Footprint Models for Atmospheric Boundary Layer

8.2.1 Analytical Footprint Models

The first concept to estimate a two-dimensional source weight distribution has been proposed by Pasquill (1972), using a simple Gaussian model to describe the transfer function between sources and measurement point. Schmid and Oke (1988, 1990) improved Pasquill's approach by including a diffusion model based on the Monin-Obukhov similarity theory, with an analytical solution of the latter proposed by van Ulden (1978). The first paper, describing a simple analytical model to the diffusion equation using a constant velocity profile and neutral conditions, was presented by Gash (1986). The same approach was later adapted by Schuepp et al. (1990) in a companion paper to Leclerc and Thurtell (1990) to describe the concept of "flux footprint." Flux footprint is the assessment of the individual signatures from a particular source either on the ground, in the understory, or in the canopy crown to a point flux measurement.

With the addition of realistic velocity profiles and stability dependence, Horst and Weil's analytical models (1992, 1994) further expanded the scope of this approach. Again, their analytical solution was based on van Ulden (1978). The analytical footprint models by Horst and Weil (1992, 1994) are not explicit and require numerical solution, although Horst and Weil (1994) have proposed an approximate analytical solution. To date, Schmid's flux and concentration footprint models (1994, 1997) have been widely used. The two-dimensional extension of these models has generated additional insight into the interpretation of experimental data collected over patchy surfaces.

It should be mentioned that the above models, however compact in their formulation, suffer from numerical instabilities and generally perform poorly in stable conditions.

Later, Haenel and Grünhage (2001) and Kormann and Meixner (2001) have proposed explicit analytical expressions for flux footprint functions. Haenel and Grünhage (2001) used power law profiles for wind speed and eddy diffusivity to obtain an analytical solution. Monin-Obukhov similarity relationships were only introduced in a later stage of their derivation. Kormann and Meixner (2001) followed a similar approach, starting with power law profiles for wind speed and eddy diffusivity and introducing Monin-Obukhov similarity profiles by fitting the power law profiles to similarity profiles at later stage. As summarized by Schmid (2002), physical accuracy was sacrified for simplifications in the derivation of explicit analytical expressions. Therefore, the model by Horst and Weil (1992, 1994) is suggested for Atmospheric Surface Layer (ASL) conditions.

Analytical footprint models, as all other footprint models described here, are based on the assumption of steady-state conditions during the course of the flux period analyzed. They furthermore assume that no contribution to a point flux is possible by downwind sources and are unable to include the influence of nonlocal forcings to flux measurements. The latter point has been shown to be incorrect (Kljun et al. 2002; Leclerc et al. 2003a). Implicit in the use of these equations are the assumptions of (1) a horizontally homogeneous turbulence field; (2) no vertical advection; (3) the Monin-Obukhov similarity theory being applicable to the layer of air above the tower; and (4) all eddy contributions from the flux being contained within a sampling period. Recent findings for nocturnal atmospheric boundary layer (Karipot et al. 2006, 2008a, b) and by Prabha et al. (2007, 2008b) have shown that vertical advection is modulating the flux response. This fact is currently not included in footprint formulations.

The original footprint concept and its analytical solutions assigned the unit source strength to upwind surface sources. Most of the analytical solutions used have been one-dimensional with the implicit assumption that the sources are infinite in crosswind direction. In practice, this is certainly an issue of relevance as few sources/sinks cover a large enough area to allow neglecting the lateral component of the flow. The lateral diffusion gains significance with decreasing windspeed, that is, the lateral turbulence intensities become larger as the wind meanders.

8.2.2 Lagrangian Stochastic Approach

The Lagrangian stochastic (LS) models describe the diffusion of a scalar by means of a stochastic differential equation, a generalized Langevin equation,

$$d\mathbf{X}(t) = \mathbf{V}(t)dt$$
$$d\mathbf{V}(t) = \mathbf{a}(t, \mathbf{X}(t), \mathbf{V}(t))dt + \sqrt{C_0\bar{\varepsilon}(\mathbf{X}(t), t)}d\mathbf{W}(t), \quad (8.2)$$

where $\boldsymbol{X}(t)$ and $\boldsymbol{V}(t)$ denote trajectory coordinates and velocity as a function of time t, C_0 is the Kolmogorov constant, $\bar{\varepsilon}$ is the mean dissipation rate of turbulent kinetic energy (TKE), and $\boldsymbol{W}(t)$ describes the three-dimensional Wiener process. This equation determines the evolution of a Lagrangian trajectory in space and time by combining the evolution of trajectory as a sum of deterministic drift $\mathbf{a}$ and random terms. The drift term is to be specified for each LS model constructed for specific flow regime (Thomson 1987).

The Lagrangian stochastic approach can be applied to any turbulence regime, thus allowing footprint calculations for various atmospheric boundary-layer flow regimes. For example, in the convective boundary layer, turbulence statistics are typically non-Gaussian and for realistic dispersion simulations, a non-Gaussian trajectory model has to be applied. An indication of the departure from Gaussianity is often obtained using the turbulence velocity skewness; for instance, in convective boundary layers, the vertical velocity skewness is typically 0.3 while a neutral canopy layer can exhibit negative vertical velocity skewness as large as −2.0 (Leclerc et al. 1991; Finnigan 2000). However, most Lagrangian trajectory models fulfill the main criterion for construction of Lagrangian stochastic models, the well-mixed condition (Thomson 1987), for only one given turbulence regime.

It should be noted, however, that the Lagrangian stochastic models are not uniquely defined for atmospheric flow conditions. Even in the case of homogeneous but anisotropic turbulence, there are several different stochastic models which satisfy the well-mixed condition (Thomson 1987; Sabelfeld and Kurbanmuradov 1998). This is often called the uniqueness problem (for details, see the discussion in Kurbanmuradov et al. 1999, 2001; Kurbanmuradov and Sabelfeld 2000). In addition to well-mixed condition by Thomson (1987), trajectory curvature has been proposed as the additional criterion to select the most appropriate Lagrangian stochastic model (Wilson and Flesch 1997), but this additional criterion does not define the unique model (Sawford 1999).

The stochastic Lagrangian method is, nevertheless, very convenient in footprint application: once the form of the parameterization is chosen, the stochastic Langevin-type equation is solved by a very simple scheme (e.g., Sawford 1985; Thomson 1987; Sabelfeld and Kurbanmuradov 1990). The approach needs only a one-point probability density function (pdf) of the Eulerian velocity field. The Lagrangian stochastic trajectory model together with appropriate simulation methods and corresponding estimators for concentration or flux footprints are usually merged into a Lagrangian footprint model. For a detailed overview of the estimation of concentrations and fluxes by the Lagrangian stochastic method, the concentration and flux footprints in particular, see Kurbanmurdov et al. (2001).

The rather long computing times due to a large number of trajectories required for producing statistically reliable results is an unavoidable weakness of Lagrangian stochastic footprint models. To overcome this, Hsieh et al. (2000) proposed an analytical model derived from Lagrangian model results. More recently, a simple parameterization based on a Lagrangian footprint model was proposed by Kljun et al. (2004a). This parameterization allows the determination of the footprint from atmospheric variables that are usually measured during flux observation programs.

8.2.3 Forward and Backward Approach by LS Models

The conventional approach of using a Lagrangian model for footprint calculation is to release particles at the surface point source and track their trajectories downwind of this source toward the measurement location forward in time (e.g., Leclerc and Thurtell 1990; Horst and Weil 1992; Rannik et al. 2000). Particle trajectories and particle vertical velocities are sampled at the measurement height. In case of horizontally homogeneous and stationary turbulence, the mean concentration at the measurement location (x, y, z) due to a sustained surface source Q located at height z_0 can be described as

$$\langle c_s(x, y, z)\rangle = \frac{1}{N}\sum_{i=1}^{N}\sum_{j=1}^{n_i}\frac{1}{|w_{ij}|}Q(x - X_{ij}, y - Y_{ij}, z_0), \tag{8.3}$$

where N is number of released particles and n_i the number of intersections of particle trajectory i with the measurement height z; w_{ij}, X_{ij} and Y_{ij} denote the vertical velocity and the coordinates of particle i at the intersection moment, respectively. Similarly, the mean flux is given by

$$F_s = \langle w(x, y, z)c_s(x, y, z)\rangle = \frac{1}{N}\sum_{i=1}^{N}\sum_{j=1}^{n_i}\frac{w_{i0}}{|w_{ij}|}Q(x - X_{ij}, y - Y_{ij}, z_0). \tag{8.4}$$

The above equations apply identically also to elevated sources located at arbitrary height.

The concentration footprint and the flux footprint can be determined as follows:

$$\phi_C = \frac{1}{Q}\frac{\partial^2 \langle c_s \rangle}{\partial x \partial y} \tag{8.5}$$

$$\phi_F = \frac{1}{Q}\frac{\partial^2 F_s}{\partial x \partial y} = \frac{1}{Q}\frac{\partial^2 \langle w c_s \rangle}{\partial x \partial y}. \tag{8.6}$$

Alternatively, it is possible to calculate the trajectories of a Lagrangian model in a backward time frame (cf. Thomson 1987; Flesch et al. 1995; Flesch 1996; Kljun et al. 2002). In this case, the trajectories are initiated at the measurement point itself and tracked backward in time, with a negative time step, from the measurement point to any potential surface source. The particle touchdown locations and touchdown velocities are sampled and mean concentration and mean flux at the measurement location can be described as

$$\langle c_s(x,y,z) \rangle = \frac{2}{N}\sum_{i=1}^{N}\sum_{j=1}^{n_i}\frac{1}{|w_{ij}|}Q(X_{ij}, Y_{ij}, z_0) \tag{8.7}$$

and

$$F_s = \langle w(x,y,z) c_s(x,y,z) \rangle = \frac{2}{N}\sum_{i=1}^{N}\sum_{j=1}^{n_i}\frac{w_{i0}}{|w_{ij}|}Q(X_{ij}, Y_{ij}, z_0), \tag{8.8}$$

where w_{i0} is the initial (release) vertical velocity of the particle i and w_{ij} is the particle touchdown velocity. Again, the concentration footprint and the flux footprint are determined using Eqs. 8.5 and 8.6. Note that in case of an elevated plane source with strength Q at arbitrary height Eqs. 8.7 and 8.8 are also applicable with the following modifications: the factor 2 is removed and the touchdown velocities are replaced by the vertical crossing velocities of the trajectories with the source level (both directions).

The forward and backward footprint estimates are theoretically equivalent. In practice, the forward LS models are applicable under horizontally homogeneous conditions since the method can be efficiently employed only using horizontal coordinate transformation. The backward estimators for concentration and flux do not assume homogeneity and stationarity of the turbulence field. The calculated trajectories can be used directly without a coordinate transformation. Therefore, if inhomogeneous probability density functions of the particle velocities are applied, backward Lagrangian footprint models hold the potential to be applied efficiently over inhomogeneous terrain.

In theory, the forward and backward footprint estimates are equivalent (Flesch et al. 1995). However, certain numerical errors must be avoided. Cai and Leclerc (2007) show that the concentration footprint inferred from backward simulation can be erroneous due to discretization error close to surface where turbulence is strongly inhomogeneous and proposed an adjustment numerical scheme to eliminate the error. In addition, the backward footprint simulation can violate the well-mixed condition at the surface when perfect reflection scheme is applied to skewed or inhomogeneous turbulence (Wilson and Flesch 1993). This numerical problem can be also avoided by a suitable numerical scheme (Cai and Leclerc 2007; Cai et al. 2008).

Lagrangian footprint models require a predefined turbulence field. Those can be obtained as parameterizations from atmospheric scaling laws such as Monin-Obukhov similarity theory or convective and stable atmospheric boundary-layer scaling laws. Alternatively, the parameterizations can be obtained from measurements or numerical modeling of atmospheric flow.

Closure models of any order can be applied to flow and footprint modeling, including horizontally inhomogeneous flow (see Sect. 8.4.1). Since computing costs may be high for three-dimensional calculations, a way to minimize the calculation time is to use flow statistics derived by an Atmospheric Boundary Layer (ABL) model for LS backward approach. Combined with closure model results, the LS approach has been applied to study the influence of transition in surface properties on the footprint function. The first attempt was done by Luhar and Rao (1994) and by Kurbanmuradov et al. (2003), later Hsieh and Katul (2009) applied stochastic model for estimating footprint and water vapor flux over inhomogeneous surfaces. They derived the turbulence field of the two-dimensional flow over a change in surface roughness using a closure model and performed Lagrangian simulations to evaluate the footprint functions.

Also Large-Eddy Simulation (LES) (see Sect. 8.2.5) approach has been used in combination with LS modeling to infer footprints for convective boundary layers as well as for canopy flow. For example, Cai and Leclerc (2007) and Steinfeld et al. (2008) performed LS simulations for sub-grid scale turbulent dispersion. More recently, Prabha et al. (2008a) made a comparison between the in-canopy footprints obtained using a Lagrangian simulation with those obtained against a large-eddy simulation. In that model, the Lagrangian stochastic model was driven by flow statistics derived from the large-eddy simulation.

8.2.4 Footprints for Atmospheric Boundary Layer

Most footprint models have been developed for a limited atmospheric flow regime. The first footprint study to apply Lagrangian simulations to the description of footprints is attributed to Leclerc and Thurtell (1990) who applied the LS approach to ABL. That study was the first to analyze the influence of atmospheric stability on footprints; it also showed for the first time the impact of surface roughness, atmospheric stability, and measurement height on the footprint. The importance

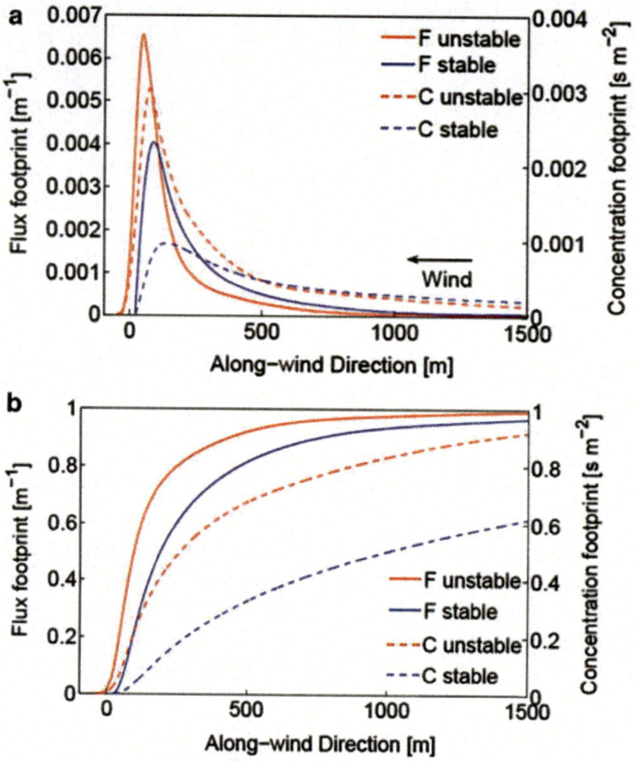

Fig. 8.2 (**a**) Crosswind-integrated flux and concentration footprints for 10 m observation height at location (0,0) and 0.01 m roughness length under unstable ($L = -30$ m, $u_* = 0.2$ m s^{-1}, $w_* = 2.0$ m s^{-1}, $z_i = 2{,}500$ m) and stable ($L = 30$ m, $u_* = 0.5$ m s^{-1}, $z_i = 200$ m) conditions. (**b**) Cumulative footprints for the same conditions

of these results is reflected in that several NASA ABLE 3-B multi-scale, multi-platform field campaigns were redesigned based on their preliminary calculations. As one of a few, Kljun et al. (2002) presented a footprint model based on a trajectory model for a wide range of atmospheric boundary-layer stratification conditions.

The stability dependence has been investigated by Kljun et al. (2002), comparing crosswind-integrated footprints predicted for different stability regimes by a three-dimensional Lagrangian simulation. In the example in Fig. 8.2, measurement height and roughness length were fixed to 10 and 0.01 m, respectively, whereas the friction velocity, vertical velocity scale, Obukhov length, and boundary-layer height were varied to represent convective, neutral, and stable conditions. In unstable conditions, the turbulence intensity is high, resulting in the upward transport of any compound and a shorter travel distance/time. Correspondingly, the peak location is closer to the receptor in unstable conditions. This is in agreement with the findings of Leclerc and Thurtell (1990) and with experimental validation of these models (Finn et al. 1996;

Leclerc et al. 1997). Stability affects strongly the footprint peak location and its maximum value. Concentration footprints tend to be longer (Fig. 8.2).

Flux and concentration footprints differ significantly in spatial extent. In Lagrangian framework, this can be explained as follows: The flux footprint value over a horizontal area element is proportional to the difference of the numbers or particles (passive tracers) crossing the measurement level in the upward and downward directions. Far from the measurement point, the number of upward and downward crossings of particles or fluid elements across an imaginary *x*–*y* plane typically tends to be about the same and thus the up- and downward movements are counterbalanced decreasing the respective fractional flux contribution of those source elements to the flux. In contrast to the flux footprint, each crossing contributes positively to the concentration footprint independently of the direction of the trajectory. This increases the footprint value at distances further apart from the receptor location.

The cumulative footprint function presented in Fig. 8.2b indicates the fraction of flux (or concentration) contributed by uniform surface sources to the measured flux. Note that the concept of cumulative effective fetch was introduced by Gash (1986) before the footprint function in differential form was proposed by Schuepp et al. (1990). The cumulative footprint function is especially useful in determining the necessary horizontally homogeneous upwind distance for the measured flux to represent certain fraction of surface flux under investigation. Depending on the requirement of representativeness of the measured flux and contrast of the surface types, the cumulative fetch can be determined for different levels of homogeneous fetch. For example, if 80% of the flux should originate from the surface of interest, the homogeneous fetch must extend up to 250 and 500 m in unstable and stable conditions, respectively, for the observation conditions in Fig. 8.2.

The crosswind-integrated footprint function is useful when the assumption of surface homogeneity in crosswind direction applies. In case of patchy surface and also for some applications of footprints (see Sect. 8.5) two-dimensional footprint functions are needed (Fig. 8.3). Again, the flux and concentration footprints exhibit significantly different spatial extent for the same height and roughness conditions.

Flux and concentration footprint functions depend on measurement level, wind speed and wind direction, atmospheric stability, and surface characteristics. Figure 8.4 illustrates the distance at which footprint peak occurs as a function of measurement height and surface roughness. The footprint peak

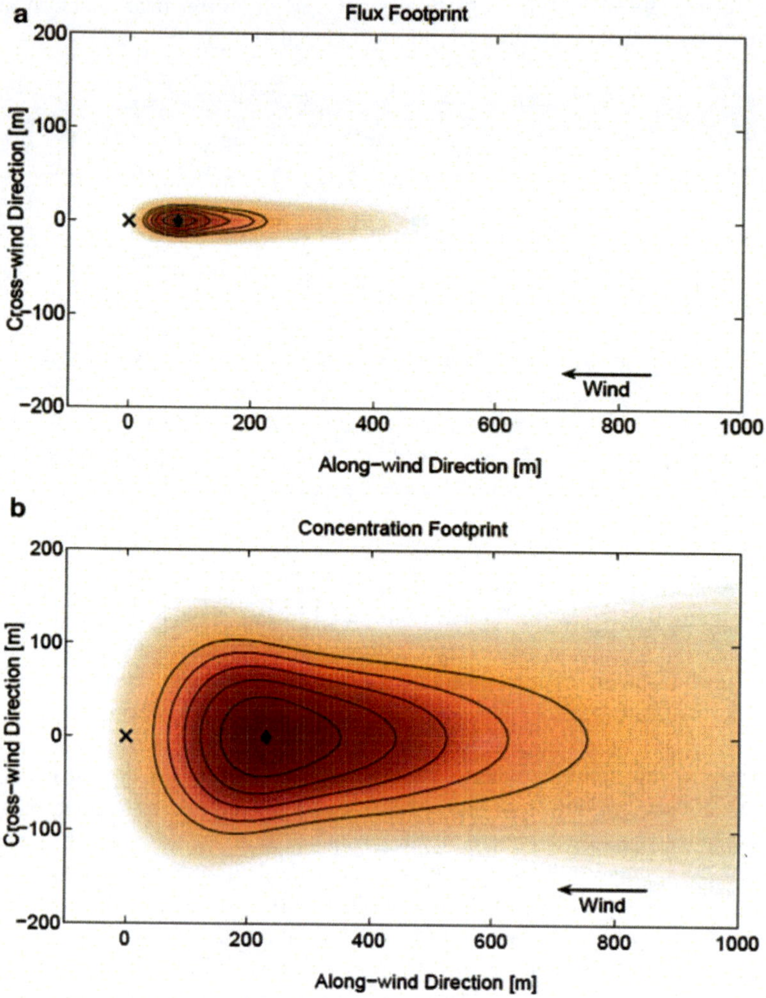

Fig. 8.3 Footprint functions for neutral atmospheric stratification conditions ($u_* = 0.8$ m s^{-1}, $z_i = 1{,}500$ m) at 10 m height and 0.01 m roughness length for (**a**) flux and (**b**) concentration. The *isolines* represent 10–50% source area. *Cross* denotes the tower location

location increases almost linearly with observation height. Surface roughness has strong impact on peak location. In unstable stratification, the footprint peaks are much closer to observation point than in stable stratification; compare the curves for the same surface roughness of 0.01 m.

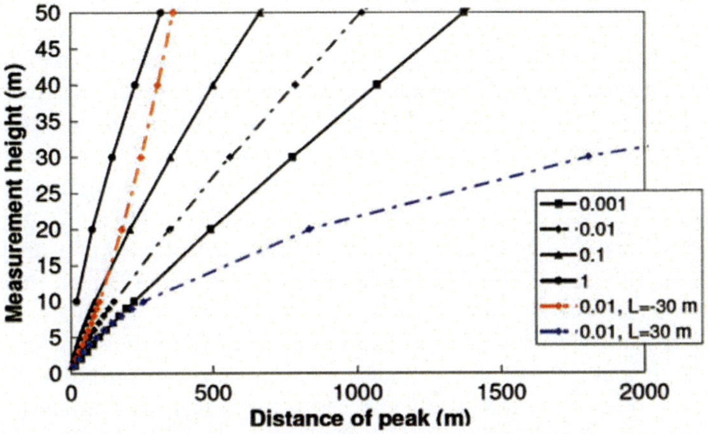

Fig. 8.4 Footprint peak distance depending on measurement height. *Curves* are presented for range of roughness lengths under neutral stratification conditions and for two stability length values for comparison with neutral case for $z_0 = 0.01$ m. ASL conditions are assumed

8.2.5 Large-Eddy Simulations for ABL

The Large-Eddy Simulation (LES) approach is free of the drawback of a predefined turbulence field. Using Navier-Stokes equations, LES resolves the large eddies with scales equal to or greater than twice the grid size, while parameterizing sub-grid scale (SGS) processes. This approach presupposes that most of the flux is contained in the large eddies: since these are directly resolved, this method provides a high level of realism to the flow despite complex boundary conditions (e.g., Hadfield 1994). The Large-Eddy Simulation is a sophisticated model which directly computes the three-dimensional, time-dependent turbulence motions, and only parameterizes the SGS motions. The choice of lateral and surface/upper boundary conditions is one of the aspects of this technique that is critically important and which depends on the application. In addition, in stable boundary layers, the errors due to an imperfect SGS parameterization become more important as the characteristic eddy size is smaller in stable conditions. This technique, applied for the first time to the atmosphere by Moeng and Wyngaard (1988), is considered the technique of choice for many cases not ordinarily studied using simpler models and can include the effect of pressure gradient.

Typically, LES predicts the three-dimensional velocity field, pressure, and turbulent kinetic energy. Depending on the purpose, it can also simulate the turbulent transport of moisture, carbon dioxide, and pollutants. There are several parameterizations available in treating the sub-grid scales. One of the most widely used simulations is that originally developed by Moeng (1984) and Moeng and Wyngaard (1988) and modified by Leclerc et al. (1997), Su et al. (1998), and by Patton et al. (2001) for adaptation to include canopy and boundary-layer scalar transport.

Often, the SGS is parameterized using the 1.5 order of closure scheme. Depending on the research interest, the LES can contain a set of cloud microphysical equations, thermodynamic equation, and can predict the temperature, concentrations, and pressure. Some LES also include a terrain-following coordinate system. A spatial cross-average and temporal average is applied to the simulated data once the simulation has reached quasi steady-state equilibrium. Typical boundary conditions are periodic with a rigid lid applied to the top of the domain so that waves are absorbed and reflection from the upper portion of the domain is decreased. The LES is computationally very expensive and limited to relatively simple flow conditions by the number of grid points in flow simulations.

This powerful type of simulations has been used extensively in atmospheric flow modeling and in particular in convective boundary layers (Mason 1988). The technique has been used successfully to describe the influence of surface patchiness on the convective boundary layers at different scales (Hadfield 1994; Shen and Leclerc 1995).

The first attempt to apply LES approach for footprint modeling was made by Hadfield (1994). Further, the LES method has been applied to simulate footprints in the convective boundary layer (Leclerc et al. 1997; Guo and Cai 2005; Peng et al. 2008; Steinfeld et al. 2008; Cai et al. 2010). In some of the recent studies (Cai and Leclerc 2007; Steinfeld et al. 2008; Cai et al. 2010), the LES was used in conjunction with the Lagrangian simulation of SGS turbulent dispersion to reproduce convective boundary-layer turbulence and infer concentration footprints. Steinfeld et al. (2008) used LES to describe the footprint in boundary layers of different complexities. They documented positive and negative flux footprints in the convective boundary layer in a manner analogous to Prabha et al. (2008a) in a forest canopy. This is consistent with Finnigan's (2004) conclusion that the flux footprint function is a functional of the concentration footprint function and in complex flows there is no guarantee that the flux footprint is positive, bounded by zero and one. Wang and Rotach (2010) applied LES with backward Lagrangian stochastic approach over undulating surface and observed impact of flow divergence and convergence on footprint function for near-surface receptors. They observed that crosswind-integrated footprint function peak was located closer to receptor in the area with surface-wind convergence and was opposite in the area with wind divergence, respectively.

8.3 Footprint Models for High Vegetation

8.3.1 Footprints for Forest Canopy

The study by Baldocchi (1997) was first to address the footprint behavior inside a forest canopy by using LS modeling approach. He used literature-based parameterizations for turbulence vertical profiles inside the canopy and similarity relationships above the canopy (within this section we use "canopy" to refer to "forest canopy"). The influence of higher-order velocity moments on footprint prediction was not included in this study. However, one of the benefits of Lagrangian models is their capability to consider both Gaussian and non-Gaussian turbulence. While the flow within the surface layer is nearly Gaussian, non-Gaussianity characterizes flow fields of both canopy layer and convective mixed layer. Another benefit of Lagrangian stochastic models over analytical ones is their applicability in near-field conditions, that is, in conditions when fluxes of constituents are disconnected from their local gradients, providing thus proper description for within canopy dispersion. This makes it possible to locate trace gas sources/sinks within a canopy. Baldocchi (1997), Rannik et al. (2000, 2003), Mölder et al. (2004), and Prabha et al. (2008a) studied qualitative effects of canopy turbulence on the footprint function. In the case of tall vegetation, the footprint prediction depends primarily on two factors: canopy turbulence and the source/sink levels inside the canopy. These factors become of particular relevance for observation levels close to the treetops (Shen and Leclerc 1997; Rannik et al. 2000; Lee 2003; Markkanen et al. 2003; Göckede et al. 2007; Sogachev and Lloyd 2004).

Lee (2003, 2004) adopted a different approach for inside-canopy scalar advection modeling based on localized near-field theory and applied the model to footprint prediction over a forest canopy. The near-field effect had an impact on footprint prediction inside the roughness sublayer but could be neglected inside the inertial sublayer.

The wind statistics necessary for LS footprint simulations originate from similarity theory, experimental data, or an output from a flow model capable to produce wind statistics. However, the description of wind statistics inside a canopy becomes uncertain due to poor understanding of stability dependence of the canopy flow as well as of Lagrangian correlation time. In terms of parameterization of the value of the Kolmogorov constant C_0 it has been shown that the LS model results are sensitive to the absolute value of the constant (Mölder et al. 2004; Rannik et al. 2003). Poggi et al. (2008) revealed that C_0 may vary nonlinearly inside the canopy while the LS model predictions were not sensitive to gradients of C_0 inside canopy.

In addition to LS approach closure modeling (cf. Sect. 8.4) and LES have been successfully applied to footprints inside and above a forest canopy. The clear benefit of these models is their ability to simulate complex canopy flows.

The versatility of the LES has been recognized as a potential tool to describe the flow over (Chandrasekar et al. 2003) and near (Shen and Leclerc 1997) or inside very strongly sheared atmospheric flows such as within plant canopies (Su et al. 1998; Shen and Leclerc 1997; Watanabe 2009) and urban canopies (Tseng et al. 2006). Recently, LES studies have been applied to canopy turbulence and been shown to reproduce many observed characteristics of airflow within and immediately above a plant canopy, including skewness, coherent structures, and two-point statistics (Su et al. 1998; Shen and Leclerc 1997; Prabha et al. 2008a).

Concentrations and flux footprints have been studied using the LES, by examining the behavior of tracers released from multiple sources inside a forest canopy.

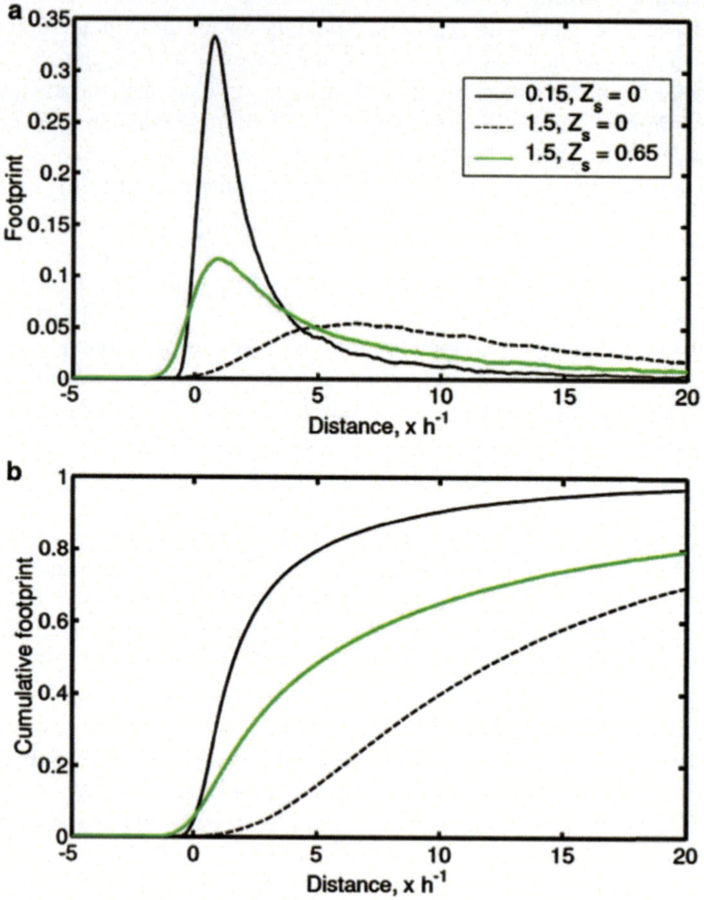

Fig. 8.5 (**a**) Flux footprints predicted for within-canopy wind statistics according to Launianinen et al. (2007) by assuming source locations at the forest floor ($Z_s = 0$) or at height 0.65× canopy height. (**b**) Cumulative footprints corresponding to (**a**). Observation levels $z\,h^{-1} = 0.15, 1.5$

Recently, the flux footprint over or inside the forest canopy using the LES has been modeled by Su and Leclerc (1998), Prabha et al. (2008a), and by Mao et al. (2008).

8.3.2 Footprint Dependence on Sensor and Source Heights

Rannik et al. (2000), Markkanen et al. (2003), and Prabha et al. (2008a) highlighted the dependence of the footprint function on the vertical source location. This is of relevance in case of flux measurements over high vegetation, where exchange of many atmospheric consituents of wide interest (CO_2) occurs mainly at the higher part of canopy. Figure 8.5 examines the influence of source height on footprint

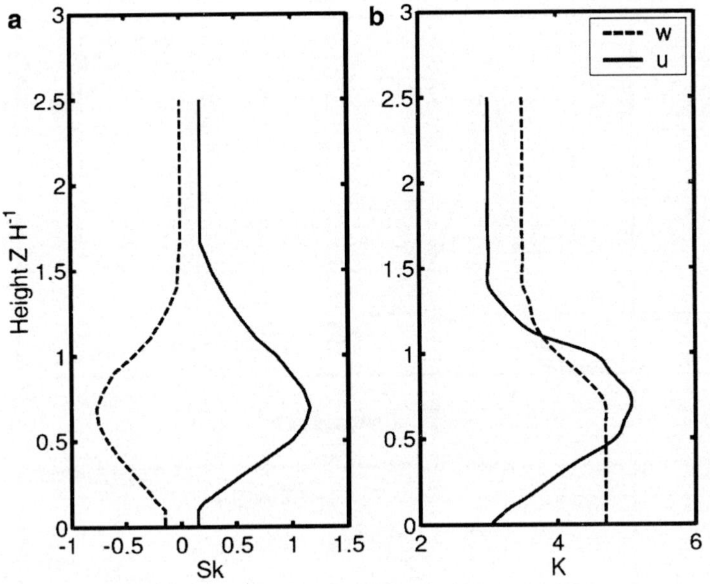

Fig. 8.6 Vertical profiles of higher moments: (**a**) skewness (*Sk*), and (**b**) kurtosis (*K*) of vertical (*w*) and along-wind (*u*) components (Rannik et al. 2003)

function. For this illustration, turbulence profiles in LS simulation of footprint functions were parameterized for pine forest according to measurements reported in Launiainen et al. (2007). It can be seen that the footprint function peak is higher for elevated sources inside the canopy (Fig. 8.5). The footprint funtion for measurements over forest at a typical height varies significantly depending on source location either on the forest floor or in the upper part of forest canopy. The footprint function for flux measurements above the forest floor inside trunk space is much more constrained.

8.3.3 Influence of Higher-Order Moments

The velocity distribution inside canopy is significantly skewed (Fig. 8.6). Leclerc et al. (1991) examined the behavior of the vertical velocity skewness inside and above a forest canopy for a wide range of atmospheric stabilities, defined as the stability above the canopy, and found that non-dimensionalized vertical velocity skewness can be as large as −2. The trajectory model of Thomson (1987) enables to account only for Gaussian turbulence statistics. Flesh and Wilson (1992) developed a two-dimensional trajectory model able to account also for third and fourth moments. Since more than 1D Lagrangian trajectory models are

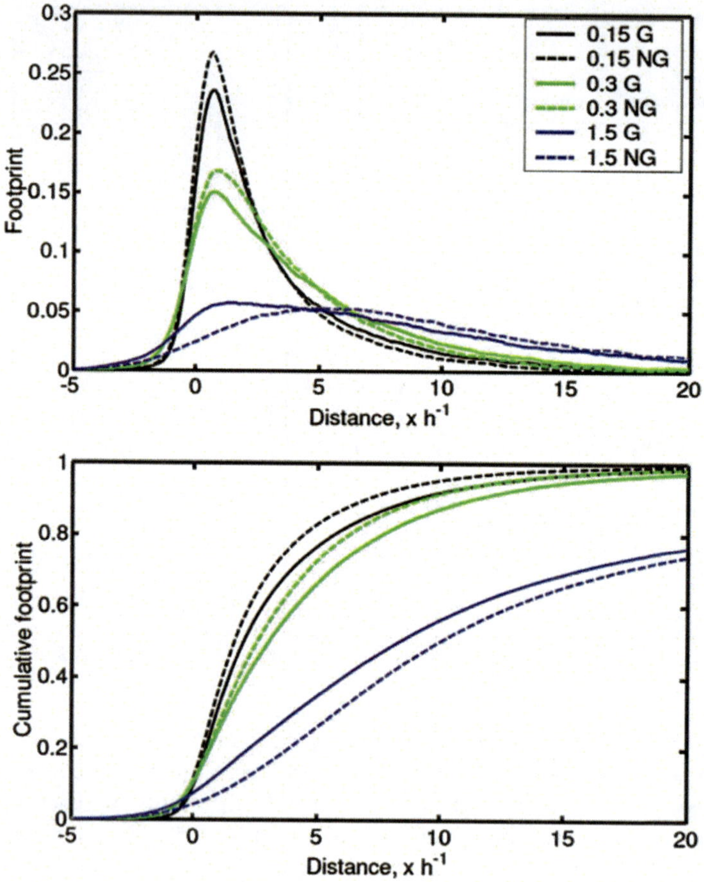

Fig. 8.7 Prediction of flux footprint pdf with Lagrangian stochastic trajectory model of Flesch and Wilson (1992), parameterized with Gaussian (*G*) and non-Gaussian (*NG*) turbulence profiles. 0.15, 0.3, and 1.5 refer to observation heights above forest surface normalized to forest height *h*, profiles parameterized according to Rannik et al. (2003) and skewness and kurtosis as presented in Fig. 8.6

not uniquely defined, the model of Flesh and Wilson (1992) was run for the comparison also with Gaussian parameterization of velocity distribution function. Non-Gaussian turbulence statistics tend to move the footprint peak further away from the measurement point, reducing the contribution from very close sources from below and around the observation point (Fig. 8.7). However, the integrals over horizontal distance (representing the fraction of flux contributed by the given horizontal distance) converge and the choice between the two trajectory models does hardly affect the estimate of the footprint extent.

8.4 Complicated Landscapes and Inhomogeneous Canopies

8.4.1 Closure Model Approach

Often the estimation methods of ecosystem–atmosphere exchange rely on horizontal homogeneity. Nevertheless, the assumption of spatial homogeneity is rarely met within most natural ecosystems and airflow passing through and over them is essentially two- or three-dimensional, leading to advective transport occurring besides the turbulent transfer. The large and often undetermined uncertainty of ecosystem–atmosphere exchange derived by single-point micrometeorological measurements has become one of the most important topics of methodological micrometeorology (e.g., Rannik et al. 2006). Capturing of advection and horizontal flux components at imperfect sites requires auxiliary experiments and cannot yet be routinely performed (e.g., Aubinet et al. 2003, 2005). Numerical modeling has been recognized as an effective and flexible tool in the investigation of spatially dependent complex processes, providing supplementary information on variables of interest, generally overlooked in field measurements.

As airflow mediates the biosphere–atmosphere exchange and coupling, the first step toward understanding the role of advection in exchange processes over complex terrain is characterizing wind flow. Over the last 30 years, different modeling approaches to simulate vegetation–atmosphere interaction have been applied to horizontally homogeneous canopies, and these form a basis for more complex flows. It has became clear that for any model that aims to adequately simulate the airflow over heterogeneous surfaces, the turbulence length scale, l, must be calculated as a dynamic variable (e.g., Ayotte et al. 1999; Finnigan 2007). For practical applications (such as footprint estimation), where information on higher-order statistics of turbulent flows is superfluous, the approach based on two-equation closure (see below) seems to be the optimal choice for modeling of such flows since second- and higher-order closure models (e.g., Rao et al. 1974; Launder et al. 1975) or Large-Eddy Simulation (e.g., Deardorff 1972; Moeng 1984) providing a practical framework for computing these statistics are computationally more demanding. The approach based on differential transport equations for the turbulent kinetic energy (TKE) E, and for a length scale determining a variable related to E (that is more often one of the following parameters: El, ε, or ω – the product of E and l, the dissipation rate of E, or the specific dissipation (ε/E), respectively), provides the minimum level of complexity that is capable of simulating l without any additional speculation (e.g., Launder and Spalding 1974; Wilcox 2002; Kantha 2004). Although having a number of well-known deficiencies, two-equation closure has still been used in industrial computations for a long time and has proved to be an excellent compromise between accuracy and computational effort (see Hanjalić 2005 or Hanjalić and Kenjereš 2008 for a review). During the last two decades, models using two-equation closure have attracted great attention in the geophysical modeling community and a number of authors have found it is sufficient for most practical tasks (Wang and Takle 1995; Umlauf and Burchard 2003; Castro et al.

2003; Hipsey et al. 2004; Katul et al. 2004). Applications of this approach to atmospheric and oceanic flows have highlighted, however, serious uncertainties in the treatment of buoyancy and plant drag effects (e.g., Duynkerke 1988; Svensson and Häggkvist 1990; Apsley and Castro 1997; Wilson et al. 1998; Baumert and Peters 2000; Kantha 2004; Sogachev and Panferov 2006). Recently, Sogachev (2009) showed how different sources/sinks appearing in the turbulent kinetic energy equation due to these effects can be treated in the supplementary equation in such a way as to minimize the uncertainty. This gives new opportunities in the use of two-equation closure models for environment problems. However, some types of models (e.g., *E–El*) have problems with properly reproducing the log-law region near wall unless extra terms are included (e.g., Kantha 2004). Application of such models to the canopy and planetary boundary layer could be limited; for example, determination of the near-wall term in the presence of vegetation could be difficult similarly to determination of l (see, for discussion, Sogachev and Panferov 2006).

A natural question demanding more careful consideration is still an ability of such models based on gradient-diffusion scheme to describe adequately turbulence under conditions of unstable stratification and inside of vegetation. Discussions on this question with reference to vegetation repeatedly rose in scientific literature (Sogachev et al. 2002; Katul et al. 2004; Sogachev et al. 2008). Here we summarize the main points. Central to any first or one-and-half order closure model is a simple relationship used for the description of the turbulent exchange within the vegetation, namely K-theory where the mean turbulent flux (F_s) is related to the mean concentration (c_s) gradient as follows:

$$F_s = -K_s(z)\frac{d\bar{c}_s}{dz}. \tag{8.9}$$

Here z is the height and $K_s(z)$ is the local eddy diffusivity for c_s. A number of investigators have noted, however, that K-theory may be inadequate for description of turbulent fluxes from local gradients within the canopy due to strong variability in the sources and sinks of any scalar s, and due to the possible occurrence of countergradient transfer (Denmead and Bradley 1985; Raupach 1988; Finnigan 2000). Nevertheless, researchers still consider models based on gradient-diffusion approximation to explore disturbed flows (Gross 1993; Wilson et al. 1998; Wilson and Flesch 1999; Pinard and Wilson 2001; Katul et al. 2004, 2006; Sogachev and Lloyd 2004; Foudhil et al. 2005; Sogachev and Panferov 2006). This is in part due to the fact that keeping the number of equations and necessary constants to a minimum provides a significant computing profitability over other methods which can reproduce nonlocal, nondiffusive behavior in the Eulerian framework such as Large-Eddy Simulation (LES) (Shaw and Schumann 1992; Shen and Leclerc 1997) and higher-order closure (Wilson and Shaw 1977, Meyers and Paw 1986) models. Most importantly, however, there is a distinct dynamical support to describe the behavior of strongly perturbated canopy flows as is the case for flows near the transition between a forest edge and an open forest gap (Wilson et al. 1998; Belcher et al. 2003) or on hills (Finnigan and Belcher 2004).

Thus, near the forest edge, most of the flow distortion initially is dominated by inertial effects, resulting in large advective terms (Belcher et al. 2003). These lead to reduction in K which is not offset by the new energetic small-scale eddies generated as the flow encounters the foliage. Hence, these eddies have a small integral length scale and the "near-field" effect (a nondiffusive contribution from nearby sources) associated with them is localized. Thus the basic requirement of K-theory – that the length scale of the mixing process needs to be substantially smaller than that of the inhomogeneity in the mean scalar or momentum gradient – is not violated here (Corrsin 1974). Airflow over hill is different from that near forest edge but it also leads to distortion and breaking up of large eddies and using K-theory is admissible (Wilson et al. 1998; Katul et al. 2004).

A common conclusion from above was expressed by Gross (1993), who found that the application of the flux-gradient approach by two-dimensional and three-dimensional-modeling is admissible, in particular, in simulations for which advective processes are of greater importance than diffusive processes. Such situations are typical for inhomogeneous vegetation and complex terrain. Regarding diffusion process that is always present irrespective of advection, we note that for the forward problem, which is considered when we are looking for flux footprint, the objective is to calculate fluxes from the canopy and underlying surface to a reference point. In this case, "near-field" dispersion provides distortions to the local concentration profiles within the canopy but does not contribute substantially to the transport between the canopy layers and the reference point (Raupach 1989; Katul et al. 1997; Leuning et al. 2000).

8.4.2 Model Validation

All numerical results presented below were derived using ABL model SCADIS based on one-and-half order closure with different closure schemes during different stages of model development. The last version of model is based on E–ω closure scheme, modified according to Sogachev (2009). There exists a variety of experimental data about airflow characteristics inside the vegetation canopy. As a rule, such data have been derived from single-point measurements. In the literature one can find many models of different levels of complexity (including analytical ones) for the canopy flow that is mainly validated by using such data. Applicability of those models is justified for homogeneous conditions but is rather questionable for heterogeneous ones. There are few natural experiments exploring turbulence characteristics spatially, that is, in vicinity of forest edge (Gash 1986; Kruijt 1994; Irvine et al. 1997; van Breugel et al. 1999; Flesch and Wilson 1999; Morse et al. 2002). The lack of the experimental data limits seriously a development of high-resolution flow models capable to take into account the natural heterogeneity. Nevertheless, the results of recent model tests over a wide range of canopy architectures by Sogachev and Panferov (2006) suggest that the model SCADIS can

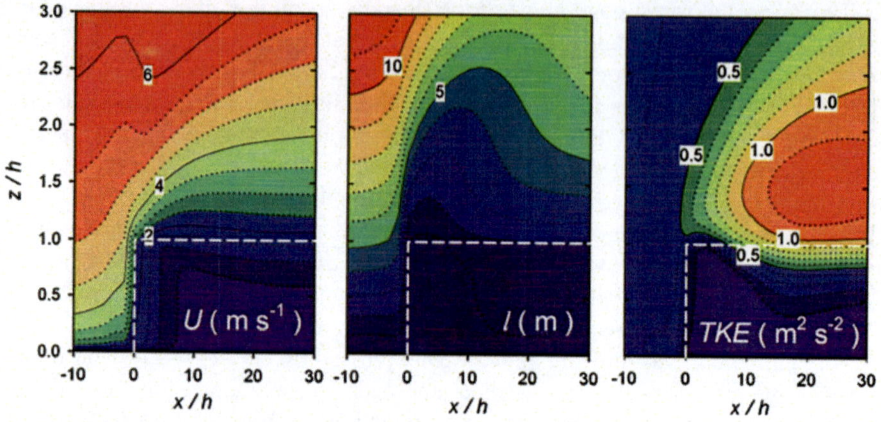

Fig. 8.8 Two-dimensional fields of horizontal wind velocity (U), mixing length (l), and turbulent kinetic energy (*TKE*) near the leading edge of a forest derived by $E - \omega$ model. The *thick dashed line* encloses the forest approximated by vertically uniform vegetation with a height of 15 m and LAI = 3. The horizontal distance is normalized by the tree height, x/h. Here and in figures below the airflow from the *left to the right* (After Sogachev and Panferov 2006)

adequately reproduce the interaction between the flow and the forest edge. Thus, the behavior of the turbulence scale and the turbulence field as predicted by our two-equation model is in qualitative agreement with the description suggested by Belcher et al. (2003) (see above) and corresponds to that experimentally obtained by Krujit (1994) and by Morse et al. (2002) (see Fig. 8.8).

Comparison of model results with observations of Chen et al. (1995) for turbulent kinetic energy in wide gap downwind of the model forest derived from wind tunnel study shows that the model also deals well with the readjustment of the turbulence field on the lee side of a forest (see Fig. 8.9).

There are differences between airflow above smooth and rough ridge. Belcher and Hunt (1998) pointed out that higher roughness of the ridge or larger wind shear of the approaching flow enhances the stress perturbation so that separation tends to occur at smaller slopes. Model results for airflow over two different ridges – one with relatively smooth surface and another covered by homogeneous forest – are demonstrated in Fig. 8.10. Comparing the left-side and the right-side panels of Fig. 8.10, it can be seen that separation occurs for a ridge with a large surface roughness, whereas there was no separation for the ridge with small surface roughness. This is in good agreement with the conclusion of Belcher and Hunt (1998). As is seen, the model reproduces qualitatively the most significant flow features of hilly terrain (Raupach and Finnigan 1997) and is therefore suitable for preliminary investigation of both scalar dispersion and footprint behavior in complex terrain.

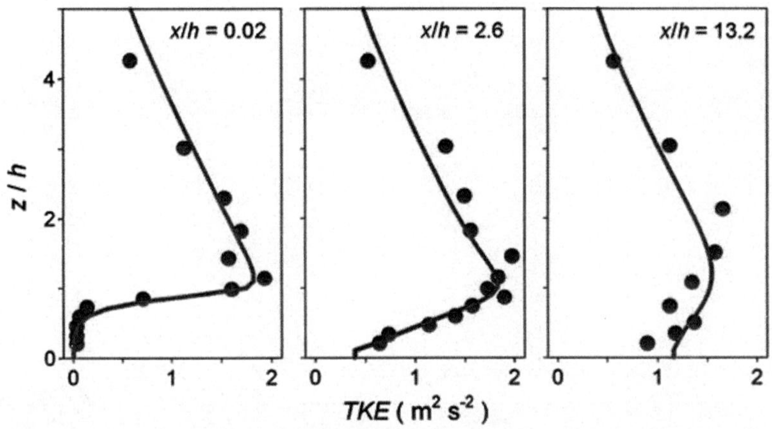

Fig. 8.9 Comparison between vertical profiles of measured (*symbols*) and modeled (*lines*) turbulent kinetic energy (*TKE*) downwind the model forest edge. The position at $x/h = 0$ corresponds to the beginning of the open place (After Sogachev and Panferov 2006)

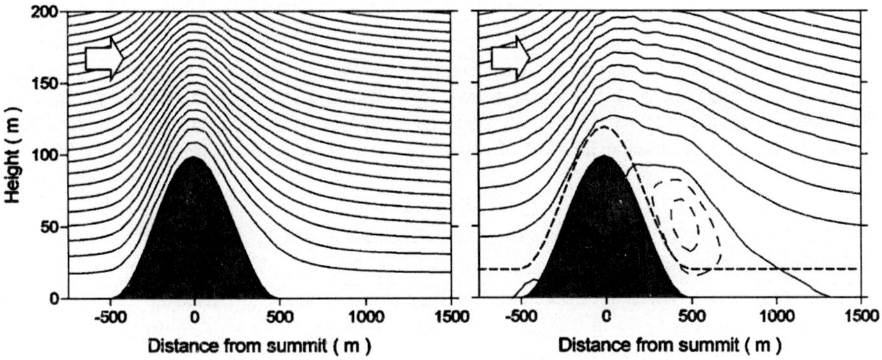

Fig. 8.10 Isolines of the stream function for neutral stability airflow over a ridge having a relatively smooth surface (soil with surface roughness assumed to be $z_0 = 0.3$ m) (*the left panel*) and having a rough surface (forest) (*the right panel*). The height of the forest was assumed to be 20 m (denoted by the *dashed line*) with LAI = 2.4 m^2 m^{-2}. Aerodynamic drag of the forest and the flow through the forest were considered. The topography variations are shown by *black area*. *Arrows* show the direction of the airflow (After Sogachev et al. 2004)

8.4.3 Footprint Estimation by Closure Models

The spatial distribution of sources and sinks within plant canopies is strongly heterogeneous and depends on vegetation properties and prevailing meteorological conditions. However, such details regarding the distribution of local sources and sinks are not needed for many practical tasks. To interpret experimental data correctly it is often sufficient to know the footprint of the measurement with some

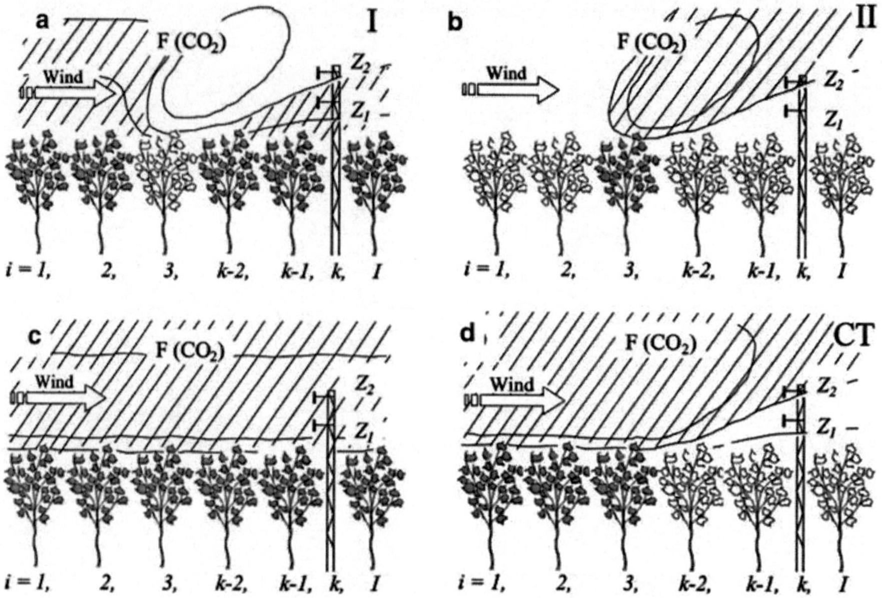

Fig. 8.11 Methods of estimation of source weight function by means of the numerical model; "*i*" indicates a model grid cell within a domain of I gridcells, "*k*" is the investigated grid cell (measurement point), "Z1" and "Z2" are the heights for which the footprint is estimated. The *dashed areas* depict high-intensity areas of vertical scalar flux (After Sogachev and Lloyd 2004)

finite horizontal resolution; this being sufficient to identify the contribution of the main vegetation types to the measured flux.

Thus assuming that the vertical scalar flux measured by a sensor at a given point can be estimated by Eq. 8.9, we can then find the integral contribution of each model cell to that measurement from modeled fields of scalar concentration and turbulent diffusion. When using SCADIS there are two nearly equivalent techniques (difference can be caused by boundary conditions at simulation domain) to estimate the contribution of any model cell to the measured vertical flux at a prescribed location. These are presented schematically in Fig. 8.11.

According to the first technique (I) the contribution of a given cell to the measured vertical flux at point (*k*,*Z*) is determined by excluding all sources and sinks in the investigated cell (e.g., $i = 3$ in Fig. 8.11a). The alternative approach (II) is complementary where all sources and sinks in the model domain are excluded (1,*I*) except for those within the investigated cell (e.g., $i = 3$ in Fig. 8.11b). The bulk vertical flux at point (*k*,*Z*) is then calculated by summing up the result of the individual calculations for each cell (Fig. 8.11c). Taking the total contribution of all cells to bulk flux as unity it is then possible to estimate the influence (or weight) of each cell and, therefore, define the flux footprint function.

In the current modeling approach, it is difficult to predefine equal source strength inside all grid cells, especially over complex terrain. This is because complex topography and varying tree compositions with different height and density will change aerodynamic resistance and stomatal conductivity in unpredictable manner. Therefore, modeling approach is needed for normalization of sources for each grid cell to get uniform distribution of sources for footprint estimation. The major problems with this approach occur when the cells next to inflow lateral border have significantly different source/sink strengths to each other or if the inflow lateral border of the model is not far enough from point (k). This is because the source/sink from the inflow border cell ($i = 1$) mostly defines the model background flux as the contribution to point (k,Z) from outside of the model domain. So any sudden changes in inflow conditions can result in uncertain footprint prediction.

These problems can, however, be overcome by imposing the mean canopy properties onto several inflow cells or by having the inflow border at a sufficient distance from estimated measurement point. Some guidance for the appropriate distance can be obtained from analytical footprint models. An irregular horizontal grid with a model step that increases as one moves away from the measurement point also helps to solve the problem with lateral border conditions, especially for two-dimensional model domains, and without significantly increasing computational requirements.

It should be noted that the footprint estimation for fluxes where the source or sink strength is dependent on specific surrounding conditions (e.g., photosynthetic activity and ambient CO_2 concentration) are, however, slightly incorrect as advective terms are ignored. Footprint estimation taking into account the upwind influences is relatively simple for the two-dimensional model when using the cumulative technique (CT). This approach is illustrated in Fig. 8.11d. The contributions of model cells to the flux at the investigated measurement point are estimated by this approach as follows. First the source/sink influence of an inflow border cell ($i = 1$) is estimated when for all other cells ($i = 2,I$) all sources/sinks are not active. Then the sources/sinks in the next downwind cell are activated and the joint influence of the two cells is estimated. Then the sources/sinks in the next downwind cell are activated ($i = 1,3$) and so on until the value of bulk flux in investigated point is reached as a result of the joint influence of all upwind sources/sinks ($i = 1,k$). After that it is easy to derive the cumulative flux for each upwind cell from numerical data. The derivative of this cumulative flux function is the footprint. This technique is much more difficult or even impossible to implement for three-dimensional conditions because of very complicated upwind conditions. So for full three-dimensional simulations it is assumed that the source/sink strengths of different cells are independent of each other with the exception of the upwind boundary cell. The resultant flux at the investigated (measurement) point is then a result of superposition of flux fields produced by all cells. According to this assumption the first two techniques of footprint modeling are equivalent.

According Sogachev and Lloyd (2004) the "footprint function" as calculated by above-described techniques does not strictly adhere to the footprint definition, for which footprint function should depend only on turbulent diffusion and source-receptor location. Rather, it represents a normalized contribution function (or "source weight function"), where variations in the horizontal distributions of

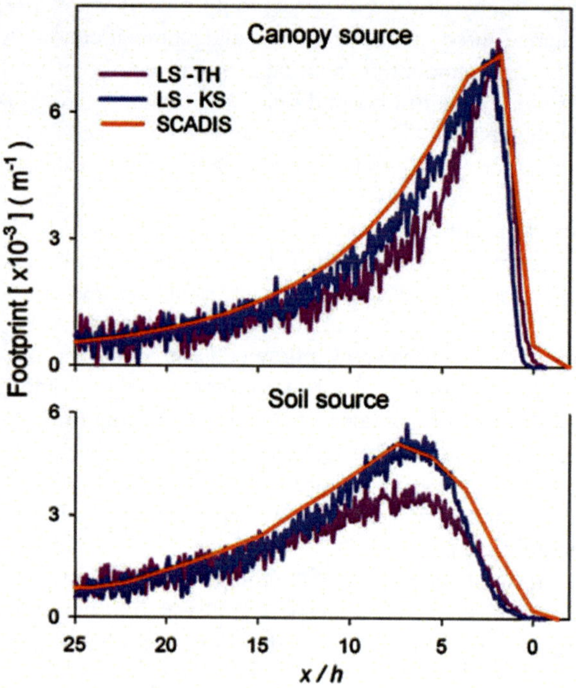

Fig. 8.12 Predictions of flux footprint with the Lagrangian stochastic trajectory simulation of Thomson (1987) (LS-TH) and Kurbanmuradov and Sabelfeld (2000) (LS-KS), and SCADIS model estimations of flux footprints above a managed forest plantation in Florida ($z = 1.4$ h) in neutral conditions (After Sogachev et al. 2005a)

fluxes will by definition also give rise to a variation in estimated footprint function. In case of horizontally homogeneous source/sink field our normalized contribution function is effectively equivalent to a footprint function and is thus referred to as such.

There are no general criteria guiding the validation of footprint models. Only a handful of validation experiments are available (see Foken and Leclerc 2004). Therefore, the approach of footprint estimation based on SCADIS was mainly validated using comparison with other approaches. Footprint functions modeled by SCADIS were compared with footprints derived from both analytical and Lagrangian stochastic approaches for condition of uniform surface (e.g., Schuepp et al. 1990; Leclerc and Thurtell 1990; Kormann and Meixner 2001). The best agreement was obtained in neutral conditions. In Sogachev et al. (2005a), additional proofs of credibility of the closure approach were given by a comparison of footprints predicted by SCADIS and two different LS models (Thomson 1987; Kurbanmuradov and Sabelfeld 2000) (see Fig. 8.12). Figure displays footprint predictions derived by different models for the same flow conditions over homogeneous vegetation. The vegetation was presented by slash pine managed forest in Florida (Leclerc et al. 2003a). The forest has a closed canopy with an average height of 13.5 m and leaf area index (LAI) of about 3. SCADIS footprints exhibit very close values compared to Lagrangian stochastic (LS) model results.

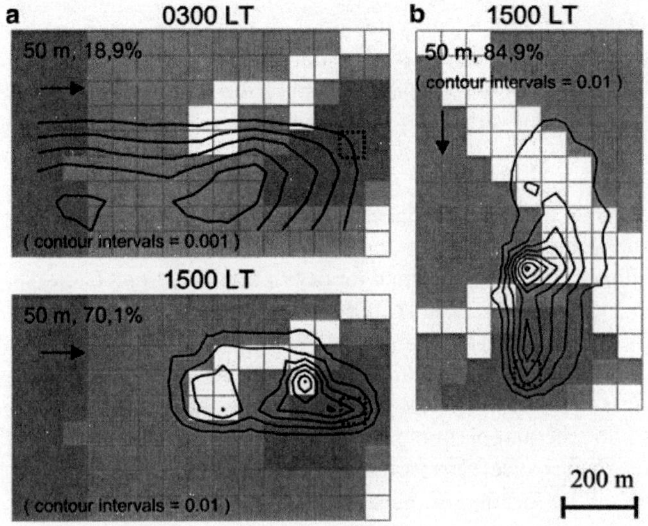

Fig. 8.13 Examples of footprint predictions for two different directions of surface wind and time points at three heights above forest canopy. (**a**) – west wind; upper panel for 3.00 LT ($L \approx 50$ m), down panel – 15.00 LT ($L \approx -120$ m). (**b**) – north wind, 15.00 LT ($L \approx -120$ m). The *dashed* quadrate at all panels indicates the location of measuring tower. *Arrow* indicates the surface wind direction. Numbers of per cent indicate total contribution from model domain to measured flux. Colors of each type correspondent to colors of vegetation type approximation in domains (*white* – *birch*; different grey for different spruce stands) (After Sogachev and Lloyd 2004)

8.4.4 Footprints over Complex Terrain

The main advantage of the approach for footprint estimation based on closure models is that it does not rely on the assumption of spatially homogeneous vegetation. Therefore, it could be successfully applied for a wide range of practical tasks like the choice of optimal sensor position for flux measurements over complex terrain, or for the data interpretation from existing measurement sites.

The approach has been applied to estimate footprints for existing flux measurement sites in Tver region (European Russia) (Sogachev and Lloyd 2004) and Hyytiälä (Finland) (Sogachev et al. 2004), taking into account mainly the vegetation heterogeneities in the first case and complex topography in the second. Applications of the method to real sites lead us to several interesting observations. For example, for a mixed coniferous forest in European Russia on a plain relief a marked asymmetry of the footprint in the crosswind direction was observed, this being especially pronounced for nonuniform plant distributions involving vegetation types with different morphological and physiological properties (see Fig. 8.13). It was also found that, other factors being equal, for above-canopy measurement sensor, the footprint peak for forest soil respiration is typically over twice the distance as compared to that for canopy photosynthesis. This result has important consequences for interpretation of annual ecosystem carbon balance estimations with the eddy-covariance method. The study of the Hyytiälä site revealed the effects of topography on scalar concentration and flux fields within the atmospheric surface layer. The fluxes at a fixed height vary as a function of position in respect of topography. The fluxes tend to be larger at the upwind foot of the ridge and at the downwind side of the ridge crest, being smaller downhill. Correspondingly, the flux footprints depend on the location of the flux measurement point and may significantly deviate from those for a flat terrain.

Vertical fluxes and footprint behavior over a few simplified landscape types were investigated by Sogachev et al. (2005b). Hypothetical heterogeneous vegetation patterns – forest with clear-cuts as well as a hypothetical heterogeneous relief, a bell-shaped valley, and a ridge covered by forest – were considered. The disturbances induce changes in scalar flux fields within the atmospheric surface layer compared to fluxes for homogeneous conditions: at a fixed height the fluxes vary as a function of distance from disturbance. Correspondingly, the flux footprint estimated from model data depends on the location of the point of interest (flux measurement point). This study demonstrated mainly that any generalization of the footprint and flux behavior as a function of landscape heterogeneity is still a challenging task due to their site specificity.

The behavior of both scalar fluxes and flux footprints near a forest edge were investigated in detail for the Florida AmeriFlux site (Sogachev et al. 2005a) and Bankenbosch forest in the Netherlands (Klaassen and Sogachev 2006). The former study examined the influence of bare soil patch located upwind of the eddy-covariance tower on fluxes in a forest plantation. Scalar fluxes and flux footprints from a clear-cut–forest transect with swaths of logged land with dimensions varying with wind direction were modeled (see Fig. 8.14). In sharp contrast with momentum fluxes, the magnitudes of CO_2 and scalar fluxes were found to be sensitive to clear-cut width. The adjustment to new underlying scalar flux values as a function of distance from the leading forest edge appeared to be far greater for scalar fluxes than for momentum fluxes. This result is consistent for all modeled clear-cut swaths – forest canopy interfaces, suggesting that CO_2 flux measurements using the eddy-covariance technique require a larger fetch for forest flux towers than previously thought. The footprint analysis indicated flux contributions from the clear-cut, forest floor, and forest canopy to the tower flux hundreds of meters downwind of the clear-cut–forest interface and highlighted the need for caution in the interpretation of data away from the leading forest edge (up to 30 canopy heights) (Fig. 8.14). This is especially true when the strengths of both surface and in-canopy sources are of comparable magnitude.

The study of Klaassen and Sogachev (2006) showed in addition that with increasing forest density, atmospheric fluxes deviate even more strongly from surface fluxes, but over shorter fetches. It was concluded that scalar fluxes over forests are commonly affected by inhomogeneous turbulence over large fetches downwind of an edge. It is recommended to take horizontal variations in turbulence into account when the footprint is calculated for atmospheric flux measurements

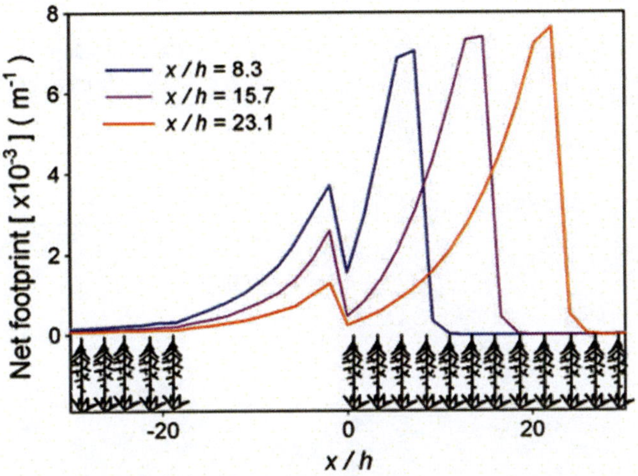

Fig. 8.14 Examples of net footprints (joint contribution of sources located within the canopy layer and on the soil surface are considered) derived by the model for a case of 17 h wide clear-cut for sensors located at various normalized distances, *x/h* downwind of the forest edge at a height of 1.4 h (After Sogachev et al. 2005a)

downwind of a forest edge. The spatially integrated footprint is recommended for describing the ratio between the turbulent flux above forest and the average surface flux in the source area.

The knowledge of the footprint itself considerably improves our ability to decompose a flux signal into its different source signatures. However, Sogachev et al. (2005b) pointed out that for establishing and locating the flux towers, the information provided by the footprint function is more convenient if presented in a different form. They introduced fractional flux function describing the contribution of given source into a signal at that imaginary flux tower. Figure 8.15 compares these fractional flux functions for measurement height $z = 1.4\,h$ obtained for the different modeled clear-cut sizes. The behavior of these functions depends on the flow structure in the clear-cut–forest transition zone, which in turn is defined by the canopy structure. The flow acceleration in the lower canopy and above, the flow deceleration in the upper canopy region together with the vertical air motions, all occurred in this zone resulting in a complicated distribution of the scalar field and vertical fluxes. With information on fluxes from the soil in clear-cut and forest areas (as might be seen during nighttime conditions with upward CO_2 fluxes, for example) and from the forest canopy, net fluxes at given height downwind of the forest edge can be estimated.

Both studies suggested that, to improve our current assessment of net carbon uptake, attention should be given to the importance of careful tower location selection in a landscape characterized by a mosaic of surface properties as observed in most natural ecosystems. For towers located in complex terrain, the approach based on two- and three-dimensional flow model capable of taking into account

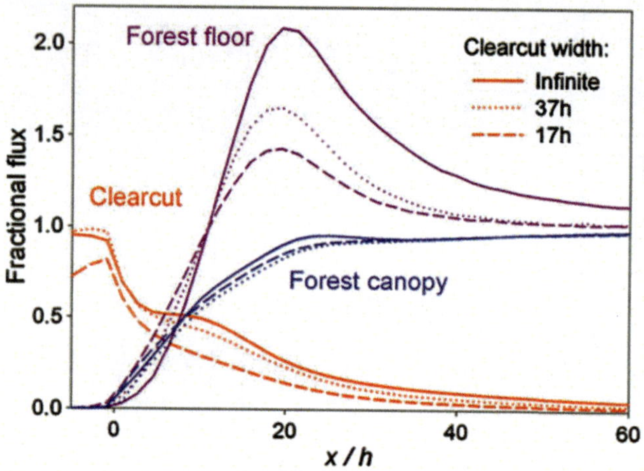

Fig. 8.15 Variation of the fractional flux functions at a height of 1.4 h with normalized distance, *x/h* downwind of the forest edge, derived by footprint modeling for sources on forest floor, inside a tree layer and on the clear-cut. These functions describe the contribution of corresponding sources to a measured signal at an arbitrary location downwind of the clear-cut–forest edge (After Sogachev et al. 2005a)

heterogeneity of surface is strongly recommended for footprint estimation. The interpretation of the eddy-covariance flux measurements over Lake Valkea-Kotinen in the framework of Helsinki Environment Research Centre (HERC) project (Vesala et al. 2006) is a practical example confirming the adequacy and usefulness of this approach.

Quantitative behavior of scalar fluxes near a forest edge depends strongly on forest structure and surrounding conditions. Nevertheless, several general conclusions can be drawn which should be taken into account when interpreting data observed at short and moderate fetches (less than 40 tree heights downwind of a forest edge):

An adjustment in the momentum flux does not necessarily mean an adjustment in scalar flux. It is recommended that more stringent fetch requirements for scalar flux observations be used.

The adjustment rates for scalar fluxes originating from sources inside canopy are faster than that for soil fluxes for any forest structure. As a consequence, in addition to upwind clear-cut dimension, the adjustment rate of the flux from combined sources/sinks depends also on the ratio of canopy/ground source strengths.

The ground source plays a major role in the formation of wave-like vertical scalar flux behavior over the forest downwind of a forest edge, despite the fact that the contribution of foliage sources/sinks changes monotonously. Such a variation is caused by scalar advection in the trunk-space. The effect was more pronounced in model forests with leaf area concentrated in the upper part of the canopy.

Overall, the interpretation of data from eddy flux measurement systems located downwind of a clear-cut–forest discontinuity always needs a more in-depth examination to ensure that correct footprints are calculated and that the measured fluxes are properly interpreted relative to net ecosystem exchange.

8.4.5 Modeling over Urban Areas

Recently, Vesala et al. (2008a) successfully implemented this method for estimation of footprint for measuring tower surrounded by complex urban terrain. Besides the above example for Tver region (European Russia) (Sogachev and Lloyd 2004), it is a second attempt of footprint prediction in three-dimensional landscape reported in the literature. Performed footprint analysis allowed for discrimination of the influence of surface and canopy sinks/sources and complex topography on observed fluxes. The heterogeneity of urban surface results in complex transport from sources to receptor and the footprint signature was asymmetric along prevailing wind direction. Thus, any two-dimensional footprint models (especially based on analytical solutions) should be avoided for urban surrounding even with flat topography. Jarvi et al. (2009) applied also the ABL model for estimation of footprint over urban areas including the effect of real urban structure on the flow. In simulations, land use was classified into nine different types including roads, parking areas, soil, and trees with two different height classes, and buildings with four different height classes. Buildings were considered to be impenetrable. The footprint calculation was made for the road sector with the surface wind from a direction perpendicular to the road, and a geostrophic wind speed of 10 ms^{-1}. Neutral stratification of the atmosphere was assumed. The cell size used in the simulation was 20×20 m^2. The airflow at the height of 10 m above surface and flux footprint for ground sources and for the sensor located at the height of 31 m are presented in Fig. 8.16.

The flow pattern was strongly affected by buildings, and therefore the footprint function of the surface fluxes showed a complex pattern, unlike the smooth pattern characteristic of horizontally homogeneous conditions. In fact, the function had two local maxima, one close to the measurement tower and another at a distance further upwind. Model simulations also indicated that the footprint function was highly sensitive to wind direction.

There are only a few attempts presented above to estimate footprint over urban area. However, over complex topography and heterogeneous terrain, the only possible way to estimate the influence of surface sources on the measured flux is through the use of numerical calculations.

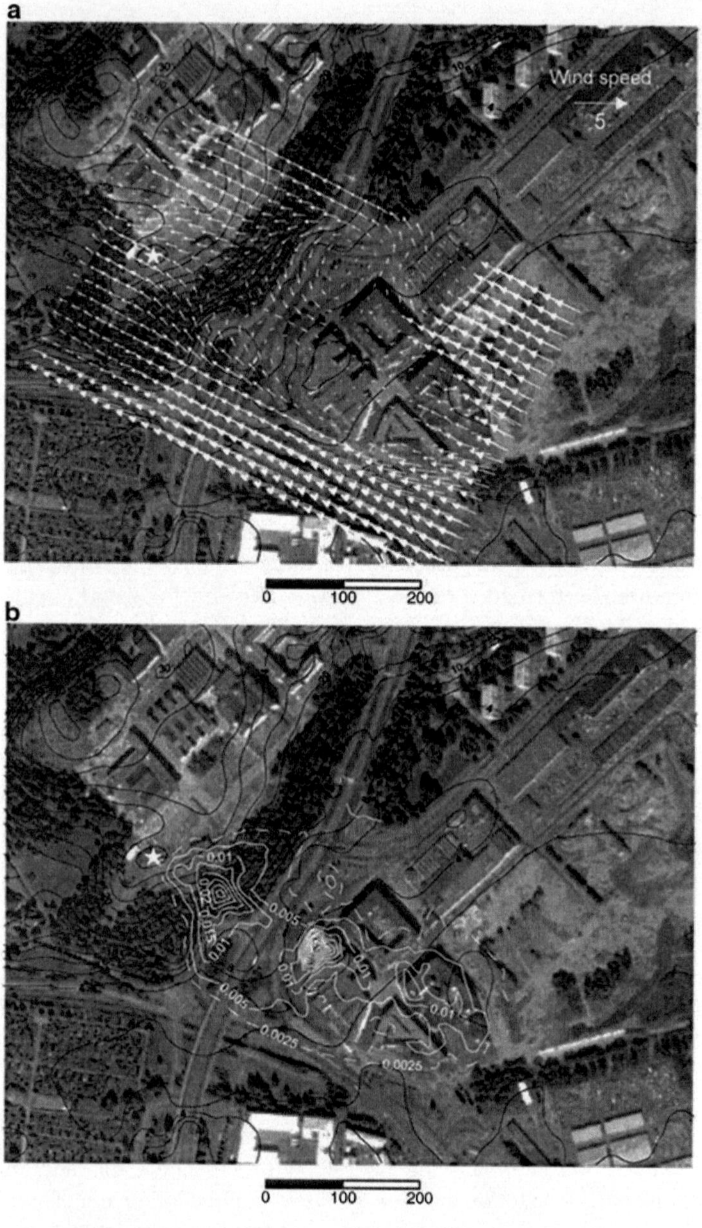

Fig. 8.16 Aerial photograph of the measurement location. Topography of the measurement site (relative to sea level) is denoted by *black contours*. Wind vector plots (**a**) and the flux footprint function (**b**) (scale 10^{-6}, the unit of flux footprint is m^{-2}) are shown when the wind direction is perpendicular to the road (117°), Geostrophic wind speed is 10 m s^{-1} and the boundary layer is neutrally stratified. The location of the measurement tower is marked by a white star, and its distance to the edge of the road is around 150 m (After Jarvi et al. 2009)

8.5 Quality Assessment Using Footprint Models

The application of the eddy-covariance technique to monitor turbulent exchange processes between surface and atmosphere is restricted to basic theoretical assumptions, the most important of which are steady-state flows, a mean vertical wind component of zero, and non-advective conditions (e.g., Foken et al. 2004; Foken 2006; Kaimal and Finnigan 1994). Deviations from these assumptions will increase measurement uncertainty, and thus have a negative impact on overall data quality (see also Sect. 4.3, Chap. 4). Heterogeneity in the area surrounding an eddy-covariance measurement site, such as clearings in a forest, fields with different crop types in an agricultural area, or obstacles like buildings or trees in an otherwise open grassland, holds the potential to disturb the atmospheric flow, and trigger the above-mentioned deviations from ideal conditions that cause data quality to decrease (e.g., Baldocchi et al. 2005; Panin and Tetzlaff 1999; Schmid and Lloyd 1999). Evaluating the influence of such terrain heterogeneity on eddy-covariance measurements through footprint modeling can, therefore, serve as an important component in the overall eddy-covariance data quality assessment strategy (Foken et al. 2004).

In recent years, the growing number of eddy-covariance sites organized in networks such as FLUXNET (Baldocchi et al. 2001), CarboEurope (Valentini et al. 2000), or Ameriflux (Law 2005), lead to a shift from ideal, homogeneous sites to complex and heterogeneous conditions (e.g., Schmid 2002). To facilitate coverage for a wide range of ecosystems many sites had to be established in heterogeneous areas with variable land cover types, since there had to be a compromise between the ecological importance of a new site and the suitability of the surrounding environment for high-quality eddy-covariance measurements. Accordingly, there is a strong interest in methods and applications that can link quality features in the measured data with characteristics of the surrounding terrain. Such efforts are particularly valuable for the increasing number of FLUXNET synthesis studies (Grant et al. 2009; Luyssaert et al. 2008; Stoy et al. 2009) that pool observations from multiple sites to generate, for example, products representative for larger scales.

As a diagnostic quality assessment tool for existing databases, footprint analyses can generally be applied in three different areas:

Testing the spatial representativeness of the measured fluxes. Footprint model results can reveal the composition of different land cover types, different forest age classes, etc., in the fetch of a measurement (Göckede et al. 2004, 2006). This information can be used to characterize the variability in the flux time series that is caused by a changing field of view of the sensors, and ideally the total flux can be decomposed into flux contributions from different biomes (Barcza et al. 2009; Soegaard et al. 2003; Wang et al. 2006). If data from a homogeneous flux source are required, for example, to train a model for a specific biome like conifer forest, the footprint filter can indicate which measurements provide the "true" forest signal, and which are "contaminated" by, for example, clearings or water bodies

(Göckede et al. 2008; Rebmann et al. 2005). A test for spatial representativeness is also necessary to link eddy-covariance measurements to data at different spatial resolution, such as upscaling to remote sensing information grids (Chen et al. 2008; Kim et al. 2006; Reithmaier et al. 2006) or aircraft data (Kustas et al. 2006; Ogunjemiyo et al. 2003), or downscaling for comparison to soil chamber measurements (Davidson et al. 2002; Myklebust et al. 2008; Reth et al. 2005).

Linking data quality to terrain features. Eddy-covariance data quality assessment results, as, for example, outlined in Sect. 4.3, can be linked with footprint analyses to produce spatial maps of the data quality (Göckede et al. 2004, 2006, see below for details) These maps hold the potential of identifying general instrumentation problems, disturbed wind sectors under different conditions of atmospheric stability, or even the influence of single obstacles in the near field of a sensor. Potential effects will show up as structures in the spatial maps, for example, a single wind sector with reduced data quality for a specific atmospheric stability regime. Such structures are often caused by subtle trends which might easily be missed in a standard database filter. Such "bad" situations can be flagged to strengthen the database.

Visualize spatial structures in ancillary parameters. In the same way as outlined above for the data quality, in principle any measured parameter (scalars and fluxes) can be linked with the footprint analyses to produce spatial maps. A classic example for this application would, for example, be the visualization of spatial structures in the mean vertical wind component (Göckede et al. 2008). Other examples include visualizing the flux fields of sensible or latent heat, which may indicate spatially variable sources for these parameters.

In addition to analyzing existing datasets in a diagnostic way, footprint modeling can also be applied in a "predictive" way to assist in the planning of new meteorological experiments. Using either hypothetical or measured wind climatology datasets, the instrument position can be optimized by, for example, maximizing the influence of fluxes from the biome intended to monitor, and/or minimizing the influence of potential obstacles in the fetch of the sensors.

8.5.1 Quality Assessment Methodology

A comprehensive quality assessment framework to include footprint analyses into eddy-covariance data quality assessment schemes was first introduced by Göckede et al. (2004). Their approach, which built on an analytic flux footprint model (FSAM, Schmid 1994, 1997), addressed all three general quality assessment areas listed above, and was successfully applied by Rebmann et al. (2005) to 18 sites of the CARBOEUROFLUX network. An upgraded version of this framework (Göckede et al. 2006), which aimed at a more reliable performance and broader applicability, replaced the analytic footprint model by a forward Lagrangian stochastic (LS) trajectory model (Rannik et al. 2003). This software tool provided the results for an extensive quality control study of CarboEurope-IP data (Göckede et al. 2008) that summarized findings from 25 forested sites.

To ensure representative findings, footprint analyses for data quality assessment should use a database of several months (at least 2–3) of meteorological measurements, so that several thousand half-hourly averaged observations are available. The correct interpretation of the findings relies on a good sample of the local wind climatology, and sufficient coverage of different atmospheric stability conditions for all wind sectors. The analysis will be strengthened by choosing a database that covers a period of the year with high absolute values of exchange fluxes between surface and atmosphere. Concerning the required gridded maps of the terrain characteristics such as land cover type or stand age, the spatial resolution as well as the number of classes assigned only play a minor role as long as the map resolves those details in the surrounding terrain the specific study is aiming at (Reithmaier et al. 2006). For example, coarse resolution maps might be sufficient for studies that simply differentiate between generic forest and the non-forest areas beyond the forest edge, while finer resolution maps will be required if also patches of coniferous, deciduous, and mixed forest need to be resolved, or the forest is interspersed by small clearings. Overall, the quality of the footprint results tend to improve through the use of more detailed, remote sensing based map material.

An integral part of footprint-based quality assessment approaches is the average source weight function for a longer measurement period, the so-called footprint climatology (e.g., Amiro 1998). It is obtained by summing up the source weight functions of individual 30-min measurements over a longer period of time, and thus reflects the local wind climatology and the distribution of different classes of atmospheric stability on the long-term measurement conditions. Figure 8.17 demonstrates the variability of footprint climatologies under different stratification regimes, highlighting also the change in the composition of land cover types within the footprint with varying conditions. In these two-dimensional visualizations, the white effect level rings indicate the three-dimensional topography of the footprint climatologies, with the most influential terrain areas located in the center of the concentric rings.

For the evaluation of the spatial representativeness of an eddy-covariance flux dataset, footprint climatologies merged with land cover maps as shown in Fig. 8.17 already provide a first impression on the potential impact of terrain heterogeneity on the observations. The most prominent land cover classes within the area encircled by the white isolines will also dominate the flux measurements. "Disturbance elements," such as clearings in a forest, will have a higher impact on the dataset the closer they are situated to the center of the concentric effect level rings. For a more detailed analysis, each 30-min source weight function has to be projected onto the gridded land cover map, assigning a weighting factor to each grid cell that represents its relative contribution to the actual measurement. The distribution of flux contributions from the different land cover types can then be obtained by accumulating these weights sorted by land cover type. Application on a larger dataset reveals patterns in the composition of the footprint that are dependent on wind sector and stability regime. This information is particularly valuable in case

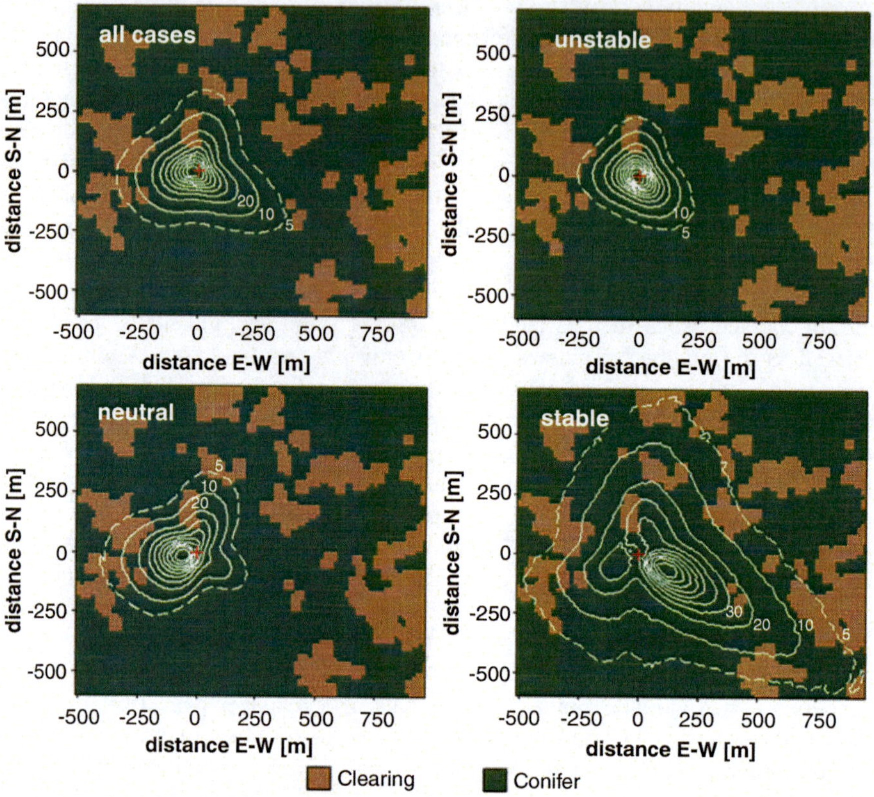

Fig. 8.17 Top-down view on footprint climatologies (*white lines*), accumulated for different regimes of atmospheric stability, obtained for the Weidenbrunnen tower in southeastern Germany. Panels give footprint climatologies for all cases (*top left*), unstable (*top right*), neutral (*bottom left*) and stable (*bottom right*) stratification. Values are in percentages to the peak of the function, with *solid lines* ranging from 90% to 10%, and the *dashed line* as 5% of the maximum. High values indicate a high relative contribution of the specific area to the fluxes measured in the given observation period. *Colors* in the background indicate land cover classes. Distances to the tower position (*red cross*) are given in [m]

a dataset is supposed to represent a certain "target land cover type," for example, for a site intercomparison, or the training of an ecophysiological model. For such applications, the footprint results can be used to provide the percentage contribution of the specified target land cover type to the total flux, and measurements that fail to reach a user-specified minimum threshold can be discarded from the database (e.g., Nagy et al. 2006).

For network intercomparison studies such as presented by Rebmann et al. (2005) or Göckede et al. (2008), it is recommended to classify the homogeneity of flux sources within the source area by defining thresholds of target area flux contributions. Göckede et al. (2008) defined four different classes:

Homogeneous measurements, with 95% or more of the flux emitted by the target land cover type
Representative measurements (80–95%)
Acceptable measurements (50–80%)
Disturbed measurements (<50%)

For a site intercomparison, it can then be determined what percentage of the total dataset at each site could, for example, be classified as homogeneous or representative measurements, which can serve as an indicator of how well the sites could be compared, or how suitable they are for model training focusing on a specific biome.

For the second way of applying footprints in eddy-covariance quality assessment, linking data quality to terrain features, the footprint results need to be coupled to approaches evaluating the flux data quality of the measurements. The specific method to assign flux data quality, as well as the definition and resolution of quality classes, can be chosen by the user and customized for each study, as long as the quality ratings are numeric to allow aggregation. Göckede et al. (2006, 2008) applied a scheme proposed by Foken and Wichura (1996) in the revised version as presented by Foken et al. (2004), which assigns quality flags between 1 (best) and 9 (worst) for the fluxes of momentum, sensible and latent heat, and CO_2 (see Sect. 4.3.3). To create spatial maps of the data quality, the quality flags for each individual 30-min measurement are projected onto a discrete grid, storing relative influence and quality flag results for each grid cell in a database. After processing the entire dataset, this information can be converted into a frequency distribution of data quality for each cell, which in turn yields the overall quality rating as the median of the distribution (see Göckede et al. 2004, 2006 for details). Visualization of the results helps in revealing spatial patterns in data quality, such as isolated wind sectors with significantly reduced quality ratings compared to neighboring regions (Fig. 8.18). Such patterns may, for example, hint at terrain structures in that specific wind sector which has a negative impact on atmospheric measurement conditions, or might be caused by flow distortion induced by the instrumental setup. Observations of multidirectional reduction in data quality for specific subsets of the measurements (Fig. 8.19) can indicate instrumental problems, such as water in the tubing of a closed-path infrared gas analyzer that only precipitates during lower temperatures at night. Whatever the cause of the reduced data quality, affected wind sectors or stability regimes can be flagged and removed from the database to improve overall data quality.

For the visualization of spatial structures in ancillary parameters, the procedure resembles the one described above for quality flag analysis, only that observational data such as the mean vertical wind speed or the friction velocity replace the data quality ratings. This application allows exploring spatial effects for a large number of parameters which hold the potential to help interpret cases of low data quality, or identify instrumental problems. An example of this type of analysis included into the framework by Göckede et al. (2006) is the visualization of spatial structures in the vertical wind component before and after application of the

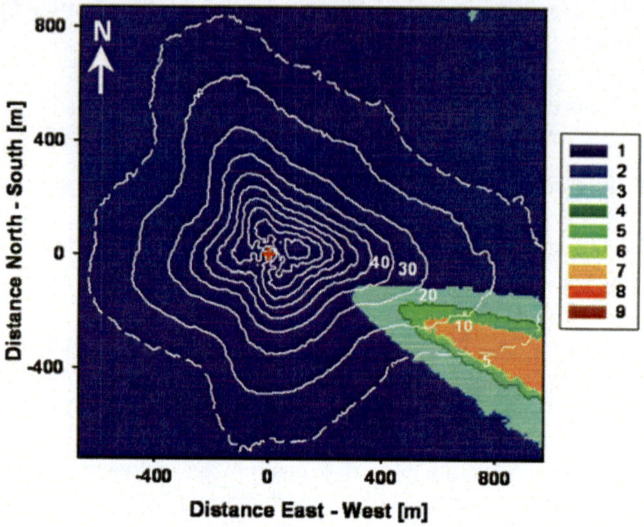

Fig. 8.18 Example for an isolated wind sector with reduced data quality, taken from Göckede et al. (2008). Background colors give the median quality rating (1 = best) of the momentum flux during stable stratification ($z/L > 0.0625$; z: measurement height [m]; L: Obukhov length [m]) at the Wetzstein site in central Eastern Germany (See caption of Fig. 8.17 for further details)

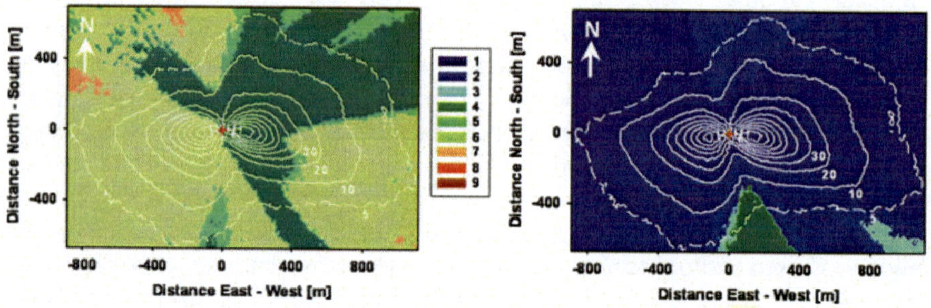

Fig. 8.19 Comparison of the spatial data quality of the latent heat flux (*left panel*) and the CO_2 flux (*right panel*) during stable stratification, taken from Göckede et al. (2008). *Background colors* give the median quality rating (1 = best) obtained for the Soroe site in Denmark (See captions of Fig. 8.17 for further details)

Planar-Fit coordinate rotation (Wilczak et al. 2001). These results indicate tilt and distortion of the initial wind field, and the effectiveness of the coordinate rotation to correct the flow conditions to a mean vertical wind of zero, as required for eddy-covariance measurements. Figure 8.20 gives an example of structures in the vertical wind field before and after rotation. In this case, the absolute deviations from the ideal value of zero could be significantly reduced through Planar-Fit, but spatial patterns still remain in the corrected dataset, because the complex terrain at this site

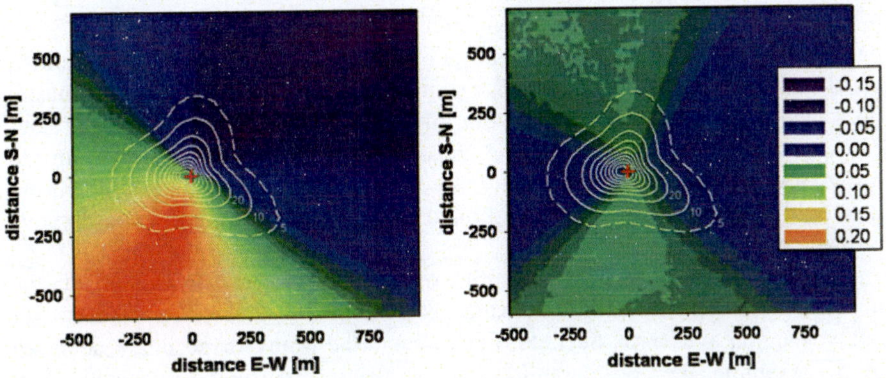

Fig. 8.20 Spatial map of the mean vertical wind component before (*left panel*) and after (*right panel*) application of the Planar-Fit coordinate rotation. Results taken from site analysis of the Weidenbrunnen site in southeastern Germany (See captions of Fig. 8.17 for further details)

produces a slightly curved wind field that cannot be completely corrected for with a single set of rotation angles. In a similar fashion, sectors with particularly low friction velocities during nighttime could be identified to highlight advection-prone conditions. The visualization of heterogeneity in the sources for momentum, heat, or CO_2, surrounding the site is also possible; however, for this application the impact of external drivers such as temperature or radiation on the flux variability needs to be taken into account through additional filters.

8.5.2 *Site Evaluation with Analytical and LS Footprint Models*

Analytical footprint models have been widely applied to characterize the "field of view" of eddy-covariance measurements. Their popularity is mainly based on their relative mathematical simplicity (e.g., Schmid 2002) that allows integrating them into eddy-covariance processing software packages without high additional computational expense, or even estimate fetch lengths through spreadsheet applications. This simplicity is what makes them attractive as a component in site evaluation tools, since particularly network studies require the processing of tens of thousands of footprint estimates. Analytic footprint models are often restricted to rather narrow ranges of input parameters like aerodynamic roughness length or stability of atmospheric stratification, which reduces the dataset that can actually be processed, and/or calls for adaptations in the assignment of effective roughness lengths. Rebmann et al. (2005) demonstrated the potential of the site evaluation tool by Göckede et al. (2004), which is based on an analytical footprint model, for network-wide studies on a large number of sites. Even though their study was impacted by certain conceptual limitations of the analytical footprint model such as the exclusion of stable stratification cases beyond a certain stability threshold, they produced a comprehensive survey on fetch conditions and data quality among the participating 18 observation sites, and pointed out footprint-related measurement problems. Since analytical footprint estimates tend to be larger than stochastic ones due to the neglect of along-wind diffusion, and at the same time the land cover structure is usually more heterogeneous with increasing distance from the tower (tower locations are commonly selected to be homogeneous at least in the near field), the site evaluations based on analytical models provide a conservative estimate of quality results.

Lagrangian stochastic (LS) footprint models offer more ways of adaptation to local measurement conditions, which is particularly valuable for studies over tall vegetation (see also Sect. 8.3). However, the gains in accuracy achieved by, for example, the consideration of within-canopy transport (Baldocchi 1997; Rannik et al. 2003), sources at multiple levels (Markkanen et al. 2003), or along-wind diffusion (Rannik et al. 2000), come along with significantly increased computational expense, which plays a major role for site evaluation concepts that cover multiple sites over a time frame of several months. Also, the quality of the simulations is dependent on a reliable description of vertical turbulence profiles under various atmospheric conditions (Göckede et al. 2007), while high-quality datasets to describe these profiles are only available for a small subset of sites within FLUXNET. Application of LS models in extensive site evaluation studies, therefore, usually calls for simplifications in the setup, such as the use of generic turbulence profiles that are not customized for each specific forest stand, and the pre-calculation of source weight functions for specific combinations of atmospheric stability, measurement height, and terrain roughness (Göckede et al. 2006). Using this strategy, the network study by Göckede et al. (2008) processed 30-min footprints for 76 data months of flux measurements from 25 sites, demonstrating the applicability of LS footprints as a standard quality assessment tool for eddy-covariance databasing.

8.5.3 *Applicability and Limitations*

Quality assessment using footprint models, like most other areas where footprints are commonly used, is subject to an inherent contradiction that needs to be taken into consideration when interpreting the results. The application of most footprint models is theoretically restricted to horizontally homogeneous flow conditions, which can only be obtained if the tower is surrounded by perfectly uniform terrain with respect to topography, aerodynamic roughness, and sources for sensible and latent heat. This is particularly the case for analytic and forward Lagrangian stochastic models which are easy to use and flexible enough to be applied to multiple sites over longer time frames, as required for the network studies cited above. The overriding objective of footprint-based site evaluation, however, is to characterize the influence of heterogeneity in the surrounding terrain on the flux measurements. Therefore, such tools will always violate the area of applicability that has been defined for the employed footprint model (Vesala et al. 2008b), except for the unusual case that the terrain is perfectly flat, and the analyzed heterogeneities only affect the sources of "passive" scalars like CO_2 but not the flow conditions. These problems can only be avoided through the use of backward LS footprint models (Kljun et al. 2002) or closure approaches (Sogachev et al. 2005a, b) (see also Sects. 8.2 and 8.4) that can explicitly handle inhomogeneous flow conditions, but their requirements for setting up the model domain might make extensive network studies impossible.

All footprint results obtained outside the area of applicability of the underlying models will be subject to increased uncertainty. Complex topography and step changes in roughness or heat flux source strength alter the atmospheric flow conditions (Foken and Leclerc 2004; Klaassen et al. 2002; Leclerc et al. 2003a; Schmid and Oke 1988), so source area predictions based on the assumption of homogeneous transport will be biased (Finnigan 2004). This uncertainty will only slightly affect qualitative site evaluation results like the identification of a wind sector with reduced data quality, but quantitative findings like the percentage flux contribution of a certain land cover type have to be evaluated carefully. A general error estimate cannot be provided, since the deviations from ideal flow conditions depend on the relative location of "disturbing" terrain elements with respect to the sensor position, and the local wind climatology, so that they need to be reassessed for every case study.

Footprint studies that aim at the assessment of long-term averaged properties, like a representative footprint climatology or the mean data quality for a specific sector, are likely to be biased by problems related to simulating source weight at night. Analytical models, like, for example, the FSAM model (Schmid 1994, 1997) used in the framework by Göckede et al. (2004), are often restricted to input parameter ranges that exclude parts of the stable stratification range, effectively discriminating against nighttime measurements which often have large source areas and tend toward lower flux quality ratings. LS models are less numerically unstable than analytical ones in situations as such, but also the representativeness of LS footprints is questionable in case of weak and intermittent turbulence, or even flow conditions dominated by wave motions. Exclusion of these situations leads to a systematic shift toward higher data quality and smaller footprint climatologies, compared to a treatment of the complete dataset; however, at least in case of the LS models, the major part of the excluded data does not fulfill the theoretic assumptions for eddy-covariance data processing either, so these data also would not be considered for the assessment of the net carbon budget, for example.

In an attempt to better characterize the problems linked to the application of footprint models in heterogeneous flow conditions, Markkanen et al. (2009) classified the agreement of different footprint models with an LES study (Steinfeld et al. 2008). Their results categorized the correlation between models in relation to the contribution of different flux sources to the total flux, and also considered the location of "disturbing" grid elements relative to the location of the peak of the source weight function. Factors influencing the model output, such as the horizontal grid resolution, or the measuring height of the sensors, were taken into account in a sensitivity study. Data quality analyses such as described in the previous sections can generally be improved by comparing the general accuracy of employed footprint models with a reference model that is better suited for application in heterogeneous conditions. Such an approach would strengthen the data quality analyses, and allow for drawing more reliable conclusions on the site characteristics

8.6 Validation of Footprint Models

Validations of footprint models are often only a comparison of different footprint models. According to Foken and Leclerc (2004), only a few experimental datasets of tracer experiments are available for validation purposes. While analytical footprint predictions were often evaluated using results of Lagrangian footprint models, there is no such simple possibility for the evaluation of Lagrangian footprint models. LS footprint models consist of a dispersion model and an estimation scheme for the footprint function. LS dispersion models have been tested against dispersion experiments in numerous cases for different turbulence regimes (Reynolds 1998; Kurbanmuradov and Sabelfeld 2000; Kljun et al. 2002); therefore, the ability of LS models to reproduce dispersion statistics for several flows is well established. Only a few footprint results from Lagrangian models were compared with experimental data: Leclerc et al. (1988) first compared a Lagrangian simulation against a tracer released at different depths above and inside a short alfalfa canopy. The results suggest that the influence of thermal stability inside plant canopies played an important role on the turbulent diffusion. Subsequently, Finn et al. (1996) performed a tracer experiment in the convective boundary layer at Hanford over a short sagebrush canopy in unstable conditions against which they tested both a Lagrangian simulation used in Leclerc and Thurtell (1990) and the Horst and Weil (1994) analytical solution to the diffusion equation. Leclerc et al. (1997) compared both LES and Lagrangian simulations against tracer flux data in the convective boundary layer with good results. Leclerc et al. (2003a, b) have compared the footprint models for fluxes over forest canopy with tracer flux measurement experiments. Mölder et al. (2004) dealt with the validation of footprint models as a way to compare both analytical and Lagrangian models against experimental data. Kljun et al. (2004b) compared forward and backward Lagrangian models against data from tracer release experiments in a wind tunnel. Such independent comparisons between models and experimental validation enable a robust assessment of model sensitivity to various environmental variables. In general, the investigated footprint models agree well with the tracer experiment. Even though these comparisons gave promising results, there is still a need for further experimental data allowing for validation of footprint models, the LS models in particular.

Foken and Leclerc (2004) pointed out that complex validation experiments are expensive, and hence difficult to achieve. Nevertheless, the authors show that ongoing experiments can also be used to validate footprint models, when two or more well-defined and neighboring surfaces with significantly different fluxes can be studied. These issues are important if footprint models are to be used as a tool to define experimental requirements and validate experimental data. An application of this method was successfully made by Göckede et al. (2005) with two flux stations over bare soil and a meadow. A third flux station with a footprint area covering both surfaces was used to validate the footprint model, because the contributions of both surfaces changed with the stability and wind velocity. Earlier investigations used a similar approach: Soegaard et al. (2003) operated five ground-level EC systems over five different crop fields together with a sixth set up on top of a higher mast to enable landscape-wide flux measurements. The agreement between high-level values and those integrated from ground-level using a reformulated version of the models of Gash (1986) and Schuepp et al. (1990) was good. Hsieh et al. (2000) developed an analytical model based on Lagrangian dispersion model and dimensional analysis. They found a good agreement with model predictions and measured fluxes over a transect from a desert to an irrigated potato field. More recently, Marcolla and Cescatti (2005) compared three analytical footprint models over a meadow with different surface characteristics and one of the models (Schuepp et al. 1990) overestimated the footprint.

The LES approach provides a valuable "dataset," with much of the flow complexities inherent in a true atmospheric flow against which simpler footprint models can be verified. Recently, Prabha et al. (2008a) made a comparison between the in-canopy footprints obtained using a Lagrangian simulation with those obtained against a large-eddy simulation. In that model, the Lagrangian stochastic model was driven by flow statistics derived from the large-eddy simulation. Markkanen et al. (2009) published a review of conventional stochastic models, that is, that of Kljun et al. (2002) and that of Rannik et al. (2000) tested against LES-driven footprint estimates for ABL and ASL conditions, respectively. They concluded that the models agreed well for most of the measurement heights. The two conventional flux footprint models agreed best under near-neutral conditions, whereas the agreement between LES and LS model for ABL was better for intermediate measurement heights and for the convective case.

Acknowledgments Support from ACCENT-BIAFLUX, IMECC EU project and ICOS EU project is acknowledged together with the Academy of Finland Center of Excellence program (project number 1118615).

References

Amiro BD (1998) Footprint climatologies for evapotranspiration in a boreal catchment. Agric For Meteorol 90(3):195–201

Apsley DD, Castro IP (1997) A limited-length-scale $k-\varepsilon$ model for the neutral and stably-stratified atmospheric boundary layer. Bound Layer Meteorol 83:75–98

Aubinet M, Heinesch B, Yernaux M (2003) Horizontal and vertical CO2 advection in a sloping forest. Bound Layer Meteorol 108(3):397–417

Aubinet M, Berbigier P, Bernhofer Ch, Cescatti A, Feigenwinter C, Granier A, Grünwald Th, Havrankova K, Heinesch B, Longdoz B, Marcolla B, Montagnani L, Sedlak P (2005) Comparing CO_2 storage and advection conditions at night at different Carboeuroflux sites. Bound Layer Meteorol 116:63–94

Ayotte KW, Finnigan JJ, Raupach MR (1999) A second-order closure for neutrally stratified vegetative canopy flows'. Bound Layer Meteorol 90:189–216

Baldocchi DD (1997) Flux footprints within and over forest canopies. Bound Layer Meteorol 85:273–292

Baldocchi D et al (2001) FLUXNET: a new tool to study the temporal and spatial variability of ecosystem-scale carbon dioxide, water vapor, and energy flux densities. Bull Am Meteorol Soc 82(11):2415–2434

Baldocchi DD et al (2005) "Wet/dry Daisyworld": a conceptual tool for quantifying the spatial scaling of heterogeneous landscapes and its impact on the subgrid variability of energy fluxes. Tellus B 57(3):175–188

Barcza Z et al (2009) Spatial representativeness of tall tower eddy covariance measurements using remote sensing and footprint analysis. Agric For Meteorol 149(5):795–807

Baumert H, Peters H (2000) Second-moment closures and length scales for weakly stratified turbulent shear flows. J Geophys Res 105:6453–6468

Belcher SE, Hunt JCR (1998) Turbulent air flow over hills and waves. Annu Rev Fluid Mech 30:507–538

Belcher SE, Jerram N, Hunt JCR (2003) Adjustement of a turbulent boundary layer to a canopy of roughness elements. J Fluid Mech 488:369–398

Cai XH, Leclerc MY (2007) Forward-in-time and backward-in-time dispersion in the convective boundary layer: the concentration footprint. Bound Layer Meteorol 123:201–218

Cai X, Peng G, Guo X, Leclerc MY (2008) Evaluation of backward and forward Lagrangian footprint models in the surface layer. Theor Appl Climatol 93:207–233

Cai X, Chen J, Desjardins RL (2010) Flux footprints in the convective boundary layer: large-eddy simulation and lagrangian stochastic modelling. Bound Layer Meteorol 137:31–47

Castro FA, Palma JMLM, Silva LA (2003) Simulation of the Askervein flow: Part 1: Reynolds averaged Navier-Stokes equations (k-ε turbulence model). Bound Layer Meteorol 107:501–530

Chandrasekar A, Philbrick CR, Clark R, Doddridge B, Georgopoulos P (2003) A large-eddy simulation study of the convective boundary layer over Philadelphia during the 1999 summer NE-OPS campaign. Environ Fluid Mech 3:305–329

Chen JM, Black TA, Novak MD, Adams RS (1995) A wind tunnel study of turbulent air flow in forest clearcuts. In: Coutts MP, Grace J (eds) Wind and trees. Cambridge University Press, London, pp 71–87, chap. 4

Chen BZ, Chen JM, Mo G, Black A, Worthy DEJ (2008) Comparison of regional carbon flux estimates from CO2 concentration measurements and remote sensing based footprint integration. Glob Biogeochem Cycles 22(2)

Corrsin S (1974) Limitations of gradient transport models in random walks and turbulence. Adv Geophys 18A:25–60

Davidson EA et al (2002) Minimizing artifacts and biases in chamber-based measurements of soil respiration. Agric For Meteorol 113(1–4):21–37

Deardorff JW (1972) Numerical investigations of neutral and unstable planetary boundary layers. J Atmos Sci 18:495–527

Denmead OT, Bradley EF (1985) Flux-gradient relationships in a forest canopy. In: Hutchison BA, Hicks BB (eds) The forest-atmosphere interaction. Reidel, Dordrecht, pp 421–442

Duynkerke PG (1988) Application of the E - ε turbulence closure model to the neutral and stable atmospheric boundary layer. J Atmos Sci 45:865–880

Finn D, Lamb B, Leclerc MY, Horst TW (1996) Experimental evaluation of analytical and Lagrangian surface-layer flux footprint models. Bound Layer Meteorol 80:283–308

Finnigan JJ (2000) Turbulence in plant canopies. Annu Rev Fluid Mech 32:519–571

Finnigan JJ (2004) The footprint concept in complex terrain. Agric For Meteorol 127:117–129

Finnigan JJ (2007) Turbulent flow in canopies on complex topography and the effects of stable stratification. In: Gayev YA, Hunt JCR (eds) Flow and transport processes with complex obstructions. Springer, Dordrecht, pp 199–219

Finnigan JJ, Belcher SE (2004) Flow over a hill covered with a plant canopy. Q J R Meteorol Soc 130:1–29

Flesch TK (1996) The footprint for flux measurements, from backward Lagrangian stochastic models. Bound Layer Meteorol 78:399–404

Flesch TK, Wilson JD (1992) A two-dimensional trajectory-simulation model for non-Gaussian, inhomogeneous turbulence within plant canopies. Bound Layer Meteorol 61:349–374

Flesch TK, Wilson JD (1999) Wind and remnant tree sway in forest cutblocks: I. Measured winds in experimental cutblocks. Agric For Meteorol 93:229–242

Flesch TK, Wilson JD, Yee E (1995) Backward-time Lagrangian stochastic dispersion models and their application to estimate gaseous emissions. J Appl Meteorol 34:1320–1332

Foken T (2006) Angewandte Meteorologie, Mikrometeorologische Methoden, 2. überarb. u. erw. Aufl. Springer, Berlin/Heidelberg/New York, 326 pp

Foken T (2008) Micrometeorology, Springer-Verlag, Berlin, Heidelberg

Foken T, Leclerc MY (2004) Methods and limitations in validation of footprint models. Agric Forest Meteorol 127:223–234

Foken T, Wichura B (1996) Tools for quality assessment of surface-based flux measurements. Agric For Meteorol 78:83–105

Foken T et al (2004) Post-field data quality control. In: Lee X et al (eds) Handbook of micrometeorology: a guide for surface flux measurements. Kluwer Academic Publishers, Dordrecht, pp 181–208

Foudhil H, Brunet Y, Caltagirone J-P (2005) A fine-scale k-ε model for atmospheric flow over heterogeneous landscapes. Environ Fluid Mech 5:247–265

Gash JHC (1986) A note on estimating the effect of a limited fetch on micrometeorological evaporation measurements. Bound Layer Meteorol 35:409–413

Göckede M et al (2004) A combination of quality assessment tools for eddy covariance measurements with footprint modelling for the characterisation of complex sites. Agric For Meteorol 127(3–4):175–188

Göckede M, Markkanen T, Mauder M, Arnold K, Leps J-P, Foken T (2005) Validation of footprint models using natural tracer measurements from a field experiment. Agric For Meteorol 135 (1–4):314–325

Göckede M et al (2006) Update of a footprint-based approach for the characterisation of complex measurement sites. Bound Layer Meteorol 118:635–655

Göckede M, Thomas C, Markkanen T, Mauder M, Ruppert J, Foken T (2007) Sensitivity of Lagrangian Stochastic footprints to turbulence statistics. Tellus B 59:577–586

Göckede M et al (2008) Quality control of CarboEurope flux data – Part 1: coupling footprint analyses with flux data quality assessment to evaluate sites in forest ecosystems. Biogeosciences 5:433–450

Grant RF et al (2009) Interannual variation in net ecosystem productivity of Canadian forests as affected by regional weather patterns - a Fluxnet-Canada synthesis. Agric For Meteorol 149(11):2022–2039

Gross G (1993) Numerical simulation of canopy flows. Springer, Berlin, 168 pp

Guo XF, Cai XH (2005) Footprint characteristics of scalar concentration in the convective boundary layer. Adv Atmos Sci 22:821–830

Hadfield MG (1994) Passive scalar diffusion from surface sources in the convective boundary layer. Bound Layer Meteorol 69:417–448

Haenel HD, Grünhage L (2001) Reply to the comment on 'footprint analysis: a closed analytical solution based on height-dependent profiles of wind speed and eddy viscosity' by T. W. Horst. Bound Layer Meteorol 101:449–458

Hanjalić K (2005) Will RANS survive LES? A view of perspectives. ASME J Fluid Eng 27:831–839

Hanjalić K, Kenjereš S (2008) Some developments in turbulence modeling for wind and environmental engineering. J Wind Eng Ind Aerodyn 96:1537–1570

Hipsey MR, Sivapalan M, Clement TP (2004) A numerical and field investigation of surface heat fluxes from small wind-sheltered waterbodies in semi-arid Western. Environ Fluid Mech 4:79–106

Horst TW, Weil JC (1992) Footprint estimation for scalar flux measurements in the atmospheric surface layer. Bound Layer Meteorol 59:279–296
Horst TW, Weil JC (1994) How far is far enough? The fetch requirements for micrometeorological measurement of surface fluxes. J Atmos Ocean Technol 11:1018–1025
Hsieh C-I, Katul G (2009) The Lagrangian stochastic model for estimating footprint and water vapor flux over inhomogeneous surfaces. Int J Biometeorol 53:87–100
Hsieh C-I, Katul G, Chi T (2000) An approximate analytical model for footprint estimation of scalar fluxes in thermally stratified atmospheric flows. Adv Water Resour 23:765–772
Irvine MR, Gardiner BA, Hill MK (1997) The evolution of turbulence across a forest edge. Bound Layer Meteorol 84:467–496
Järvi L, Rannik Ü, Mammarella I, Sogachev A, Aalto PP, Keronen P, Siivola E, Kulmala M, Vesala T (2009) Annual particle flux observations over a heterogeneous urban area. Atmos Chem Phys 9:7847–7856
Kaimal JC, Finnigan JJ (1994) Atmospheric boundary layer flows: their structure and measurement. Oxford University Press, New York, 289 pp
Kantha LH (2004) The length scale equation in turbulence models. Nonlinear Process Geophys 11:83–97
Karipot A, Leclerc MY, Zhang G, Martin T, Starr G, Hollinger D, McCaughey JH, Hendrey GR (2006) Nocturnal CO_2 exchange over a tall forest canopy associated with intermittent low-level jet activity. J Theor Appl Climatol 85:243–248
Karipot A, Leclerc MY, Zhang G (2008a) Climatology of the nocturnal low-level jets observed over north Florida. Mon Weather Rev 137:2605–2621
Karipot A, Leclerc MY, Zhang G, Lewin K, Nagy J, Starr G (2008b) Influence of nocturnal low-level jet on turbulence structure and CO_2 flux measurements over a forest canopy. J Geophys Res 113, D10102
Katul GG, Oren R, Ellsworth D, Hseih CI, Phillips N, Lewin K (1997) A Lagrangian dispersion model for predicting CO_2 sources, sinks and fluxes in uniform loblolly pine (Pinus taeda L.) stand. J Geophys Res 102:9309–9321
Katul GG, Mahrt L, Poggi D, Sanz C (2004) One- and two-equation models for canopy turbulence. Bound Layer Meteorol 113:81–109
Katul GG, Finnigan JJ, Poggi D, Leuning R, Belcher SE (2006) The influence of hilly terrain on canopy-atmosphere carbon dioxide exchange. Bound Layer Meteorol 118:189–216
Kim J et al (2006) Upscaling fluxes from tower to landscape: overlaying flux footprints on high-resolution (IKONOS) images of vegetation cover. Agric For Meteorol 136(3–4):132–146
Klaassen W, Sogachev A (2006) Flux footprint simulation downwind of a forest edge. Bound Layer Meteorol 121:459–473
Klaassen W et al (2002) Increased heat fluxes near a forest edge. Theor Appl Climatol 72 (3–4):231–243
Kljun N, Rotach MW, Schmid HP (2002) A 3-D backward Lagrangian footprint model for a wide range of boundary layer stratifications. Bound Layer Meteorol 103:205–226
Kljun N, Calanca P, Rotach MW, Schmid HP (2004a) A simple parameterisation for flux footprint predictions. Bound Layer Meteorol 112:503–523
Kljun N, Kastner-Klein P, Fedorovich E, Rotach MW (2004b) Evaluation of a Lagrangian footprint model using data from a wind tunnel convective boundary layer. Special issue on footprints of fluxes and concentrations. Agric For Meteorol 127:189–201
Kormann R, Meixner FX (2001) An analytic footprint model for neutral stratification. Bound Layer Meteorol 99:207–224
Kruijt B (1994) Turbulence over forest downwind of an edge. PhD thesis, University of Groningen, Groningen
Kurbanmuradov OA, Sabelfeld KK (2000) Lagrangian stochastic models for turbulent dispersion in the atmospheric boundary layer. Bound Layer Meteorol 97:191–218
Kurbanmuradov O, Rannik Ü, Sabelfeld K, Vesala T (1999) Direct and adjoint Monte Carlo algorithms for the footprint problem. Monte Carlo Methods Appl 5:85–112

Kurbanmuradov O, Rannik Ü, Sabelfeld KK, Vesala T (2001) Evaluation of mean concentration and fluxes in turbulent flows by Lagrangian stochastic models. Math Comput Simul 54:459–476

Kurbanmuradov O, Levykin AI, Rannik Ü, Sabelfeld K, Vesala T (2003) Stochastic Lagrangian footprint calculations over a surface with an abrupt change of roughness height. Monte Carlo Methods Appl 9:167–188

Kustas WP et al (2006) Using a remote sensing field experiment to investigate flux-footprint relations and flux sampling distributions for tower and aircraft-based observations. Adv Water Resour 29:355–368

Launder BE, Spalding DB (1974) The numerical computation of turbulent flows. Comput Methods Appl Mech Eng 3:269–289

Launder BE, Reece GJ, Rodi W (1975) Progress in the development of a Reynolds-stress turbulent closure. J Fluid Mech 68:537–566

Launiainen S, Vesala T, Mölder M, Mammarella I, Smolander S, Rannik Ü, Kolari P, Hari P, Lindroth A, Gatul G (2007) Vertical variability and effect of stability on turbulence characteristics down to the floor of a pine forest. Tellus 59B:919–936

Law BE (2005) Carbon dynamics in response to climate and disturbance: recent progress from multiscale measurements and modeling in AmeriFlux. In: Omasa K et al (eds) Plant responses to air pollution and global change. Springer, Tokyo, pp 205–213

Leclerc MY, Thurtell GW (1990) Footprint prediction of scalar fluxes using a Markovian analysis. Bound Layer Meteorol 52:247–258

Leclerc MY, Thurtell GW, Kidd GE (1988) Measurements and Langevin simulations of mean tracer concentration fields downwind from a circular line source inside an alfalfa canopy. Bound Layer Meteorol 43:287–308

Leclerc MY, Beissner KC, Shaw RH, den Hartog G, Neumann HH (1991) The influence of buoyancy on third-order turbulent velocity statistics within a decidious forest. Bound Layer Meteorol 55:109–123

Leclerc MY, Shen S, Lamb B (1997) Observations and large-eddy simulation modeling of footprints in the lower convective boundary layer. J Geophys Res 102(D8):9323–9334

Leclerc MY, Karipot A, Prabha T, Allwine G, Lamb B, Gholz HL (2003a) Impact of non-local advection on flux footprints over a tall forest canopy: a tracer flux experiment. Agric For Meteorol 115:19–30

Leclerc MY, Meskhidze N, Finn D (2003b) Comparison between measured tracer fluxes and footprint model predictions over a homogeneous canopy of intermediate roughness. Agric For Meteorol 117:145–158

Lee X (2003) Fetch and footprint of turbulent fluxes over vegetative stands with elevated sources. Bound Layer Meteorol 107:561–579

Lee X (2004) A model for scalar advection inside canopies and application to footprint investigation. Agric For Meteorol 127(3–4):131–141

Leuning R, Denmead OT, Miyata A, Kim J (2000) Source-sink distributions of heat, water vapour, carbon dioxide and methane in rice canopies estimated using Lagrangian dispersion analysis. Agric For Meteorol 104:233–249

Luhar AK, Rao KS (1994) Source footprint analysis for scalar fluxes measured over an inhomogeneous surface. In: Gryning SE, Milan MM (eds) Air pollution modeling and its applications. Plenum Press, New York, pp 315–323

Luysseart S et al (2008) Old-growth forests as global carbon sinks. Nature 455(11):213–215

Mao S, Leclerc MY, Michaelides EE (2008) Passive scalar flux footprint analysis over horizontally inhomogeneous plant canopy using large-eddy simulation. Atmos Environ 42:5446–5458

Marcolla B, Cescatti A (2005) Experimental analysis of flux footprint for varying stability conditions in an alpine meadow. Agric For Meteorol 135:291–301

Markkanen T, Rannik Ü, Marcolla B, Cescatti A, Vesala T (2003) Footprints and fetches for fluxes over forest canopies with varying structure and density. Bound Layer Meteorol 106:437–459

Markkanen T, Steinfeld G, Kljun N, Raasch S, Foken T (2009) Comparison of conventional Lagrangian stochastic footprint models against LES driven footprint estimates. Atmos Chem Phys 9:5575–5586

Mason PJ (1988) Large-eddy simulation of the convective atmospheric boundary layer. J Atmos Sci 46:1492–1516

Meyers T, Paw UKT (1986) Testing of a higher-order closure model for modeling airflow within and above plant canopies. Bound Layer Meteorol 37:297–311

Moeng C-H (1984) A large-eddy simulation model for the study of planteray boundary-layer turbulence. J Atmos Sci 41:252–2061

Moeng C-H, Wyngaard JC (1988) Spectral analysis of large-eddy simulations of the convective boundary layer. J Atmos Sci 45:3575–3587

Mölder M, Klemedtsson L, Lindroth A (2004) Turbulence characteristics and dispersion in a forest—tests of Thomson's random-flight model. Agric For Meteorol 127(3–4):203–222

Morse AP, Gardiner BA, Marshall BJ (2002) Mechanisms controlling turbulence development across a forest edge. Bound Layer Meteorol 103:227–251

Myklebust MC et al (2008) Comparison of eddy covariance, chamber, and gradient methods of measuring soil CO2 efflux in an annual semi-arid grass, Bromus tectorum. Agric For Meteorol 148(11):1894–1907

Nagy MT et al (2006) Footprint-adjusted net ecosystem CO2 exchange and carbon balance components of a temperate forest. Agric For Meteorol 139(3–4):344–360

Ogunjemiyo SO et al (2003) Methods of estimating CO2, latent heat and sensible heat fluxes from estimates of land cover fractions in the flux footprint. Agric For Meteorol 117(3–4):125–144

Panin GN, Tetzlaff G (1999) A measure of inhomogeneity of the land surface and parametrization of turbulent fluxes under natural conditions. Theor Appl Climatol 62(1–2):3–8

Pasquill F (1972) Some aspects of boundary layer description. Q J R Meteorol Soc 98:469–494

Pasquill F, Smith FB (1983) Atmospheric diffusion, 3rd edn. Wiley, New York

Patton EG, Davis HJ, Barth MC, Sullivan P (2001) Decaying scalars emitted by a forest canopy: a numerical study. Bound Layer Meteorol 100:91–129

Peng G, Cai X, Zhang H, Li A, Hu F, Leclerc MY (2008) Heat flux apportionment to heterogeneous surfaces using flux footprint analysis. Adv Atmos Sci 25:107–116

Pinard J-P, Wilson JD (2001) First- and second-order closure models for wind in a plant canopy. J Appl Meteorol 40:1762–1768

Poggi D, Katul GG, Cassiani M (2008) On the anomalous behavior of the Lagrangian structure function similarity constant inside dense canopies. Atmos Environ 42:4212–4231

Prabha T, Leclerc MY, Karipot A, Hollinger DY (2007) Low-frequency effects on eddy-covariance fluxes under the influence of a low-level jet. J Appl Meteorol 46:338–352

Prabha T, Leclerc MY, Baldocchi D (2008a) Comparison of in-canopy flux footprints from Lagrangian simulations against wind tunnel experiments and large-eddy simulation. J Appl Meteorol Climatol 47(8):2115–2128

Prabha TV, Leclerc MY, Karipot A, Hollinger DY, Mursch-Radlgruber E (2008b) Influence of nocturnal low-level jets on eddy covariance fluxes over a tall forest canopy. Bound Layer Meteorol 126:219–236

Rannik Ü, Aubinet M, Kurbanmuradov O, Sabelfeld KK, Markkanen T, Vesala T (2000) Footprint analysis for the measurements over a heterogeneous forest. Bound Layer Meteorol 97:137–166

Rannik Ü, Markkanen T, Raittila J, Hari P, Vesala T (2003) Turbulence statistics inside and over forest: influence on footprint prediction. Bound Layer Meteorol 109:163–189

Rannik Ü, Kolari P, Vesala T, Hari P (2006) Uncertainties in measurement and modelling of net ecosystem exchange of a forest ecosystem at different time scales. Agric For Meteorol 138:244–257

Rao KS, Wyngaard JC, Coté OR (1974) Local advection of momentum, heat, and moisture in micrometeorology. Bound Layer Meteorol 7:331–348

Raupach M (1988) Canopy transport processes. In: Flow and transport in the natural environment: advances and applications. Springer, Berlin, pp 95–127

Raupach MR (1989) Applying Lagrangian fluid mechanics to infer scalar source distributions from concentration profiles in plant canopies. Agric For Meteorol 47:85–108

Raupach MR, Finnigan JJ (1997) The influence of topography on meteorology variables and surface-atmosphere interactions. J Hydrol 190:182–213

Rebmann C et al (2005) Quality analysis applied on eddy covariance measurements at complex forest sites using footprint modelling. Theor Appl Climatol 80(2–4):121–141

Reithmaier LM et al (2006) Use of remotely sensed land use classification for a better evaluation of micrometeorological flux measurement sites. Theor Appl Climatol 84(4):219–233

Reth S, Göckede M, Falge E (2005) CO2 efflux from agricultural soils in Eastern Germany – comparison of a closed chamber system with eddy covariance measurements. Theor Appl Climatol 80(2–4):105–120

Reynolds AM (1998) A two-dimensional Lagrangian stochastic dispersion model for convective boundary layers with wind shear. Bound Layer Meteorol 86:345–352

Sabelfeld KK, Kurbanmuradov OA (1990) Numerical statistical model of classical incompressible isotropic turbulence. Sov J Numer Anal Math Model 5:251–263

Sabelfeld KK, Kurbanmuradov OA (1998) One-particle stochastic Lagrangian model for turbulent dispersion in horizontally homogeneous turbulence. Monte Carlo Methods Appl 4:127–140

Sawford BL (1985) Lagrangian statistical simulation of concentration mean and fluctuation fields. J Clim Appl Meterol 24:1152–1166

Sawford BL (1999) Rotation of trajectories in Lagrangian stochastic models of turbulent dispersion. Bound Layer Meteorol 93:411–424

Schmid HP (1994) Source areas for scalar and scalar fluxes. Bound Layer Meteorol 67:293–318

Schmid HP (1997) Experimental design for flux measurements: matching scales of observations and fluxes. Agric For Meteorol 87:179–200

Schmid HP (2002) Footprint modeling for vegetation atmosphere exchange studies: a review and perspective. Agric For Meteorol 113:159–183

Schmid HP, Lloyd CR (1999) Spatial representativeness and the location bias of flux footprints over inhomogeneous areas. Agric For Meteorol 93(3):195–209

Schmid HP, Oke TR (1988) Estimating the source area of a turbulent flux measurement over a patchy Surface. In: Proceedings of the eighth symposium on turbulence and diffusion, Preprints. American Meteorological Society, Boston, pp 123–126

Schmid HP, Oke TR (1990) A model to estimate the source area contributing to turbulent exchange in the surface layer over patchy terrain. Q J R Meteorol Soc 116:965–988

Schuepp PH, Leclerc MY, MacPherson JI, Desjardins RL (1990) Footprint prediction of scalar fluxes from analytical solutions of the diffusion equation. Bound Layer Meteorol 50:355–373

Shaw RH, Schumann U (1992) Large-eddy simulation of turbulent flow above and within a forest. Bound Layer Meteorol 61:47–64

Shen S, Leclerc MY (1995) How large must surface inhomogeneities be before they influence the connective boundary layer structure? A case study. Q J R Meteorol Soc 121:1209–1228

Shen S, Leclerc MY (1997) Modelling the turbulence structure in the canopy layer. Agric For Meteorol 87:3–25

Soegaard H, Jensen NO, Boegh E, Hasager CB, Schelde K, Thomsen A (2003) Carbon dioxide exchange over agricultural landscape using eddy correlation and footprint modelling. Agric For Meteorol 114:153–173

Sogachev A (2009) A note on two-equation closure modelling of canopy flow. Bound Layer Meteorol 130(3):423–435. doi:10.1007/s10546-008-9346-2

Sogachev A, Leclerc MY, Karipot A, Zhang G, Vesala T (2005a) Effect of clearcuts on footprints and flux measurements above a forest canopy. Agric For Meteorol 133:182–196

Sogachev A, Panferov O, Gravenhorst G, Vesala T (2005b) Numerical analysis of flux footprints for different landscapes. Theor Appl Climatol 80(2–4):169–185

Sogachev A, Leclerc MY, Zhang G, Rannik U, Vesala T (2008) CO_2 fluxes near a forest edge: a numerical study. Ecol Appl 18(6):1454–1469

Sogachev A, Lloyd JJ (2004) Using a one-and-a-half order closure model of the atmospheric boundary layer for surface flux footprint estimation. Bound Layer Meteorol 112:467–502

Sogachev A, Menzhulin G, Heimann M, Lloyd J (2002) A simple three dimensional canopy – planetary boundary layer simulation model for scalar concentrations and fluxes. Tellus 54B:784–819

Sogachev A, Panferov O (2006) Modification of two-equation models to account for plant drag. Bound Layer Meteorol 121:229–266

Sogachev A, Rannik Ü, Vesala T (2004) On flux footprints over the complex terrain covered by a heterogeneous forest. Agric For Meteorol 127:143–158

Sogachev A, Sedletski A (2006) SCADIS "Footprint calculator": operating manual. In: Kulmala M, Lindroth A, Ruuskanen T (eds) Proceedings of bACCI, NECC and FCoE activities 2005, Book B, Report Series in Aerosol Science 81B, The Finnish Association for Aerosol Research, Helsinki, Finland

Steinfeld G, Raasch S, Markkanen T (2008) Footprints in homogeneously and heterogeneously driven boundary layers derived from a Lagrangian Stochastic particle model embedded into large-eddy simulation. Bound Layer Meteorol 129:225–248

Stoy PC et al (2009) Biosphere-atmosphere exchange of CO2 in relation to climate: a cross-biome analysis across multipe time scales. Biogeosciences 6:2297–2312

Strong C, Fuentes JD, Baldocchi DD (2004) Reactive hydrocarbon flux footprints during canopy senescence. Agric For Meteorol 127:159–173

Su HB, Leclerc MY (1998) Large-eddy simulation of trace gas footprints from infinite crosswind line sources inside a forest canopy. In: Preprints, Proceedings of the 23rd conference on agriculture and forest meteorology. American Meteorological Society, Boston, pp 388–391

Su H-B, Shaw RH, Paw KT, Moeng C-H, Sullivan PP (1998) Turbulent statistics of neutrally stratified flow within and above a sparse forest from large-eddy simulation and field observations. Bound Layer Meteorol 88:363–397

Svensson U, Häggkvist K (1990) A two-equation turbulence model for canopy flows. J Wind Eng Ind Aerodyn 35:201–211

Thomson DJ (1987) Criteria for the selection of stochastic models of particle trajectories in turbulent flows. J Fluid Mech 189:529–556

Tseng Y-H, Meneveau C, Parlange MB (2006) Modeling flow around bluff bodies and predicting urban dispersion using large Eddy simulation. Environ Sci Technol 40(8):2653–2662

Umlauf L, Burchard H (2003) A generic length-scale equation for geophysical turbulence models. J Mar Res 61:235–265

Valentini R et al (2000) Respiration as the main determinant of carbon balance in European forests. Nature 404(6780):861–865

Van Breugel PB, Klaassen W, Moors EJ (1999) Fetch requirements near a forest edge. Physics and chemistry of the earth, Part B. Hydrol Oceans Atmos 24:125–131

van Ulden AP (1978) Simple estimates for vertical diffusion from sources near the ground. Atmos Environ 12:2125–2129

Vesala T, Huotari J, Rannik Ü, Suni T, Smolander S, Sogachev A, Ojala A (2006) Eddy covariance measurements of carbon exchange and latent and sensible heat fluxes over a boreal lake for a full open-water period. J Geophys Res. doi:10.1029/2005JD006365

Vesala T, Järvi L, Launiainen S, Sogachev A, Rannik U, Mammarella I, Siivola E, Keronen P, Rinne J, Riikonen A, Nikinmaa E (2008a) Surface-atmosphere interactions over complex urban terrain in Helsinki, Finland. Tellus 60B:188–199

Vesala T, Kljun N, Rannik Ü, Rinne J, Sogachev A, Markkanen T, Sabelfeld K, Foken Th, Leclerc MY (2008b) Flux and concentration footprint modelling: state of the art. Environ Pollut 152:653–666

Vesala T, Kljun N, Rannik Ü, Sogachev A, Markkanen T, Sabelfeld K, Foken Th, Leclerc MY (2010) Flux and concentration footprint modelling. In: Modelling of pollutants in complex environmental systems, vol II. ILM Publications, St Albans, pp 339–355

Wang W, Rotach M (2010) Flux footprints over an undulating surface. Bound Layer Meteorol 136:325–340

Wang H, Takle ES (1995) A numerical simulation of boundary-layer flows near shelterbelts. Bound Layer Meteorol 75:141–173

Wang WG et al (2006) Decomposing CO2 fluxes measured over a mixed ecosystem at a tall tower and extending to a region: a case study. J Geophys Res Biogeosci 111(G2):1–14
Watanabe T (2009) LES study on the structure of coherent eddies inducing predominant perturbations in velocities in the roughness sublayer over plant canopies. J Meteorol Soc Jpn 87:39–56
Wilcox DC (2002) Turbulence modeling for CFD. DCW Industries Inc, La Cañada, CA, 540 pp
Wilczak JM, Oncley S, Stage SA (2001) Sonic anemometer tilt correction algorithms. Bound Layer Meteorol 99(1):127–150
Wilson JD, Flesch TK (1993) Flow boundaries in random-flight dispersion models: enforcing the well-mixed condition. J Appl Meteorol 32:1695–1707
Wilson JD, Flesch TK (1997) Trajectory curvature as a selection criterion for valid Lagrangian stochastic dispersion models. Bound Layer Meteorol 84:411–426
Wilson JD, Flesch TK (1999) Wind and remnant tree sway in forest openings III. A windflow model to diagnose spatial variation. Agric For Meteorol 93:259–282
Wilson NR, Shaw RH (1977) A higher order closure model for canopy flow. J Appl Meteorol 16:1197–1205
Wilson JD, Swaters GE (1991) The source area influencing a measurement in the planetary boundary-layer – the footprint and the distribution of contact distance. Bound Layer Meteorol 55:25–46
Wilson JD, Finnigan JJ, Raupach MR (1998) A first-order closure for disturbed plant-canopy flows, and its application to winds in a canopy on a ridge. Q J R Meteorol Soc 124:705–732

Chapter 9
Partitioning of Net Fluxes

Markus Reichstein, Paul C. Stoy, Ankur R. Desai, Gitta Lasslop, and Andrew D. Richardson

9.1 Motivation

Eddy covariance measures the net exchange of matter and energy between ecosystems and the atmosphere. The net ecosystem exchange of CO_2 (NEE) results from two larger fluxes of opposite sign: CO_2 uptake by photosynthesis (gross ecosystem productivity – GEP) and CO_2 release from ecosystem respiration (R_{eco}) following the definition equation.

$$\text{NEE} = R_{eco} + \text{GEP} \tag{9.1}$$

with fluxes from atmosphere to biosphere considered negative per the meteorological convention. As per this definition, R_{eco} is always positive, and GEP is negative or zero at nighttime. NEE gives a valuable measure of ecosystem carbon sequestration, but by itself does not describe the processes responsible for carbon flux. Measurements or estimates of R_{eco} and GEP are necessary to

M. Reichstein (✉) • G. Lasslop
Max-Planck Institute for Biogeochemistry, 07745 Jena, Germany
e-mail: mreichstein@bgc-jena.mpg.de; gitta.lasslop@zmaw.de

P.C. Stoy
Department of Land Resources and Environmental Sciences, Montana State University, P.O. Box 173120, Bozeman, MT, 59717-3120, USA
e-mail: paul.stoy@montana.edu

A.R. Desai
Atmospheric and Oceanic Sciences, University of Wisconsin, Madison, USA
e-mail: desai@aos.wisc.edu

A.D. Richardson
Department of Orgasmic and Evolutionary Biology, Harvard University Herbaria, 22 Divinity Avenue, Cambridge, MA 02138, USA
e-mail: arichardson@oeb.harvard.edu

M. Aubinet et al. (eds.), *Eddy Covariance: A Practical Guide to Measurement and Data Analysis*, Springer Atmospheric Sciences, DOI 10.1007/978-94-007-2351-1_9,

obtain information about the processes that contribute to NEE for the purposes of ecosystem studies and modeling. Flux partitioning algorithms are necessary to estimate these fluxes over long time periods using eddy covariance data.

Inferring two dependent variables (R_{eco} and GEP) from one observation (NEE) is an ill-posed problem; the same net flux can result from an indefinite number of combinations of R_{eco} and GEP if both are simultaneously occurring or have occurred over the temporal averaging interval used to describe NEE. Hence, additional constraints or information about flux processes are needed. Most flux partitioning strategies are based on the notion that only R_{eco} occurs at night in ecosystems dominated by C_3 and/or C_4 photosynthesis, while GEP is virtually zero [but not with CAM photosynthesis, San-José et al. (2007)]. The challenge comes in extrapolating these nighttime R_{eco} measurements to daytime conditions to estimate GEP by difference using Eq. 9.1. These difficulties are compounded by the fact that nighttime flux measurements are often compromised by stable atmospheric conditions with insufficient turbulence to satisfy the assumptions of the eddy covariance measurement system. These observations must be filtered from the eddy covariance data record (Sect. 5.3), leaving incomplete information about R_{eco} and thereby GEP.

This chapter summarizes existing strategies for NEE flux partitioning and discusses their benefits and limitations, focusing on challenges of model formulation and parameterization. We describe briefly the standard flux partitioning approaches used in the FLUXNET database by Reichstein et al. (2005a) using nighttime data, and Lasslop et al. (2010) using primarily daytime data, noting that these algorithms are subject to improvement and additional algorithms may be added to FLUXNET in the future. We conclude with suggestions for future directions in flux partitioning research, including techniques for estimating assimilation, respiration, and respiratory sources directly using high-frequency eddy covariance measurements (Scanlon and Kustas 2010; Scanlon and Sahu 2008; Thomas et al. 2008) and stable isotope measurements (Zobitz et al. 2007, 2008), as well as challenges in partitioning eddy covariance-based evapotranspiration measurements into evaporation and transpiration for process-based studies in hydrology and for coupled carbon/water cycle science research. We emphasize the use of simple models for flux partitioning for a simple, data-driven understanding of the processes at hand, but also note the important contributions from other strategies including data assimilation, neural networks, and more complex process-based ecosystem models that provide a more complete picture of the processes that contribute to NEE (cf. Desai et al. 2008).

9.2 Definitions

R_{eco} is the combination of respiratory sources from autotrophic respiration, predominantly from organisms whose primary energy source is the sun (i.e., plants) and heterotrophic respiration, whose primary energy source comes from other organisms. In some ecosystems geologic CO_2 release or sequestration cannot be discounted (Emmerich 2003; Kowalski et al. 2008; Mielnick et al. 2005; Were et al. 2010), but we can consider these fluxes minor across most global ecosystems such that Eq. 9.1 represents biological processes.

Important flux quantities are defined here, to avoid ambiguities that might occur, because terms in the literature are sometimes used with different meanings. The following equations and definitions are valid throughout this chapter (see also Sect. 1.4.2),

$$\mathrm{NEE} = F_{\mathrm{C}}^{\mathrm{EC}} + F_{\mathrm{C}}^{\mathrm{STO}} = R_{\mathrm{eco}} + \mathrm{GEP} \tag{9.2}$$

where $F_{\mathrm{C}}^{\mathrm{EC}}$ is the net turbulent CO_2 flux through a horizontal plane above the canopy (conventionally positive when directed toward the atmosphere) (term IV in Eq. 1.24, where the considered component is CO_2), $F_{\mathrm{C}}^{\mathrm{STO}}$ is the change of carbon storage in the atmosphere below the horizontal plane (positive when increasing) (term I in Eq. 1.24), and NEE is the net ecosystem exchange of CO_2 (positive when emitted) (term V in Eq. 1.24). Net ecosystem CO_2 uptake (often called net ecosystem productivity – NEP) is equal to –NEE. With this definition of NEE, the ecosystem boundaries are leaf, stem, branch, (animal), and soil surfaces, which are in conformity with the models used for flux partitioning, described below. Gross ecosystem photosynthesis (GEP) is the CO_2 flux originating from primary production, and R_{eco} (ecosystem respiration) is the CO_2 flux originating from all respiring compartments of the ecosystem. Analogous to NEE and NEP having opposite signs, GEE can also be used as the negative of GEP. The eddy covariance method gives estimates of $F_{\mathrm{C}}^{\mathrm{EC}}$ (see, e.g., Sects. 1.4 and 3.3). Further, the storage term ($F_{\mathrm{C}}^{\mathrm{STO}}$) can be estimated by the integration of a vertical CO_2 concentration profile (see also Sects. 1.4.2 and 2.5), whereupon the middle term of Eq. 9.2 is determined.

Depending on research objectives, R_{eco} may be separated functionally into respiration of autotrophic and heterotrophic organisms, or spatially into above- and below-ground respiration (R_{above}, R_{soil}), where R_{soil} consists of root and microbial (i.e., edaphon) respiration. Neglected here is soil CO_2 efflux originating from inorganic processes (mainly weathering of carbonates in the soil) and from lateral transport into and out of the flux footprint, which is assumed to be minor.

Evapotranspiration (E_{tot}) is defined here as the flux of H_2O through a horizontal plane above the canopy (positive when directed toward the atmosphere, as with CO_2 flux). It consists of transpiration (E_{plant}), evaporation of intercepted water (E_{int}) and evaporation from the soil surface (E_{soil}).

$$F_{\mathrm{v}}^{\mathrm{EC}} = E_{\mathrm{tot}} = E_{\mathrm{plant}} + E_{\mathrm{int}} + E_{\mathrm{soil}} \tag{9.3}$$

Under turbulent conditions the eddy covariance method measures the total flux ($F_{\mathrm{v}}^{\mathrm{EC}} = E_{\mathrm{tot}}$) (term IV of Eq. 1.24, where the considered component is water vapor) (see also Sect. 3.3.3). Sapflow methods can be used to measure E_{plant}, which must be scaled to the volume of canopy measured by the eddy covariance flux footprint (see Sect. 11.3.4).

9.3 Standard Methods

9.3.1 Overview

Flux partitioning algorithms have been compared extensively across multiple measurement sites using multiple methods (Desai et al. 2008; Lasslop et al. 2010; Moffat et al. 2007; Reichstein et al. 2005a; Stoy et al. 2006b). Existing methods differ in: (1) the form of the model including driving variables, (2) parameterization including the cost function used to estimate parameters, (3) choices regarding temporal variability of parameters, and (4) the use of nighttime, daytime or all eddy covariance data used for model parameterization (Moffat et al. 2007).

For convenience, we classify flux partitioning approaches as those that use only filtered (Sect. 5.3) nighttime data to directly measure R_{eco} (Reichstein et al. 2005a), and those that exploit both day- and nighttime data or only daytime data, using light-response curves, to estimate R_{eco} either as the intercept parameter at zero light or a population of data points at zero light for further modeling (Table 9.1). (We note that data assimilation approaches rely on some a priori model structure rather than light- or temperature-response curves per se.) These two broad approaches have been compared by Falge et al. (2002), Stoy et al. (2006b), Lasslop et al. (2010), and others, resulting in generally good agreement, although some are prone to bias (Desai et al. 2008), and any output must be carefully interpreted and preferably compared against independent measurements or models should these exist.

9.3.2 Nighttime Data-Based Methods

Flux partitioning techniques that rely on nighttime data must first ensure that the quality of these data is reliable. The challenge is that turbulence is often suppressed at night and the assumptions of the eddy covariance system – that the transfer of mass between surface and atmosphere can be approximated as the vertical turbulent flux across a plane above the ecosystem, plus storage below this plane Eq. 9.2 – are often violated by nontrivial horizontal and vertical advective fluxes (Aubinet et al. 2010; Rebmann et al. 2010; Staebler and Fitzjarrald 2004). This issue is covered extensively in Chap. 5. Most techniques for ensuring flux data quality employ some friction velocity (u*) filter (Aubinet et al. 2000; Barford et al. 2001; Falge et al. 2001; Papale et al. 2006; Reichstein et al. 2005a) (Sects. 5.3 and 5.4), but techniques that also account for atmospheric stability, thereby including both the buoyant and mechanical terms (Novick et al. 2004; van Gorsel et al. 2009), flux footprint dimensions (Rebmann et al. 2005; Stoy et al. 2006b), and those that approach the data filtering issue from comprehensive data quality rating systems (Foken et al. 2004) are also common. After filtering for data quality, the remaining population of

Table 9.1 Classification of currently available statistical flux-partitioning approaches to separate ecosystem respiration (R_{eco}) and gross ecosystem productivity (GEP) using eddy covariance-measured net ecosystem exchange of CO_2 (NEE)

Approach		Advantages	Disadvantages
A: Nighttime data		Flux data represents R_{eco}	Extrapolation to daytime periods is necessary, nighttime eddy covariance data quality concerns
1	R_{eco} is modeled as a time-invariant function of temperature Hollinger et al. (1994)	Simplicity, global applicability	Only applicable where no other factors than temperature influence R_{eco} significantly, not generic
2	R_{eco} is modeled as a time-invariant function of temperature and other environmental drivers (Rambal et al. 2003; Reichstein et al. 2002)	Simplicity, accounts for additional drivers of R_{eco}, allows for seasonally varying temperature sensitivity	Results in selection of site specific factors that determine R_{eco}, ancillary variables (e.g., SWC) may not be measured, or not measured uniformally, across ecosystems
3	R_{eco} is modeled using temporally varying base respiration (Rref) parameters and one single temperature sensitivity derived from annual data set (Falge et al. 2002; Law et al. 2002)	Accounts for temporally varying respiration rates at reference temperature resulting from varying substrate availability	Long-term temperature sensitivity from annual data may not reflect short-term responses to environmental variability, introduction of systematic error when extrapolating to daytime
4.	R_{eco} is modeled using temporally varying functions of temperature (both Rref and temperature sensitivity varying) (this study)	Accounts for temporally varying respiration rates at reference temperature, caused by any factor, seasonally varying temperature sensitivity is accounted for (Davidson and Janssens 2006)	Often noisiness of eddy covariance data does not allow derivation of temperature sensitivity for large periods of the year, i.e., limited practical applicability

(continued)

Table 9.1 (continued)

Approach	Advantages	Disadvantages
B: R_{eco} derived from daytime NEE observations	Increases size of data set used to make inference; reduces reliance on night-time data.	Depends on specific light-response curve model; light-response curve can be confounded by other factors (e.g., vapor pressure deficit), sometimes yields unstable parameter estimates (high standard errors); R_{eco} estimate susceptible to storage flux problems, since those occur in the morning and evening during low-light conditions
1. R_{eco} as *y*-intercept from light-response curve of GEP (Falge et al. 2002)	Day-to-day variation of R_{eco} reflected	Only daily R_{eco} can be derived
2. R_{eco} (temperature driven) and GEP (radiation driven) are simultaneously modeled as parts of one fixed model equation (Gilmanov et al. 2003)	Uses all data (night- and daytime)	Resulting GEP is from a model and thus constrained by model assumptions (disallows comparison with other models), temperature sensitivity maybe confounded by response of GEP to environmental factors, that are hard to separate (e.g., is afternoon drop in NEE caused by R_{eco} as $f(T)$ or by high VPD, or even by plant-internal hydraulic constraints)
3. R_{eco} and GEP are simultaneously modeled as parts of one model equation with state dependent parameters (data-based mechanistic modeling approach)	Uses all data (night- and daytime); very flexible approach; parameters can evolve with time and state	Statistical assumptions (e.g., noncorrelated residuals) and robustness against violations may be problematic; maybe affected by confounding factors similar to B2.
4. R_{eco} (temperature driven) and GEP (radiation and VPD driven) are simultaneously modeled as parts of one model equation with state dependent parameters (Lasslop et al. 2010)	Uses nighttime data to derive the t-sensitivity and daytime data for GEP parameters and flux magnitude of R_{eco}	Equifinality and/or local minima in the cost function. Drought limitation of respiration. $-R_{eco}$ is strongly determined by morning and evening measurements where storage peaks can occur.
5. R_{eco} and GEP are derived via a posteriori analysis of an artificial neural network conditioned with all data. (cf. Papale and Valentini 2002) for gap-filling; flux-partitioning not explored	Uses all data (night- and daytime); very flexible approach; influence of different input data can be evaluated for best description of the data set	Extrapolation problem, since zero radiation has to be assumed estimating R_{eco} from neural network during the day; potentially confounded by other factors similar to B2.

nighttime data points, assumed to comprise R_{eco}, are modeled using approaches that make differing assumptions about model formulation and the temporal variability of model parameters (Reichstein et al. 2005a).

9.3.2.1 Model Formulation: Temperature – Measurements

Respiration is an enzyme-mediated biological reaction and thus depends on temperature and substrate availability. Therefore, the simplest possible mechanistic model of ecosystem respiration is a single equation that is a function of temperature and a so-called base respiration which is implicitly dependent on substrate availability.

The treatment of ecosystem respiration as a single temperature-dependent equation may be the simplest possible approach, but carries additional challenges. Which temperature should one choose given that ecosystems encompass some range of temperatures across which respiratory processes occur in the soil, roots, stems, leaves, and other organisms? How should temporal variability in respiration model parameters be treated given that a different mix of substrates with different temperature sensitivities are being respired across time and space (Fierer et al. 2005; Janssens and Pilegaard 2003)?

Despite these complexities, R_{eco} models that are a simple function of air temperature tend to explain more of the observed variance in R_{eco} models compared to models driven by soil temperature (Van Dijk and Dolman 2004), despite site-level differences (Richardson et al. 2006a), and despite the fact that few respiratory sources are at the measured temperature(s) of air at any one time. The better relationship, on average, between air temperature and R_{eco} is likely due to the fact that a larger percentage of soil respiration occurs near the surface; diurnal hysteresis effects are found for respiration when plotting R_{eco} against soil temperature at depth (Bahn et al. 2008; Vargas and Allen 2008). This indicates that soil temperatures are often measured at a level too deep for optimal correlation with ecosystem respiration. In theory, dual- or multiple source models (cf. Ciais et al. 2005; Reichstein et al. 2005b) where respiration is a multivariate function of different temperature should perform better, but empirical evidence to justify multiple source models is lacking. From the practical perspective, soil temperature measurements are lacking for some sites and site-years in the FLUXNET data record. Hence, air temperature is currently mostly used as the independent variable in R_{eco} models for flux partitioning in the FLUXNET database. Nevertheless, for studying individual sites it is recommended to analyze which temperatures correlate best with flux observations.

9.3.2.2 R_{eco} Model Formulation

A common approach to model R_{eco} using temperature as a dominant driver is the so-called Q_{10} equation:

$$R_{eco} = R_{10} Q_{10}^{\frac{\theta-10}{10}} \tag{9.4}$$

Where R_{10} is ecosystem base respiration at 10°C and Q_{10} is the temperature sensitivity parameter, here describing the amount of change in R_{eco} for a 10°C change in temperature (i.e., a Q_{10} of 2 results in a doubling of R_{eco} for every 10°C change in temperature). Base temperatures other than 10°C can be used accordingly (Ryan 1991).

Respiration is also commonly empirically modeled using the Arrhenius equation or variants thereof; for example, Lloyd and Taylor (1994) used soil respiration data from multiple sources to arrive at a popular expression following Arrhenius kinetics:

$$\mathrm{R}_{eco} = R_{10} \exp\left[E_0 \left(\frac{1}{283.15 - \theta_0} - \frac{1}{\theta - \theta_{ref}}\right)\right] \tag{9.5}$$

where E_0 is an activation energy parameter and is fitted to data, and the θ_{ref} parameter is often set to 227.13 K (−46.02°C) as recommended in the original study (see, e.g., Reichstein et al. 2005a). Numerous studies on ecosystem respiration using eddy covariance data have parameterized equations of this sort for the purposes of flux partitioning (Falge et al. 2001).

Other exponential temperature-based models derived on thermodynamic kinetics (e.g., Eyring model, Desai et al. 2005; Cook et al. 2004) or the modified Arrhenius equation (Gold et al. 1991) have also been proposed in the literature, but fundamentally they retain a functional form and sensitivity similar to the aforementioned equations.

9.3.2.3 Challenges: Additional Drivers of Respiration

R_{eco} responds to more than just temperature alone; sufficient water and nutrient levels are required for biological functioning to occur in the first place. Nutrient limitations may constrain the amount of biomass held by the ecosystem and do not tend to vary dramatically over short timescales in natural or minimally managed ecosystems. These dynamics may be best incorporated into the base respiration parameter rather than explicitly as an additional variable in R_{eco} models. The effects of soil moisture on R_{eco} are arguably more complicated to model for the purposes of flux partitioning because it is dynamic in time and space, constrains autotrophic and heterotrophic respiration differently, and quick changes related to precipitation may induce respiratory pulses, possibly in concert with changes in nutrient availability (e.g., Jarvis et al. 2007, and early references from H.F. Birch within).

Soil moisture strongly impacts R_{eco} and soil respiration by constraining biological activity under dry conditions and inhibiting oxygen availability under extremely wet conditions (Carbone et al. 2008; Irvine and Law 2002). Soil moisture effects enter models as different adjustment terms to the base respiration parameter, the temperature sensitivity parameter, or as multipliers to the entire temperature-based R_{eco} equation (Palmroth et al. 2005). To date, to our knowledge, no single model formulation that includes soil moisture has been demonstrated to perform better than others across multiple sites at the ecosystem level using eddy covariance data.

Unfortunately soil moisture is measured at a minority of FLUXNET sites to date, which limits the global applicability of soil moisture-inclusive models. Hence, in flux network-wide studies that include multiple sites, the effects of soil moisture variability and other limitations on biological functioning may be best approached by varying the parameters of the R_{eco} model in time, rather than changing model formulation given uncertainties regarding the best formulation and a lack of data availability. At the site level, it is critical to understand the effects of soil moisture on respiration from different carbon pools for a comprehensive understanding of ecosystem carbon metabolism, but from the flux network perspective, a simpler R_{eco} model formulation is preferred.

The role of photodegradation, the breakdown of organic matter by solar irradiance, on R_{eco} is beginning to be tested at eddy covariance research sites (Rutledge et al. 2010). The importance of photodegradation to R_{eco} and the best way to model this process across global ecosystems need to be explored further, but it is likely to be important across a wide range of ecosystems with exposed organic matter (Austin and Vivanco 2006; Rutledge et al. 2010).

9.3.2.4 Challenges: Photosynthesis – Respiration Coupling and Within-Ecosystem Transport

Recent research has demonstrated that much of the carbon respired as R_{eco} across many ecosystems was recently fixed as GEP (Barbour et al. 2005; Drake et al. 2008; Högberg et al. 2001; Horwath et al. 1994; Janssens et al. 2001; Knohl et al. 2005; Zhang et al. 2006). This provides an additional complication for R_{eco} modeling and partitioning: If R_{eco} is a function of GEP after some time lag (Mencuccini and Hölttä 2010), and R_{eco} is used to determine GEP by difference Eq. 9.1, a circularity ensues (Vickers et al. 2009). One may incorporate GEP estimates from previous days into an R_{eco} model following findings from isotopic studies (e.g., Table 1 in Stoy et al. 2007) but the time lags between GEP and root/soil respiration may be quite rapid if pressure/concentration waves in the phloem are considered (Mencuccini and Hölttä 2010; Thompson and Holbrook 2003).

Measuring ecosystem metabolism using the eddy covariance system is further complicated by lags due to gas transport from the location of the respiratory source to the eddy covariance instrumentation (Baldocchi et al. 2006; Stoy et al. 2007; Suwa et al. 2004). In other words, the eddy covariance system measures CO_2 efflux, which results from respiration that occurred sometime in the past, depending on the timescales of transport through the soil or plant and the atmosphere. These time lags between CO_2 production in the soil and transport to the above-canopy atmosphere often exceed the common 30-min averaging time for both flux and micrometeorological measurements. In other words, part of the CO_2 that the flux system "sees" as respiration was likely produced under different temperature conditions than measured at the time of its ejection from the ecosystem volume.

These lags decouple the measurement of temperature with the actual process of respiration. Comprehensive treatments of CO_2 production and transport in the soil

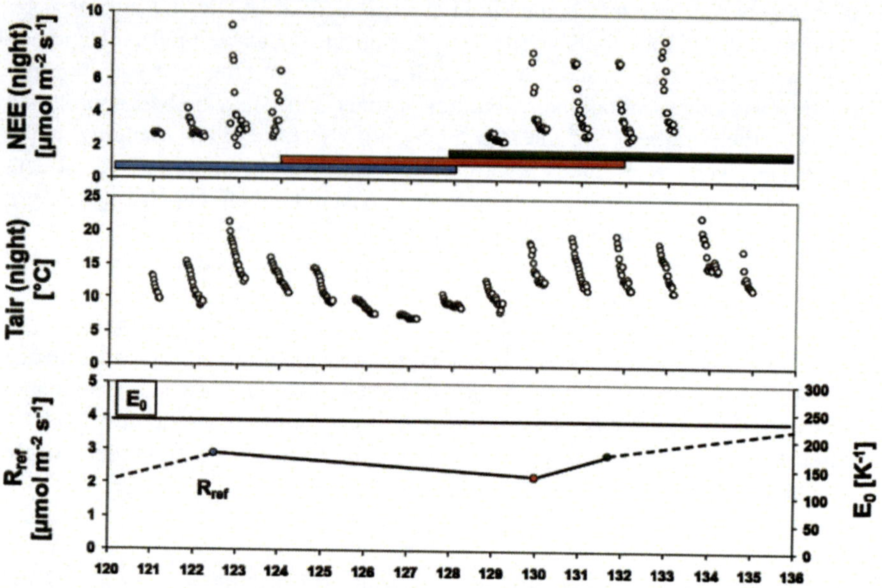

Fig. 9.1 Scheme for derivation of ecosystem respiration parameters from eddy covariance nighttime flux data. *Upper panel* shows the flux data (incl. gaps) with the bars being the 50% overlapping windows used for parameter estimation. *Lower panel* shows the estimates of the reference respiration (*Rref*) based on the data in the respective windows. The estimates of Rref are assigned to the data weighted center of the time window (*dots*) and then linearly interpolated. E_0 is kept constant here as an estimate for the whole year but that is not necessary

or whole ecosystem is commendable and advisable for elucidating the mechanisms responsible for CO_2 production and transport, but involve extensive additional measurements of CO_2 flux within the ecosystem domain (Baldocchi et al. 2006; Daly et al. 2009; Tang and Baldocchi 2005). Incorporating such knowledge into R_{eco} models for eddy covariance applications would involve making extensive assumptions about the location of respiratory sources and transport in the soil, which are not solvable using eddy covariance-based whole-ecosystem measurements alone. The aforementioned processes may be best incorporated into flux partitioning models by adding temporal variability to the R_{eco} model parameters rather than by incorporating additional processes into the model when little information about these processes exists in most cases. By estimating the reference respiration (R_{eco} at reference temperature), every few days with a moving window approach (Fig. 9.1), the reference respiration may vary implicitly as a function of any other factor not explicitly accounted for in the equation (e.g., phenology, soil moisture, substrate availability). The size of the moving window has to reflect a compromise between data availability to estimate statistical models and the necessity to have as small as possible window sizes. Desai et al. (2005) present an approach where the window size varies based on the amount of data, while Reichstein et al. (2005a) use a fixed window size. In any case, the assumption of this approach is that within the time-window used for parameter estimation, R_{ref} does not vary other than described by the linear interpolation. In particular if the reference respiration varies diurnally (e.g., because of links to GEP or short-term variation in soil moisture, or with CO_2 of geogenic origin), this is not reflected in the approach and will cause biases. Moreover, rapid response of the reference respiration, for example, to rain pulses cannot be described with this approach.

9.3.3 *Daytime Data-Based Methods*

A concern about using nighttime data for R_{eco} modeling is that the input data represent a subset of the total available data that are unlikely to be of the best quality. The alternate approach is to fit a model to daytime NEE observations that accounts for the effects of radiation and vapor pressure deficit (VPD) on GEP as well as the effects of temperature on R_{eco} (Falge et al. 2001; Gilmanov et al. 2003). This approach is to date less common than flux partitioning based on nighttime data, but has been used in earlier eddy covariance studies (Lee et al. 1999) and can complement nighttime data-based methods (Lasslop et al. 2010).

9.3.3.1 Model Formulation: The NEE Light Response

The rectangular hyperbola is a simple, common equation to model the effects of radiation (here the photosynthetically active photon flux density, PPFD) on NEE:

$$\mathrm{NEE} = \frac{\alpha_{\mathrm{RH}}\beta_{\mathrm{RH}}\mathrm{PPFD}}{\alpha_{\mathrm{RH}}\mathrm{PPFD} + \beta_{\mathrm{RH}}} + \gamma_{\mathrm{RH}} \tag{9.6}$$

R_g, the global radiation, can be used in place of PPFD in Eq. 9.6; the values and units for the fitted parameters α_{RH} (the initial slope of the light-response curve) and β_{RH} (GEP at light saturation) will change accordingly. γ_{RH}, the intercept parameter at zero light, represents R_{eco} and can be expanded using a temperature-driven equation (e.g., Gilmanov et al. 2010) (see Fig. 9.2). The rectangular hyperbola has a long history for gap-filling daytime flux data, often with slight modifications concerning the parameters (e.g., Wofsy et al. 1993).

The non-rectangular hyperbola adds a parameter that describes the degree of curvature (θ_{NRH}).

$$\mathrm{NEE} = -\frac{1}{2\theta_{\mathrm{NRH}}}\left(\alpha_{\mathrm{NRH}}\mathrm{PPFD} + \beta_{\mathrm{NRH}} - \sqrt{(\alpha_{\mathrm{NRH}}\mathrm{PPFD} + \beta_{\mathrm{NRH}})^2 - 4\alpha_{\mathrm{NRH}}\beta_{\mathrm{NRH}}\theta_{\mathrm{NRH}}\mathrm{PPFD}}\right) + \gamma_{\mathrm{NRH}} \tag{9.7}$$

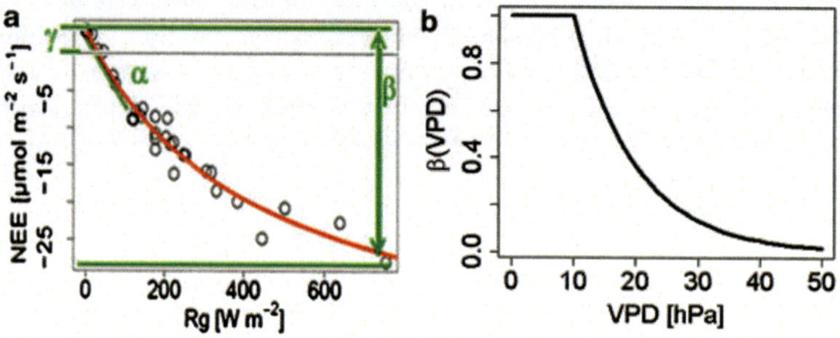

Fig. 9.2 (**a**) Observed net ecosystem exchange, as a function of global radiation, explaining the three parameters with respect to the function's shape: α the light utilization efficiency, is the initial slope, β, the maximum carbon uptake, is the range of NEE and γ, the respiration, is the offset. (**b**) The function decreasing the parameter beta as a function of VPD according to Eq. 9.9; note that the parameter k defining the steepness of the equation is estimated from the data

The non-rectangular light-response curve tends to fit measured data better than the rectangular hyperbola (Gilmanov et al. 2003; Marshall and Biscoe 1980) – as it should give the additional parameter – but the convergence of the parameter routine may be less frequent and logical parameter bounds and initial guesses are encouraged to ensure optimal parameter sets (Stoy et al. 2006b).

Lindroth et al. (2008) and Aubinet et al. (2001) used a slightly different form of a light-response function (Mitscherlich model):

$$\mathrm{NEE} = -\left(\beta_{\mathrm{M}} + \gamma_{\mathrm{M}}\right)\left(1 - \exp\left(\frac{-\alpha_{\mathrm{M}}\mathrm{PPFD}}{\beta_{\mathrm{M}} + \gamma_{\mathrm{M}}}\right)\right) + \gamma_{\mathrm{M}} \tag{9.8}$$

It is important to note that, whereas the various light-response models Eqs. 9.6–9.8 may fit the data equally well, the parameters of the equations need not take the same values (hence the different subscripts) and may not take realistic values of carbon exchange phenomena as demonstrated in Fig. 9.3 and Table 9.2. Here, 1 day of observed NEE from the Duke Hardwood forest ecosystem (US-Dk2) was modeled using Eqs. 9.6–9.8 and nonlinear least squares was chosen to find the optimum parameter values. For the rectangular hyperbola, the optimized value of β_{RH} is 0.66 mg C m^{-2} s^{-1}, far greater than the largest observed flux that day (0.34 mg C m^{-2} s^{-1}) which itself may be considered an outlier. This saturating value of β_{RH} exists at a light level that will never realistically be reached and is not the saturating value of NEE under field conditions, rather a parameter that describes the maximum value of the rectangular hyperbola fit to observations. Flux studies should take care to note this distinction, a more reasonable value of the maximum carbon uptake can be computed by using the model parameters and a radiation value that can be considered a maximum radiation. In Fig. 9.1, $\beta_{\mathrm{NRH}} = 0.29\mathrm{mgCm^{-2}s^{-1}}$, roughly the median of the points at high light. $\beta_{\mathrm{M}} = 0.39\mathrm{mgCm^{-2}s^{-1}}$, beyond the limits of what was observed but closer to a realistic value of NEE at saturation than

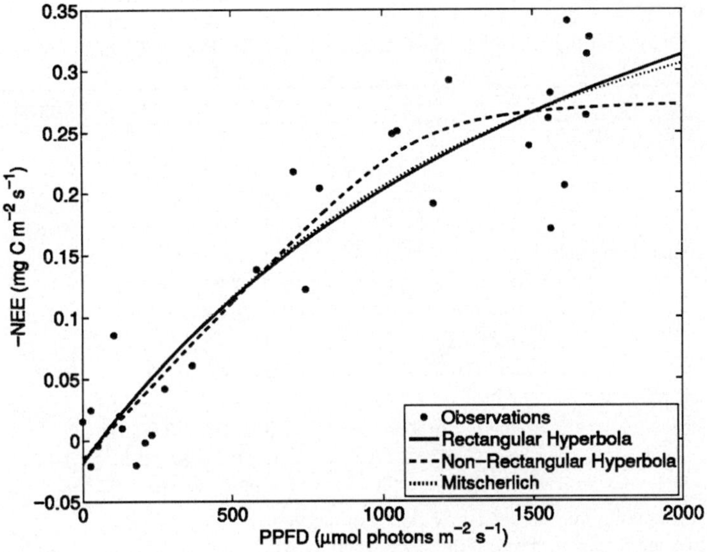

Fig. 9.3 Observed (negative) net ecosystem exchange (−NEE, i.e., net ecosystem productivity, NEP), as a function of photosynthetically active photon flux density (PPFD) for day of year 170, 2005 in the Duke Forest hardwood ecosystem fit using a rectangular hyperbola, a non-rectangular hyperbola, and the Mitscherlich model (Aubinet et al. 2001; Lindroth et al. 2008) Eqs. 9.6–9.8. Fitted parameters are listed in Table 9.2

β_{RH}. Whereas any of the above equations may result in defensible values of modeled NEE and partitioned GEP and R_{eco}, the parameter values themselves may not make physical sense.

9.3.3.2 Challenges: Additional Drivers and the FLUXNET Database Approach

Radiation is not the only driver of NEE; the photosynthetic term that dominates during the day may be constrained by stomatal closure, often modeled as a function of vapor pressure deficit (VPD) (Oren et al. 1999; Lasslop et al. (2010)). These effects are embodied in a hysteresis pattern in the light-response curve, with lower NEE values in the afternoon when temperature and vapor pressure deficit (VPD) are higher (Gilmanov et al. 2003). Stomatal behavior has been successfully explained by the so-called optimality hypothesis which assumes that stomata behave to maximize carbon gain while minimizing water loss (see, e.g., Cowan 1977; Mäkelä et al. 2002). The fundamental role of stomata in regulating both carbon and water fluxes suggests that transpiration estimates can be used to constrain GEP. From the eddy covariance perspective, such an approach would require additional modeling

Table 9.2 Parameter values, with units and logical bounds, for the light-response curves Eqs. 9.6–9.8 fit to 1 day of observed eddy covariance-measured NEE in the Duke Forest hardwood ecosystem on DOY 170, 2005 (Fig. 9.3)

Parameter	Units	Value (Fig. 9.3)	Logical bounds
Rectangular hyperbola			
α_{RH}	(mg C μmol photons^{-1})	0.00033 ± 0.00024	[a]
β_{RH}	(mg C m^{-2} s^{-1})	0.66 ± 0.52	[NEE_{min}, $R_{eco,max}$]
γ_{RH}	(mg C m^{-2} s^{-1})	0.019 ± 0.042	[0, $R_{eco,max}$]
Non-rectangular hyperbola			
α_{NRH}	(mg C μmol photons^{-1})	0.00026 ± 0.00013	[a]
β_{NRH}	(mg C m^{-2} s^{-1})	0.29 ± 0.10	[NEE_{min}, $R_{eco,max}$]
θ_{NRH}	(unitless)	0.98 ± 0.13	[0, 1]
γ_{NRH}	(mg C m^{-2} s^{-1})	0.016 ± 0.036	[0, $R_{eco,max}$]
Mitscherlich equation			
α_{M}	(mg C μmol photons^{-1})	0.00033 ± 0.00020	[a]
β_{M}	(mg C m^{-2} s^{-1})	0.39 ± 0.22	[NEE_{min}, $R_{eco,max}$]
γ_{M}	(mg C m^{-2} s^{-1})	0.020 ± 0.040	[0, $R_{eco,max}$]

$R_{eco,max}$ is the maximum observed ecosystem respiration
[a]The positive and negative values of the slope where NEE_{max} is reached at the lowest logical value of PPFD, that is, the greatest logical slope of the light-response curve

of transpiration from evapotranspiration while noting that eddy covariance-based evapotranspiration measurements are not independent from eddy covariance-based GEP estimates.

The degree to which R_{eco} is enhanced by higher temperatures and GEP is reduced by stomatal responses to VPD is uncertain. VPD is partly a function of temperature, and both R_{eco} and GEP occur simultaneously during the day when leaves are present. Despite these challenges, multiple approaches separating GEP and R_{eco} from daytime NEE observations have been tested.

Gilmanov et al. (2006, 2003) introduced an exponential function in the place of γ_{NRH} in Eq. 9.7 and added an exponential decrease of GEP with relative humidity to account for stomatal effects Lasslop et al. (2010) expanded on this approach by introducing the Lloyd and Taylor model Eq. 9.5 in place of γ_{RH} in Eq. 9.6 and added a VPD limitation on NEE that decreases β_{RH} exponentially from a maximum value β_0 for VPD higher than a limiting value (VPD_0), which was determined to be 1 kPa based on a synthesis of leaf-level findings (Körner 1995) (note also Oren et al. 1999) (see Fig. 9.3):

$$\beta_{RH} = \begin{cases} \beta_0 \, e^{-k\cdot(VPD-VPD_0)} & \text{for VPD} > VPD_0 \\ \beta_0 & \text{for VPD} < VPD_0 \end{cases} \tag{9.9}$$

Parameterizing a model that combines Eqs. 9.5, 9.6, and 9.9 is challenging and parameter equifinality is likely to occur: the decrease in GEP due to VPD has the same effect on NEE as an increase in R_{eco} due to temperature. Lasslop et al. (2010) estimated the parameters of the combined equation using a multistep process. The temperature sensitivity of R_{eco} was estimated first from nighttime data using 15-day windows after Reichstein et al. (2005a). In a second step, the temperature sensitivity was fixed and the remaining fitted parameters were estimated using 4-day windows of daytime data, noting that the base respiration parameter was fit alongside the other parameters using daytime data to ensure a degree of independence from the nighttime data. Including these five parameters in the optimization routine still results in an overparameterized model in certain situations. For instance if VPD is low, the parameter k is not well constrained, but it can influence the results if it is used for extrapolation to high VPD. Meaningless photosynthetic parameters are common for deciduous forests and polar ecosystems in winter. (Table A1 in Lasslop et al. 2010, explains how parameters were treated if they were not in a predefined range.)

The myriad choices available for modeling R_{eco} and GEP using daytime data from global ecosystems leaves open the possibility for multiple improvements to the FLUXNET flux partitioning algorithm in the future. Desai et al. (2008) demonstrated significant differences among light-response curve-based methods and showed that, whereas some methods may be more subject to biases than others, it is not possible to identify one superior method given flux observations and an unknown "true" flux. This suggests that future work on flux partitioning using multiple, complementary methods is an ideal way forward to ensure defensible partitioned estimates with conservative error bounds.

9.3.3.3 Unresolved Issues and Future Work

It has been reported that canopy assimilation is not only affected by the overall shortwave radiation flux density, but also by its direct or diffuse characteristics; higher assimilation rates have been observed at the same overall radiation flux density under conditions dominated by diffuse radiative flux (Baldocchi et al. 1997; Gu et al. 2003; Hollinger et al. 1994; Jenkins et al. 2007; Knohl and Baldocchi 2008; Niyogi et al. 2004). Diffuse radiation is measured at few FLUXNET sites to date, and incorporating the effects of diffuse radiation on NEE for global flux partitioning would require models to separate direct and diffuse radiation from net radiation measurements. This introduces the problem of using modeled data to drive a model. Diffuse radiation is also correlated with low VPD values, and the relative importance of each needs to be ascertained before modeling efforts proceed (Rodriguez and Sadras 2007; Wohlfahrt et al. 2008).

To summarize, we recommend simple, process-based R_{eco} models with varying parameters to incorporate rapid, seasonal, or interannual changes in canopy structure, soil moisture, ecosystem nutrient level, and carbon transport for the purpose of partitioning GEP and R_{eco} across the global eddy covariance tower network (Reichstein et al. 2005a). At the site level, we advocate integrating above-canopy eddy covariance instrumentation, below-canopy eddy covariance (Baldocchi et al. 1997), carefully designed respiration chambers (Bain et al. 2005; Subke et al. 2009; Xu et al. 2006), isotopic techniques (Ekblad et al. 2005; Ekblad and Hogberg 2001;

Högberg et al. 2001), laboratory analyses (Conant et al. 2008), and modeling studies (Adair et al. 2008; Thompson and Holbrook 2004) for developing a comprehensive ecosystem-level mechanistic understanding of R_{eco}.

9.4 Additional Considerations and New Approaches

9.4.1 *Oscillatory Patterns*

Circadian rhythms of stomatal conductance have not been formally considered for flux partitioning to date. They are either endogenous or caused by hydraulic limitations in the afternoon. These patterns in the diurnal cycle can persist for more than a week, independent of environmental influences (Hennessey and Field 1991). Although this effect has been widely observed (Gorton et al. 1993; Hennessey et al. 1993; Nardini et al. 2005), the degree to which they affect the carbon exchange under field conditions is less clear. Williams et al. (1998) suggested by using a modeling approach that these circadian rhythms do not significantly affect photosynthesis and stomatal conductance in field conditions. Recent laboratory-based findings have found the circadian rhythms of root functioning to be coupled to leaf function at the plant level (James et al. 2008), but ecosystem-level relationships have yet to be explored and for the moment oscillatory patterns may be best treated by model parameterization rather than changing model structure.

9.4.2 *Model Parameterization*

So far we have discussed model parameters but not methods for determining their value and associated uncertainty, which is critical for assimilating data into ecosystem models (Raupach et al. 2005; Williams et al. 2009). The form of the cost function, rather than the technique used to find the optimum parameter values, tends to be more important for accurate parameter estimation using flux data (Fox et al. 2009; Trudinger et al. 2007). It has been argued that the error in flux measurements follows a Laplace (double exponential) distribution such that least absolute deviations rather than least-squares techniques should be used for the cost function (Hollinger and Richardson 2005; Richardson et al. 2006b, see also Sect. 7.2.5), but other studies have suggested that error in eddy covariance flux measurements can be approximated as a normal distribution with nonstationary variances that are a function of flux magnitude (Lasslop et al. 2008). Rannik and Vesala (1999) presented relative systematic and random error distributions for sensible heat fluxes, which are qualitatively same for other scalars. Importantly, any method should not understate uncertainty in parameter values or resulting partitioned flux estimates.

A major theme of the discussion to this point is that half-hourly eddy covariance observations alone are not sufficient to understand the mechanisms responsible for R_{eco} and GEP fluxes. The simple models advocated to this point are but one approach for flux partitioning, albeit the most common. Additional techniques can and should be investigated to improve our understanding of ecosystem processes and the biosphere–atmosphere flux of CO_2.

9.4.3 Flux Partitioning Using High-Frequency Data

It has been argued that the high-frequency (e.g., 10 or 20 Hz) flux data contains more information about the sources of CO_2 (Thomas et al. 2008) and the assimilation/respiration dynamics (Scanlon and Kustas 2010; Scanlon and Sahu 2008) than is commonly acknowledged. To partition respiration sources into above- and below-canopy components Thomas et al. (2008) used a conditional sampling method to identify turbulent events that represented both a source of water vapor and CO_2 to the atmosphere, and attributed these events to transport from below the plant canopy. It was noted that the resultant respiratory fluxes agreed with chamber-based measurements and the intercept of eddy covariance light-response curves.

Scanlon and Kustas (2010) noted that stomatal processes (i.e., GEP and E_{transp}) and non-stomatal processes (R_{eco} and E_{soil}) each conform separately to flux-variance (Monin-Obukhov) similarity and provided an analytical expression based on the water use efficiency to partition both CO_2 and water vapor fluxes using high-frequency data (Scanlon and Sahu 2008). Seasonal patterns of these partitioned flux estimates followed closely canopy development in an agricultural ecosystem.

An obvious problem with these approaches for integration into the FLUXNET database is the lack of available or synthesized high-frequency flux data to perform these analyses globally, although for site-level studies and future research they may prove extremely valuable for not only quantifying ecosystem carbon and water dynamics, but also transport phenomena at the biosphere–atmosphere interface.

9.4.4 Flux Partitioning Using Stable Isotopes

As discussed, a fundamental problem with flux partitioning is that one measurement (NEE) is being used to infer two processes (R_{eco} and GEP). A natural solution would be to add measurements that provide additional information. Naturally abundant stable isotopes in the atmosphere provide a way forward. Stable isotope observations to better understand plant ecology and biochemistry have a long history (Dawson et al. 2002), but their use for partitioning eddy covariance-measured NEE is more recent (Bowling et al. 2001; Lloyd et al. 1996). The biochemistry of photosynthesis is such that plants prefer the lighter isotope of CO_2, thereby imprinting that signature on both organic matter (depleted in heavier isotopes) and in the atmosphere (enriched) (Yakir and da Silveira Lobo Sternberg 2000). Photosynthetic fractionation leads to atmospheric enrichment of ^{13}C in CO_2 and, through equilibration of transpired water and assimilation of CO_2, to enrichment of ^{18}O in CO_2. Additional fractionation of CO_2 isotopes during autotrophic and microbial respiration further separates the isotopic signature of respired products from assimilation (Knohl and Buchmann 2005).

An equation for isotopic fractionation by GEP and R_{eco} can be written following Ogée et al. (2004):

$$\delta_N \text{NEE} = \delta_R R_{eco} - \left(\delta_a - \Delta_{canopy}\right) \text{GEP} \tag{9.10}$$

where the first term represents the product of NEE and its isotopic composition (δ_N), commonly called the isoflux, the second term the effect of respiration on atmospheric isotopic composition (δ_R), and the latter term the discrimination by photosynthesis (Δ_{canopy}) for lighter isotopes of CO_2 in the atmosphere, which has its own isotopic composition (δ_a). Isotopic ratios are commonly expressed in units of per mil with respect to a benchmark standard. Combining Eq. 9.10 with Eq. 9.1, and observations of NEE, the isoflux, δ_R, δ_a, and a model of Δ_{canopy}, allows one to infer R_{eco} and GEP.

Currently, eddy covariance observations of the isoflux are limited by the frequency responses of instrumentation, so it is instead generally inferred from flux-gradient or relaxed (or disjunct) eddy accumulation techniques. The isotopic composition of R_{eco} is typically measured from the intercept of a Keeling plot, which plots the inverse of nighttime CO_2 versus its isotopic composition (Pataki et al. 2003). Isotopic discrimination during assimilation (Δ_{canopy}) is typically assumed from equations of stomatal conductance and leaf cellular CO_2 diffusion during the photosynthetic process.

There are a number of uncertainties in this approach that need to be propagated for defensible GEP and R_{eco} estimates. These include the mismatch between concentration profiles and flux footprints, the sensitivity of micrometeorological flux-gradient techniques to atmospheric stability and mixing, the assumptions made in Keeling plot analysis and the canopy discrimination model (which, for example, differs substantially for C3 and C4 photosynthesis), the sampling frequency of isotope observations, and assumptions made about isotopic equilibration with plant and soil water and equivalency in fractionation for autotrophic and heterotrophic respiration. For example, Ogée et al. (2004) demonstrated that uncertainty could exceed 4 μmol m^{-2} s^{-1} for half-hourly observations of GEP and R_{eco} using isotopic methods. Further, isotopic flux partitioning is strongly sensitive to the extent of isotopic disequilibrium between R_{eco} and GEP, which is relatively small for $^{13}CO_2$. Direct in situ high-frequency isotope observations (e.g., Zhang et al. 2006) and Bayesian parameterization of canopy photosynthetic and isotopic models (e.g., Zobitz et al. 2007) address some of the uncertainties associated with isotopic techniques. Isotopic partitioning of NEE is still primarily limited by the lack of stable isotope observations at most FLUXNET sites; however, these deficiencies will likely change in the future as sensor prices and stability improve.

9.4.5 Chamber-Based Approaches

Eddy covariance measurements of NEE can be partitioned to different component fluxes by upscaling chamber measurements (e.g., soil, leaf, bole, and coarse woody debris fluxes) of CO_2 uptake and release (Bolstad et al. 2004; Harmon et al. 2004; Lavigne et al. 1997; Law et al. 1999; Ohkubo et al. 2007; Wang et al. 2010). Upscaling involves extrapolation of measurements both in space (i.e., from individual chambers to the whole ecosystem) and in time (i.e., from periodic or intermittent measurements to a half-hourly time step commensurate with the tower-measured fluxes, or to an annual time step for ecosystem C budgets). Also required is information about the size of various C pools, for example, leaf area index and canopy density profiles, bole volume, and sapwood area of trees of different diameter classes, and the amount and state of decay of coarse woody debris. The overall approach to upscaling, and the way in which component fluxes interact with environmental drivers, varies among studies and is highly dependent on the data available and the assumptions that need to be made; the studies cited above provide a range of examples.

There are major uncertainties inherent in chamber-based approaches for measuring photosynthetic uptake or respiration from stems, leaves, and soil (Lavigne et al. 1997; Loescher et al. 2006). These include sampling uncertainties (representativity and spatial heterogeneity), scale mismatches between chambers and the tower footprint, as well as random and systematic measurement errors (e.g. Savage et al. 2008; Subke et al. 2009). For example, Lavigne et al. (1997) reported poor agreement between upscaled chamber measurements and nocturnal NEE measurements at six evergreen boreal field sites, largely because of the inherent noise in both estimates, but also because of a systematic bias on the order of 20–40%. These uncertainties will ideally be reduced as improved chamber designs are developed and improved spatiotemporal measurement strategies are adopted (Bain et al. 2005; Subke et al. 2009; Xu et al. 2006).

Estimating the uncertainties inherent in individual measurements, and then propagating these forward in the upscaling methodology is desirable, but is rarely done in a comprehensive manner. This is, however, a relatively straightforward task if the upscaling is conducted using a model-data fusion framework in conjunction with a process-based model of ecosystem C dynamics: posterior uncertainties in partitioned fluxes can be estimated conditional on both the model and the data used as constraints (e.g., Richardson et al. 2010). (For an alternative Monte Carlo approach conducted at the annual time step, see Harmon et al. 2004.)

9.4.6 Partitioning Water Vapor Fluxes

Eddy covariance flux partitioning need not be limited to carbon fluxes. Given the ubiquity of carbon flux investigations, and the relative paucity of water and energy flux studies to date, carbon flux partitioning has been the overwhelming focus. Process-based studies in hydrology can benefit tremendously from knowledge of the pathways by which water enters the atmosphere from the terrestrial surface.

In a similar manner to carbon fluxes, periods exist where terms of the evapotranspiration equation Eq. 9.3 are zero or negligible. For example, in deciduous forests, E_{transp} and E_{int} are near zero during leaf-off except immediately after rain events. Assuming that stem evaporation is minor, $E_{tot} \cong E_{soil}$. Stoy et al. (2006a) modeled E_{soil} as a function of radiation that penetrated the aboveground vegetation in temperate forest and grass ecosystems in southeastern USA. The model was parameterized using eddy covariance measured E_{tot} during dry periods when the respective canopies were known to be inactive. Partitioned E_{transp} estimates approximated well stand-level E_{transp} estimated by sapflux for the Duke Forest loblolly pine ecosystem (Schäfer et al. 2002). (Oishi et al. 2008) modeled E_{soil} as a function of VPD using a subset of dry, wintertime eddy covariance data from the Duke Forest hardwood ecosystem and found good agreement between annual eddy covariance-measured ET, and annual ET based on the sum of this evaporation model, stand-scaled sapflux measurements, and modeled canopy interception. Partitioning eddy covariance E_{tot} by directly using upsclaled sapflux measurements is another common technique (see Sect. 11.3.4).

Stable isotope-based approaches for partitioning evaporation and transpiration from evapotranspiration measurements have been explored (Wang and Yakir 2000) (Albertson et al. 2001) but not widely applied to date. We note that the US–based National Earth Observation Network (NEON) will use a stable isotope-based approach in conjunction with eddy covariance data to separate evaporation and transpiration and such approaches are likely to find wide applicability in the near future.

9.5 Recommendations

Extensive work on ecosystem carbon flux partitioning has been completed to date, but there is more to be done. We caution against using a single standard algorithm for partitioning R_{eco} and GEP given the potential for bias (Desai et al. 2008); multiple methods should be compared at each site to ensure that the outcome is robust. We recommend comparing both light-response curve and temperature response curve methods as quasi-independent checks (Lasslop et al. 2010; Reichstein et al. 2005a) and to develop additional flux partitioning routines to challenge and improve standard approaches.

An argument often arises: why not use more complex process-based models for the purpose of flux partitioning (Desai et al. 2008)? More complex models have the potential to deliver more accurate partitioned fluxes, but the uncertainty of the model formulation is difficult to quantify (Rastetter et al. 2010) and the partitioned estimates may be used to constrain model output or compare against model output, resulting in a circularity. By ensuring that flux estimates are data-driven to the extent that this is possible using the simplest physiologically reasonable models available, the values that are least contaminated by model assumptions can be found. Techniques that are entirely data-driven (e.g. artificial neural networks) are likewise of value but may have difficulties extrapolating observations.

We note that the techniques favored to date are not static or "final" and that ample opportunity for improvement exist. Checks of eddy covariance-derived net and partitioned fluxes against independent flux estimates continue to have the potential to improve algorithms. Given the centralized management of the FLUXNET database, new, different, and/or improved approaches can be integrated as additional derived products without extensive additional effort and will aid in the generation of conservative error bounds on NEE, GEP, and R_{eco}. We encourage continued investigations into partitioning carbon and water fluxes using the FLUXNET database.

References

Adair EC et al (2008) Simple three-pool model accurately describes patterns of long-term litter decomposition in diverse climates. Glob Chang Biol 14:2636–2660

Albertson JD, Kustas WP, Scanlon TM (2001) Large eddy simulation over heterogeneous terrain with remotely sensed land surface conditions. Water Resour Res 37:1939–1953

Aubinet M et al (2000) Estimates of the annual net carbon and water exchange of forests: the EUROFLUX methodology. Adv Ecol Res 30:113–175

Aubinet M, Chermanne B, Vandenhaute M, Longdoz B, Yernaux M, Laitat E (2001) Long-term carbon dioxide exchange above a mixed forest in the Belgian Ardennes. Agric For Meteorol 108:293–315

Aubinet M et al (2010) Direct advection measurements do not help to solve the night-time CO_2 closure problem: evidence from three different forests. Agric For Meteorol 150:655–664

Austin AT, Vivanco L (2006) Plant litter decomposition in a semi-arid ecosystem controlled by photodegradation. Nature 442:555–558

Bahn M et al (2008) Soil respiration in European grasslands in relation to climate and assimilate supply. Ecosystems 11:1352–1367

Bain WG et al (2005) Wind-induced error in the measurement of soil respiration using closed dynamic chambers. Agric For Meteorol 131:225–232

Baldocchi DD, Vogel CA, Hall B (1997) Seasonal variation of carbon dioxide exchange rates above and below a boreal jack pine forest. Agric For Meteorol 83:147–170

Baldocchi D, Tang J, Xu L (2006) How switches and lags in biophysical regulators affect spatial-temporal variation of soil respiration in an oak-grass savanna. J Geophys Res Atmos 111:G02008. doi:02010.01029/02005JG000063

Barbour MM et al (2005) Variation in the degree of coupling between δC13 of phloem sap and ecosystem respiration in two mature Nothofagus forests. New Phytol 166:497–512

Barford CC et al (2001) Factors controlling long- and short-term sequestration of atmospheric CO_2 in a mid-latitude forest. Science 294:1688–1691

Bolstad PV, Davis KJ, Martin J, Cook BD, Wang W (2004) Component and whole-system respiration fluxes in northern deciduous forests. Tree Physiol 24:493–504

Bowling DR, Tans PP, Monson RK (2001) Partitioning net ecosystem carbon exchange with isotopic fluxes of CO_2. Glob Chang Biol 7:127–145

Carbone MS, Winston GC, Trumbore SE (2008) Soil respiration in perennial grass and shrub ecosystems: linking environmental controls with plant and microbial sources on seasonal and diel timescales. J Geophys Res Biogeosci 113. doi:10.1029/2007JG000611

Ciais P et al (2005) Europe-wide reduction in primary productivity caused by the heat and drought in 2003. Nature 437:529–533

Conant RT et al (2008) Sensitivity of organic matter decomposition to warming varies with its quality. Glob Chang Biol 14:868–877

Cook BD et al (2004) Carbon exchange and venting anomalies in an upland deciduous forest in norhern Wisconsin, USA. Agric For Meteorol 126:271–295

Cowan I (1977) Stomatal behaviour and environment. Adv Bot Res 4:117–228

Daly E et al (2009) The effects of elevated atmospheric CO_2 and nitrogen amendments on subsurface CO_2 production and concentration dynamics in a maturing pine forest. Biogeochemistry 94:271–287

Davidson EA, Janssens IA (2006) Temperature sensitivity of soil carbon decomposition and feedbacks to climate change. Nature 440:165–173

Dawson TE, Mambelli S, Plamboeck AH, Templer PH, Tu KP (2002) Stable isotopes in plant ecology. Annu Rev Ecol Syst 33:507–559

Desai AR, Bolstad PV, Cook BD, Davis KJ, Carey EV (2005) Comparing net ecosystem exchange of carbon dioxide between an old-growth and mature forest in the upper Midwest, USA. Agric For Meteorol 128:33–55

Desai AR et al (2008) Cross-site evaluation of eddy covariance GPP and RE decomposition techniques. Agric For Meteorol 148:821–838

Drake JE, Stoy PC, Jackson RB, DeLucia EH (2008) Fine root respiration in a loblolly pine (Pinus taeda) forest exposed to elevated CO_2 and N fertilization. Plant Cell Environ 31:1663–1672

Ekblad A, Hogberg P (2001) Natural abundance of ^{13}C reveals speed of link between tree photosynthesis and root respiration. Oecologia 127:305–308

Ekblad A, Boström B, Holm A, Comstedt D (2005) Forest soil respiration rate and $\delta^{13}C$ is regulated by recent above ground weather conditions. Oecologia 143:136–142

Emmerich WE (2003) Carbon dioxide fluxes in a semiarid environment with high carbonate soils. Agric For Meteorol 116:91–102

Falge E et al (2001) Gap filling strategies for defensible annual sums of net ecosystem exchange. Agric For Meteorol 107:43–69

Falge E et al (2002) Seasonality of ecosystem respiration and gross primary production as derived from FLUXNET measurements. Agric For Meteorol 113:53–74

Fierer N, Craine J, McLauchlan K, Schimel JP (2005) Litter quality and the temperature sensitivity of decomposition. Ecology 86:320–326

Foken T, Göckede M, Mauder M, Mahrt L, Amiro BD, Munger JW (2004) Post-field data quality control. In: Lee X, Massman W, Law B (eds) Handbook of micrometeorology: a guide for surface flux measurement and analysis. Kluwer, Dordrecht, p 250

Fox A et al (2009) The REFLEX project: comparing different algorithms and implementations for the inversion of a terrestrial ecosystem model against eddy covariance data. Agric For Meteorol 149:1597–1615

Gilmanov TG et al (2003) Gross primary production and light response parameters of four southern plains ecosystems estimated using long-term CO2-flux tower measurements. Glob Biogeochem Cycle 17:1071. doi:1010.1029/2002GB002023

Gilmanov TG, Aires L, Barcza Z, Baron VS, Belelli L, Beringer J, Billesbach D, Bonal D, Bradford J, Ceschia E, Cook D, Corradi C, Frank A, Gianelle D, Gimeno C, Gruenwald T, Guo HQ, Hanan N, Haszpra L, Heilman J, Jacobs A, Jones MB, Johnson DA, Kiely G, Li SG, Magliulo V, Moors E, Nagy Z, Nasyrov M, Owensby C, Pinter K, Pio C, Reichstein M, Sanz MJ, Scott R, Soussana JF, Stoy PC, Svejcar T, Tuba Z, Zhou GS (2010) Productivity respiration, and light-response parameters of world grassland and agroecosystems derived from flux-tower measurements. Rangel Ecol Manage 63(1):16–39

Gilmanov TG, Svejcar TJ, Johnson DA, Angell RF, Saliendra NZ, Wylie BK (2006) Long-term dynamics of production, respiration and net CO_2 exchange in two sagebrush-steppe ecosystems. Rangel Ecol Manage 59:585–599

Gold V, Loening KL, McNaught AD, Sehmi P (1991) Compendium of chemical terminology softcover (IUPAC chemical data series). CRC Press, Boca Raton

Gorton HL, Williams WE, Assmann SM (1993) Circadian-rhythms in stomatal responsiveness to red and blue-light. Plant Physiol 103:399–406

Gu L et al (2003) Response of a deciduous forest to the mount Pinatubo eruption: enhanced photosynthesis. Science 299:2035–2038

Harmon ME et al (2004) Production, respiration, and overall carbon balance in an old-growth Pseudotsuga-tsuga forest ecosystem. Ecosystems 7:498–512

Hennessey TL, Field CB (1991) Circadian rhythms in photosynthesis: oscillations in carbon assimilation and stomatal conductance under constant conditions. Plant Physiol 96:831–836

Hennessey TL, Freeden AL, Field CB (1993) Environmental-effects of circadian-rhythms in photosynthesis and stomatal opening. Planta 189:369–376

Högberg P et al (2001) Large-scale forest girdling shows that current photosynthesis drives soil respiration. Nature 411:789–792

Hollinger DY, Richardson AD (2005) Uncertainty in eddy covariance measurements and its application to physiological models. Tree Physiol 25:873–885

Hollinger DY, Kelliher FM, Byers JN, Hunt JE, Mcseveny TM, Weir PL (1994) Carbon-dioxide exchange between an undisturbed old-growth temperate forest and the atmosphere. Ecology 75:134–150

Horwath WR, Pretziger KS, Paul EA (1994) ^{14}C allocation in tree-soil systems. Tree Physiol 14:1163–1176

Irvine J, Law BE (2002) Seasonal soil CO2 effluxes in young and old ponderosa pine forests. Glob Chang Biol 8:1183–1194

James AB et al (2008) The circadian clock in Arabidopsis roots is a simplified slave version of the clock in shoots. Science 322:1832–1835

Janssens IA, Pilegaard K (2003) Large seasonal changes in Q(10) of soil respiration in a beech forest. Glob Chang Biol 9:911–918

Janssens IA et al (2001) Productivity overshadows temperature in determining soil and ecosystem respiration across European forests. Glob Chang Biol 7:269–278

Jarvis P et al (2007) Drying and wetting of Mediterranean soils stimulates decomposition and carbon dioxide emission: the "Birch effect". Tree Physiol 27:929–940

Jenkins JP, Richardson AD, Braswell BH, Ollinger SV, Hollinger DY, Smith ML (2007) Refining light-use efficiency calculations for a deciduous forest canopy using simultaneous tower-based carbon flux and radiometric measurements. Agric For Meteorol 143:64–79

Knohl A, Baldocchi DD (2008) Effects of diffuse radiation on canopy gas exchange processes in a forest ecosystem. J Geophys Res 113:G02023. doi:02010.01029/02007JG000663

Knohl A, Buchmann N (2005) Partitioning the net CO_2 flux of a deciduous forest into respiration and assimilation using stable carbon isotopes. Glob Biogeochem Cycle 19:GB4008

Knohl A, Werner RA, Brand WA, Buchmann N (2005) Short-term variations in δ13C of ecosystem respiration reveals link between assimilation and respiration in a deciduous forest. Oecologia 142:70–82

Körner C (1995) Leaf diffusive conductances in the major vegetation types of the globe. In: Schulze E-D, Caldwell MM (eds) Ecophysiology of photosynthesis. Springer, Berlin, pp 463–490

Kowalski AS et al (2008) Can flux tower research neglect geochemical CO_2 exchange? Agric For Meteorol 148:1045–1054

Lasslop G, Reichstein M, Kattge J, Papale D (2008) Influences of observation errors in eddy flux data on inverse model parameter estimation. Biogeosci Discuss 5:751–785

Lasslop G et al (2010) Separation of net ecosystem exchange into assimilation and respiration using a light response curve approach: critical issues and global evaluation. Glob Chang Biol 16:187–208

Lavigne MB et al (1997) Comparing nocturnal eddy covariance measurements to estimates of ecosystem respiration made by scaling chamber measurements at six coniferous boreal sites. J Geophys Res Atmos 102:28977–28985

Law BE, Ryan MG, Anthoni PM (1999) Seasonal and annual respiration of a ponderosa pine ecosystem. Glob Chang Biol 5:169–182

Law BE et al (2002) Environmental controls over carbon dioxide and water vapor exchange of terrestrial vegetation. Agric For Meteorol 113:97–120

Lee X, Fuentes JD, Staebler RM, Neumann HH (1999) Long-term observation of the atmospheric exchange of CO_2 with a temperate deciduous forest in southern Ontario, Canada. J Geophys Res Atmos 104:15975–15984

Lindroth A, Klemedtsson L, Grelle A, Weslien P, Langvall O (2008). Measurement of net ecosystem exchange, productivity and respiration in three spruce forests in Sweden shows unexpectedly large soil carbon losses. Biogeochemistry 89:43–60

Lloyd J, Taylor JA (1994) On the temperature dependence of soil respiration. Function Ecol 8:315–323

Lloyd J et al (1996) Vegetation effects on the isotopic composition of atmospheric CO_2 at local and regional scales: theoretical aspects and a comparison between rain forest in Amazonia and a boreal forest in Siberia. Austr J Plant Physiol 23:371–399

Loescher HW, Law BE, Mahrt L, Hollinger DY, Campbell J, Wofsy SC (2006) Uncertainties in, and interpretation of, carbon flux estimates using the eddy covariance technique. J Geophys Res Atmos 111:D21S90

Mäkelä A, Givnish TJ, Berninger F, Buckley TN, Farquhar GD, Hari P (2002) Challenges and opportunities of the optimality approach in plant ecology. Silva Fennica 36:605–614

Marshall B, Biscoe PV (1980) A model for C_3 leaves describing the dependence of net photosynthesis on irradiance. J Exp Bot 31:29–39

Mencuccini M, Hölttä T (2010) The significance of phloem transport for the speed with which canopy photosynthesis and belowground respiration are linked. New Phytol 185:189–203

Mielnick PA, Dugas WA, Mitchell K, Havstad K (2005) Long-term measurements of CO_2 flux and evapotranspiration in a Chihuahuan desert grassland. J Arid Environ 60:423–436

Moffat AM et al (2007) Comprehensive comparison of gap-filling techniques for eddy covariance net carbon fluxes. Agric For Meteorol 147:209–232

Nardini A, Salleo S, Andri S (2005) Circadian regulation of leaf hydraulic conductance in sunflower (Helianthus annuus L. cv Margot). Plant Cell Environ 28:750–759

Niyogi D et al (2004) Direct observations of the effects of aerosol loading on net ecosystem CO2 exchanges over different landscapes. Geophys Res Lett 31:L20506. doi:20510.21029/22004GL020915

Novick KA et al (2004) Carbon dioxide and water vapor exchange in a warm temperate grassland. Oecologia 138:259–274

Ogée J et al (2004) Partitioning net ecosystem carbon exchange into net assimilation and respiration with canopy-scale isotopic measurements: an error propagation analysis with 13CO2 and CO18O data. Glob Biogeochem Cycle 18:GB2019

Ohkubo S, Kosugi Y, Takanashi S, Mitani T, Tani M (2007) Comparison of the eddy covariance and automated closed chamber methods for evaluating nocturnal CO_2 exchange in a Japanese cypress forest. Agric For Meteorol 142:50–65

Oishi AC, Oren R, Stoy PC (2008) Estimating components of forest evapotranspiration: a footprint approach for scaling sap flux measurements. Agric For Meteorol 148:1719–1732

Oren R et al (1999) Survey and synthesis of intra- and interspecific variation in stomatal sensitivity to vapor pressure deficit. Plant Cell Environ 22:1515–1526

Palmroth S et al (2005) Contrasting responses to drought of the forest floor CO_2 efflux in a loblolly pine plantation and a nearby oak-hickory forest. Glob Chang Biol 11:421–434

Papale D, Valentini R (2002) A new assessment of European forests carbon exhcanges by eddy fluxes and artificial neural network spatialization. Glob Chang Biol 9:525–535

Papale D et al (2006) Towards a standardized processing of net ecosystem exchange measured with eddy covariance technique: algorithms and uncertainty estimation. Biogeosciences 3:571–583

Pataki DE et al (2003) The application and interpretation of keeling plots in terrestrial carbon cycle research. Global Biogeochem Cycle 17:1022

Rambal S et al (2003) Drought controls over conductance and assimilation of a Mediterranean evergreen ecosystem: scaling from leaf to canopy. Glob Chang Biol 9:1813–1824

Rannik Ü, Vesala T (1999) Autoregressive filtering versus linear detrending in estimation of fluxes by the eddy covariance method. Bound Layer Meteorol 91:259–280

Rastetter EB et al (2010) Processing arctic eddy-flux data using a simple carbon-exchange model embedded in the ensemble Kalman filter. Ecol Appl 20:1285–1301

Raupach MR et al (2005) Model-data synthesis in terrestrial carbon observation: methods, data requirements and data uncertainty specifications. Glob Chang Biol 11:378–397

Rebmann C et al (2005) Quality analysis applied on eddy covariance measurements at complex forest sites using footprint modelling. Theor Appl Climatol 80:121–141

Rebmann C et al (2010) Treatment and assessment of the CO_2-exchange at a complex forest site in Thuringia Germany. Agric For Meteorol 150:684–691

Reichstein M, Tenhunen JD, Ourcival JM, Rambal S, Dore S, Valentini R (2002) Ecosystem respiration in two mediterranean evergreen holm oak forests: drought effects and decomposition dynamics. Funct Ecol 16:27–39

Reichstein M et al (2005a) On the separation of net ecosystem exchange into assimilation and ecosystem respiration: review and improved algorithm. Glob Chang Biol 11:1424–1439

Reichstein M, Subke J-A, Angeli AC, Tenhunen J (2005b) Does the temperature sensitivity of decomposition of soil organic matter depend upon water content, soil horizon, or incubation time? Glob Chang Biol 11:1754–1767

Richardson AD et al (2006a) Comparing simple respiration models for eddy flux and dynamic chamber data. Agric For Meteorol 141:219–234

Richardson AD et al (2006b) A multi-site analysis of random error in tower-based measurements of carbon and energy fluxes. Agric For Meteorol 136:1–18

Richardson AD et al (2010) Estimating parameters of a forest ecosystem model with measurements of stocks and fluxes as joint constraints. Oecologia 164:25–40

Rodriguez D, Sadras VO (2007) The limit to wheat water-use efficiency in eastern Australia. I. Gradients in the radiation environment and atmospheric demand. Aust J Agric Res 58:287–302

Rutledge S, Campbell DI, Baldocchi DD, Schipper LA (2010) Photodegradation leads to increased carbon dioxide losses from terrestrial organic matter. Glob Chang Biol 16:3065–3074

Ryan MG (1991) Effects of climate change on plant respiration. Ecol Appl 1:157–167

San-José J, Montes R, Nikonova N (2007) Diurnal patterns of carbon dioxide, water vapour, and energy fluxes in pineapple [Ananas comosus (L.) Merr. cv. Red Spanish] field using eddy covariance. Photosynthetica 45:370–384

Savage K, Davidson EA, Richardson AD (2008) A conceptual and practical approach to data quality and analysis procedures for high-frequency soil respiration measurements. Funct Ecol 22:1000–1007

Scanlon TM, Kustas WP (2010) Partitioning carbon dioxide and water vapor fluxes using correlation analysis. Agric For Meteorol 150:89–99

Scanlon TM, Sahu P (2008) On the correlation structure of water vapor and carbon dioxide in the atmospheric surface layer: a basis for flux partitioning. Water Resour Res 44:W10418

Schäfer KVR, Oren R, Lai CT, Katul GG (2002) Hydrologic balance in an intact temperate forest ecosystem under ambient and elevated atmospheric CO_2 concentration. Glob Chang Biol 8:895–911

Staebler RM, Fitzjarrald D (2004) Observing subcanopy CO_2 advection. Agric For Meteorol 122:139–156

Stoy PC et al (2006a) Separating the effects of climate and vegetation on evapotranspiration along a successional chronosequence in the southeastern U.S. Glob Chang Biol 12:2115–2135

Stoy PC, Katul GG, Siqueira MBS, Juang J-Y, Novick KA, Oren R (2006b) An evaluation of methods for partitioning eddy covariance-measured net ecosystem exchange into photosynthesis and respiration. Agric For Meteorol 141:2–18

Stoy PC et al (2007) Are ecosystem carbon inputs and outputs coupled at short time scales? a case study from adjacent pine and hardwood forests using impulse-response analysis. Plant Cell Environ 30:700–710

Subke J-A, Heinemeyer A, Reichstein M (2009) Experimental design to scale up in time and space and its statistical considerations. In: Kutsch W, Bahn M, Heinemeyer A (eds) Soil carbon dynamics: an integrated methodology. Cambridge University Press, Cambridge, 315

Suwa M et al (2004) Impact of elevated atmospheric CO_2 on forest floor respiration in a temperate pine forest. Glob Biogeochem Cycle 18:GB2013. doi:2010.1029/2003GB00218

Tang J, Baldocchi DD (2005) Spatial-temporal variation in soil respiration in an oak-grass savanna ecosystem in California and its partitioning into autotrophic and heterotrophic components. Biogeochemistry 73:183–207

Thomas C et al (2008) Estimating daytime subcanopy respiration from conditional sampling methods applied to multi-scalar high frequency turbulence time series. Agric For Meteorol 148:1210–1229

Thompson MV, Holbrook NM (2003) Application of a single-solute non-steady-state phloem model to the stufy of long-distance assimilate transport. J Theor Biol 220:419–455

Thompson MV, Holbrook NM (2004) Scaling phloem transport: information transmission. Plant Cell Environ 27:509–519

Trudinger CM et al (2007) OptIC project: an intercomparison of optimization techniques for parameter estimation in terrestrial biogeochemical models. J Geophys Res 112:G02027

Van Dijk AIJM, Dolman AJ (2004) Estimates of CO2 uptake and release among European forests based on eddy covariance data. Glob Chang Biol 10:1445–1459

van Gorsel E, Delpierre N, Leuning R, Black A, Munger JW, Wofsy S, Aubinet M, Feigenwinter C, Beringer J, Bonal D, Chen B, Chen J, Clement RR, Davis KJ, Desai AR, Dragoni D, Etzold S, Grünwald T, Gu L, Heinesch B, Hutyra LR, Jans WW, Kutsch W, Law BE, Leclerc MY, Mammarella I, Montagnani L, Noormets A, Rebmann C, Wharton S (2009) Estimating nocturnal ecosystem respiration from the vertical turbulent flux and change in storage of CO2. Agric For Meteorol 149:1919–1930

Vargas R, Allen MF (2008) Environmental controls and the influence of vegetation type, fine roots and rhizomorphs on diel and seasonal variation in soil respiration. New Phytol 179:460–471

Vickers D, Thomas CK, Martin JG, Law BE (2009) Self-correlation between assimilation and respiration resulting from flux partitioning of eddy-covariance CO_2 fluxes. Agric For Meteorol 149:1552–1555

Wang X-F, Yakir D (2000) Using stable isotopes of water in evapotranspiration studies. Hydrol Process 14:1407–1421

Wang M, Guan DX, Han SJ, Wu JL (2010) Comparison of eddy covariance and chamber-based methods for measuring CO_2 flux in a temperate mixed forest. Tree Physiol 30:149–163

Were A, Serrano-Ortiz P, de Moreno Jong C, Villagarcia L, Domingo F, Kowalski AS (2010) Ventillation of subterranean CO_2 and eddy covariance incongruities over carbonate ecosystems. Biogeosciences 7:859–867

Williams WE, Gorton HL (1998) Circadian rhythms have insignificant effects on plant gas exchange under field conditions. Physiol Plant 103:247–256

Williams M et al (2009) Improving land surface models with FLUXNET data. Biogeosciences 6:1341–1359

Wofsy SC et al (1993) Net exchange of CO2 in a mid-latitude forest. Science 260:1314–1317

Wohlfahrt G, Hammerle A, Haslwanter A, Bahn M, Tappeiner U, Cernusca A (2008) Disentangling leaf area and environmental effects on the response of the net ecosystem CO2 exchange to diffuse radiation. Geophys Res Lett 35:L16805. doi:16810.11029/12008GL035090

Xu L, Furtaw MD, Madsen RA, Garcia RL, Anderson DJ, McDermitt DK (2006) On maintaining pressure equilibrium between a soil CO_2 flux chamber and the ambient air. J Geophys Res 111:D08S10

Yakir D, da Silveira Lobo Sternberg L (2000) The use of stable isotopes to study ecosystem gas exchange. Oecologia 2000:297–311

Zhang J, Griffis TJ, Baker JM (2006) Using continuous stable isotope measurements to partition net ecosystem CO_2 exchange. Plant Cell Environ 29:483–496

Zobitz JM, Burns SP, Ogee J, Reichstein M, Bowling DR (2007) Partitioning net ecosystem exchange of CO_2 in a high-elevation subalpine forest: comparison of a Bayesian/isotope approach to environmental regression methods. J Geophys Res Biogeosci 112:G03013

Zobitz JM, Burns SP, Reichstein M, Bowling DR (2008) Partitioning net ecosystem carbon exchange and the carbon isotopic disequilibrium in a subalpine forest. Glob Chang Biol 14: 1–16

Chapter 10
Disjunct Eddy Covariance Method

Janne Rinne and Christof Ammann

10.1 Introduction

The eddy covariance method (EC) requires that all frequencies of turbulent motions and trace gas variations contributing to the flux are resolved by the measurement system. In conventional EC systems this is achieved by using anemometers and gas analyzers with fast response time and high data sampling frequency in order to catch the high-frequency end and by using long enough averaging period to catch the low-frequency end. Commonly, instruments with response times of around 0.1 s are used. However, for many atmospheric trace compounds analyzers with this short response time are not readily available or they do not provide continuous time series.

One possibility to reduce the requirements for analysers is the disjunct eddy covariance method (DEC). In DEC only a subset of the full continuous concentration and wind data series is used to obtain the flux. The reduced number of samples allows a noncontinuous (i.e., disjunct) sampling which gives the opportunity of a slower trace gas analysis or of sequential measurement (scanning) of multiple compounds with the same instrument.

10.2 Theory

Conventional EC systems are operated with a typical data sampling frequency of 10 Hz (sample interval of 0.1 s). For a typical flux calculation period of half an hour, this results in a total of $n = 18{,}000$ recorded values. From the recorded time series

J. Rinne (✉)
Department of Physics, University of Helsinki, FI-00014 Helsinki, Finland
e-mail: Janne.Rinne@helsinki.fi

C. Ammann
Agrosocope Reckenholz Tanikon Res Stn ART, CH-8046 Zurich, Switzerland
e-mail: christof.ammann@art.admin.ch

M. Aubinet et al. (eds.), *Eddy Covariance: A Practical Guide to Measurement and Data Analysis*, Springer Atmospheric Sciences, DOI 10.1007/978-94-007-2351-1_10,

of vertical wind speed w_j and trace gas concentrations c_{sj}, the EC flux is calculated as (by analogy with Eq. 3.9a).

$$\overline{c'_s w'} = \frac{1}{N}\sum_{j=1}^{N}\left[\left(c_{sj} - \bar{c}_s\right)\left(w_j - \bar{w}\right)\right] = \frac{1}{N}\sum_{j=1}^{N} c'_{sj} w'_j \qquad (10.1)$$

The basic idea of DEC is that the same flux (Eq. 10.1) may be calculated from a random-like subsampling of the full w_j and c_{sj} series. Since well-developed turbulence is not periodic, subsamples taken at regular intervals can be considered as random. Thus, disjunct samples are typically taken at a constant interval of 1–30 s. Accordingly, the number of DEC samples (N) taken within half an hour varies between 70 and 1,800 in published studies.

The DEC method has been validated against conventional EC method by data simulations and field experiments (Lenschow et al. 1994; Rinne et al. 2000, 2008; Bosweld and Beljaars 2001; Ammann et al. 2006; Turnipseed et al. 2009). The results confirm the assumptions that the reduced sample number of the DEC method does not cause systematic error to the measured flux values but it increases the random uncertainty of the fluxes. The fact that the subsampling of the time series in the DEC method does not bias the fluxes is caused by aliasing, which will be discussed in Sect. 10.4. However, the lower sampling rate of DEC method leads to increased random uncertainty in the measured fluxes. As will be shown in Sect. 10.5, this uncertainty is in many cases a function of number of samples used for flux calculations.

As there exists some confusion in the terminology of DEC measurement technique in the literature, we list here some recommendations for terminology and justify the suitability of these terms.

10.2.1 Sample Interval

The sample interval Δ is the constant time interval between samples (data recordings) of both w and c_s. In conventional EC, Δ is typically 0.1 s ($\ll$integral turbulence timescale, cf. Lenschow et al. 1994). In DEC, Δ is limited by the sampling regime of the concentration data and is in the range 1–30 s. The inverse of the sample interval is called the sampling frequency: $f_s = 1/\Delta$.

10.2.2 Response Time

The response time τ_R of a measurement system is a characteristic scale describing the time it takes to adjust the output signal as the input (measured quantity) is changing. This typically depends on the internal architecture of the analyzer, for

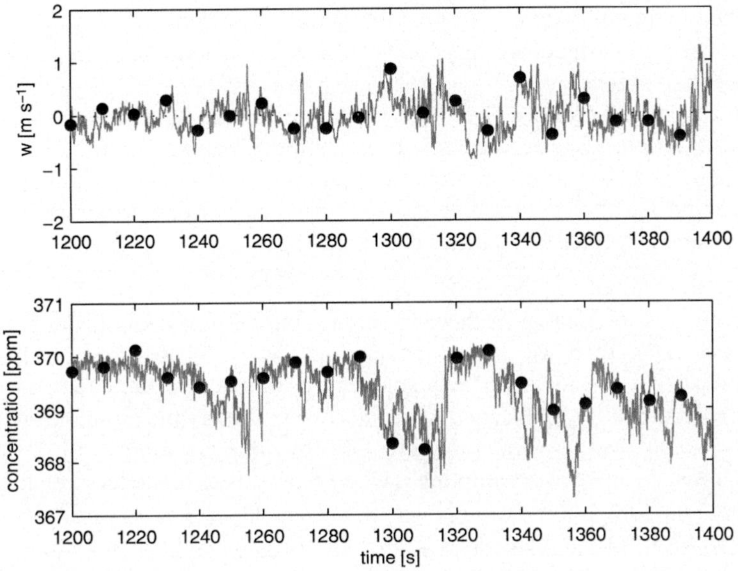

Fig. 10.1 Illustration for the basic idea of disjunct subsampling of time series. The *gray line* represents the original continuous 10 Hz time series while the *black dots* are 0.1 s long sub-samples taken from these data

example, dimensions of the measurement cell and inlet tube, sample flow rate, length of the internal integration time needed for a reliable measurement, etc. (see Sect. 4.1.3.2). The response time defines the high-frequency cut-off of the EC or DEC system (rather than the sampling frequency). In conventional eddy covariance systems the sampling interval Δ is typically close to τ_R, while in the DEC systems it is considerably longer.

10.2.3 Definition of DEC

We define any eddy covariance system with a *sample interval* Δ *considerably longer than the system response time* τ_R as disjunct eddy covariance (DEC). This term was coined by Lenschow et al. (1994), and justified over some other terms used at that time.

There are presently two main application types of the DEC method in use:

1. DEC by grab sampling (DEC-GS): Air is sampled near-instantaneously (during typically 0.1 s) into a reservoir at intervals $\Delta = 1$–30 s. This procedure allows a slower trace gas analysis of the grab sample compared to conventional EC measurements, and it results in a disjunct time series of near instantaneous concentration values (see Fig. 10.1). In this case the system response time is

defined by the time it takes to grab the sample. The vertical wind speed, which is measured continuously in general, has to be synchronized to the disjunct concentration time series.

2. DEC by mass scanning (DEC-MS): This technique is typically suitable for quadrupole mass spectrometers with a continuous sample flow, which provide fast response detection of multiple masses (trace gases). However, only a sequential detection of different masses is possible, and therefore each individual compound is recorded as a disjunct time series.

The term "virtual disjunct eddy covariance" (vDEC) often used for the second type of DEC is misleading as there is nothing virtual in it being disjunct or eddy covariance. Therefore, we do not recommend the use of this term. Instead, our suggestion is to use the term disjunct eddy covariance method for all systems fulfilling the definition above, and, when necessary specify the technique to sample air: for example, DEC by grab sampling, DEC by mass scanning.

The term "disjunct eddy sampling" (DES) refers to the procedure of taking grab samples of air with relatively long interval between them. The sampled air can then be processed in various ways, leading to either disjunct eddy covariance technique or disjunct eddy accumulation technique (see, e.g., Rinne et al. 2000, 2001).

Practical examples of both DEC application types are presented in Sect. 10.3. The outlines for measurement systems are based on published ones and are not necessarily the best possible.

10.3 Practical Applications of DEC

10.3.1 DEC by Grab Sampling

A gas analyzer able to conduct a concentration measurement within a time of 1–60 s can be used for DEC measurements if one is able to take a suitable air sample in a fraction of a second. We call such a sample a "grab sample." Here we present one basic design for disjunct eddy covariance measurements with grab sampling (DEC-GS), based on a design presented by Rinne et al. (2001). This design employs so-called intermediate storage reservoirs (ISR) which can be evacuated by a vacuum pump, and subsequently filled by opening a fast reacting high flow conductance valve (α-valve, Fig. 10.2). The sample interval, typically between 10 and 60 s, is used to analyze the trace gas concentrations inside the ISRs and to evacuate it for the next sample. Thus the operational sequence of this system consists of (1) evacuating the ISR, (2) taking a grab sample by opening the α-valve for a fraction of a second, and (3) analyzing the content of the ISR.

Operating sequence of a DEC system with two grab samples is shown in Table 10.1. The sampling time of this example would be 0.1 s and the sample interval 20.5 s. The short time (0.2 s) both sampling valve (S1 and S2) and analysis valve (A1 and A2) are closed between sampling and analysis is to ensure that the

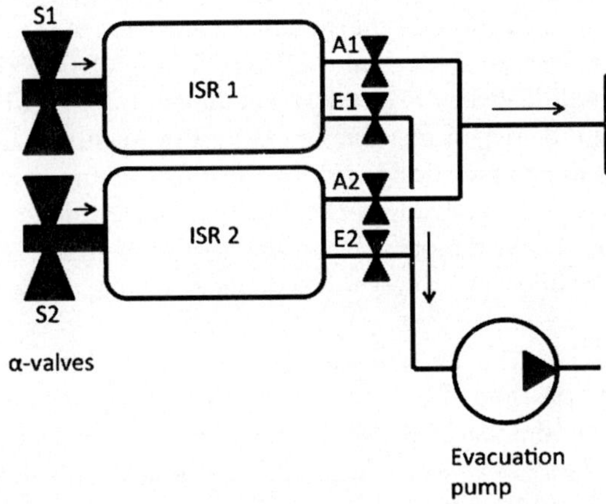

Fig. 10.2 Schematic of DEC system with two grab samplers; ISR is intermittent storage reservoir; The valve numbering refers to Table 10.1

Table 10.1 Operating sequence of a DEC system with grab samplers

	0.2 s	0.1 s	0.2 s	20 s	0.2 s	0.1 s	0.2 s	20 s
S1	x	o	x	x	x	x	x	x
S2	x	x	x	x	x	o	x	x
A1	x	x	x	o	x	x	x	x
A2	x	x	x	x	x	x	x	o
E1	x	x	x	x	x	o	o	o
E2	x	o	o	o	x	x	x	x

o signifies open valve, x closed valve. S1 and S2 are sampling valves leading to ISRs 1 and 2, respectively, A1 and A2 are valves leading to analyzer from ISRs 1 and 2, respectively, and E1 and E2 are valves leading to evacuation pump from ISRs 1 and 2, respectively

valves are not simultaneously open while switching. The time allocated for analysis depends on the needs of the analytical instrument and the ISRs can be evacuated in much shorter time with suitable pump and large diameter tubing and valves.

Most of the parameters causing uncertainty or bias to the fluxes measured by this kind of system are discussed in Sect. 10.5. Here we discuss how to minimize the effects of sampling time and sample interval in this kind of measurement system. In addition, a feature causing error and uncertainty for fluxes, typical for a DEC-GS, is sample carryover. This will also be discussed below.

The sampling time is the time during which the ISR is opened to grab an ambient air sample and it defines the response time of the system (τ_R in Sect. 10.2). A response time much longer than about 0.1 s would lead to a loss of higher frequencies and thus to systematic underestimation of the turbulent flux. Thus, the sampling time must be minimized to approach this value. For this purpose the used α-valve must have a high flow conductance and it must open and close in tens of milliseconds. The flow conductance, that is, flow coefficient required, depends on the size of the ISR as this defines the amount of air that needs to pass the valve for one grab sample. The α-valves used by Rinne et al. (2001) (Skinner 71215SN33N00N0L111P3) to fill ISRs of 1 l had a flow coefficient $C_v = 2$ and their opening and closing times were below 15 ms. The limitations in readily available valves pose limitations for grab sampler design. Only the direct acting solenoid valves have short enough opening and closing times, whereas, for example, pilot operated valves are much slower. On the other hand, the commercially available direct acting solenoid valves tend to have lower flow conductance than large pilot operated valves.

Also to reduce the flow resistance, the tube connecting the α-valve and ISR should be as short and wide as possible. The filling of the ISR slows down as the pressure difference between the ambient and ISR is reduced. Thus, typically the sampling time is terminated when the pressure inside the ISR reaches 80–90% of the ambient pressure.

The sample interval defines how many samples are taken within the eddy covariance averaging period and thus partly defines the uncertainty of the measured flux values. In the case of a simple grab sampler consisting just of one ISR, the sample interval depends on the time needed to analyze the contents of the ISR and that needed to evacuate it prior to sampling. However, most realized grab samplers use two parallel ISRs to minimize sample interval. While the contents of one ISR are being analyzed the other one is being evacuated. In this way, the sample interval can be significantly shortened. In such a system, a pump able to evacuate the ISR within the analysis time is sufficient and diaphragm pumps are often used.

A typical problem for DEC-GS, sample carryover, is due to the fact that there remains always some air from the previous sample in the ISR. This tends to lower the variance of concentrations and thus leads to systematic underestimation in the flux. This underestimation can be estimated and corrected for by a simple mixing consideration as presented by Langford et al. (2009):

$$c_{\mathrm{s,corr}} = \frac{c_{\mathrm{s}}\,p_{\mathrm{f}} - c_{\mathrm{s,old}}\,p_{\mathrm{e}}}{p_{\mathrm{f}} - p_{\mathrm{e}}} \tag{10.2}$$

where $c_{\mathrm{s,corr}}$ is the corrected concentration, c_{s} is the current measured concentration in the ISR, $c_{\mathrm{s,old}}$ is the concentration of the previous measurement, p_{f} is the pressure of the ISR when full and p_{e} is the pressure after evacuation. As the sample carryover reduces the concentration differences it also lowers the signal-to-noise ratio of the concentration variations and thus increases the uncertainty of the measured flux values. To minimize the sample carryover, the ISR should be evacuated to a pressure as low as possible. For this not only a sufficient pump performance but also low flow restriction of the evacuation line is needed. This can be achieved by minimizing the length of the line and using wide diameter tubing. In the system presented by Rinne et al. (2001), the evacuation pressure below 10% of the ambient pressure was achieved by using a diaphragm pump with 2 m of one half in tubing.

The grab samples should represent ambient trace gas concentrations within the sonic measurement volume at the moment the valve is opened. Data processing is made easier if the valve opening times are flagged into the sonic data record by, for example, sending an analogue signal to the input of the sonic, or by recording the wind values at the time of opening into the trace gas data.

Since there must be no inlet tube in front of the α-valve, the grab sampler has to be positioned close to the sonic anemometer. This may lead to obstructions to the free flow of air around the sonic anemometer, as the grab sampler is necessarily a somewhat bulky object, the dimensions of a sampler being typically around $15 \times 30 \times 50$ cm (H × W × L). To reduce flow distortion effects in the sonic anemometer measurement, the sampler has to be situated some tens of centimeters from the sonic anemometer, which can lead to underestimation of the flux (sensor separation effect). However, placing the sampler below the sonic anemometer leads to minimal flux underestimation, even for low measurement heights of 2–3 m (Kristensen et al. 1997). The underestimation of the flux caused by displaced sonic and sampler can, when the latter is below the former, be estimated by the equation:

$$\frac{F_{\mathrm{vd}}}{F} = \left(1 - 0.1\left[\frac{h_{\mathrm{mw}}}{h_{\mathrm{mc}}} - 1\right]\right) \qquad (10.3)$$

where F_{vd} is the flux measured with vertically displaced sensors, F is the true flux. h_{mw} and h_{mc} are the measurement height of vertical wind speed and concentration, respectively.

It has to be noted that the effect of vertical sensor separation is strongly asymmetric. If the sampler is situated above the anemometer the flux underestimation is much larger than the one estimated by the above equation. Also a lateral displacement leads to higher flux underestimation.

The choice of materials for grab sampler DEC systems depends on the compounds targeted. For reactive hydrocarbons Teflon is commonly used for sample lines and ISRs. Also Teflon bodied valves are preferable over stainless steel or brass bodied ones. However, suitable α-valves with sufficient response time and flow conductance are not necessarily available with Teflon body. As the air is in contact with this valve for a very short time, the use of stainless steel can be justified. The surfaces of the ISRs can have an effect on chemically active compounds. Thus it is wise to minimize the surface-to-volume ratio of the ISR, if possible, by choosing a reservoir with suitable geometry. Moreover, the length of the sample line between the ISR and the analyzer should be as short as possible in order to minimize the transfer time, and potential surface effects.

10.3.2 DEC by Mass Scanning

Online mass spectrometers with response times well below 1 s can be used to measure, for example, hydrocarbon fluxes by conventional EC method (Karl et al. 2001). A typical example is the proton transfer reaction – quadrupole mass spectrometer (PTR-QMS). However, quadrupole mass spectrometers can only measure one compound/mass at a time. Thus, the capability of these instruments to measure a suite of compounds is not exploited with the conventional EC method. One can, however, use the DEC method by scanning through a set of masses creating a disjunct time series for each individual compound. The sample interval in this DEC-MS method is the length of the QMS measurement cycle. This depends on the number of masses measured and on the integration or dwell time of each concentration measurement. The sample intervals of DEC-MS measurements reported have been between 1 and 10 s and thus the sample number n has been higher than with DEC-GS systems.

The response time of the PTR-QMS is defined by the longer of either the nominal response time of the instrument or the integration/dwell time used for concentration measurements. The response time of the instrument depends on the dimensions and flow rates in the analyzer and in the inlet tube. The dwell time used for the concentration measurement is set by the user. While for the DEC measurement one would like to choose as short a dwell time as possible, low concentrations may require the use of longer dwell times for the signal to be above detection limit. Thus, dwell times of 0.5 s have been used. At higher measurement heights, for example, above forest, this causes typically less than 10% flux underestimation. However, the lower the measurement height is, the higher the underestimation.

The measurement setup for DEC-MS (Fig. 10.3) is simple as it does not require a complicated grab sampler. In many ways this setup is similar to the setup for conventional EC measurements. Also many of the issues of the measurements are similar. One must, for example, have high enough sample flow in the sample line to have turbulent flow (high Reynolds number). However, as the PTR-QMS analyzers are typically heavy (over 100 kg) and they need protection from elements, the sample line needed especially in forest ecosystems can be 30 m or more.

The specifications of PTR-QMS analyzers lead to some restrictions in system setup. The sample flow into the analyzer is very small. Thus, to minimize the high-frequency losses in this part of the sample line it should be made as short as possible. Also the pressure in the inlet must not be lower than the set point of the pressure controller of the PTR-QMS inlet. Thus, wide diameter tubing should be used in the main sample line to ensure high flow with minimal pressure drop. For longer sample lines (30 m or so) inner diameter of 8 mm is needed. A class of pumps suitable for high flows with near ambient pressure are side canal blowers.

In this kind of measurement setup, as in any closed path EC measurement system, the sample line causes a time lag between the wind and concentration measurement. Thus, in order to be able to calculate the flux, one needs to be able to align the two time series, that is, to find the time lag (Sect. 3.2.3.2). This is a common problem for EC measurements of low fluxes, further increased by the reduced statistic of the DEC method. Often the automatic algorithms based on identification of the maximum of the covariance function do not work as there is much noise in the covariance function. A further complication arises from the fact that often the wind data from sonic anemometer and the concentration data from QMS are stored in different computers. Thus, the drift of computer clock causes an additional lag time to the two time series.

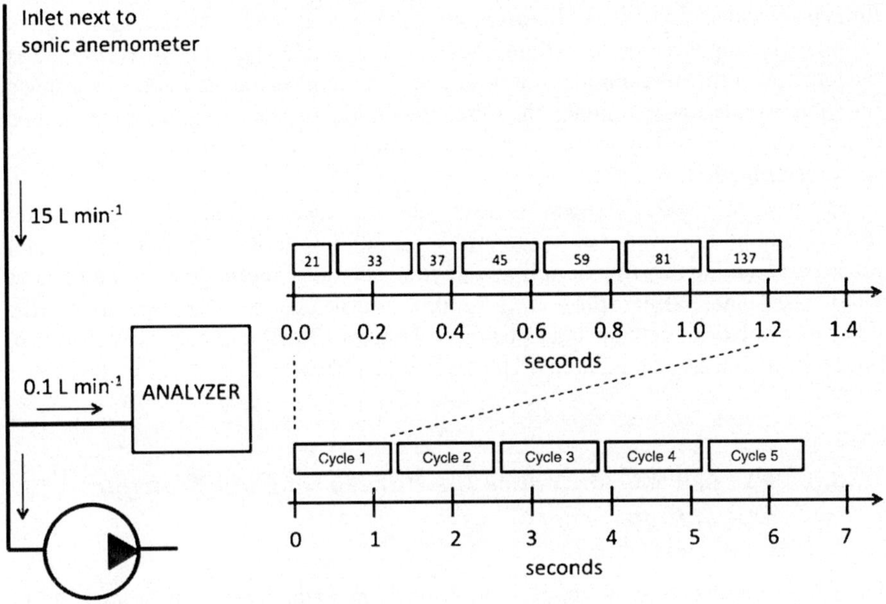

Fig. 10.3 *Left panel*: Schematic of scanning DEC system. The sample inlet should be as close to the sonic as possible. *Right panel*: Example of mass scanning cycle for measurement

Typically one need to guide the selection of the lag time by visually analyzing the covariance functions. There are a couple of tricks to ease the detection of the lag time. To remove the lag time due to the drifting computer clocks one can do one of two things. One can read the vertical wind velocity signal into the PTR-QMS analogue input once a measurement cycle and then use that signal to define the time lag due to the clock drift. This is done by correlating the disjunct vertical wind speed signal in the QMS data file with the continuous vertical wind speed signal in the original sonic data files. For this one needs a sonic anemometer with analogue output. The other option is to read the QMS raw signal into the sonic anemometer analogue input. As two signals are needed, the mass and cps signal, one needs two analogue inputs in the sonic anemometer.

To help the determination of the lag time due to the sample tubing with PTR-QMS measurements, one can use the M37 water cluster signal. As the water vapor flux is orders of magnitudes higher than the fluxes of VOCs, the peak in the covariance function between w and M37 is much clearer. Thus, one can first find the peak of the covariance function for the M37 signal and then find the other peaks in a small window around that. However, wall sorption effects can cause the lag time to be different for compounds with, for example, different solubility or polarity (Ammann et al. 2006).

To reduce the possible biasing effect of the conventional maximum covariance method for lag-time determination, one can also use smoothing of the covariance function (Taipale et al. 2010). In this approach, the covariance function is smoothed using, for example, a five-point running mean filter, and the lag time is determined as the location of the maximum of the smoothed function. As the smoothed covariance function would underestimate the flux, the value of the original un-smoothed covariance function has to be used to determine the flux at the lag time of the smoothed function.

Spirig et al. (2005) suggested easing the processing of the data created by the disjunct eddy covariance by interpolating, or imputating, the trace gas data, recorded at a relatively low sampling rate, to match the sampling rate of the vertical wind speed data. While Spirig et al. (2005) used nearest point interpolation, also other interpolation methods could be used. However, such approaches lead to flux underestimation as pointed out by Hörtnagl et al. (2010).

10.3.3 Use of DEC to Reduce the Burden on Data Transfer and Storage

DEC can also be used to relieve the burden on data transfer or storage as in Boulder Atmospheric Observatory in the 1980s (Kaimal and Gaynor 1983). In this application measurements are conducted as in conventional EC method with fast response instruments, but only a subset of data is stored or transmitted. While the data storage typically is not a limiting factor in most EC measurement sites, it may become an important factor in more remote sites with no line power and limited access. If the data logger at such sites does not allow the storage of all the raw data between site visits, one can of course decide to save only the processed flux values. However, any reprocessing of raw data becomes then impossible. Another option is to use the disjunct eddy covariance approach. By storing, for example, only every tenth data point of a 10 Hz measurement, there are still 1,800 data points per half hour for the flux calculation. In this way a specific logger will be able to store data of a ten-time longer measurement period as compared to conventional EC data. This application of DEC can only be used for systems with colocated sensors with no lag time due to tubing, etc., as the lag time between the trace gas and vertical wind speed measurement can generally not be determined and adjusted with disjunct data. Thus one should use open path gas analyzers located near, or preferably below the sonic anemometer.

10.4 DEC in Spectral Space

Any non-biased eddy covariance flux measurement must resolve atmospheric motions and variations in trace gas concentration in all temporal scales which are involved in turbulent transport of the substances in the surface layer. The

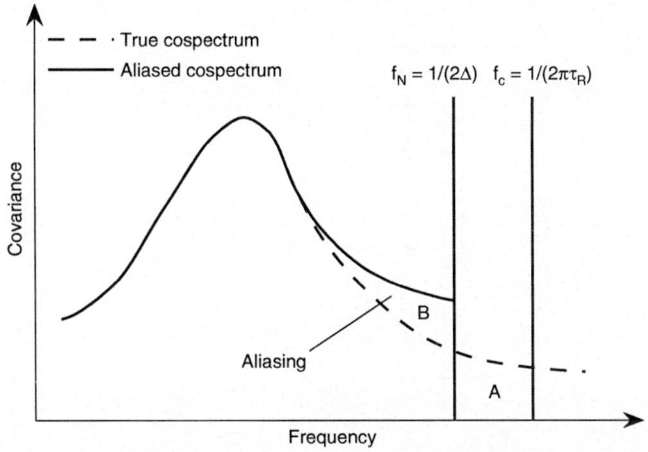

Fig. 10.4 Schematic on co-spectrum of a trace gas flux with half-power frequency f_c due to analyzer response time and Nyquist frequency f_0 due to DES

contribution of the various scales is shown by the co-spectrum and ogive of vertical wind and trace gas concentration obtained by Fourier-transformation (see Sects. 1.5.1 and 4.1.3.3 for co-spectrum and ogive respectively). While most of the flux is carried by relatively large eddies, that is, low frequencies, the higher frequencies make a significant contribution as well. For conventional EC and DEC systems, the ability to detect the flux contributions of all relevant frequencies is limited by the response time τ_R of the measurement system. The contribution of frequencies above the half-power frequency $f_c = 1/(2\pi\tau_R)$ is mostly lost (Kaimal and Finnigan 1994).

However, due to the reduced sampling frequency f_s of DEC systems, the corresponding co-spectrum is also reduced to frequencies up to the Nyquist frequency, $f_N = f_s/2$. According to the definition in Sect. 10.2.3, for DEC systems f_N is lower than f_c. However, the reduced DEC co-spectrum does not mean that frequencies above f_N (and below f_c) are lost, but they are aliased (mirrored) to frequencies below f_N (Kaimal and Finnigan 1994) as shown in Fig. 10.4. The integral of this aliased spectrum equals the integral of the full spectrum obtained from the non-disjunct time series, as can be seen by the ogive, and thus no bias is introduced to the flux. Figure 10.5 shows an example of ogives obtained by time series sampled with 10 Hz and the same time series which has been subsampled with 6.4 s sample interval. We can clearly see the aliasing in both the co-spectrum and in the ogive. Due to the aliasing, both the high-frequency ogive and the ogive from subsampled time series reach the same value. This is due to the fact that each individual concentration and wind measurement still represents a fast-response (0.1 s) value. One drawback of DEC cospectra is that the aliased and non-aliased contributions cannot easily be

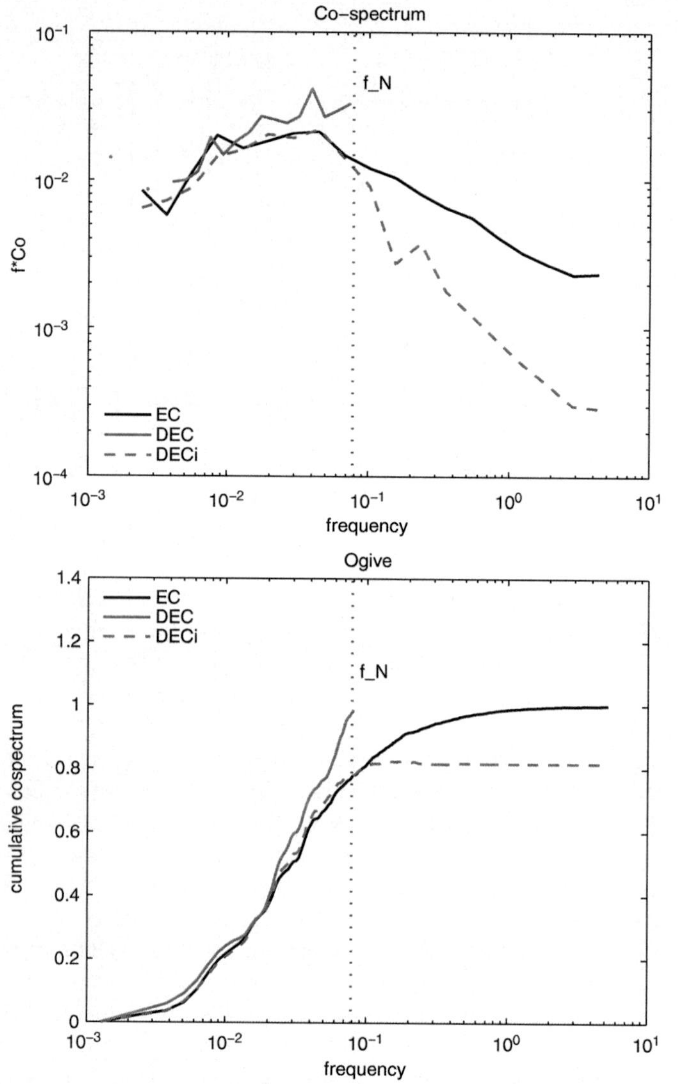

Fig. 10.5 Example of co-spectrum and ogive for sensible heat flux. EC indicates conventional eddy covariance method with 10 Hz sampling rate. DEC is disjunct eddy covariance, simulated by taking samples with 6.4 s interval from full 10 Hz time series. DECi is the imputed disjunct eddy covariance, simulated by interpolating the DEC time series to 10 Hz using nearest point interpolation. *Dashed* vertical line is the Nyquist frequency for DEC measurement

separated. However, a high-frequency damping correction based on spectral or ogive similarity in the low-frequency range is still possible in many cases (see, e.g., in Fig. 10.5).

Hörtnagl et al. (2010) showed that the data interpolation proposed by Spirig et al. (2005) leads to underestimation of the flux. The co-spectrum and its ogive (Fig. 10.5) show that this is due to (unnecessary) effective loss of higher frequencies.

It has to be considered that a potential time lag between w and c_s may be a fraction (not a full multiple) of the DEC sample interval. Therefore, it has to be determined using the full EC time series (10 Hz) of w. For an effective lag determination via FFT calculation of the cross-covariance function, the disjunct concentration time series has to be expanded to a full 10 Hz time series (by filling the gaps with zeroes).

10.5 Uncertainty Due to DEC

It can be concluded from theoretical considerations as well as from experimental results that while the reduced sampling rate seriously distorts the spectra, it does not lead to a systematic error in the fluxes. However, it adds another source of random uncertainty. This is illustrated by the reduction of the correlation between EC and DEC fluxes as the number of samples decreases (Fig. 10.6, Turnipseed et al. 2009).

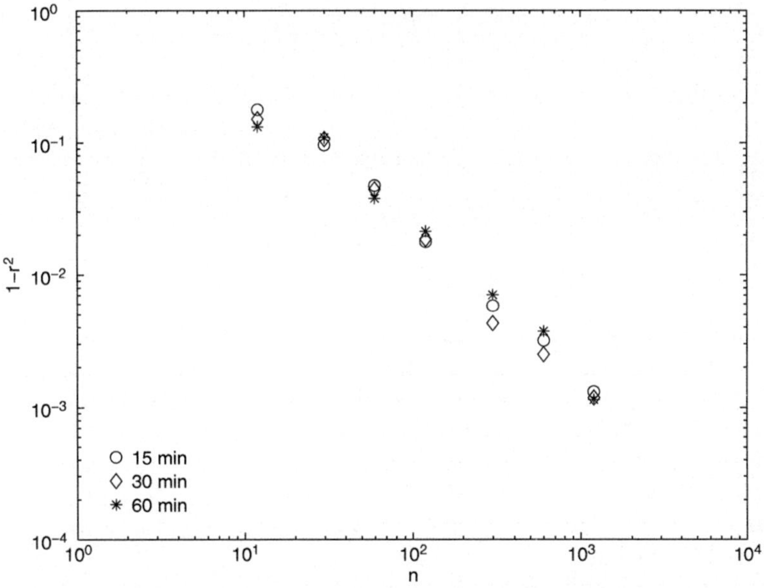

Fig. 10.6 Correlation (plotted as $1 - r^2$ for convenience) between EC and simulated DEC sensible heat fluxes as a function of number of samples (n) for different flux averaging times (15, 30, 60 min). The 10 Hz data was recorded during 4 days in Morgan County, CO (Warneke et al. 2002) and covers 91 h. The simulated DEC fluxes were realized by subsampling the full time series with suitable sample interval

Lenschow et al. (1994) formulated the random uncertainty of fluxes measured by the DEC method as

$$\frac{\sigma_F^2}{\sigma_{w'c'}^2}\frac{T}{\tau_f} = \frac{\Delta}{\tau_f}\coth\left(\frac{\Delta}{2\tau_f}\right) \tag{10.4}$$

where σ_F^2 is the resulting error variance of the flux, $\sigma_{w'c'}^2$ is the variance of the product time series $w'c'$, T is the flux averaging time, and τ_f is the integral timescale of $w'c'$. Equation 10.4 can be solved for σ_F which represents an estimate of the flux uncertainty. In the case when the sample interval Δ is much shorter than this integral timescale, Eq. 10.4 can be simplified to yield

$$\frac{\sigma_F^2}{\sigma_{w'c'}^2} = 2\frac{\tau_f}{T} \tag{10.5}$$

which is the same as for the flux measured by the conventional EC method and depends only on the ratio of integral timescale to the flux averaging period.

In the opposite case when the sample interval is much longer than the integral timescale, Eq. 10.4 reduces to

$$\frac{\sigma_F^2}{\sigma_{w'c'}^2} = \frac{\Delta}{T} \tag{10.6}$$

By noting that $\Delta/T = 1/n$ this leads to the same result as obtained when considering the subsequent records of the $w'c'$ time series to be statistically independent and writing an expression for the statistical uncertainty of their average (Rinne et al. 2008):

$$\sigma_F = \frac{\sigma_{w'c'}}{\sqrt{n}} \tag{10.7}$$

Thus the error of the flux measured by the DEC method, when the sample interval is longer than the integral timescale of $w'c'$, is proportional to the inverse of the square root of the number of samples. As the uncertainty of the flux in this case is independent from the sample interval, one can decrease the uncertainty for a given sample interval by increasing the flux averaging period, leading to increased number of samples. However, as in the case of the conventional EC method, the requirement of stationarity poses limitations to this approach. In many cases an averaging period of 1 h is still practical.

The integral timescale τ_f is a parameter that cannot be obtained from disjunct time series. It can be estimated by assuming a similarity with another scalar for which continuous high-frequency measurements exist, such as often is the case for carbon dioxide or water vapor. Typically, the surface layer integral timescales of scalars (trace gas concentrations) are longer than that of the vertical wind speed and thus is constrained by integral timescale of w and the latter can be used to give a lower limit/proxy for τ_f. The integral timescale of w scales with the measurement height and as a rule of the thumb one can assume τ_w (in units of s) $\approx z$ (in units of m). Thus, especially for measurement over low vegetation, the sample interval is mostly longer than the integral timescale and the number of samples rather than the sample interval is the parameter controlling the uncertainty of the fluxes.

The variance of $w'c'$ is due to the variation in w and in c. The variation in both quantities is caused partly by the atmospheric variability and partly by the instrument noise. For the measurements of trace gas fluxes, the instrument noise of the gas analyzer is usually the one mostly influencing the uncertainty of the flux. If in Eq. 10.7 only the influence of instrumental noise on the concentration measurement is considered, $\sigma_{w'c'}$ can be replaced by

$$\sigma_{w'c'}^2 = \sigma_w^2 \cdot \sigma_{c,\text{noise}}^2 \tag{10.8}$$

since the variations in w' and c'_{noise} are uncorrelated. This leads to

$$\sigma_F = \frac{\sigma_w \sigma_{c,\text{noise}}}{\sqrt{n}} \tag{10.9}$$

Thus the uncertainty of the flux is directly proportional to the uncertainty of the concentration due to instrument noise. The uncertainty of a concentration measurement can usually be decreased by increasing the time used for the analysis. However, as the uncertainty of the concentration measurement depends usually inversely on the square root of the measurement time, the increase of concentration measurement time does not lead to decrease of the uncertainty of the flux value.

A method for estimation of uncertainty of the eddy covariance measurements, based on the covariance function, is suitable also for DEC methods. In this method, proposed by Wienhold et al. (1994), the uncertainty of the flux is determined from the variance of the covariance function far away from its peak value.

One must note that periodic sampling can be considered statistically non-biased only for random process such as fully developed turbulence. In the case of wave motions this kind of regular subsampling may lead to systematic errors in statistics. Thus, the use of DEC with regular sampling interval in strongly stable conditions, where such wave motions can exist, can be problematic.

10.6 On the History of the DEC Approach

The effects of reduced sampling rate on measured turbulent fluxes were first discussed by Duane Haugen (1978). He pointed out that the important parameter determining the uncertainty of turbulent statistic obtained from noncontinuous data is sample size rather than sample interval. He also noted that while reduced sampling rate seriously distorts the spectra due to aliasing, it does not lead to systematic error in turbulence statistics. However, his estimates on the magnitude of the uncertainty due to the disjunct eddy sampling on the turbulent fluxes were pessimistic compared to later studies.

The first practical application suggested for the DEC approach was to reduce the burden on data storage and transmission of the Boulder Atmospheric Observatory, Colorado, in the early 1980s (Kaimal and Gaynor 1983). The application of DEC method for trace gas flux measurements was first suggested in the early 1990s by Dabbert et al. (1993) and Lenschow et al. (1994). Lenschow et al. also coined the term disjunct eddy covariance. Rinne et al. (2000) developed a disjunct eddy accumulation system utilizing grab sampler in the end of 1990s and suggested that similar sampler could be used for eddy covariance measurements.

The first practical applications of DEC method for trace gas flux measurements were realized in the first years of the new millennium (Rinne et al. 2001; Karl et al. 2002; Warneke et al. 2002). The DEC method with grab sampler was tested at an alfalfa field in Morgan County, Colorado, to measure VOC fluxes with slow response PTR-QMS by Rinne et al. (2001). They also suggested the DEC method with mass scanning, later to be applied by Karl et al. (2002) to measure VOC fluxes at an alpine coniferous forest site at Niwot Ridge, Colorado. They used a prototype of the fast response time PTR-QMS as the analyzer. During the first decade of this millennium, the DEC-MS has been a method of choice for flux measurements of volatile organic compounds using fast response PTR-QMS instruments, and it has been applied at forest, wetland, and grassland ecosystems. As the PTR-QMS is basically a laboratory instrument needing protection from the elements, and the total power consumption of the measurement system is often around 1 kW, the measurements have usually been conducted at sites with reasonably good infrastructure.

While the development of the new proton transfer reaction – time of flight mass spectrometer (PTR-ToFMS) enables flux measurements of VOCs using the conventional EC method (Müller et al. 2010), it is currently still very expensive and complicated to operate in field conditions.

Acknowledgments We thank Risto Taipale for drawing Fig. 10.4.

References

Ammann C, Brunner A, Spirig C, Neftel A (2006) Technical note: water vapour concentration and flux measurements with PTR-MS. Atmos Chem Phys 6:4643–4651

Bosveld FC, Beljaars ACM (2001) The impact of sampling rate on eddy-covariance flux estimates. Agric For Meteorol 109:39–45

Dabberdt WF, Lenschow DH, Horst TW, Zimmerman PR, Oncley SP, Delany AC (1993) Atmosphere-surface exchange measurements. Science 260:1472–1481

Haugen DA (1978) Effects of sampling rates and averaging periods on meteorological measurements. In: Proceedings of the fourth symposium on meteorological observations and instrumentation, American Meteorological Society, Denver, CO, pp 15–18

Hörtnagl L, Clement R, Graus M, Hammerle A, Hansel A, Wohlfahrt G (2010) Dealing with disjunct concentration measurements in eddy covariance applications: a comparison of available approaches. Atmos Environ 44:2024–2032

Kaimal JC, Finnigan JJ (1994) Atmospheric boundary layer flows their structure and measurement. Oxford University Press, New York, p 289

Kaimal JC, Gaynor JE (1983) The boulder atmospheric observatory. J Clim Appl Meteorol 22:863–880

Karl T, Guenther A, Jordan A, Fall R, Lindinger W (2001) Eddy covariance measurement of biogenic oxygenated VOC emissions from hay harvesting. Atmos Environ 35:491–495

Karl TG, Spirig C, Rinne J, Stroud C, Prevost P, Greenberg J, Fall R, Guenther A (2002) Virtual disjunct eddy covariance measurements of organic trace compound fluxes from a subalpine forest using proton transfer reaction mass spectrometry. Atmos Chem Phys 2:279–291

Kristensen L, Mann J, Oncley SP, Wyngaard JC (1997) How close is close enough when measuring scalar fluxes with displaced sensors? J Atmos Ocean Technol 14:814–821

Langford B, Davison B, Nemitz E, Hewitt CN (2009) Mixing ratios and eddy covariance flux measurements of volatile organic compounds from an urban canopy (Manchester, UK). Atmos Chem Phys 9:1971–1987

Lenschow DH, Mann J, Kristensen L (1994) How long is long enough when measuring fluxes and other turbulence statistics? J Atmos Ocean Technol 11:661–673

Müller M, Graus M, Ruuskanen TM, Schnitzhofer R, Bamberger I, Kaser L, Titzmann T, Hörtnagl L, Wohlfahrt G, Karl T, Hansel A (2010) First eddy covariance flux measurements by PTR-TOF. Atmos Meas Tech 3:387–395

Rinne HJI, Delany AC, Greenberg JP, Guenther AB (2000) A True eddy accumulation system for trace gas fluxes using disjunct eddy sampling method. J Geophys Res 105:24791–24798

Rinne HJI, Guenther AB, Warneke C, de Gouw JA, Luxembourg SL (2001) Disjunct eddy covariance technique for trace gas flux measurements. Geophys Res Lett 28:3139–3142

Rinne J, Douffet T, Prigent Y, Durand P (2008) Field comparison of disjunct and conventional eddy covariance techniques for trace gas flux measurements. Environ Pollut 152:630–635

Spirig C, Neftel A, Ammann C, Dommen J, Grabmer W, Thielmann A, Schaub A, Beauchamp J, Wisthaler A, Hansel A (2005) Eddy covariance flux measurements of biogenic VOCs during ECHO 2003 using proton transfer reaction mass spectrometry. Atmos Chem Phys 5:465–481

Taipale R, Ruuskanen TM, Rinne J (2010) Lag time determination in DEC measurements with PTR-MS. Atmos Meas Tech 3:853–862

Turnipseed AA, Pressley SN, Karl T, Lamb B, Nemitz E, Allwine E, Cooper WA, Shertz S, Guenther AB (2009) The use of disjunct eddy sampling methods for the determination of ecosystem level fluxes of trace gases. Atmos Chem Phys 9:981–994

Warneke C, Luxembourg SL, de Gouw JA, Rinne HJI, Guenther AB, Fall R (2002) Disjunct eddy covariance measurements of oxygenated volatile organic compound fluxes from an alfalfa field before and after cutting. J Geophys Res 107(D8):4067, 10.1029/2001JD000594

Wienhold FG, Frahm H, Harris GW (1994) Measurements of N2O fluxes from fertilized grassland using a fast response tunable diode laser spectrometer. J Geophys Res 99(D8):16557–16568

Chapter 11
Eddy Covariance Measurements over Forests

Bernard Longdoz and André Granier

11.1 Introduction

In the 1970s, flux estimation over tall vegetation, like forests, using flux-gradient relationships were found impracticable (Raupach 1979). The roughness of the exchanging surface boosted turbulent mixing, reducing the concentration gradient and invalidating Monin-Obukhov similarity theory (Lenschow 1995). In the 1990s, the eddy covariance (EC) method was developed and turned out to be very promising for CO_2, latent, and sensible heat exchange quantification over these tall ecosystems. When the first networks of EC measurements were implemented (EuroFlux, Valentini et al. 2000; Ameriflux, Running et al. 1999), they included then a majority of forest sites. The other reasons for this historical forest leading position were their large terrestrial cover (FAO 2005 report) and their potentiality to store carbon over long periods (Valentini 2003).

EC over forest presents some particularities in (1) the methodology for flux computation, selection and determination of flux dependence, (2) the complementary measurements requested to interpret correctly the EC data, and (3) the interference created by ecosystem management. In this chapter, we propose to detail these different particularities.

B. Longdoz (✉) • A. Granier
INRA, UMR1137 Ecologie et Ecophysiologie Forestières, Centre de Nancy, F-54280 Champenoux, France
e-mail: longdoz@nancy.inra.fr; agranier@nancy.inra.fr

M. Aubinet et al. (eds.), *Eddy Covariance: A Practical Guide to Measurement and Data Analysis*, Springer Atmospheric Sciences, DOI 10.1007/978-94-007-2351-1_11,

11.2 Flux Computation, Selection, and Dependence

11.2.1 Correction for High Frequency Losses

When closed-path infrared gas analyzers (IRGA) are used (Sect. 2.4.2), high-frequency fluctuations of the gas concentration is attenuated during tubing transport (Sect. 4.1.3.2). This provokes losses in the high-frequency component of the co-spectra, needing implementation of correction factors. The amplitude of these losses is principally linked to tube length, airflow rate, measurement height above the canopy surface, roughness of this surface, and wind conditions (Sect. 4.1.3.2).

In tall forests, the gas transfer may be very long (more than 30 m) as the closed-path analyzer cannot be installed near the sampling point but in a shelter close to the tower basis. On one hand, this configuration facilitates the frequent calibration operations when performed manually and improves the comfort when technical interventions are requested. On the other hand, the high-frequency losses and then the correction factors could become very significant, leading to large uncertainties. One way to reduce them is to hold a relatively large airflow rate in the tube. This requests electric power and could decrease the life span of the pumps. In this case, it is suggested to employ two or more pumps in serial with lower voltage than the nominal one. The installation of the IRGA at the top of the tower to reduce high-frequency losses and the impact of the correction factor is advisable when the injection of a calibration gas could be performed from the ground or driven by an automatic system.

As mentioned before, the high-frequency losses depend on the canopy surface roughness. So, for deciduous forests, the correction factor can vary between leafless and leafy periods. The difference has to be tested by determining this correction factor for each of these periods. Similarly, if during the measurement campaign the EC system height above the top of the forest is significantly reduced (due to significant tree growth), the eddy size viewed by the EC system is smaller, enhancing the high-frequency component of the co-spectra. The impact of high-frequency attenuation becomes more important. It is then recommended to reevaluate the correction factor. The possibility to lift up the EC system is another method to overcome this problem but it implies a new footprint analysis to certify that fluxes measured are still coming from the targeted ecosystem (Chap. 8).

11.2.2 Rotation Method

In forests, the support of the EC system is quasi-systematically a tower with a structure larger than masts that can be utilized for grasslands or crops (Sect. 2.2). The sonic anemometer has to be put aside from the tower using an arm with a length equivalent to two or three times the tower diameter (see Sect. 2.2.2.5) which may lead to sensor stability and horizontality problems. In addition, the presence of other measurement devices may constitute additional obstacles (other sensors, supplementary tower element, etc.). In these conditions, the distortion created in the wind streamline flow can have a significant impact and the choice of the Planar Fit Sector approach (Sect. 3.2.4.3) as rotation method becomes necessary.

11.2.3 Friction Velocity Threshold

When turbulence decreases, a significant part of the trace gas studied can be stored in the canopy air or migrate out by advection (Sect. 5.1.3, Aubinet et al. 2005). Then, EC system may underestimate the flux exchanged by the ecosystem. Measurements of vertical temperature or concentration profile (see Sect. 11.3.1) allow correcting EC data for storage but, presently, the only way to overcome advection problem is to apply u^* filtering (Sect. 5.3). The high density of some forest canopies, which reduces eddy penetration, combined with the large internal air space, can lead to relatively frequent and important storage and advection events. Consequently, the vertical profile should be determined with care (see Sect. 11.3.1) and the friction velocity threshold for data filtering ($u_{*\mathrm{crit}}$, see Sect. 5.3.2) could be high, leading to an important number of rejected data. This creates large gaps in the data sets covering up to 50% of the time (Papale et al. 2006). The accuracy of data gap filling method (Chap. 6) is then crucial for determining correctly the net exchange integrated over long periods (month, season, year).

11.2.4 Selection Based on Footprint

The choice of measurement height should be a trade-off between the necessity to reduce the high-frequency losses (Sect. 4.1.3.2) and those to delimit the footprint extent to measure fluxes mainly coming from the targeted ecosystem (Sect. 8.3.2). Unfortunately, in forests, the problem of access to the material (existence of platforms, supporting arms, etc.) can limit the possibilities in the choice of system height, giving sometimes non adequate positioning. Then it is important to perform footprint studies (Chap. 8) also as an additional tool for data screening and selection (Göckede et al. 2008). This procedure is also necessary when some plots need to be excluded from the accepted footprint area because they become very specific compared to the surrounding forest. This happens, for example, when the thinning is operated by different persons (different plot adjudications in state forests) leading to large spatial heterogeneity.

11.3 Additional Measurements

Some additional measurements to the EC fluxes are necessary to obtain the net ecosystem exchange, to partition it between its main components, and to interpret them. Beside the characterization of the climatic conditions (radiation, air temperature, humidity, etc.), soil efflux (topic is developed in details for CO_2 in Kutsch et al. 2010), and reflectance indexes (NDVI, PRI, see Grace et al. 2007), some of the complementary measurements have some specificities when they are performed on forests.

11.3.1 Vertical Profile of Concentration in Canopy Air

The quantification of storage in the canopy air can be necessary for some gases like CO_2 for better estimation of the half-hour NEE when turbulence is relatively low and ecosystem is tall. This situation is very usual in forests (up to half of the time, Longdoz et al. 2008) with tall trees and dense canopy that limit penetration of eddies. The storage is calculated as the difference between successive estimates of gas content in the air canopy, themselves calculated from vertical profile of concentration (Xu et al. 1999). This profile includes sampling levels not only in the free air below EC system but also in the soil, as CO_2 can also be stored in the soil pore air. In forests, the distance between the higher and lower sampling levels can be large when trees are tall. As gas concentration can change rapidly, all the levels of the profile have to be sampled within a short period in order to estimate accurately the total air canopy content. The fast purge of the different tubes is then performed by one large pump but it induces too large depression in the gas analyzer to correctly measure concentration. Consequently, another small pump should suck air from the main tube into the analyzer.

Vertical profile in soil pore air should be measured when gas concentrations have significant temporal fluctuations. In the forest, the number of sampled points in the soil (different depths) is larger than in the free air as concentration gradients are steeper because mixing processes are less active. Different methods have been tested to measure this gradient (Risk et al. 2002; Tang et al. 2003; Jassal et al. 2005). It seems that porous tubing (Gut et al. 1998) inserted horizontally and connected in close loops with gas analyzer gives the best results (Flechard et al. 2007), presenting the advantages of larger spatial representativeness and/or shorter response time and/or less expensive compared to other techniques (syringe sampling, sensors buried or located in vertical tube holed at the sampled depth).

11.3.2 Leaf Area Index

In contrast to crops, leaf area index (LAI) is relatively constant in forests during the growing season and its determination is essential to analyze interannual variability in fluxes. Different nondestructive methods exist (radiation transmission, LAI meters, litter collection, etc.; see Bréda 2003). Each of them present advantages and limitations and their combination remains the best way to have the more accurate and representative estimation. LAI estimated from radiation interception by canopies can be performed with permanent or moving sensors. One is located above the canopy and several are set below. Obviously, data have to be recorded simultaneously. The number of these sensors depends on the spatial heterogeneity of the canopy but ten is a minimum (Widlowski 2010). The same comment can be made about the number of hemispherical pictures. Optical methods do not require frequent visits at the studied forest but necessitate specific equipments (LAI meter, radiation sensors, etc.) and important assumption about the leaf angle distribution. This last can be estimated by various ways according to the species and tree density (Beta distribution function, ellipsoidal function, rotated-ellipsoidal function, Verhoef's algorithm, and de Wit's functions; see Wang et al. 2007). About the litter collection method, leaves or needles fallen in bags (could be hanged or laid on the soil and micro perforated to evacuate water) have to be brought back to laboratory for area (with an area meter) and dry mass determination. Collections have to be frequent to avoid leaves or needles decomposition in the bags and modification of their area and mass before measurement. This method is time consuming because of frequent site visits and of the large amount of leaves/needles to analyze with the area meter but it also gives quantitative information about the litter production biomass.

11.3.3 Biomass Estimates

When biomass increment and biomass carbon (or other element) content are known, they can be associated for comparison and validation of EC net ecosystem exchange (Granier et al. 2008). In forests, over medium-term periods (typically 1 year), when neglecting the variations of soil carbon content and wood carbon density, carbon sequestration estimated with EC can be compared to annual tree biomass increment. The latter takes into account estimates of mortality and exports (resulting from management, thinning or clear-cut) and temporal evolution of the whole trees and understory biomasses. Tree biomass is often estimated with allometric relationships from tree diameter at breadth height (DBH) and tree height (Van Laar and Akça 2007). The main source of uncertainty comes from the estimation of the belowground biomass but, more and more, root system excavations are performed with this goal (Peichl and Arain 2007), reducing this uncertainty for the most investigated tree species. The tree biomass estimation from DBH requires DBH inventories of a representative tree sample within the footprint area. This selection includes trees from different diameter classes and with different status (dominant, codominant, intermediate, suppressed) over the different soil types in the footprint area. In consequence, the number of trees selected can be large and thus the DBH manual measurement can be time-consuming, which explains the limited number of the campaigns (season to year). In addition, estimate of mortality (trees and branches) results also from important field campaigns and accuracy on exported wood quantity depends on information given by forest managers (see Sect. 11.4). Due to all these limitations, a relatively low number of forest sites can address the comparison between NEE and biomass increment (Granier et al. 2008). For example, it can be observed in Fig. 11.1 that at the end of the growing season for the

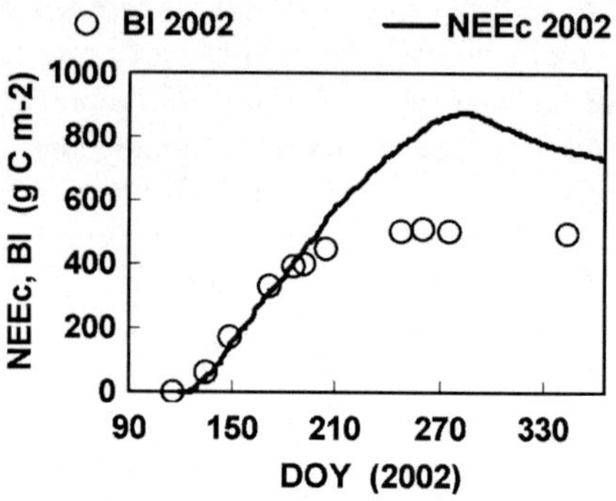

Fig. 11.1 Temporal variation of the cumulated net ecosystem exchange (NEEc) computed from EC data (*black line*) and carbon biomass increment (BI) estimated from successive DBH inventory campaigns at the Hesse forest (*open circles*) in 2002. The starting point of NEE is set at the beginning of radial tree growth (DOY 120)

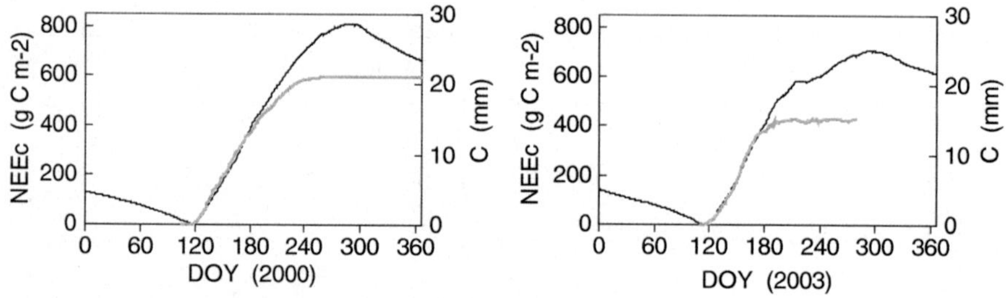

Fig. 11.2 Temporal variation of the cumulated net ecosystem exchange (NEEc) computed from the Hesse EC data (*black line*) and mean tree circumference (*gray line*) measured with dendrometer bands (up to steady state). The starting point of NEE is set at the beginning of radial tree growth (DOY 119)

Hesse forest (beech stand in the North-East of France), a deviation appears between the cumulated NEE, as estimated with EC data, and the biomass increment from manual tree growth measurements. Tree growth stops several weeks before that ecosystem turns from a CO_2 sink to a source. This deviation can be explained by the switch from structural carbon production to carbon storage (in sugars, starch, amino acids, lipids, etc.).

On shorter timescale (day, up to season), automatic dendrometer bands give the variation of tree diameter or circumference. For technical and cost reasons, dendrometer bands can be installed only on a small number of trees. The comparison of the estimated increase in biomass from dendrometer band measurement with EC carbon sequestration is, therefore, often only relative, but can bring very interesting results. Figure 11.2 shows that from 1 year to another this deviation occurs at different dates, due to different environmental conditions (an exceptional drought has been experimented by the Hesse forests in 2003). There is therefore an interannual variability in the amount of carbon stored in the trees that will impact the budburst date, LAI, and growth in the following year.

11.3.4 Sap Flow

Stand-scaled sap flow measurements can be compared to EC latent heat flux in order to separate transpiration from soil evaporation and understory transpiration, as the deviation between both fluxes can reach up to 25% of the total water vapor emission even in closed forests (Granier et al. 1996). The transpiration corresponds to the loss of water in the root zone and its determination is essential to complete the soil hydraulic balance. One of the most used methods for measuring sap flow, the heat dissipation (Granier method; Lu et al. 2004), has been developed for trees. Consequently, most of sap flow data sets concern forests.

When the objective is the estimation of transpiration at stand scale (on the footprint area), the maximal accuracy is obtained by measuring sap flow on the trees belonging to classes having the larger weight in the transpiration flux. The amount of sampled trees is a trade-off between the necessity to cover the heterogeneity in age, diameter, soil composition, foliage structure, and the restriction imposed by the material available. Most of the time, three tree status are considered (dominant, codominant, suppressed) and three (for suppressed) to five (for dominant) sap flow sensors are requested according to the status. The sap flow density of the stand (E_{SF}, m^3 of water $m^{-2}$of soil s^{-1}) is given by

$$E_{SF} = \sum_i (u_{SF,i} \cdot A_{SF,i}) \tag{11.1}$$

where the index i refers to tree classes, u_{SFi} is the sap flux density ($m^3$of water $m^{-2}$of sapwood s^{-1}), and A_{SFi} is the sapwood area (m^2 of sapwood $m^{-2}$of soil). This last can be estimated for each class by combining sapwood depth determination (from analysis of cores sampled in trunks) and DBH measurements.

11.3.5 Extractable Soil Water, Throughfall, and Stem Flow

Soil water content (SWC) in the root zone is an important factor regulating stomatal opening and then helps analyzing transpiration and carbon assimilation deduced from EC data (Granier et al. 2007). SWC is often expressed as relative water content (REW), varying between 0 and 1, corresponding to the ratio between the actual extractable soil water EW and the maximum extractable soil water EW_{max}. EW and EW_{max} are the difference between, respectively, actual or field capacity soil water contents and the permanent wilting point (−1.6 MPa). In forests, spatial variability of soil water and therefore of REW can be large. This variability is partly due to that of throughfall, consecutive to gaps in the foliage and to stem flow accumulation at trunk base. It can reach up to 30% of the incident precipitation for deciduous (Andre et al. 2011), but is generally lower for coniferous (Levia and Frost 2003). The experimental setup for soil water measurement at the plot level has to be designed to capture both temporal and spatial variability. Temporal variations are monitored using automatic sensors buried in the soil (often deducing SWC from measurement of soil dielectric constant by time or frequency domain reflectometry; see Prichard 2010) and installed at different depths. Those sensors have a fast response, but their installation necessitates digging trenches and/or holes in the soil leading to perturbations of the soil structure (layer mixing, compaction, creation of preferential flow chimney). Moreover, their small sampling volume (few tens to hundreds of cubic centimeters) limits the measurement of spatial variability. In the ideal situation, they are combined with movable systems used to perform measurements manually on a large number of locations at different depths. These systems use one probe circulating in vertical tubes inserted in the soil and able to measure SWC without direct contact with the soil (presence of the tube wall). Neutron probes and sticks with frequency domain reflectometry rings are the main systems employed (see Prichard 2010).

Measurements of throughfall and stem flow complete the water balance database (with incident precipitation and soil water content). The throughfall measurements are performed with collectors located at the ground level. When focusing on the short-term (hourly to daily) components of the water balance, the collectors have to be connected to automatic tipping buckets in order to be able to estimate the time lag with incident precipitation. These buckets should tip over for relatively low water amount (0.1 or 0.2 mm) as the throughfall quantity could be quite small in case of weak precipitation. The problem of the collector number required to cover the spatial heterogeneity is similar to one of the radiation interceptions by the canopy (see Sect. 11.3.2) and a minimum of ten is also recommended.

The stem flow collection is performed with channels stuck on the trunk and the water is driven into a rain gauge with automatic tipping buckets when short-term quantification is required. In this case and to overcome a too-rapid bucket filling (leading to non counted water losses), the volume of the buckets has to be larger than for throughfall. Indeed, even if stem flow is lower at minimum third time, the surface of interception (a tree canopy for stem flow compared to the collector surface for the throughfall) is about ten times larger.

11.3.6 Heat Storage

The ecosystem heat storage is one of the fluxes involved in the energy balance closure problem (Hendricks Franssena et al. 2010) and its knowledge is critical to evaluate the impact of the climatic changes on soil and vegetation temperature. Heat transfer in the soil compartment, can be measured using soil heat flux plates (Mayocchi and Bristow 1995). Moreover, in forests, heat storage in tree stems and canopies can be important. Its quantification can be realized using thermocouples inserted at different depths in stems, at different heights (bottom and top of trunks, main and secondary branches), on different azimuths (north and south for variation in sun exposition) and in trees among different status (dominant, codominant, suppressed). In addition to this relatively heavy experimental design, another crucial point is the determination of the wood-specific heat. This can become complicated because wood-specific heat is species-dependent and it may vary significantly during the day due to wood water changes (Čermák et al. 2007).

11.4 Impact of Ecosystem Management and Manipulation

In managed forests, stand structure is modified by thinning and clear-cuts. A good cooperation with the forest manager is necessary to collect quantitative information about the woody biomass exported and the part left on the ground. This information is essential to establish the complete carbon ecosystem balance. Unfortunately, in forests, this information is sometimes difficult to get, as several owners could share the surface representing the footprint area, with different ways to select the harvested trees and to operate the thinning. Management creates new heterogeneities (LAI, biomasses) in the footprint area which makes a regular map update necessary. The quite large surface of EC system footprint area in forests makes difficult any ecosystem manipulation (fertilization, water exclusion or addition, root exclusion, etc.), as it should be performed on the whole footprint area to analyze its impact in an unambiguous way on EC data.

References

Andre F, Jonard M, Jonard F, Ponette Q (2011) Spatial and temporal patterns of throughfall volume in a deciduous mixed-species stand. J Hydrol 400(1–2):244–254

Aubinet M, Berbigier P, Bernhofer CH et al (2005) Comparing CO_2 storage and advection conditions at night at different CarboEuroflux sites. Bound Layer Meteorol 116:63–94

Bréda N (2003) Ground-based measurements of leaf area index: a review of methods, instruments and current controversies. J Exp Bot 54(392):2403–2417

Čermák J, Kučera J, Bauerle WL et al (2007) Tree water storage and its diurnal dynamics related to sap flow and changes in stem volume in old-growth Douglas-fir trees. Tree Physiol 27:181–198

FAO (2005) Chapter 2: Extent of forest resources. In: Global forest resources assessment 2005, progress towards sustainable forest management. FAO, Rome, pp 11–12

Flechard CR, Neftel A, Jocher M et al (2007) Temporal changes in soil pore space CO_2 concentration and storage under permanent grassland. Agric For Meteorol 142:66–84

Göckede M, Foken T, Aubinet M et al (2008) Quality control of CarboEurope flux data – part 1: coupling footprint analyses with flux data quality assessment to evaluate sites in forest ecosystems. Biogeosciences 5:433–450

Grace J, Nichol C, Disney M et al (2007) Can we measure terrestrial photosynthesis from space directly, using spectral reflectance and fluorescence? Glob Chang Biol 13:1484–1497

Granier A, Biron P, Köstner B et al (1996) Comparison of xylem sap flow and water vapour flux at the stand level and derivation of the canopy conductance for Scots Pine. Theor Appl Climatol 53:115–122

Granier A, Breda N, Reichstein M et al (2007) Evidence for soil water control on carbon and water dynamics in European forests during the extremely dry year: 2003. Agric For Meteorol 143:123–145

Granier A, Breda N, Longdoz B et al (2008) Ten years of fluxes and stand growth in a young beech forest at Hesse, North-eastern France. Ann For Sci 64:704–726

Gut A, Blatter A, Fahrni M et al (1998) A new membrane tube technique (METT) for continuous gas measurements in soils. Plant Soil 198:79–88

Hendricks Franssena HJ, Stöcklid R, Lehner I et al (2010) Energy balance closure of eddy-covariance data: a multisite analysis for European FLUXNET stations. Agric For Meteorol 150:1553–1567

Jassal R, Black A, Novak M et al (2005) Relationship between soil CO_2 efflux concentrations and forest-floor CO_2 effluxes. Agric For Meteorol 130:176–192

Kutsch WL, Bahn M, Heinemeyer A (2010) Integrated methodology on soil carbon flux measurements. Cambridge University Press, Cambridge

Lenschow DH (1995) Micrometeorological techniques for measuring biosphere-atmosphere trace gas exchange. In: Matson PA, Harriss RC (eds) Biogenic trace gas: measuring emissions from soil and water. Blackwell, London, pp 126–163, Great-Britain

Levia DF Jr, Frost EE (2003) A review and evaluation of stemflow literature in the hydrologic and biogeochemical cycles of forested and agricultural ecosystems. J Hydrol 274:1–29

Longdoz B, Gross P, Granier A (2008) Multiple quality tests for analysing CO_2 fluxes in a beech temperate forest. Biogeosciences 5:719–729

Lu P, Urban L, Zhao P (2004) Granier's Thermal Dissipation Probe (TDP) method for measuring Sap flow in trees: theory and practice. Acta Botanica Sin 46(6):631–646

Mayocchi CL, Bristow KL (1995) Soil surface heat flux: some general questions and comments on measurements. Agric For Meteorol 75:43–50

Papale D, Reichstein M, Aubinet M et al (2006) Towards a standardized processing of Net ecosystem exchange measured with eddy covariance technique: algorithms and uncertainty estimation. Biogeosciences 3:571–583

Peichl M, Arain MA (2007) Allometry and partitioning of above and below ground tree biomass in an age-sequence of white pine forests. For Ecol Manag 253:68–80

Prichard TL (2010) Document. Soil moisture measurement technology. http://cecentralsierra.ucanr.org/files/96233.pdf

Raupach MR (1979) Anomalies in flux-gradient relationships over forests. Bound Layer Meteorol 16:467–486

Risk D, Kellman L, Beltrami H (2002) Carbon dioxide in soil profiles: production and temperature dependence. Geophys Res Lett 29(6):1029–2001

Running SW, Baldocchi DD, Turner D et al (1999) A global terrestrial monitoring network, scaling tower fluxes with ecosystem modelling and EOS satellite data. Remote Sens Environ 70:108–127

Tang J, Baldocchi DD, Qi Y et al (2003) Assessing soil CO_2 efflux using continuous measurements of CO2 profiles in soils with small solid-state sensors. Agric For Meteorol 118:207–220

Valentini R (2003) An integrated network for studying the long-term responses of biospheric exchanges of carbon, water and energy of European forests. Ecol Stud 163:1–8

Valentini R, Matteucci G, Dolman AJ et al (2000) Respiration as the main determinant of carbon balance in European forests. Nature 404:861–864

Van Laar A, Akça A (2007) Tree volume tables and equation. In: Managing forest ecosystems, forest mensuration. Springer Edition, Dordrecht

Wang W-M, Li Z-L, Su H-B (2007) Comparison of leaf angle distribution functions: effects on extinction coefficient and fraction of sunlit foliage. Agric For Meteorol 143(1–2):106–122

Widlowski J-L (2010) On the bias of instantaneous FAPAR estimates in open-canopy forests. Agric For Meteorol 150:1501–15022

Xu L-K, Matista AA, Hsiao TC (1999) A technique for measuring CO_2 and water vapour profiles within and above plant canopies over short periods. Agric For Meteorol 94:1–12

Chapter 12
Eddy Covariance Measurements over Crops

Christine Moureaux, Eric Ceschia, Nicola Arriga, Pierre Béziat, Werner Eugster, Werner L. Kutsch, and Elizabeth Pattey

12.1 Introduction

Croplands are managed ecosystems with rapid development over the course of the growing season under nearly optimal growth conditions with respect to nutrient availability (fertilization), water availability (possible irrigation in dry conditions), competition (monocultures where herbicide and fungicides applications keep other competitors off the plot) and plant health (insecticides minimize herbivory by insects).

C. Moureaux (✉)
Gembloux Agro-Bio Tech, University of Liege, Gembloux, Belgium
e-mail: christine.moureaux@ulg.ac.be

E. Ceschia • P. Béziat
Centre d'Etudes Spatiales de la BIOsphère (CESBIO), Toulouse, France
e-mail: Eric.ceschia@cesbio.cnes.fr

N. Arriga
DIBAF, University of Tuscia, Viterbo, Italy
e-mail: arriga@unitus.it

W. Eugster
Department of Agricultural and Food Sciences, Institute of Agricultural Sciences, ETH Zurich, Zurich, Switzerland
e-mail: werner.eugster@agrl.ethz.ch

W.L. Kutsch
Institute for Agricultural Climate Research, Johann Heinrich von Thünen Institute (vTI), Braunschweig, Germany
e-mail: werner.kutsch@vti.bund.de

E. Pattey
ECORC, Agriculture and Agri-Food Canada, Ottawa, Canada
e-mail: elizabeth.pattey@agr.gc.ca

M. Aubinet et al. (eds.), *Eddy Covariance: A Practical Guide to Measurement and Data Analysis*, Springer Atmospheric Sciences, DOI 10.1007/978-94-007-2351-1_12,

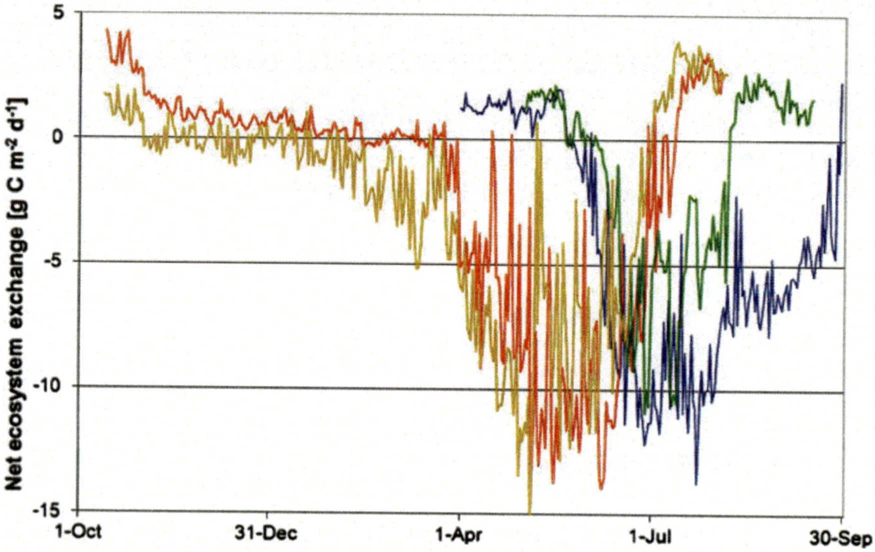

Fig. 12.1 Evolution of the daily net ecosystem exchange between sowing and harvest for four crops in Belgium (Lonzée site). The red and orange lines correspond to winter wheat crops in 2005 and 2007. Sugar beet crop (2004) is represented by the blue line and seed potato crop (2006) by the green line

Seeding and harvest take place within a time span of a few months, most often less than 1 year. In warm climates, especially in the subtropic and tropic zones, two or even three crops and harvest cycles can be accomplished over 1 year.

Between the cropping periods, the soil could remain bare or covered with crop residues with a possible development of either crop regrowths or weeds. In other cases, a fallow crop could be seeded.

So, during a year, very different and contrasted conditions are observed on a specific crop field: from bare soil to the crop maximum development. This implies large variations in canopy height, canopy structure, leaf area index (LAI) and vegetation area index (VAI). As a consequence, the structure of the turbulence and the albedo evolves during the cropping period and large variations of heat and net CO_2 fluxes are measured above these ecosystems, including a sign reversal for CO_2 fluxes.

Another crop specificity is that planting and harvest dates depend both on crop species and pedoclimatic conditions of the field. For example, in Europe and North America, winter crops are usually seeded between September and December (see Eugster et al. 2010) while spring crops are planted around April-May (e.g., spring wheat, rapeseed, potato, maize, sunflower). Consequently, the active growth and high CO_2 assimilation rate periods vary according to the crop type as illustrated in Fig. 12.1 for Belgian crops of a 4-year rotation. The duration of the period between cropping seasons depends on the crop succession and may range from a few weeks to several months. For monocropping the duration is relatively constant, while for

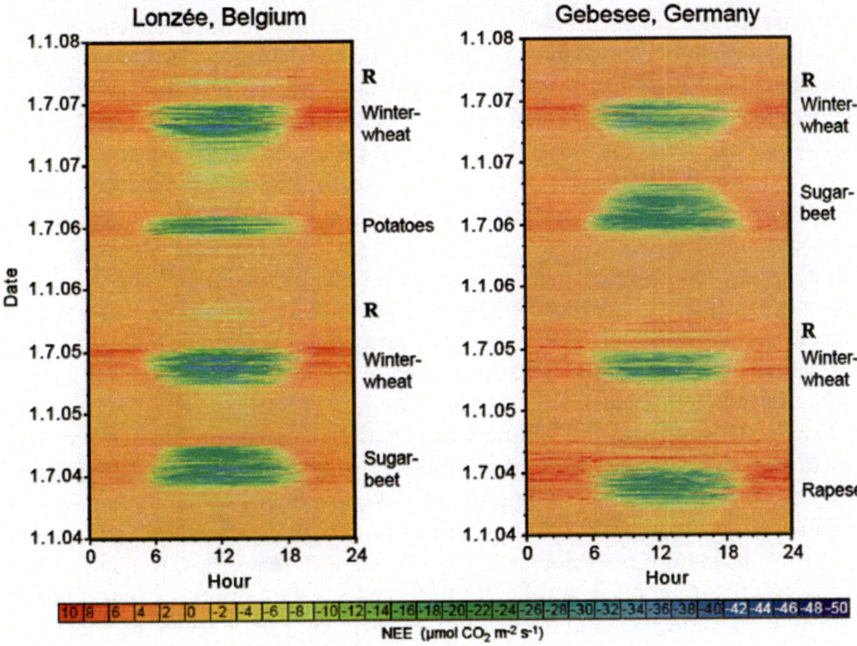

Fig. 12.2 Evolution of the net ecosystem exchange for rotation of four crops in Belgium (Lonzée site) and Germany (Gebesee site). R symbolizes crop regrowth

crop rotation, it may vary. This is illustrated in Fig. 12.2 for Belgian and German 4-year rotations. For both sites, the periods between harvest and next seeding range from less than one month between spring crops and winter wheat crops to 8 or even 9 months, when a winter wheat crop is followed by a spring crop.

During the period between seeding and harvest, the accumulated carbon in the biomass could reach high values, for example, 0.810 ± 0.311 kg C m^{-2} for shoot biomass of maize in southwest of France (Béziat et al. 2009), 0.88 ± 0.05 kg C m^{-2} and 1.01 ± 0.09 kg C m^{-2} for total biomass of winter wheat and sugar beet in Belgium (Aubinet et al. 2009). The dry biomass of these last two crops is 2.6 kg m^{-2} and 1.97 kg m^{-2}, respectively (Moureaux et al. 2006, 2008). For maize, reported values of shoot biomass ranged from 1.7 to 2.5 kg DM m^{-2} in North America (Pattey et al. 2001; Suyker et al. 2004, 2005).

The net CO_2 ecosystem exchange (NEE) measured over crops with the eddy covariance technique could reach high daily values. Net assimilation fluxes between -9 and -13 g C m^{-2} day^{-1} were observed for winter wheat (Baldocchi 2003; Soegaard et al. 2003; Anthoni et al. 2004; Moureaux et al. 2008; Béziat et al. 2009). Similar values were reported for soybean (Hollinger et al. 2005), rapeseed (Béziat et al. 2009), and sugar beet (Moureaux et al. 2006). In North America, reported maximum net uptake values in maize crops reached -18 to -20 g C m^{-2} day^{-1} (Pattey et al. 2001; Hollinger et al. 2005; Verma et al. 2005).

Another specific feature of crop ecosystems is the numbers of management practices: tillage, planting, applications of fertilizer, herbicides, fungicides, insecticides, and eventually of defoliant, irrigation, harvest, etc. The management activities are largely influenced by the cultivated crops, the pedoclimatic conditions and the crop rotation (e.g., in Belgium, generally a reduced tillage is carried out for a winter wheat crop after a potato crop). In addition to the impact of these practices on the NEE, CO_2 is emitted by the machinery, which could affect CO_2 concentration and flux measurements.

In general, crop canopies have a more homogeneous spatial composition compared to either forest or grassland canopies. They are often located on flat or gently rolling topography and surrounded by other agricultural fields reducing part of the potential issues that could affect flux measured over "natural" ecosystems using the eddy covariance technique.

This chapter discusses the specificities of eddy covariance measurements performed over cultivated areas. It deals with the aspects of setup, measurements, and data processing that are specific to the agroecosystems, while a more general presentation is provided in Chaps. 2–9. Ancillary measurements required to interpret CO_2 fluxes during crop development, to compare different crops, to quantify the net ecosystem carbon balance (NECB) and to assess the impacts of management practices are also presented in this chapter. Finally, we discuss recent development of flux measuring systems that can be deployed in subplots, in order to compare management practices and to quantify their impact on carbon fluxes and its budget.

12.2 Measurement System

12.2.1 Choice of the Site and Communication with the Farmer

The challenge in measuring fluxes from croplands comes from (1) the potential interference between the management practices and the instruments and (2) the rapid crop development. The choice of a representative site is discussed in Chap. 2. In addition to these general aspects, site investigators have to establish a collaboration agreement with the land owner/producer. This either formal or informal agreement should include the following aspects: (1) common agreement on the measurement site location, (2) communication in a timely manner between the producer and the research staff of next management practice for protecting/removing temporarily the equipment when deemed necessary, (3) access to detailed information on the management practices for the research team, (4) potential compensation for destructive plant sampling (see Sect. 12.7), access to electricity, etc. Information on management practices need to be documented as they might impact the fluxes during and following their occurrence. This can also be done by the research staff that needs to go on site every 1 or 2 weeks for the maintenance of the measuring system.

12.2.2 Flux Tower and Meteorological Station Configuration

General criteria to position the tower on the site such as predominant wind direction, fetch, and site homogeneity, as discussed in Sect. 2.2.2.1, are also crucial for measurement in agricultural systems but some additional aspects have to be considered. One of which is the mast, which represents an obstacle for the tractors. A way to reduce disruption is to establish the tower and station at the field border in such a way that they do not obstruct the work by the farmer, but still close enough to have most of the flux footprint in the field of interest. If the tower is established within the field, farmer will have to adopt a smart driver strategy.

In response to the challenge of measuring an intensively managed ecosystem, some investigators adopted a mobile measuring system (Fig. 12.3a) rather than a permanent installation (Fig. 12.3b–d). The permanent system consists of a fixed mast supporting the eddy covariance system surrounded by the meteorological sensors in a fixed enclosure. In this configuration, the soil is not tilled under the mast and the area surrounding the installation is either not cultivated or is managed by hand. This configuration allows having soil sensors and equipment enclosures

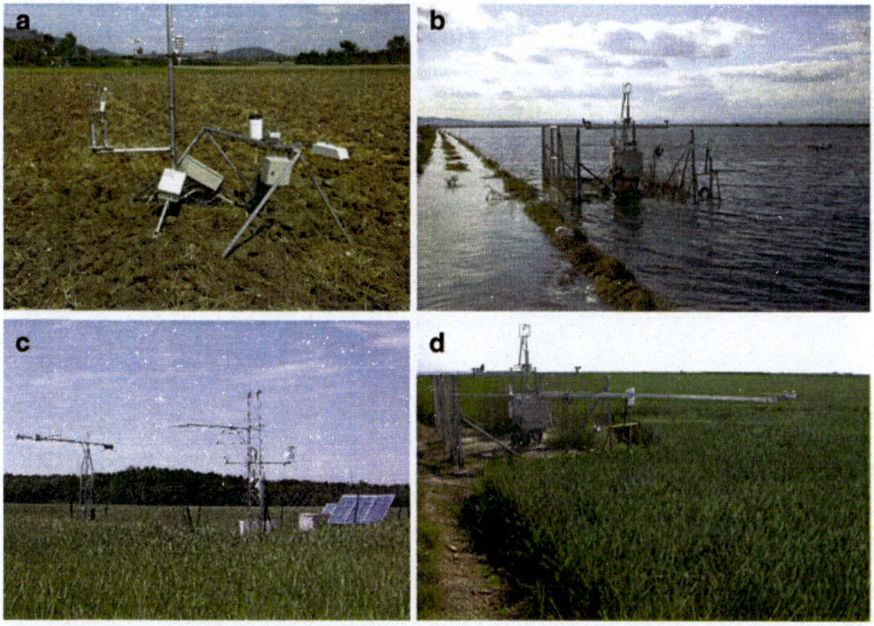

Fig. 12.3 Example of eddy covariance cropland sites: (**a**) portable lightweight towers where only the power outlet/battery box (white box above black PVC tube) is a fixed structure (CH-Oen2); (**c**) fixed position in the center of the field, where the crop inside the fence is managed by hand similarly to the main crop outside the fence (FR-Lam). The example of a rice paddy from Spain (**b, d**) shows that special planning will be necessary for sites that are seasonally flooded (Photo credits: Eugster (**a**), Carrara (**b, d**), Béziat (**c**))

installed permanently. In addition, this type of installation allows continuous measurements even during management practices giving the possibility to monitor the ecosystem responses right after their completion. However, as the equipped mast area is not cultivated in the same way as the rest of the field, it may become not representative of the rest of the field and can create a chimney effect as described in Sect. 2.2.1.2. Soil temperature, moisture, and heat flux as well as net radiation could be biased if located in this area. However, the unmanaged area can be minimized in such a way that the flux footprint area remains mostly unaffected. Soil sensors – at least the ones below the ploughing depth – can be installed elsewhere under the managed area as long as electrical cables are guided to the data logger at a depth below the ploughing depth. The same can also be recommended for the other cables such as power supply and connection wires that should be buried at a depth greater than the tillage depth but above the drains depth, in case the field is drained.

A mobile or roving measurement system is often a less invasive solution. The eddy covariance system and the meteorological sensors could be fixed on light masts (e.g., tripods or guy-wired masts) which are installed in the crop field after seeding. However, the deep soil sensors could be installed before sowing. In this way, the whole field is cultivated and the crop is less disturbed. Nevertheless, the eddy covariance system and the meteorological sensors have to be removed before the harvest or other cultivation practices and reinstalled as soon as possible. As a consequence, the flux measurements will be interrupted and some key measurement periods will be missing. Moreover, the installation of soil sensors disturbs the soil profile. It is recommended to dig a hole to insert the sensors on an undisturbed side of the hole. The hole needs to be refilled by respecting the soil horizons. A good contact between the soil and the sensors is required to ensure good quality measurements. It might take several days, depending on soil texture and precipitation to get representative soil measurements.

Mixed configurations could also be considered using, for example, a fixed eddy covariance mast and weather station in combination with temporary soil and radiation sensors installed in the field.

12.2.3 Measurement Height

Here again, the rapid development of the crops and more particularly its evolving height impacts the measurement height. How close can flux instruments be from the canopy? Several considerations need to be taken into account. The first one is the path-length between the transducers that determines the response to small-scale turbulence through line averaging the wind velocity along the path, especially at the low height of measurement as usually encountered in agricultural studies (Pattey et al. 2006). For a path-length of 0.1 m, at 0.5 m above the displacement height d, a reduction of 5% in vertical wind speed variance (σ_w) is observed, while at 2.5 m, this reduction is less than 3% (Wamser et al. 1997). The second one is the sampling frequency that should be higher closer to the ground (See Sect. 1.5.4).

Similarly, for the covariance determination, high-frequency underestimations occur for combination of high wind speed and low measuring height. By using a threshold value of 5 for the normalized frequency ($f_s(h_m-d)/u$; where f_s, is the sampling frequency = 10 Hz, h_m the measurement height in m, d the displacement height in m, and u the horizontal wind velocity in m s^{-1}), underestimation of high frequency will take place for the following conditions: at $h_m-d=0.5$ m for $u>1$ m s^{-1}, at $h_m-d=1.0$ m for $u>2$ m s^{-1}, at $h_m-d=1.5$ m for $u>3$ m s^{-1}, at $h_m-d=2.0$ m for $u>4.5$ m s^{-1} (Pattey et al. 2006). By locating eddy covariance sensors within the inertial sublayer (also called equilibrium boundary layer or well-mixed layer), near-field influence associated with roughness sublayer heterogeneity can be avoided. The inertial sublayer depth increases with the fetch and is a function of canopy architecture (Munroe and Oke 1975). For a fetch of 200 m, the inertial sublayer depth varies between 2.4 and 3.4 m with the canopy height of maize, while for a fetch of 100 m, depth varies between 0.1 and 1.7 m, which shows that the fetch is too limited in the latter case for monitoring fluxes over a maize field for the entire growing season (Pattey et al. 2006). The bottom of the inertial sublayer can be approximated as 1.66–2.16 h_c (where h_c is the canopy height).

Moreover, the confounding effects of surrounding areas have to be minimized and this is relevant for crops since in some regions fields have limited size. It is however less of a concern if the surrounding fields contain the same crop or one with a similar phenology, such that the division into individual fields is rather a logistical than a plant physiological issue. Moreover, similar adjacent fields contribute increasing the fetch and accommodate footprint increase at night. As discussed in Chap. 8 and later in this chapter, the footprint area is related to the aerodynamical displacement, and the closer to the canopy the measurement system is, the smaller is the footprint area (Sect. 8.3.2). However, to obtain flux measurements representative of field areas, sensors have to remain in the limit of the inertial sublayer.

At some sites, the measurement height is adapted according to the crop height, that is, by means of a telescopic tower, or by vertically moving a horizontal boom with the instruments on a solid mast. In this way, measurements are performed in the inertial sublayer and the footprint area is minimized.

12.2.4 Maintenance

Harvest and soil tillage during dry conditions could generate a lot of dust. In case of closed-path analyzer, this dust could rapidly obstruct the filter at the inlet of airflow to the analyzer. Furthermore, as the measurement systems over crops are quite close to the soil (generally less than 4 m) and as some agricultural sites may be located close to residential area, pollution could also rapidly block this filter, especially during winter. For this reason it is crucial to continuously control the filters (or monitor the air flow) and change them when dirty.

Similarly open-path analyzers may suffer from dust deposits on the optical windows. Rain following such dust events may clean the optical windows under some circumstances, whereas in other cases manual cleaning is required.

Further complications are birds and rodents. Since cropland areas are often lacking natural elements for birds of prey (e.g., hedges, single trees), particularly in intensively managed croplands, birds tend to perch on the tallest element in the landscape, which often is the eddy covariance flux system. A T-pole next to the system, which is taller than the system itself, would help to solve this problem, but there have been reports within southern localities where more elaborate structures were needed to keep the birds away from scientific equipment. Rodents also may be present in crops and might chew wires or enter in cabinets placed in the fields and defecate on electronic components. Steel wool could be placed around the wires and in all possible entries of the cabinets.

12.3 Flux Calculation

The dynamic height variation of crops has an impact on coordinate rotations, which are applied on the raw means and second moments. Therefore, the half-hourly 2D rotation (See Sect. 3.2.4) is recommended for measurements over crops. First rotation aligns the coordinate system with the mean wind, second rotation accounts for the inclination of streamlines to yield zero mean vertical wind speed. The Planar Fit method (See Sect. 3.2.4.3) for tilt correction is not appropriate since it requires several weeks of measurement during which the setup conditions remain constant. This is rarely the case when measuring above a crop.

12.4 Flux Corrections

12.4.1 Storage Term

In short ecosystems like croplands, the storage term ($F_C{}^{STO}$, see Eqs. 1.24b and 1.25b) is expected to be small compared to forest ecosystem (Sect 2.5) and consequently is frequently computed on the basis of one single CO_2 concentration measurements at eddy covariance flux measurement height (Anthoni et al. 2004; Moureaux et al. 2006; Suyker et al. 2005; Verma et al. 2005; Wohlfahrt et al. 2005; Xu and Baldocchi 2004; Béziat et al. 2009). The storage term estimated from the single point method was compared to a multiple point profile. For low turbulent periods, Saito et al. (2005) reported a 22% underestimation of the storage term with the single point method in comparison with a six point profile, while Moureaux et al. (2008) found a 6% overestimation of the single point method during turbulent periods which suggests that the single height method could be used under turbulent conditions. Whenever half-hourly CO_2 fluxes are discussed, CO_2 storage flux should be taken into account.

12.4.2 Nighttime Flux Data Screening

Micrometeorological techniques based on turbulent transfer frequently underestimate CO_2 fluxes during nighttime conditions when turbulence is low. A filtering procedure is proposed in Sect. 5.3 for screening data into either windy or calm condition. Two selection criteria were proposed: one based on the friction velocity (u_*) and the other based on the standard deviation of the vertical wind speed (σ_w). Pattey et al. (2002) found that σ_w was a more robust criterion, independent of the sonic anemometer head configuration. The threshold value allowing the filtering could be dependent on the crop species and on the presence or absence of the crop (Moureaux et al. 2008; Béziat et al. 2009).

For this reason the threshold in cropland should be calculated for the different management periods that are a function of seeding and harvest dates, as well as regrowth events. The year could even been subdivided according to the intensity of crop development or soil tillage. However, the length of the different periods has to allow a reliable determination of the threshold. Béziat et al. (2009) defined crop functioning periods between the dates of sowing, maximum crop development, harvest, and tillage and determined a u_{*crit} threshold for each crop functioning period.

12.5 Data Gap Filling and Footprint Evaluation

Similar to the u_{*crit} determination, gap-filling (Chap. 6) and footprint evaluation (Chap. 8) in crop ecosystems require attention related to the fast or even abrupt changes in the ecosystem status due to rapid crop development and management practices.

12.6 Cumulated Carbon Exchange

Commonly, fluxes from eddy covariance measurements are integrated and compared over 1 year time. However, the calendar year is not adequate for crops.

In order to compare CO_2 fluxes of different crops, the integration period should start at seeding and finish either at harvest or prior to the next seeding. In several synthesis studies of European crops (Kutsch et al. 2010; Ceschia et al. 2010), the selected integration period ranged from early October to end of September. This includes sowing and harvest of the spring and winter studied crops. However, in this way, the carbon degradation of crop residues occurring after the harvest, that is, during autumn, winter, and even spring, will be included in the following crop period and the impact of this degradation will be attributed to the next crop. For spring crops, starting the integration period in the spring at sowing and finishing it prior to the next seeding allow including the initial residue degradation.

Therefore, the best approach to derive the cumulated fluxes or the carbon balance of crop rotations is to integrate from seeding to prior the next seeding of the crop rotation for the entire sequence of crop rotation, which normally means that the integration limits are not aligned with the Gregorian calendar. This allows taking into account the crop sequence, the impact of management practices and periods between harvest and seeding. This was performed for 2-year rotations in North America (maize/soybean) by Hollinger et al. (2005), Suyker et al. (2004, 2005), and Verma et al. (2005), for a Belgian 4-year rotation by Aubinet et al. (2009), and integrating six full crop rotations of different European agricultural sites by Kutsch et al. (2010).

12.7 Additional Measurements

The need for supplemental measurements depends on the objectives of the research. However, the knowledge of the sowing, harvest and tillage dates, plant density, LAI, and biomass distribution dynamics is important to understand the fluxes. Extensive sampling might be required to cover the flux footprint area. Because of its influences on photosynthetic radiation interception, latent and sensible heat fluxes, LAI is important to measure over space and time. Recently, digital color photography was proposed to measure LAI from crops (Liu and Pattey 2010) in addition to conventional methods, since the approach is less limited by radiation conditions and the protocol can easily be implemented for extensive sampling. For crop comparisons, the produced biomass is a key element. In the frame of the carbon balance assessment, to compute either the net ecosystem carbon balance (NECB) or net biome productivity (NBP), the imported and exported biomasses have to be known.

In order to obtain reliable dry biomass assessment and the associated uncertainties, it is recommended to collect several samples in representative areas of the field. To follow closely the vegetation dynamics, sampling can be performed every week or 2 weeks in relation to the dynamics of the crop. The biomass of the various organs could be estimated by separating samples into seeds/fruits, green and dead stems and leaves. Root biomass is very challenging to measure and carries a lot of uncertainty and for this it is usually not routinely measured.

The harvested biomass assessment by the farmer by weighting some of the wagons containing the exported part of the crop (i.e., grain) might not be very accurate. An alternative way is to destructively measure the dry biomass to be exported (e.g., grain, shoot) right before the harvest. Another way is to assess total biomass before the harvest and crops residues remaining thereafter and to subtract them. Finally, yield monitor installed on board of the combine can also be used, provided they are calibrated. They offer the advantage to provide a yield map.

In any case, attention should be paid to reduce the uncertainties on biomass sampling since uncertainties on those estimations might be bigger than uncertainties on other flux measurements (Béziat et al. 2009). In order to obtain reliable assessment of carbon inputs in case of organic manure application, several buckets of known area have to be placed on the field during the application and the carbon content of the collected samples has to be measured.

12.8 Future Experimentations

The agricultural management practices are expected to impact the carbon fluxes and the carbon budget. In the frame of carbon mitigation opportunities, these practices have to be evaluated in terms of C fluxes and budgets. An attractive way to compare the agricultural practices is to divide a crop area into subplots managed in different ways and use several EC masts (e.g., Pattey et al. 2006; Davis et al. 2010). In order to reduce source areas of scalar fluxes, while measuring "representative" data, the flux systems may be placed at the bottom of the inertial sublayer. If the flux measuring system is located in the roughness sublayer, flux measurement detects the near-field contribution, at the expenses of a more average contribution. Moreover, there are technical and theoretical issues limiting the eddy flux–canopy top minimum distance as discussed in Sect. 12.3. Using instruments with small-size transducers and with higher sampling frequency should allow reducing the measurement height requirements.

Experimental test involving natural tracers release and comparison of results coming from different technical setups should provide data that are presently missing to better understand how to perform small-scale fluxes with actual technologies and what are the most significant drawbacks when measuring fluxes at a short distance from the surface. Experiments involving multiple deployments of eddy flux systems at various heights above the crop canopy could allow the agrometeorologists to find suitable empirical corrections when placing systems in the roughness sublayer. These experiments would also benefit the footprint models, such as those based on large eddy simulations that need adequate parameterization of small-scale turbulent dynamics, to better predict the source areas for scalar concentrations and fluxes.

Acknowledgments CM acknowledges financial support by the EU (FP 6), the Belgian Fonds de la recherche Scientifique (FNRS-FRS), and the Communauté française de Belgique (Action de Recherche Concertée).

References

Anthoni PM, Freibauer A, Kolle O, Schulze E-D (2004) Winter wheat carbon exchange in Thuringia, Germany. Agric For Meteorol 121:55–67

Aubinet M, Moureaux C, Bodson B, Dufranne D, Heinesch B, Suleau M, Vancutsem F, Vilret A (2009) Carbon sequestration by a crop over a 4-year sugar beet/winter wheat/seed potato/winter wheat rotation cycle. Agric For Meteorol 149:407–418

Baldocchi DD (2003) Assessing the eddy covariance technique for evaluating carbon dioxide exchange rates of ecosystems: past, present and future. Glob Chang Biol 9:479–492

Béziat P, Ceschia E, Dedieu G (2009) Carbon balance of a three crop succession over two cropland sites in South West France. Agric For Meteorol 149:1628–1645

Ceschia E, Béziat P, Dejoux JF, Aubinet M, Bernhofer C, Bodson B, Buchmann N, Carrara A, Cellier P, Di Tomasi P, Elbers JA, Eugster W, Grünwald T, Jacob CMJ, Jans WWP, Jones M, Kutsch W, Lanigan G, Magliulo E, Marloie O, Moors EJ, Moureaux C, Olioso A, Osborne B, Sanz MJ, Saunders M, Smith P, Soegaard H, Wattenbach M (2010) Management effects on net ecosystem carbon and GHG budgets at European crop sites. Agric Ecosyst Environ 139:363–383

Davis PA, Clifton BJ, Saunders M, Lanigan G, Wright E, Fortune T, Burke J, Connolly J, Jones MB, Osborne B (2010) Assessing the effects of agricultural management practices on carbon fluxes: spatial variation and the need for replicated estimates of net ecosystem exchange. Agric For Meteorol 150:564–574

Eugster W, Moffat A, Ceschia E, Aubinet M, Ammann C, Osborne B, Davis PA, Smith P, Jacobs C, Moors E, Dantec VL, Béziat P, Saunders M, Jans W, Grünwald T, Rebmann C, Kutsch W, Czerný R, Janouš D, Moureaux C, Dufranne D, Carrara A, Magliulo V, Tommasi PD, Olesen JE, Schelde K, Olioso A, Bernhofer C, Cellier P, Larmanou E, Loubet B, Wattenbach M, Marloie O, Sanz M-J, Søgaard H, Buchmann N (2010) Management effects on European cropland respiration. Agric Ecosyst Environ 139:346–362

Hollinger SE, Bernacchi CJ, Meyers TP (2005) Carbon budget of mature no-till ecosystem in north central region of the United States. Agric For Meteorol 130:59–69

Kutsch WL, Aubinet M, Buchmann N, Smith P, Osborne B, Eugster W, Wattenbach M, Schulze ED, Tomelleri E, Ceschia E, Bernhofer C, Béziat P, Carrara A, Di Tommasi P, Grünwald T, Jones M, Magliulo V, Marloie O, Olioso A, Sanz MJ, Saunders M, Soegaard H, Ziegler W (2010) The net biome production of full crop rotations in Europe. Agric Ecosyst Environ 139:336–345

Liu J, Pattey E (2010) Measuring agricultural crop canopy structural descriptors using digital photography. Agric For Meteorol 150:1485–1490 (GreenCropTracker software to process the color photos can be downloaded for free from Flintbox www.flintbox.com)

Moureaux C, Debacq A, Bodson B, Heinesch B, Aubinet M (2006) Annual net ecosystem carbon exchange by a sugar beet crop. Agric For Meteorol 139:25–39

Moureaux C, Debacq A, Hoyaux J, Suleau M, Tourneur D, Vancutsem F, Bodson B, Aubinet M (2008) Carbon balance assessment of a Belgian winter wheat crop (Triticum aestivum L.). Glob Chang Biol 14:1353–1366

Munro DS, Oke TR (1975) Aerodynamic boundary-layer adjustment over a crop in neutral stability. Bound Layer Meteorol 9:53–61

Pattey E, Strachan IB, Boisvert JB, Desjardins RL, McLaughlin NB (2001) Detecting effects of nitrogen rate and weather on corn growth using meteorological and hyperspectral reflectance measurements. Agric For Meteorol 108:85–99

Pattey E, Strachan IB, Desjardins RL, Massheder J (2002) Measuring nighttime CO_2 flux over terrestrial ecosystems using eddy covariance and nocturnal boundary layer methods. Agric For Meteorol 113(1–4):145–153

Pattey E, Edwards G, Strachan IB, Desjardins RL, Kaharabata S, Wagner RC (2006) Towards standards for measuring greenhouse gas flux from agricultural fields using instrumented towers. Can J Soil Sci 86:373–400

Saito M, Miyata A, Nagai H, Yamada T (2005) Seasonal variation of carbon dioxide exchange in rice paddy field in Japan. Agric For Meteorol 135:93–109

Soegaard H, Jensen NO, Boegh E, Hasager CB, Schelde K, Thomsen A (2003) Carbon dioxide exchange over agricultural landscape using eddy correlation and footprint modelling. Agric For Meteorol 114:153–173

Suyker AE, Verma SB, Burba GG, Arkebauer TJ, Walters DT, Hubbard KG (2004) Growing season carbon dioxide exchange in irrigated and rainfed maize. Agric For Meteorol 124:1–13

Suyker A, Verma S, Burba G, Arkebauer TJ (2005) Gross primary production and ecosystem respiration of irrigated maize and irrigated soybean during a growing season. Agric For Meteorol 131:180–190

Verma SB, Dobermann A, Cassman KG, Walters DT, Knops JM, Arkebauer TJ, Suyker AE, Burba GG, Amos B, Yang H (2005) Annual carbon dioxide exchange in irrigated and rainfed maize-based agroecosystems. Agric For Meteorol 131:77–96

Wamser C, Peters G, Lykossov VN (1997) The frequency response of sonic anemometers. Bound Layer Meteorol 84:231–246

Wohlfahrt G, Anfang C, Bahn M, Haslwanter A, Newesely C, Schmitt M, Drosler M, Pfadenhauer J, Cernusca A (2005) Quantifying nighttime ecosystem respiration of a meadow using eddy covariance, chambers and modelling. Agric For Meteorol 128:141–162

Xu L, Baldocchi DD (2004) Seasonal variation in carbon dioxide exchange over a Mediterranean annual grassland in California. Agric For Meteorol 123:79–96

Chapter 13
Eddy Covariance Measurements over Grasslands

Georg Wohlfahrt, Katja Klumpp, and Jean-François Soussana

In this chapter we first provide a historic overview of – and outline some of the peculiarities associated with – grassland eddy covariance flux measurements, elaborate on the additional terms that need to be quantified when estimating the grassland net ecosystem carbon balance and finally discuss some of the challenges associated with upcoming nitrous oxide and methane flux measurements in managed grasslands.

13.1 Historic Overview of Grassland Eddy Covariance Flux Measurements

One of the first latent/sensible heat flux measurements by means of the eddy covariance method were done in 1950 by Swinbank (1951) over a grassland. At that time and during the following years, experimental evaluations of the eddy covariance method were focused on understanding the turbulent properties of surface layer and on mastering the technical challenges posed by the available equipment. Technical advances in sonic anemometers and scalar sensors then led to more routine applications of the eddy covariance method by the mid-1980s when the scientific interest was already to determine the source/sink strength of grassland ecosystems for latent/sensible heat and trace gases such as nitrogen oxides, ozone and carbon dioxide (CO_2) (Delany et al. 1986; Kim and Verma 1990; Verma et al. 1989; Zeller et al. 1989). These studies, however, were usually

G. Wohlfahrt (✉)
Institute of Ecology, University of Innsbruck, Innsbruck, Austria
e-mail: Georg.Wohlfahrt@uibk.ac.at

K. Klumpp • J.-F. Soussana
INRA, Grassland Ecosystem Research (UREP), Clermont-Ferrand, France
e-mail: katja.klumpp@clermont.inra.fr; Jean-Francois.Soussana@clermont.inra.fr

M. Aubinet et al. (eds.), *Eddy Covariance: A Practical Guide to Measurement and Data Analysis*, Springer Atmospheric Sciences, DOI 10.1007/978-94-007-2351-1_13,

confined to a few months and it was not until 1996 when the first year-round carbon dioxide and energy eddy covariance flux measurements over grasslands were initiated (Meyers 2001; Suyker and Verma 2001). During those early years of the FLUXNET project, the majority of sites were situated in forest ecosystems (only 3 out of 34 sites listed in Baldocchi et al. (2001) were non-forest ecosystems). This changed dramatically at the beginning of the twenty-first century when two EU projects dedicated to grassland ecosystems were initiated: CarboMont (Cernusca et al. 2008) and GreenGrass (Soussana et al. 2007). At about the same time, eddy covariance flux measurements were also started at several sites in the large grassland areas of northern America (e.g. Flanagan et al. 2002; Hunt et al. 2004) and central Asia (e.g. Kato et al. 2004; Li et al. 2005). By now, the relative number of grassland flux towers within FLUXNET is approximately equal to the global percentage land cover of grasslands (http://www.fluxdata.ornl.gov).

13.2 Peculiarities of Eddy Covariance Flux Measurements over Grasslands

In the following we aim at complementing the previous chapters by identifying and giving examples for issues specific to eddy covariance flux measurements over grassland ecosystems.

A major difference between eddy covariance flux measurements over grasslands as compared to forests, or more generally between tall and short canopies like cropland and wetland, is that flux measurements over grasslands are made closer to the ground surface. This entails the advantage of the storage flux usually being small as compared to the eddy flux (see Eq. 1.24 in Sect. 1.4.2) and that any error in the quantification of the storage flux will have comparably small numerical consequences for the derived net ecosystem CO_2 exchange (NEE). For example, for a measurement height of 3 m and a typical averaging period of 30 min, a uniform 1 ppm change in CO_2 mole fraction translates to a storage flux of only 0.07 μmol CO_2 m^{-2} s^{-1} (at 20°C and a static air pressure of 101.3 kPa). At the grassland study site Neustift (Austria; Wohlfahrt et al. 2008a) the long-term storage flux (sign removed) amounts to 0.03 ± 0.04 μmol CO_2 m^{-2} s^{-1}, which is more than two orders of magnitude smaller than the corresponding average night-time (6 μmol CO_2 m^{-2} s^{-1}) and midday (-10 μmol CO_2 m^{-2} s^{-1}) NEE. The lower measurement height at grassland sites, however, causes co-spectra to shift to higher frequencies, as compared to forest sites (see also Sects. 1.5.4 and 4.1.3). This in turn has implications for flux loss due to low- and high-pass filtering (Sect. 4.1.3.2) associated to a certain degree with any eddy covariance system and the corrections required for eliminating this bias (Massman 2000). As a consequence, frequency response corrections tend to be higher for grassland as compared to forest sites (Fig. 13.1), in particular at high wind speeds and/or with stable stratification when co-spectra are characterised by larger high-frequency contents (Kaimal and Finnigan 1994). Differences between typical grassland and

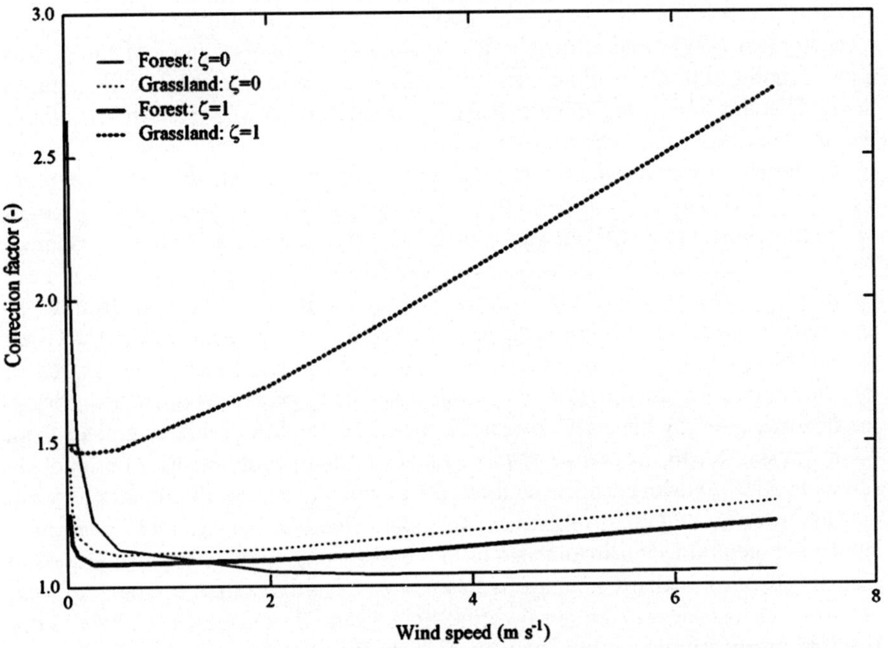

Fig. 13.1 Frequency response correction factors as a function of horizontal wind speed for hypothetical forest (measurement height = 30 m, zero-plane displacement height = 14 m) and grassland (measurement height = 3 m, zero-plane displacement height = 0.7 m) site for near-neutral ($\zeta = 0$) and stable ($\zeta = 1$) conditions. Calculations have been performed for a 30 min averaging period (no de-trending) and for the combination of a sonic anemometer (negligible time response, path-length = 0.15 m) and an open-path instrument (response time = 0.1 s, path-length = 0.15 m) separated laterally by 0.2 m and assumed a co-spectral reference model according to Kaimal and Finnigan (1994)

forest frequency response correction factors diminish or even reverse at very low wind speeds, when high-pass filtering associated with block averaging and any detrending operations leads to a relatively larger flux loss for forests due to the larger low-frequency content (Fig. 13.1).

A major challenge for conducting eddy covariance flux measurements in managed grasslands is the fact that the footprint is often heterogeneous due to different land uses present (e.g. a mix of grass- and croplands), different management intensities (e.g. number of grazing animals, cutting frequency, fertiliser type and quantity), temporal asynchrony in management activities (e.g. cutting events occurring at different dates), as well as landscapes with complex topography. Under these circumstances, a careful site selection, which takes the surface heterogeneity into account, is required. As an alternative to the EC method, chamber-based measurements may be suitable for monitoring NEE from short-plant ecosystems in complex topography (Risch and Dougas 2005; Li et al. 2005; Schmitt et al. 2010).

Footprint models (see Chap. 8) may be used to determine, for any given measurement height and atmospheric conditions, the likely extent of the source area (Schmid 2002). An illustrative example is given in Fig. 13.2, which shows the NEE at the study site Neustift (Austria) for a situation where the first 100 m of the daytime source area had regrown after a cut 11 days earlier and were followed further upwind by grass which had been cut 3 weeks before and thus did have more time to regrow. The NEE in these two areas was measured concurrently by means of transparent chambers (Wohlfahrt et al. 2005). The source area weighted chamber flux (Fig. 13.2) corresponds nicely with the NEE measured by eddy covariance (slope and *y*-intercept of a linear regression not significantly different from unity and zero, respectively) confirming the validity of the footprint model by Hsieh et al. (2000). During the morning hours, the young grass, which as compared to the older grass exhibits a relatively modest net CO_2 uptake, contributes most to the flux measured by the eddy covariance tower (up to 95% of flux originates from young grass). During the course of the day this contribution diminishes continuously (down to 41%), however; because the NEE of the older grass diminishes as well, measured and modelled CO_2 fluxes change relatively little. This example is meant to show the potential footprint analysis offers in analysing eddy covariance grassland data in situations where management leads to heterogeneities in the footprint. However, except for very simple situations (e.g. Marcolla and Cescatti 2005) and/or if additional measurements are available as in the example shown in Fig. 13.2, it will usually be difficult to partition fluxes based on footprint models. Rather footprint models can be used for guiding site selection and tower setup for maximising data capture from the grassland of interest and as a post-processing quality control criterion (Novick et al. 2004) for excluding flux measurements contaminated by heterogeneities in the footprint.

The low measurement height of grassland flux towers and associated relatively small footprint offer the advantage of making grasslands study sites amenable to manipulation at the field scale. Adopting treatment and control footprints, factorial experiments can be designed which combine the strengths of the EC method, that is, near-continuous spatially integrated flux monitoring, with the explanatory power of causal analysis offered by classical ecological factorial experiments. With identical equipment, land-use history and near-identical environmental conditions, systematic uncertainties are minimised (Ammann et al. 2007), allowing accurate determination of treatment effects. To date such experiments have been limited to changes in management (Allard et al. 2007; Ammann et al. 2007); however, manipulations of environmental factors (e.g. additional precipitation) seem possible as well, and we in fact expect major new developments in this direction in the near future.

Another issue with, in particular managed, grasslands is their rapid growth or regrowth after grazing/cutting, which is reflected in rapid changes of the NEE and energy fluxes (Hammerle et al. 2008; Wohlfahrt et al. 2008a). Imputation of inevitable gaps in EC time series (Falge et al. 2001), a standardised step in the processing of FLUXNET data (see also Chap. 6), needs to take these rapid changes into account (Ammann et al. 2007). Thereby a compromise between the length of the time window, which should be a short as possible in order to capture the dynamic

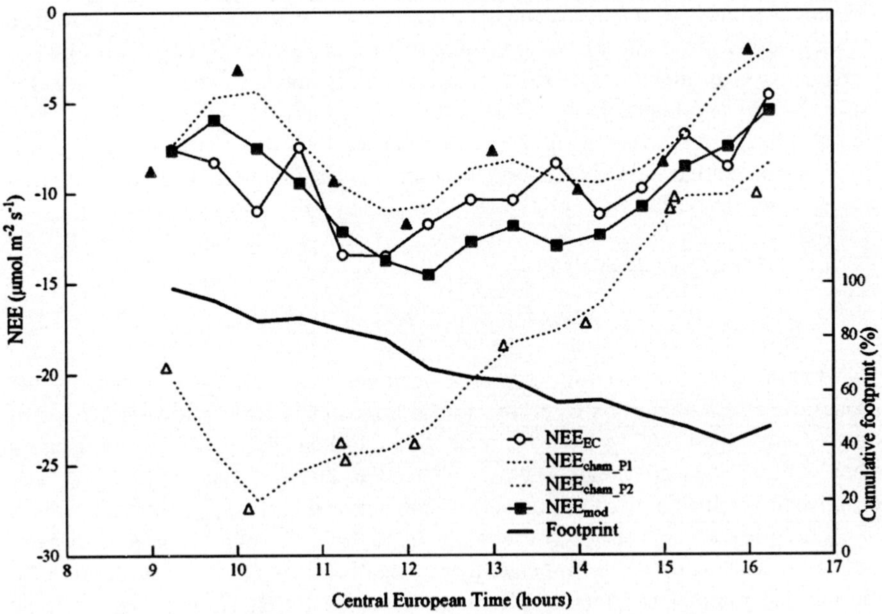

Fig. 13.2 Example illustrating the effect of a bimodal CO_2 sink strength distribution in the footprint of an eddy covariance tower on the measured net ecosystem CO_2 exchange (NEE_{EC}). The area immediately upwind of the tower (Plot 1) had been cut 11 days before measurements and was followed further upwind by an area (Plot 2) which had been cut 3 weeks before. NEE on these two plots (NEE_{cham_P1}, NEE_{cham_P2}) was measured by means of transparent chambers as described in Wohlfahrt et al. (2005). Based on the footprint model by Hsieh et al. (2000) the expected NEE at the EC tower (NEE_{mod}) was then calculated as the source-area weighted average of the NEEs measured by chambers. Unpublished data by Wohlfahrt G. and Drösler M. from the study site Neustift (Austria)

behaviour, and the number of data upon which gap-filling is based, which should be as large as possible for statistical significance, has to be achieved (see Sect. 6.3.2.1 for further details).

13.3 Estimating Grassland Carbon Sequestration from Flux Measurements

An alternative to the direct measurement of carbon (C) stock changes in grasslands (Conant et al. 2001) is to measure the net balance of C fluxes (i.e. net ecosystem carbon balance, NECB; Chapin et al. 2006) exchanged at the system boundaries. Using this approach changes in C stocks can be detected within 1 year. In contrast, direct measurements of stock changes by soil coring require several years or even several decades to detect significant effects given the high variability amongst samples (Arrouays et al. 2003). The main drawback of flux measurements, however, is that several C fluxes may need to be quantified (Soussana et al. 2010): (1) gaseous C exchange with the atmosphere, (2) particulate organic C fluxes, (3) dissolved C flux in waters and lateral transport of soil C through erosion (see also Sect. 13.4). The NECB (gC m^{-2} $year^{-1}$) is the mass balance of all these fluxes (Eq. 13.1). It should be noted that NBP (Net Biome Productivity) is another term frequently used when scaling up NECB from plot to regional scale (e.g. Schulze et al. 2009).

$$NBP = NECB = NEP + F_{CH4-C} + F_{VOC} + F_{fire} + F_{manure} + F_{harvest} + F_{animal-products} + F_{leach} + F_{erosion} \tag{13.1}$$

Adopting a sign convention where a positive C flux indicates a gain by the ecosystem and a negative flux a loss to the atmosphere, the following groups of fluxes (units of gC m^{-2} $year^{-1}$) can be distinguished: fluxes which constitute a clear gain by the ecosystem, such as C imports through manure (F_{manure}), and negative fluxes such as carbon exports through harvesting ($F_{harvest}$), animal products ($F_{animal\text{-}products}$), erosion ($F_{erosion}$), leaching ($F_{leach}$; organic and/or inorganic carbon) and fire emissions (F_{fire}). Both positive and negative fluxes have been reported for the net ecosystem CO_2 uptake (NEP, see Sect. 9.2), CH_4 ($F_{CH4\text{-}C}$) and volatile organic compound (F_{VOC}) exchange (Soussana et al. 2007; Wohlfahrt et al. 2008a; Ruuskanen et al. 2011).

Depending on the system studied and its management, some of these fluxes can be neglected for NECB calculation. For instance, fire emissions by grasslands are very low in temperate regions like Europe (i.e., below 1 gC m^{-2} $year^{-1}$ over 1997–2004), whilst they reach 10 and 100 gC m^{-2} $year^{-1}$ in Mediterranean and in tropical grasslands, respectively (Van der Werf et al. 2006). Erosion ($F_{erosion}$) is also rather insignificant in permanent grasslands (e.g. in Europe), but can be increased by tillage in the case of sown grasslands. The global map of $F_{erosion}$ created by Van Oost et al. (2007) indicates that grassland C erosion rates are usually below 5 gC m^{-2} $year^{-1}$, even in dry tropical grasslands (Van Oost et al. 2007). VOC emissions by grassland systems are increased in the short term by cutting (Ruuskanen et al. 2011) and tend to be higher with legumes than with grass species (Davison et al. 2008). However, VOC fluxes from grasslands, even if some VOC compounds include several C atoms, are usually at least 1 order of magnitude smaller and can usually be neglected in terms of carbon exchange. Therefore, for temperate managed grasslands, Eq. 13.1 can be simplified as (Allard et al. 2007)

$$NECB = NEP + F_{CH4-C} + F_{manure} + F_{harvest} + F_{animal-products} + F_{leach} \tag{13.2}$$

Since eddy flux covariance measurements use a free air technique, as opposed to enclosures, there is no disturbance of the measured area, which can be freely accessed by herbivores. CO_2 belched by ruminants during grazing (digestive + metabolic CO_2), which can be measured by the SF_6 method (Pinares-Patiño et al. 2007), is thus included in NEP measurements. It has no direct effect on the atmospheric CO_2 concentration, because it is ‘short-cycling’ carbon, which has been fixed by plants earlier.

Quite often not all components of the NECB budget are measured. For example, DOC/DIC losses as well as C exports in milk and meat products are sometimes neglected. Siemens and Janssens (2003) have estimated at the European scale the average DOC/DIC loss at 11 ± 8 gC m^{-2} $year^{-1}$. This flux tends to be highly variable depending on soil (pH, carbonate) and climate (rainfall, temperature) factors and it could reach higher values in wet tropical grasslands, especially on calcareous substrate. Assuming a value at the upper range of this estimate, would reduce the grassland NBP by 20%. In contrast, the role of organic C exports is relatively small with meat production systems (e.g. 1.6% of NBP, Allard et al. 2007) but can be higher with intensive dairy production systems.

13.4 Additional Measurements

Calculation of the NECB (Eq. 13.2) requires several additional carbon fluxes in addition to the CO_2 and CH_4 fluxes, treated in Sects. 13.2 and 13.5, respectively, to be quantified. In a cutting regime, a large part of the primary production is exported from the plot (as hay or silage), and may be compensated by organic C imports through cattle slurry and liquid manure. To determine a complete C budget requires quantification of the amount of harvested dry-matter and organic fertiliser as well as the corresponding C content (gC m^{-2} $year^{-1}$). Under grazing, up to 60% of the above-ground dry-matter production is ingested by domestic herbivores (Lemaire and Chapman 1996), whereas a large part of the ingested C is emitted as CO_2 and CH_4 shortly after intake. The magnitude of these fluxes largely depends on quality of ingested biomass and number, weight and type of animals (i.e., sheep, heifers, dairy cows, etc.). For example, with the SF_6 dual tracer technique (e.g. Pinares-Patiño et al. 2007), methane emission comprised between 0.33 and 0.45 g CH_4 kg^{-1} LW day^{-1} for heifers and bulls and reached 0.68–0.97 g CH_4 kg^{-1} LW day^{-1} for lactating cows (Soussana et al. 2007). Quantity and quality of herbage biomass offered to animals can be determined by grazing exclusion cages. To closely follow the vegetation dynamics, cages should be moved and cut at least on a monthly basis. Detailed analyses of the harvested biomass offers further insights into the fraction of senescent plant material and plant functional groups (i.e., grass, forbs and legumes). The potential herbage production of the pasture, which is closely related to gross primary productivity, can be estimated, by measuring the monthly biomass regrowth (g DM m^{-2} day^{-1}).

The non-digestible C (25–40%) of the intake is usually returned to pastures/hay meadows in excreta (mainly as faeces) and urine for fertilisation. Manure, urine and mineral fertiliser in turn are known to stimulate N_2O emissions, their magnitude depending on soil humidity conditions at the time of application. Detailed knowledge of stocking rate (mean number of livestock units per unit area), weight and growth of livestock, grazing period, as well as date of fertilisation and meteorological soil conditions (i.e., soil water content, soil temperature) are thus helpful for understanding and interpreting eddy covariance data (see also Sect. 13.5).

Because particular managed grasslands exhibit a very dynamic canopy growth, knowledge of the amount of above-ground biomass is crucial for interpreting flux measurements (Hammerle et al. 2008; Wohlfahrt et al. 2008b). The amount of above-ground biomass or the leaf area index (LAI) should thus be determined at least episodically, for example, using destructive harvesting and leaf area determination. Indirect measurements of LAI, by inverting models of within-canopy radiative transfer (Wohlfahrt et al. 2001), or measurements of proxies for above-ground biomass and LAI, for example, canopy height, can be done in an automated fashion using line PAR (Wohlfahrt et al. 2010) and snow height (Jonas et al. 2008) sensors, respectively, and can be combined with direct measurements to yield quantitative high-resolution time series information about vegetation development (Wohlfahrt et al. 2008a).

13.5 Other Greenhouse Gases

During the past two decades, a large body of literature has been accumulated on grassland carbon cycling based on eddy covariance flux measurements (e.g. Gilmanov et al. 2007, 2010; Soussana et al. 2007, 2010; Wohlfahrt et al. 2008b). In the near future, we anticipate that these CO_2 flux measurements will be increasingly accompanied by flux measurements of the other two important greenhouse gases originating from grasslands, that is, methane (CH_4) and nitrous oxide (N_2O), whose emissions may easily offset any carbon gains due to the larger warming potential (Soussana et al. 2007, 2010). However, due to a combination of source complexity (i.e., spatial and temporal variation), limitations in equipment and methodology, measurements of CH_4 and N_2O fluxes are accompanied by significant uncertainties. N_2O emissions in soils usually occur in 'hot spots' associated with urine spots and particles of residues and fertiliser (Flechard et al. 2007). Nitrous oxide emissions from grasslands tend to occur in short-lived bursts following the application of fertilisers (Leahy et al. 2004). Temporal and spatial variations contribute large sources of uncertainty in N_2O fluxes at the field and annual scales (Flechard et al. 2005). Methane emissions by cattle show temporal and spatial variations, depending upon stocking rate and the quantity and quality of feed consumed (Pinares-Patiño et al. 2007) (see also Sect. 13.4). Additionally, a large variability results from animal behaviour, as animals do not behave at random. Grazing and ruminating is separated in time and space. The area of interest (i.e., paddock) is in most cases larger than measured footprint, which may make it necessary to track animals (e.g. using webcams or laser systems). Moreover, ruminating will create plumes of CH_4 that may appear as spikes in the concentration time series, but are in fact 'natural phenomena' which may be unintentionally removed by automated despiking algorithms (Vickers and Mahrt 1997).

Instrumentation for CH_4 and N_2O EC measurements, have to satisfy four criteria: (Nelson et al. 2002): (1) continuity, that is, the system should be able to operate unattended on a continuous basis; (2) the analyser response time should be on the order of 0.1 s to capture also the smallest eddies (Monteith and Unsworth 1990); however, sample interval may be much longer as long as the system time response is short enough (see Sect. 10.2.3); (3) a minimal drift should occur during a period of atmospheric stationarity (i.e., 30 min); and (4) a precision of 4 and 0.3 ppb for CH_4 and N_2O, given an average ambient concentration of 1,800 and 320 ppb. Instruments that meet those requirements for CH_4 and N_2O EC measurements are now becoming available. A limited number of closed-path EC measurements have been published using lead salt tunable diode laser (TDL) spectrometers (e.g. Smith et al. 1994; Wienhold et al. 1994; Laville et al. 1999; Hargreaves et al. 2001; Werle and Kormann 2001), quantum cascade (QC) lasers for (CH_4 and N_2O; Kroon et al. 2007; Neftel et al. 2007; Eugster et al. 2007), off-axis integrated cavity output spectroscopy (CH_4; Hendriks et al. 2008; Smeets et al. 2009) and as of this writing even open-path CH_4 analysers are becoming commercially available. In those studies, CH_4 and N_2O fluxes are measured over time periods ranging from 1 week to several years, yielding averaged emission rates with their standard deviations. The standard deviation, however, is mainly an indication of the temporal variability and does not represent the uncertainty associated with the mean flux (Kroon et al. 2009). Based on 30 min EC fluxes, the relative uncertainty is mainly attributed to relatively small EC fluxes and one-point sampling. On average this term contributes to more than 90% to the total uncertainty. The other 10% comprise the uncertainty in correction algorithms for systematic errors (e.g. inadequate footprint, non-stationarity, advection, storage, low– and high-pass filtering, etc.; see also previous chapters). Nevertheless, because the EC method integrates over a large spatial area and provides near-continuous data, EC flux measurements can contribute to more accurate estimates of the net ecosystem exchange of N_2O and CH_4 than estimates based on chamber measurements only.

Acknowledgements Authors acknowledge financial support by the EU (FP 5, 6 and 7) as well as Austrian and French National Science fund, the Tyrolean Science fund, the Austrian Academy of Sciences, the Österreichische Forschungsgemeinschaft and the help of numerous persons who have helped in keeping the eddy covariance flux measurements going over the years.

References

Allard V, Soussana JF, Falcimagne R (2007) The role of grazing management for the net biome productivity and greenhouse gas budget (CO_2, N_2O and CH_4) of semi-natural grassland. Agric Ecosyst Environ 121:47–58

Ammann C, Flechard CR, Leifeld J et al (2007) The carbon budget of newly established temperate grassland depends on management intensity. Agric Ecosyst Environ 121:5–20

Arrouays D, Jolivet CI, Boulonne L et al (2003) Le Réseau de Mesures de la Qualité des Sols (RMQS) de France. Étude et Gestion des Sols 10:241–250

Baldocchi DD, Falge E, Gu L et al (2001) FLUXNET: a new tool to study the temporal and spatial variability of ecosystem-scale carbon dioxide, water vapor and energy flux densities. Bull Am Meteorol Soc 82:2415–2435

Cernusca A, Bahn M, Berninger F et al (2008) Preface to CarboMont special feature: effects of land-use changes on sources, sinks and fluxes of carbon in European mountain grasslands. Ecosystems 11:1335–1337

Chapin FS III, Woodwell GM, Randerson JT et al. (2006) Reconciling carbon cycle concepts, terminology, and methods. Ecosystems 9:1041–1050

Conant RT, Paustian K, Elliott ET (2001) Grassland management and conversion into grassland: effects on soil carbon. Ecol Appl 11:343–355

Davison B, Brunner A, Ammann C, Spirig C (2008) Cut-induced VOC emissions from agricultural grasslands. Plant Biol 10:76–85

Delany AC, Fitzjarrald DR, Lenschow DH et al (1986) Direct measurement of nitrogen oxides and ozone fluxes over grassland. J Atmos Chem 4:429–444

Eugster W, Zeyer K, Zeeman M (2007) Methodical study of nitrous oxide eddy covariance measurements using quantum cascade laser spectrometry over a Swiss forest. Biogeosciences 4:927–939

Falge E, Baldocchi D, Olson R et al (2001) Gap filling strategies for defensible annual sums of net ecosystem exchange. Agric For Meteorol 107:43–69

Flanagan LB, Wever LA, Carlson PJ (2002) Seasonal and interannual variation in carbon dioxide exchange and carbon balance in a northern temperate grassland. Glob Change Biol 8:599–615

Flechard CR, Ambus P, Skiba U et al. (2007) Effects of climate and management intensity on nitrous oxide emissions in grassland systems across Europe. Agr Ecosyst Env 121:135–152

Flechard CR, Neftel A, Jocher M et al (2005) Bi-directional soil/atmosphere N_2O exchange over two mown grassland systems with contrasting management practices. Glob Change Biol 11:2114–2127

Gilmanov T, Soussana JF, Aires L et al (2007) Partitioning European grassland net ecosystem CO_2 exchange into gross primary productivity and ecosystem respiration using light response function analysis. Agric Ecosyst Environ 121:93–120

Gilmanov TG, Aires L, Barcza Z et al (2010) Productivity, respiration, and light-response parameters of world grassland and agroecosystems derived from flux-tower measurements. Rangel Ecol Manage 63:16–39

Hammerle A, Haslwanter A, Tappeiner U et al (2008) Leaf area controls on energy partitioning of a temperate mountain grassland. Biogeosciences 5:421–431

Hargreaves KJ, Fowler D, Pitcairn CER, Aurela, M (2001) Annual methane emission from Finnish mires estimated from eddy covariance campaign measurements. Theor Appl Climatol 70:203–213

Hendriks DMD, Dolman AJ, Van der Molen MK et al (2008) A compact and stable eddy covariance set-up for methane measurements using off-axis integrated cavity output spectroscopy. Atmos Chem Phys 8:1–13

Hsieh CI, Katul G, Chi TW (2000) An approximate analytical model for footprint estimation of scalar fluxes in thermally stratified atmospheric flows. Adv Water Resour 23:765–772

Hunt JE, Kelliher FM, McSeveny TM, Ross DJ et al (2004) Long-term carbon exchange in a sparse, seasonally dry tussock grassland. Glob Change Biol 10:1785–1800

Jonas T, Rixen C, Sturm M, Stoeckli V (2008) How alpine plant growth is linked to snow cover and climate variability. J Geophys Res 113:G03013. doi:10.1029/2007JG000680

Kaimal JC, Finnigan JJ (1994) Atmospheric boundary layer flows. Oxford University Press, Oxford, 289 pp

Kato T, Tang Y, Gu S, Cui X et al (2004) Carbon dioxide exchange between the atmosphere and an alpine meadow ecosystem on the Qinghai-Tibetan Plateau, China. Agric For Meteorol 124:121–134

Kim J, Verma SB (1990) Components of surface energy balance in a temperate grassland ecosystem. Bound Layer Meteorol 51:401–417

Kroon PS, Hensen A, Jonker HJJ et al (2007) Suitability of quantum cascade spectroscopy for CH_4 and N_2O eddy co-variance flux measurements. Biogeosciences 4:715–728

Kroon PS, Hensen A, Jonker HJJ et al (2009) Uncertainties in eddy covariance flux measurements assessed from CH_4 and N_2O Observation. Agric For Meteorol. doi:10.1016/j.agrformet.2009.08.008

Laville P, Jambert C, Cellier P (1999) Nitrous oxide fluxes from a fertilized maize crop using micrometeorological and chamber methods. Agric For Meteorol 96:19–38

Leahy P, Kiely G, Scanlon TM (2004) Managed grasslands: a greenhouse gas sink or source? Geophys Res Lett 31:L20507. doi:10.1029/2004GL021161

Lemaire G, Chapman D (1996) Tissue flows in grazed plant communities. In: Hodgson J, Illius AW (eds) The ecology and management of grazing systems. CABI, Wallingford

Li S-G, Asanuma J, Eugster W (2005) Net ecosystem carbon dioxide exchange over grazed steppe in Mongolia. Glob Change Biol 11:1941–1955

Marcolla B, Cescatti A (2005) Experimental analysis of flux footprint for varying stability conditions in an alpine meadow. Agric For Meteorol 135:291–301

Massman WJ (2000) A simple method for estimating frequency response corrections for eddy covariance systems. Agric For Meteorol 104:185–198

Meyers T (2001) A comparison of summertime water and CO_2 fluxes over rangeland for well-watered and drought conditions. Agric For Meteorol 104:185–198

Monteith JL, Unsworth MH (1990) Principles of environmental physics, Edward Arnold, London, 1990

Neftel A, Flechard C, Ammann C (2007) Experimental assessment of N_2O background fluxes in grassland systems. Tellus B 59:470–482

Nelson DD, Shorter JH, McManus JB (2002) Sub-part-per-billion detection of nitric oxide in air using a thermoelectrically cooled mid-infrared quantum cascade laser spectrometer. Appl Phys B 75:343–350

Novick KA, Stoy PC, Katul GG, Ellsworth DS, Siqueira MBS, Juang J, Oren R (2004) Carbon dioxide and water vapor exchange in a warm temperate grassland. Oecologia 138:259–274

Pinares-Patiño CS, Dhour P, Jouany JP (2007) Effects of stocking rate on methane and carbon dioxide emissions from grazing cattle. Agric Ecosyst Environ 121:30–46

Risch A, Frank DA (2005) Carbon dioxide fluxes in a spatially and temporally heterogeneous temperate grassland. Oecologia 147:291–302

Ruuskanen TM, Müller M, Schnitzhofer R, Karl T, Graus M, Bamberger I, Hörtnagl L, Brilli F, Wohlfahrt G, Hansel A (2011) Eddy covariance VOC emission and deposition fluxes above grassland using PTR-TOF. Atmos Chem Phys 11:611–625

Schmid HP (2002) Footprint modeling for vegetation atmosphere exchange studies: a review and perspective. Agric For Meteorol 113:159–183

Schmitt M, Bahn M, Wohlfahrt G, Tappeiner U, Cernusca A (2010) Land use affects the net ecosystem CO_2 exchange and its components in mountain grasslands. Biogeosciences 7:2297–2309

Schulze ED, Ciais P, Luyssaert S et al (2009) The greenhouse gas balance of Europe: methane and nitrous oxide compensate the carbon sink of EU-25. Nat Geosci 2:842–850

Siemens J, Janssens IA (2003) The European carbon budget: a gap. Science 302:1681

Smeets CJPP, Holzinger R, Vigano R et al (2009) Eddy covariance methane measurements at a Ponderosa pine plantation in California. Atmos Chem Phys 9:8365–8375

Smith KA, Clayton H, Arah JRM et al. (1994) Micrometeorological and chamber methods for measurement of nitrous oxide fluxes between soils and the atmosphere: Overview and conclusions. J Geophys Res 99: 541–16

Soussana JF, Allard V, Pilegaard K et al (2007) Full accounting of the greenhouse gas (CO_2, N_2O, CH_4) budget of nine European grassland sites. Agric Ecosyst Environ 121:121–134

Soussana JF, Tallec T, Blanfort V (2010) Mitigating the greenhouse gas balance of ruminant production systems through carbon sequestration in grasslands. Animal 4:334–350

Suyker AE, Verma SB (2001) Year-round observations of the net ecosystem exchange of carbon dioxide in a native tallgrass prairie. Glob Change Biol 7:279–289

Swinbank WC (1951) The measurement of vertical transfer of heat and water vapor by eddies in the lower atmosphere. J Meteorol 8:135–145

Van der Werf GR, Randerson JT, Giglio L et al (2006) Interannual variability in global biomass burning emissions from 1997 to 2004. Atmos Chem Phys 6:3423–3441

Van Oost K, Quine TA, Govers G et al (2007) The impact of agricultural soil erosion on the global carbon cycle. Science 318:626–629

Verma SB, Kim J, Clement RJ (1989) Carbon dioxide, water vapor and sensible heat fluxes over a tallgrass prairie. Bound Layer Meteorol 46:53–67

Vickers D, Mahrt L (1997) Quality control and flux sampling problems for tower and aircraft data. J Atmos Ocean Technol 14:512–526

Werle P, Kormann R (2001) Fast chemical sensor for eddy correlation measurements of methane emissions from rice paddy fields. Appl Opt 40:846–858

Wienhold FG, Klemedtsson L, Galle B (1994) Micrometeorological and chamber methods for measurement of nitrous oxide fluxes between soils and the atmosphere: overview and conclusions. J Geophys Res 99:541–548

Wohlfahrt G, Sapinsky S, Tappeiner U, Cernusca A (2001) Estimation of plant area index of grasslands from measurements of canopy radiation profiles. Agric For Meteorol 109:1–12

Wohlfahrt G, Anfang Ch, Bahn M et al (2005) Quantifying nighttime ecosystem respiration of a meadow using eddy covariance, chambers and modelling. Agric For Meteorol 128:141–162

Wohlfahrt G, Anderson-Dunn M, Bahn M et al (2008a) Biotic, abiotic and management controls on the net ecosystem CO_2 exchange of European mountain grasslands. Ecosystems 11:1338–1351

Wohlfahrt G, Hammerle A, Haslwanter A et al (2008b) Seasonal and inter-annual variability of the net ecosystem CO_2 exchange of a temperate mountain grassland: effects of weather and management. J Geophys Res 113:D08110. doi:10.1029/2007JD009286

Wohlfahrt G, Pilloni S, Hörtnagl L, Hammerle A (2010) Estimating carbon dioxide fluxes from temperate mountain grasslands using broad-band vegetation indices. Biogeosciences 7: 683–694

Zeller K, Massman W, Stocker D et al (1989) Initial results from the Pawnee eddy correlation system for dry acid deposition research, USDA Forest Service Research Paper RM-282. U.S. Dept. of Agriculture, Forest Service, Rocky Mountain Forest and Range Experiment Station, Fort Collins

Chapter 14
Eddy Covariance Measurements over Wetlands

Tuomas Laurila, Mika Aurela, and Juha-Pekka Tuovinen

14.1 Introduction

Wetland ecosystems can be classified according to various systems, one of which defines three major groups: (1) northern peatlands (with a total area of 350×10^6 ha), (2) freshwater swamps and marshes (204×10^6 ha), and (3) coastal wetlands (36×10^6 ha) (Mitsch et al. 2009). Depending on the definition, wetlands cover 3–6% of the Earth's land surface. This chapter concentrates on northern peatlands, which constitute a highly important component of the global biogeochemical cycling, as these boreal and arctic mires have accumulated about one-third of the global organic soil carbon (Gorham 1991). Turunen et al. (2002) estimated the size of this carbon pool as 270–370 Tg C, while tropical peatlands, which are also a very important source of greenhouse gases (GHGs) to the atmosphere, are estimated to store about 50 Tg C in peat (Hooijer et al. 2006).

Peatlands in the boreal zone may originate from different processes, but the main prerequisite for their development is a surplus of water. Mires start to grow in lowlands where draining is poor, and usually precipitation exceeds evaporation (Kuhry and Turunen 2006). The poorly aerated conditions due to the high water table result in a slow decay of plant litter. The accumulation of carbon results mainly from the inhibited decomposition in anoxic conditions, rather than high photosynthetic uptake rates. On the other hand, the reduction reactions prevailing in these conditions lead to the microbially mediated production of CH_4 (Limpens et al. 2008).

Mires affect the radiative forcing of the atmosphere in two opposite ways: (1) cooling induced by the uptake of CO_2 from the atmosphere, acting on a timescale of millennia, and (2) warming induced by CH_4 emissions on a timescale of decades

T. Laurila (✉) • M. Aurela • J.-P. Tuovinen
Finnish Meteorological Institute, P.O. Box 503, FI-00101 Helsinki, Finland
e-mail: tuomas.laurila@fmi.fi; mika.aurela@fmi.fi; juha-pekka.tuovinen@fmi.fi

M. Aubinet et al. (eds.), *Eddy Covariance: A Practical Guide to Measurement and Data Analysis*, Springer Atmospheric Sciences, DOI 10.1007/978-94-007-2351-1_14,

(Frolking et al. 2006). The stability of these large organic carbon masses largely depends on the hydrological conditions, which may alter as result of climate change (Griffis et al. 2000; Lafleur et al. 2003), and which also in turn depend upon the scale of human intervention, such as the draining of mires for agriculture and forestry.

The GHG exchange of mires has interested the scientific community for a long time, and different approaches have been used to measure the exchange rates. The long-term apparent accumulation of carbon in peat during the Holocene or over a shorter period reflects the CO_2 exchange and its net uptake (Clymo et al. 1998; Turunen et al. 2002; Schulze et al. 2002). Chamber measurements have provided important information on the GHG exchange on the scale of plant communities and on the environmental responses of these fluxes (Moore and Knowles 1990; Alm et al. 1999; Riutta et al. 2007). Utilization of the micrometeorological eddy covariance (EC) method has been a great step forward for the understanding of the present-day GHG exchange of wetlands, because EC measurements provide continuous nonintrusive observations on an ecosystem scale. It is nowadays feasible to make year-round EC measurements and thus obtain direct observations of the current annual GHG balances.

This chapter first provides a brief history of EC-based GHG studies and summarizes the current network of wetland measurement sites. As throughout the chapter, the emphasis of this site survey is placed on the natural northern mires, but examples are also shown of wetlands located in lower latitudes. After that, some special features related to the application of the EC technique in mire ecosystems are highlighted. This is followed by an outline of the ancillary and complementary measurements that would support the interpretation of EC flux data and a discussion of the additional challenges of carrying out flux measurements in the winter conditions encountered in northern latitudes. Finally, both the determination of the total carbon balance of an ecosystem and its related climate impacts are discussed as examples of the application of EC-based flux data.

14.2 Historic Overview

For the measurement of CO_2 fluxes using the EC technique, fast-response nondispersive infrared (NDIR) sensors have been available for about three decades. In present-day EC systems, the CO_2 fluxes are measured using perfected open- or closed-path instruments (Sect. 2.4). Compared to CO_2, the measurement of CH_4 is technically more difficult, due to its lower atmospheric concentration and less favorable absorption spectrum. Thus, the development of user-friendly analyzers for the EC field measurements of CH_4 fluxes has been slower, and has involved more diverse (laser absorption) techniques. The first of these instruments was based on the Zeeman-split HeNe laser (Fan et al. 1992), but the tunable diode laser (TDL) enabled a better selection of the absorption peak (Verma et al. 1992; Zahniser et al. 1995). While lead-salt TDL spectrometers were already commercially available in the 1990s, they suffered from instability and were laborious to maintain, as they require cooling by liquid nitrogen. The quantum cascade laser (QCL) spectrometers offered an alternative that was more stable and accurate (Faist et al. 1994; Kroon et al. 2007). More recently, with the introduction of the cavity ring-down method, the sensitivity and stability of the analyzers has improved markedly. Instruments based on QCL or narrow-band industrial lasers have also proved feasible for CH_4 detection at room temperature, making the maintenance of field measurement sites much more convenient (Hendriks et al. 2008).

The first micrometeorologists to study the ecosystem–atmosphere exchange of GHGs were originally already attracted by the remote wetlands that guaranteed the acquisition of original data. The first micrometeorological CO_2 flux measurements were conducted over wet meadow tundra in Alaska in 1971 by means of the gradient method (Coyne and Kelly 1975). The first wetland measurements of CO_2 and CH_4 fluxes with the EC method were made in 1988 at a tundra site in Alaska by Fan et al. (1992), who used an NDIR instrument for CO_2 and both a HeNe laser spectrometer and a total hydrocarbon detector for CH_4 concentrations. The first EC measurements of CO_2 above a raised bog were made in the Hudson Bay lowlands in July 1990 by Neumann et al. (1994). Within the same study, Edwards et al. (1994) measured CH_4 fluxes using a TDL-based instrument. On a more southern bog in Minnesota, Verma et al. (1992) demonstrated the applicability of their newly developed TDL sensor for the EC measurements of CH_4 fluxes. Multiyear measurements at the same site showed that an ecosystem can act either as a CO_2 sink or a source, depending on the meteorological and hydrological conditions during the growing season (Shurpali et al. 1995).

In Europe, EC measurements on mires started later than in North America. The first measurements of CO_2 fluxes were probably carried out on a disturbed bog in the Netherlands in 1994–1995 using commercial instruments (Nieveen et al. 1998). The first European measurements on a pristine wetland were conducted on a subarctic fen in 1995. The CO_2 fluxes from this measurement campaign were reported by Aurela et al. (1998) and the CH_4 fluxes by Hargreaves et al. (2001). Measurements at this site, Kaamanen, located in northern Finland, have continued since 1997. Another long-term (since 1998) EC-based time series of CO_2 fluxes has been collected on an ombrotrophic bog near Ottawa, Canada (Lafleur et al. 2001). These sites demonstrate the feasibility of the EC technique for continuous multiyear measurements in wetland ecosystems, providing data that have proved most useful for studying the environmental responses, also enabling consideration of climate change effects on the GHG exchange (Lafleur et al. 2003; Aurela et al. 2004).

During recent years, a growing number of EC measurements have been commenced on different wetland ecosystems. The first measurements within the vast wetland areas of northern Russia were conducted in 2000 by Arneth et al. (2002). The importance of these areas, especially in relation to the possible melting of permafrost soil, has subsequently served as a motivation for several EC-based studies (Corradi et al. 2005; Kutzbach et al. 2007; van der Molen et al. 2007; Sachs et al. 2008; Laurila et al. 2010). Figure 14.1 and Table 14.1 present a survey of wetland sites at which EC measurements have been conducted. The majority

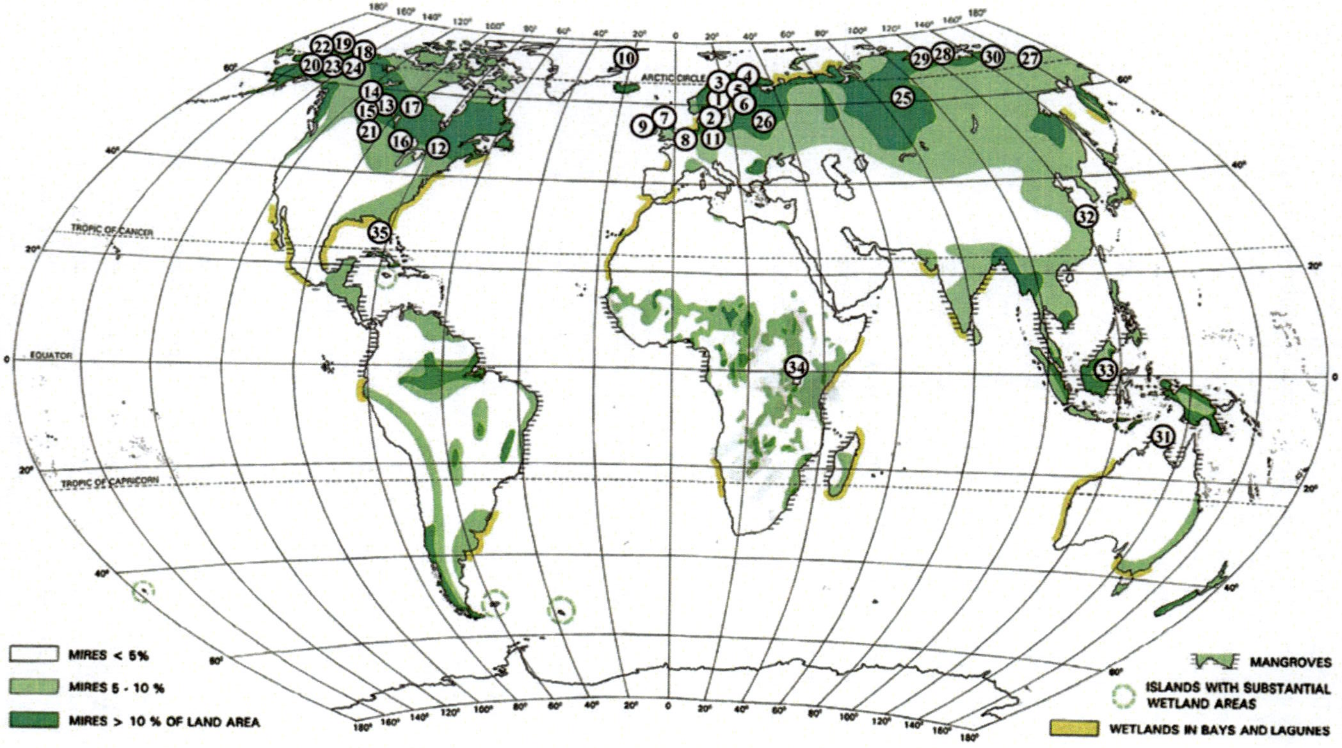

Fig. 14.1 Global distribution of mires (Lappalainen 1996) together with the locations of present and earlier eddy covariance measurement sites (*numbered circles*) in the world's various wetland ecosystems. The numbers refer to Table 14.1, in which additional information is presented for each site

Table 14.1 Eddy covariance measurement sites in various wetland ecosystems

ID	Site name	Coordinates	Seasonal coverage	Wetland type	Measured GHGs	Gas analyzer	Sonic anemometer	Reference
Europe								
1	Degerö	64°11′N, 19°33′E	A	Fen	CO_2	LI6262	Gill R2	Sagerfors et al. (2008)
2	Fäjemyr	56°15′N, 13°33′E	A	Bog	CO_2	LI6262	Gill R3	Lund et al. (2007)
3	Stordalen	68°20′N, 19°03′E	A	Fen	CO_2, CH_4	Aerodyne TDL, LI7500	Gill R3	Jackowicz-Korczyński et al. (2010)
4	Kaamanen	69°08′N, 27°17′E	A	Fen	CO_2, CH_4	LI7000, Aerodyne TDL, LGR RMT-200	ATI SWS-211, METEK USA-1	Aurela et al. (2004), Hargreaves et al. (2001)
5	Siikaneva	61°50′N, 24°11′E	A	Fen	CO_2, CH_4	LI7000, Campbell TGA100, LGR RMT-200	METEK USA-1	Rinne et al. (2007), Aurela et al. (2007)
6	Lompolojänkkä	67°60′N, 24°13′E	A	Fen	CO_2, CH_4	LI7000, LGR RMT-200	METEK USA-1	Aurela et al. (2009)
7	Auchencorth Moss	55°48′N, 3°15′W	A	Bog	CO_2, CH_4	LI7000, LGR RMT-200	Gill R2	Dinsmore et al. (2010)
8	Fochtelooer area	53°01′N, 6°24′E	A	Bog	CO_2	LI6262, LI7000	Gill R2	Nieveen et al. (1998)
9	Ireland	51°55′N, 9°55′W	A	Bog	CO_2	LI7500	Campbell 81000	Sottocornola and Kiely (2005)
10	Zackenberg	74°28′N, 20°34′W	GS	Tundra	CO_2	LI6262	Gill R2	Soegaard and Nordstroem (1999)
11	Rzecin	52°46′N, 16°31′E	A	Fen	CO_2	LI7500	Gill R3	Chojnicki et al. (2007)

(continued)

Table 14.1 (continued)

ID	Site name	Coordinates	Seasonal coverage	Wetland type	Measured GHGs	Gas analyzer	Sonic anemometer	Reference
North America								
12	Mer Bleue	45°25′N, 75°31′W	A	Bog	CO_2	LI6262, LI7000	Gill R2	Lafleur et al. (2003)
13	Canada-WP1	54°57′N, 112°28′W	GS	Treed fen	CO_2	LI7000	Gill R3	Syed et al. (2006)
14	Canada-WP2	55°32′N, 112°20′W	GS	Fen	CO_2	LI7500	Campbell CSAT3	Glenn et al. (2006)
15	Canada-WP3	54°28′N, 113°19′W	GS	Fen	CO_2	LI7500	Campbell CSAT3	Glenn et al. (2006)
16	Minnesota	47°32′N, 93°28′E	GS	Bog	CO_2, CH_4	LI6251, Unisearch Associates TDL	Kaijo-Denki, DA-600	Shurpali et al. (1995), Verma et al. (1992)
17	Manitoba	55°54′N, 98°24′W	GS	Fen	CO_2	LI6252	ATI SWS-211	Joiner et al. (1999)
18	Atqasuk	70°28′N, 157°25′W	A	Tundra	CO_2	LI7500	Gill R3	Kwon et al. (2006)
19	Barrow[a]	71°19′N, 156°38W	A	Tundra	CO_2	LI7500, Advanet E009a	Gill R3, Kaijo-Denki DA-600	Harazono et al. (2003), Kwon et al. (2006)
20	Ivotuk	68°29′N, 155°45′W	A	Tundra	CO_2	LI7500	Gill R3	Walter Oechel and Olaf Vellinga, personal communication (2011)
21	Saskatchewan	53°57′N, 105°57′E	GS	Fen	CO_2	LI6262, LI7000	ATI Sx	Suyker et al. (1997)
22	Happy Valley	69°09′N, 148°51′W	GS	Tundra	CO_2	LI6262	ATI SWS-211	Vourlitis and Oechel (1999)
23	U-Pad	70°17′N, 148°53′W	GS	Tundra	CO_2	LI6262	ATI SWS-211	Vourlitis and Oechel (1997)
24	24-Mile	69°56′N, 148°48′W	GS	Tundra	CO_2	NOAA/ATDD open-path	ATI SWS-211	Vourlitis and Oechel (1997)

Russia								
25	Zotino	60°45′N, 89°23′E	GS	Bog	CO_2	LI6262	Gill R3	Arneth et al. (2002)
26	Fyodorowskoye	57°27′N, 32°55′E	GS	Bog	CO_2	LI6262	Gill R3	Arneth et al. (2002)
27	NE Siberia	68°37′N, 161°20′E	GS	Tundra	CO_2	LI6262	Gill R3	Corradi et al. (2005)
28	Lena River Delta	72°22′N, 126°30′E	GS	Tundra	CO_2, CH_4	LI7000, Campbell TGA100	Gill R3	Kutzbach et al. (2007), Sachs et al. (2008)
29	Tiksi	72°22′N, 126°30′E	A	Tundra	CO_2, CH_4	LI7000, LGR RMT-200	METEK USA-1	Laurila et al. (2010)
30	Kytlyk	70°50′N, 147°30′E	GS	Tundra	CO_2	LI7500	Gill R3	van der Molen et al. (2007)
Tropical wetlands								
31	Fogg Dam	12°54′S, 131°31′E	A	Floodplain	CO_2	LI7500	Campbell CSAT3	Jason Beringer, personal communication (2011)
32	Dongtan	31°31′N, 121°58′E	A	Estuarine	CO_2	LI7500	Campbell CSAT3	Yan et al. (2008)
33	Indonesia	2°20′S, 114°2′E	A	Swamp	CO_2	LI7500	Campbell CSAT3	Hirano et al. (2007)
34	Jinja	0°24′N, 33°11′E	A	Papyrus	CO_2	LI7500	Campbell CSAT3	Saunders et al. (2007)
35	Everglades	25°22′N, 81°04′W	A	Mangrove	CO_2	LI7500	Gill RS50	Barr et al. (2010)

The ID number of the site refers to the symbols on the wetland map (Fig. 14.1). The annual coverage of measurements is presented as two categories: (A) annual cycle and (GS) growing season.

[a]At Barrow there have been several eddy covariance systems for CO_2 fluxes on different tundra ecosystems

of these sites are located on boreal and arctic peatlands, which are the focus of this chapter, but examples are also shown of the measurements in tropical wetland ecosystems.

14.3 Ecosystem-Specific Considerations

Many of the special features identified in connection with EC measurements over grasslands (Chap. 15) apply to open mires as well. The microtopography of mires is typically not as even as that of grasslands, but a relatively low (3–5 m) measurement height is normally sufficient, especially as the typical mire vegetation is rather short, consisting of mosses, shrubs, grasses, and sedges. On the one hand, this reduces the importance of the storage flux and the inherent uncertainties related to its determination, but on the other hand increases the importance of the corrections required for the imperfect frequency response of the measurement system (Sects. 1.5.4 and 4.1.3). Since mires are usually located in a flat landscape, advection problems generated by a sloping terrain are diminished. However, in small mires the turbulent flow field may be influenced by land cover types surrounding the area of interest; this disturbance should be evaluated with footprint models, as described in Chap. 8. The low measurement height also significantly simplifies the design of the measurement tower/mast (Sect. 2.2). In the case of forested wetlands, such as the treed fens and tropical swamps that fall outside the scope of this chapter, many of the EC specifics are common with those of forests in general (Chap. 11).

While the design and operation of measurement sites are discussed in great detail in Chap. 2, there are certain questions specific to wetlands that require further attention. A large proportion of wetlands are located in remote areas, which introduces additional requirements for logistics and site infrastructure. Mires often form extensive complexes, within which the most attractive sites, from the micrometeorological point of view, are difficult to access by car. It is advantageous, if a strip of land having mineral soil extends close to the measurement site. The installation of the measurement system and access to the site are complicated by the fact that peat soil is not firm and occasionally may become inundated. To ensure the stability of the measurement mast, it should be erected on a steady foundation, which can be constructed as a platform with supporting poles extending deep into the peat or, if possible, down to the mineral soil. Similarly, boardwalks are typically necessary for accessing the site and minimizing perturbations to the ecosystem during the maintenance. As mire vegetation is fragile, walking should be restricted to the boardwalks only. This also prevents disturbances by ebullition. An example of a measurement site established on wetland is shown in Fig. 14.2.

One of the main practical issues is the availability of mains power. Mains power is the ideal solution when CO_2 fluxes are measured with a closed-path analyzer in cold conditions, in which case the heating of the inlet tubes and the instrument cabin are necessary for most of the year. Mains power is usually also needed for the pump of a closed-path CO_2 analyzer, and practically always for the present closed-path CH_4

Fig. 14.2 Flux measurement site at the Lompolojänkkä fen located in northern Finland (67°59.832′N, 24°12.551′E). The micrometeorological mast is erected on a supported platform. Radiation sensors are mounted on a separate mast. The horizontal support carries a heated inlet tube from the sonic anemometer to the analyzer

analyzers, which require more powerful pumps. Nowadays, open-path instruments with lower power requirements are also available for both CO_2 and CH_4. However, solar radiation may not be sufficient throughout the year, which compromises the seasonal coverage of the measurements. In addition, in practice it is necessary to use a sonic anemometer that is equipped with sensor-head heating, in order to obtain year-round gap-free data in cold conditions. A generator-powered system requires constant maintenance, which is often difficult to accomplish at remote sites. If a generator is used, the effect of its exhaust gases on the trace gas measurements must be minimized.

Some of the specific solutions suggested above have implications for the quality assurance of the measurements. The necessary constructions at the site, especially the instrument shelter or cabin, potentially disturb the flux and other measurements, which should be taken into account when situating the sensors (Fig. 14.2). Wind-direction-based screening of the flux data may be necessary to minimize such disturbances. A similar screening procedure may also be needed for fetch considerations, if the measurement mast is located close to the edge of the mire, for example, to optimize the flux footprint in certain flow directions, or for the logistic reasons mentioned above (e.g., Aurela et al. 1998). The same applies to the avoidance of generator exhaust.

The northern location of many mires (Fig. 14.1) implies specific winter-related measurement questions, including the detection of small GHG fluxes, which are discussed below in a separate section. However, as a summer-specific final note on practicalities, it should be emphasized that additional filtering of the inlet flow is crucial in order to avoid a rapid blocking of the standard inlet filters due to an abundance of small insects, such as black flies, in wetland areas.

14.4 Complementary Measurements

Net radiation, global and reflected solar radiation, and global and reflected photosynthetically active photon flux density (PPFD) are the basic radiation components that should be measured as ancillary data for the EC fluxes. At boreal and arctic sites, the albedo decreases rapidly as the snow melts and the dark mire surface is exposed. With the appearance of new plants, the albedo gradually increases, but the reflected PPFD remains low. The ratios between the outgoing and incoming shortwave radiation, as well as the corresponding PPFD terms, can be used to trace the emergence and senescence of vegetation defining the growing season (Huemmrich et al. 1999).

In biological terms, boreal and arctic mires are characterized by a variation of habitats having differing plant communities, often corresponding to hydrological conditions that vary with microtopography and result in a mosaic of wet hollows, drier hummocks, and intermediate zones. Ancillary measurements should represent these different surfaces. Thus, the environmental measurements complementing the EC data should include more than one soil temperature profile. In northern regions, soil temperatures remain close to freezing point for extended periods. The soil temperature measurement system, including the data logger as well as the sensors, should be able to resolve variations of 0.1 K, and physically should tolerate inundated and icy conditions.

Continuous measurement of the water table depth (WTD) by a submerged pressure sensor is usually needed at one point at least. The reference level for WTD is usually at the peat surface, but it should be noted that the level of this surface may itself vary with a fluctuating WTD. Thus, it is useful if an additional WTD sensor can be anchored to the mineral soil to measure the absolute WTD changes. Above the WTD, the peat moisture content can be measured by time-domain reflectometer (TDR) probes. The TDR probes also serve as sensors for detecting the freezing of the soil, because dielectricity drops strongly when soil freezes, while it is difficult to distinguish the phase transitions from temperature measurements alone. Snow-depth measurements should also be made, employing continuously registering instruments.

Soil heat flux is normally measured using heat flux plates, which are basically designed for mineral soils. A poor or varying contact between the plate and the surface, as well as variations in the soil moisture, can cause erroneous signals. In mires, the WTD varies, and the heat flux sensor may be above or below it, but in either case the conditions are difficult for a heat flux plate. Below the WTD, the heat transport due to water flow may totally contaminate the measurement, while in the other case problems are caused by the large variations in the peat moisture as well as the limited contact between porous peat and the plate. To some degree, these problems may be overcome using so-called self-calibrating heat flux plates (Ochsner et al. 2006). However, a more reliable way of estimating the soil heat flux would be to employ high-resolution temperature profiles and to calculate the heat flux from temperature variations.

In addition to the physical parameters discussed above, measurements of biochemical variables would provide useful information for both the characterization of the soil and for understanding the soil processes. For example, ombrotrophic bogs are characterized by acidic soil water with low pH (3.0–4.5), while minerotrophic fens usually have a higher pH (4.5–8.0); oxic acrotelm and anoxic catotelm show opposite redox potential characteristics; O_2 concentration and redox potential vary seasonally responding to possible ice and snow cover in winter and flooding in spring; heavy rain events cause variation in acidity and nutrient status. For detecting variations of this kind, sensor packages are currently available for continuous measurements of O_2, redox, soil water pH, temperature, and pressure; these sensors can be permanently installed in the peat. These instruments provide exciting information on the temporal variation of biogeochemical processes, complementing the data on ecosystem–atmosphere exchange.

The atmospheric fluxes of CO_2 and CH_4 vary markedly between the microrelief elements described above. CO_2 exchange is most intense at the hummocks, while CH_4 emissions are highest from the wet surfaces. For this reason, it is laborious to obtain spatially representative flux data on the scale of a mire ecosystem by employing the chamber measurement technique. In contrast, the EC method provides a weighted measurement of the surface exchange within the observation footprint or the "field of view" of the flux sensor (Chap. 8), thus averaging over the small-scale heterogeneity related to habitat variability. However, for understanding the functioning of the ecosystem, it may be necessary to determine the response of a plant community to environmental variables, such as temperature, WTD, and VPD, separately for each microrelief. While a footprint analysis provides additional information for the interpretation of the EC measurements (e.g., Aurela et al. 2009; Laine et al. 2006), this small-scale variation cannot be fully extracted from the spatially averaged EC flux data, but entails the use of additional chamber measurements and ancillary meteorological measurements representing the individual surface types.

Measurements based on the chamber technique have previously been considered as an alternative to the micrometeorological flux measurements. However, the complementary nature of the EC and chamber measurements over low-vegetation mires is evident from those studies in which both these techniques have been employed. For example, the measurements on a homogeneous blanket bog (Laine et al. 2006), an even boreal fen (Riutta et al. 2007) and a subarctic fen with a pronounced microtopography (Maanavilja et al. 2011) all showed similar results. All these sites, while being highly variable in terms of plant communities, can be considered rather homogeneous at the ecosystem scale resolved with the EC method. However, the chamber measurements indicated that the CO_2 exchange varied markedly (twofold to sixfold) between the different plant communities of these mires, and that the variation was similar for respiration and photosynthesis.

In addition to revealing the differences between the microsites, the chamber data facilitate the partitioning of the measured net ecosystem exchange (NEE) into its photosynthesis and respiration terms. For the EC measurements, this partitioning can be calculated using different approaches described in Sect. 9.3. However, chambers enable a more direct partitioning based on the simultaneously measured photosynthesis and respiration fluxes, although introducing a potential representativeness problem with respect to EC data. Thus, it is recommended to integrate the two partitioning approaches.

The studies discussed above illustrate the role of chamber measurements as a complementary source of information supporting the EC flux measurements. However, it must be kept in mind that the gas exchange of porous peat measured in an enclosure may differ greatly from that taking place under the influence of turbulent atmospheric flow, which calls for nonintrusive flux measurement methods such as EC (Sachs et al. 2008). It should also be noted that a vegetation inventory and continuous leaf area measurements are essential, even if no chamber measurements are performed. A special feature of mires is the abundance of mosses, which in such an inventory should be characterized by their coverage and green biomass.

14.5 EC Measurements in the Wintertime

Mires are predominantly to be found in the northern high latitudes (Fig. 14.1). In these boreal and arctic areas, the winter season is relatively long, and snow covers the ecosystems for a significant part of the year. During the wintertime, GHG fluxes are typically small as compared to the growing season. The soil temperature under the insulating snow cover remains close to zero irrespective of the possible harshness of the air temperature. In several studies it has been found that the microbial activity in the soil continues even in subfreezing temperatures (Flanagan and Bunnell 1980; Coxon and Parkinson 1987; Zimov et al. 1993), keeping the formation of CO_2 and CH_4 going throughout the winter. Several long-term studies have shown that these wintertime emissions contribute significantly to the annual CO_2 balances. For example, in a southern boreal bog in Canada, the wintertime CO_2 flux was estimated to represent 25–35% of the annual CO_2 balance (Lafleur et al. 2003), while in a subarctic fen in northern Finland the wintertime efflux is actually greater than the annual uptake (Aurela et al. 2002). For CH_4, the wintertime fluxes are not as important for the annual sum. At a southern boreal fen in Finland, the wintertime CH_4 emission was about 5% of the annual total efflux (Rinne et al. 2007). An example of the seasonal variation of CO_2 and CH_4 fluxes observed at a subarctic fen is shown in Fig. 14.3.

The wintertime decrease in the GHG formation, and consequently in the magnitude of the fluxes, results in an increased noise in the data measured with the EC technique. This is typically manifested by negative values in the flux time

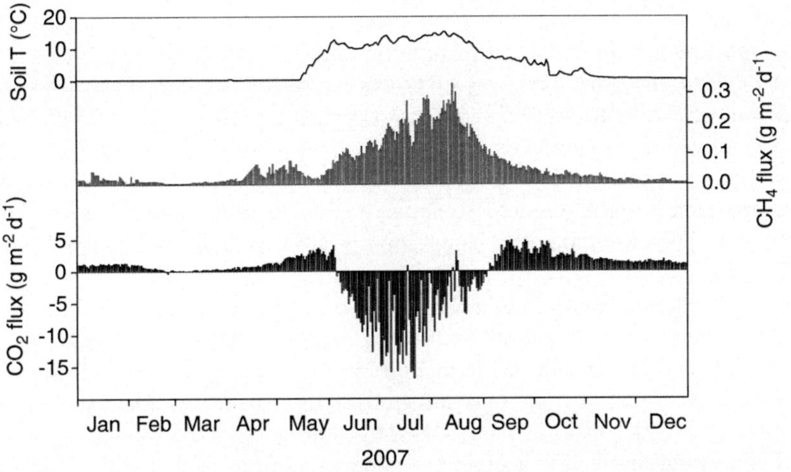

Fig. 14.3 Mean daily exchange of CO_2 (*black bars*) and CH_4 (*gray bars*) together with the soil temperature (at −10 cm) (*line*) at the Lompolojänkkä fen in northern Finland (67°59.832′N, 24°12.551′E)

series, corresponding to apparent uptake in conditions in which no uptake can be expected. As this variation can be considered to be random noise, averaging these data over longer periods produces unbiased estimates. Therefore, it is important to pay attention to the data screening protocols, in order to avoid systematic errors due to data selection. For example, discarding all the negative values as nonphysical would definitely bias the estimated averages and, consequently, the long-term GHG balances.

The decrease of the signal-to-noise ratio also has an indirect influence, especially on the measurement of GHG fluxes with a closed-path instrument, for which the lag time related to the air travel in the tubes must be taken into account in the flux calculations (Sect. 3.2.3.2). When the fluxes are sufficiently high and the turbulent flow is well-defined, the lag-covariance relationship shows a clear (absolute) maximum. As the fluxes decrease, however, this relationship becomes noisier, with many local maxima within the lag search window. If this variation is even partly due to random noise, a systematic error is induced into the long-term mean flux, if the highest value is invariably selected. Thus, the lag algorithm should be chosen with care when the fluxes to be measured are small, such as the GHG emissions from northern mires in winter. In this case, it is recommended to define a rather narrow search window, and apply a predefined constant lag value, if a window limit is exceeded.

In addition to diminished fluxes, the snow cover also brings about other effects that complicate the interpretation of the data measured in winter. As the snowpack acts as a buffer, the observed GHG fluxes are partly decoupled from the actual formation of these gases related to the microbial activity taking place in the different peat layers. This has an influence on the parameterization of the relationships between fluxes and their drivers, needed for the partitioning of the fluxes into the productivity and respiration components, for example (Sect. 9.3). During the summertime, the observed fluxes can be related to concurrent soil or air temperatures, while during the snow-cover season air temperature variations have a lesser influence on the decomposition processes, and thus a weaker correlation with the observed fluxes. However, as the thermal conditions beneath the snow are rather constant, it is still possible to capture the longer-term variations with a simple temperature-response model by using appropriate soil temperatures (Aurela et al. 2002).

Another phenomenon complicating the winter fluxes is their dependence on the atmospheric flow, appearing in practice as a correlation with wind speed or friction velocity. Events where the measured CO_2 efflux increases with increasing wind speed are often observed over snow-covered ecosystems (Goulden et al. 1996; Aurela et al. 2002). This is usually caused by ventilation of the snowpack, in which GHG concentrations tend to accumulate under conditions of slow diffusion into the atmosphere (i.e., with low wind speeds) (Massman et al. 1997). In principle, it would be possible to include this phenomenon in the CO_2 flux parameterizations, but the dependence is often nonlinear, even in the short term. While ventilation increases the fluxes, it decreases the storage, and thus the effect gets weaker in time. Usually, this flow-dependency is not taken into account in gap-filling models, as its effect becomes negligible over longer averaging periods.

14.6 Carbon Balances and Climate Effects

In terms of their GHG balances, wetlands differ from many other ecosystems in that CH_4 emissions play a central role, while N_2O fluxes are typically insignificant. As CH_4 analyzers have become more affordable and easier to deploy in the field, an increasing number of flux stations have started measuring CH_4 alongside with CO_2 fluxes. Figure 14.3 illustrates a typical annual cycle of CO_2 and CH_4 exchanges observed at a northern fen, with the characteristic seasonal dynamics that collectively define the annual balances. As discussed above, in winter the fluxes are small but clearly non-negligible. The increased soil temperature after the snowmelt enhances both the CO_2 and CH_4 effluxes. During the melting process, CH_4 fluxes often show an additional enhancement that cannot be directly related to soil temperature variations or CH_4 formation processes. This so-called spring pulse is caused by the release of a CH_4 reservoir that has accumulated beneath the frozen peat layer during winter (Hargreaves et al. 2001). The growing-season fluxes of CO_2 are co-controlled by the simultaneous photosynthetic uptake and respiration processes. The seasonal cycle of CH_4 appears simpler, being controlled mainly by soil temperature and vegetation phenology. It is thus much easier to fit a simple site-specific temperature-response function to the CH_4 fluxes; this function can then be used for gap-filling the measurement time series. CH_4 fluxes can even be parameterized as daily averages, something that is not possible for CO_2 fluxes.

In addition to the net exchanges of CO_2 and CH_4 that can be measured with the EC technique, there are also other components in the net carbon balance of a mire ecosystem: the export and import by lateral water flow of total organic carbon (TOC), dissolved inorganic carbon (DIC) and CH_4, as well as the import of carbon through precipitation. In practice, not all of these components are significant, but some of them may play an important role in the total balance. For example, at a boreal oligotrophic minerogenic mire in Sweden, the largest term of the balance was carbon gain by net CO_2 uptake (48 g C m^{-2} in 2005), followed by CH_4 emission (14 g C m^{-2}) and the TOC export by a stream (12 g C m^{-2}) (Nilsson et al. 2008). On the other hand, the stream export of DIC and CH_4 (3.1 and 0.1 g C m^{-2}, respectively) and TOC deposition (1.4 g C m^{-2}) were of lesser importance. These components add up to a total annual carbon accumulation of 20 g C m^{-2}, which is 42% of the NEE of CO_2.

It is interesting to compare an estimate of the current carbon balance with the carbon accumulation in the peat profile over a longer time horizon, which is possible if either a certain peat layer (using, for example, the initiation of tree growth; Schulze et al. 2002) or the peat bottom is dated. The latter provides the long-term apparent carbon accumulation (LARCA) over the life span of the mire during the Holocene. The current carbon accumulation rate can be estimated from the LARCA based on an accumulation model (e.g., Clymo et al. 1998), which for the Swedish mire discussed above resulted in an accumulation rate that is very close to the current balance derived from the EC and other measurements (Nilsson et al. 2008).

The climate effects of GHG fluxes are profoundly altered, if the natural ecosystems are subjected to an intervention by human management (e.g., Lohila et al. 2010). For example, the draining of a natural mire for agriculture or forestry typically stops CH_4 emissions, while for CO_2 the peat soil may turn from a sink into a source. While a knowledge of both the CO_2 and CH_4 fluxes is essential for the carbon balance and climate effects to be estimated, the measurement of the third major GHG, that is, N_2O, is usually not needed, because cool, inundated conditions do not favor its production (Regina et al. 1996). However, N_2O emissions are usually high, if the peat soil has been prepared for agriculture (Augustin et al. 1998). This effect may last a very long period. For example, a bog that had been afforested three decades earlier (Lohila et al. 2007), after being first drained for agriculture, still emitted large amounts of N_2O (Mäkiranta et al. 2007). As practically all the histosol sites in Europe, excluding the outskirts of the continent, have experienced human influence during the past centuries, the management history of a potential flux measurement site should be explored to identify measurement needs, and to facilitate the subsequent interpretation of the results obtained. The importance of the non-CO_2 GHGs is highlighted by the fact that, compared to CO_2, they have much higher radiative forcing efficiencies per unit emission, and also different atmospheric lifetimes, complicating the analysis of climate effects of these ecosystems.

14.7 Concluding Remarks

This chapter has concentrated on natural mires in boreal and arctic environments. However, it is important to briefly note the significance of wetlands located in lower latitudes, where high fluxes with sometimes unexpected variations may be observed, as well as the part played by peatlands that have experienced management or are influenced by climate change effects. For example, 3 years of EC data from a drained tropical swamp ecosystem showed very high respiration (3,870 g C m^{-2} $year^{-1}$), resulting in a significant net loss of CO_2 to the atmosphere (430 g C m^{-2} $year^{-1}$) (Hirano et al. 2007). A mangrove forest in Florida was a large sink of CO_2 (1,170 g C m^{-2} $year^{-1}$) due to the low respiration in comparison with the high photosynthetic uptake (2,340 g C m^{-2} $year^{-1}$) (Barr et al. 2009). At these sites, the potential technical problems, such as those related to the impact of high humidity and temperature, thunderstorms and hurricanes, certainly differ from those discussed in this chapter. Thus, these measurements should be considered outstanding accomplishments, further demonstrating the applicability of the EC technique.

Another environment worth paying more attention to is the permafrost region that is thawing due to the warming climate. In the degrading permafrost area, the landscape consists of flarks, lawns, and strings with different vegetation types and non-vegetated peat molds and ponds. This comprises an extremely heterogeneous surface with hotspots of fluxes with opposite directions. This variability is true not only for CH_4 and CO_2, but also for N_2O, for which emissions have been observed from a bare peat surface in a subarctic wetland (Repo et al. 2009). As another example, the area of wet flarks has been observed to increase in a thawing wetland in alpine northern Sweden, enhancing the CH_4 emissions (Christensen et al. 2004). In these environments, EC measurements provide an invaluable instrument for tracing the impacts of climate change on GHG exchange on a long-term basis.

References

Alm J, Schulman L, Walden J, Nykänen H, Martikainen PJ, Silvola J (1999) Carbon balance of a boreal bog during a year with an exceptionally dry summer. Ecology 80:161–174

Arneth A, Kurbatova J, Kolle O, Shibistova O, Lloyd J, Vygodskaya NN, Schulze ED (2002) Comparative ecosystem-atmosphere exchange of energy and mass in a European Russian and a central Siberian bog II. Interseasonal and interannual variability of CO_2 fluxes. Tellus 54B: 514–530

Augustin J, Merbach W, Rogasik J (1998) Factors influencing nitrous oxide and methane emissions from minerotrophic fens in northeast Germany. Biol Fertil Soils 28:1–4

Aurela M, Tuovinen J-P, Laurila T (1998) Carbon dioxide exchange in a subarctic peatland ecosystem in northern Europe measured by eddy covariance technique. J Geophys Res 103:11289–11301

Aurela M, Laurila T, Tuovinen J-P (2002) Annual CO_2 balance of a subarctic fen in northern Europe: importance of the wintertime efflux. J Geophys Res 107:4607. doi: 10.1029/2002JD002055

Aurela M, Laurila T, Tuovinen J-P (2004) The timing of snow melt controls the annual CO_2 balance in a subarctic fen. Geophys Res Lett 31:L16119. doi:10.1029/2004GL020315

Aurela M, Riutta T, Laurila T, Tuovinen J-P, Vesala T, Tuittila E-S, Rinne J, Haapanala S, Laine J (2007) CO_2 balance of a sedge fen in southern Finland – the influence of a drought period. Tellus 59B:826–837

Aurela M, Lohila A, Tuovinen J-P, Hatakka J, Riutta T, Laurila T (2009) Carbon dioxide exchange on a northern boreal fen. Boreal Environ Res 14:699–710

Barr JG, Fuentes JD, Engel V, Zieman JC (2009) Physiological responses of red mangroves to the climate in the Florida Everglades. J Geophys Res 114:G02008. doi:10.1029/2008JG000843

Barr JG, Engel V, Fuentes JD, Zieman JC, O'Halloran TL, Smith TJ III, Anderson GH (2010) Controls on mangrove forest-atmosphere carbon dioxide exchanges in western Everglades National Park. J Geophys Res 115:G02020. doi:10.1029/2009JG001186

Chojnicki BH, Urbaniak M, Józefczyk D, Augustin J, Olejnik J (2007) Measurements of gas and heat fluxes at Rzecin wetland. In: Okruszko T, Maltby E, Szatylolowicz J, Świątek D, Kotowski W (eds) Wetlands: monitoring, modeling and management. Taylor & Francis Group, London, pp 125–131

Christensen TR, Johansson T, Åkerman HJ, Mastepanov M, Malmer N, Friborg T, Crill P, Svensson BH (2004) Thawing sub-arctic permafrost: effects on vegetation and methane emissions. Geophys Res Lett 31:L04501. doi:10.1029/2003GL018680

Clymo RS, Turunen J, Tolonen K (1998) Carbon accumulation in peatland. Oikos 81:368–388

Corradi C, Kolle O, Walter K, Zimov SA, Schulze ED (2005) Carbon dioxide and methane exchange of a north-east Siberian tussock tundra. Glob Change Biol 11:1910–1925

Coxon DS, Parkinson D (1987) Winter respiratory activity in aspen woodland forest floor litter and soils. Soil Biol Biochem 19:49–59

Coyne PI, Kelly JJ (1975) CO_2 exchange in the Alaskan tundra: meteorological assessment by the aerodynamic method. J Appl Ecol 12:587–611

Dinsmore KJ, Billett MF, Skiba UM, Rees RM, Helfter C (2010) Role of the aquatic pathway in the carbon and greenhouse gas budgets of a peatland catchment. Glob Change Biol 16:2750–2762

Edwards GC, Neumann HH, den Hartog G, Thurtell GW, Kidd G (1994) Eddy correlation measurements of methane fluxes using a tunable diode laser at the Kinosheo Lake tower site during the Northern Wetlands Study (NOWES). J Geophys Res 99:1511–1517

Faist J, Capasso F, Sivco DL, Sirtori C, Hutchinson AL, Cho AY (1994) Quantum cascade laser. Science 264:553–556

Fan SM, Wofsy SC, Bakwin PS, Jacob DJ, Anderson SM, Kebabian PL, McManus JB, Kolb CE, Fitzjarrald DR (1992) Micrometeorological measurements of CH_4 and CO_2 exchange between the atmosphere and subarctic tundra. J Geophys Res 97:16627–16643

Flanagan PW, Bunnell FL (1980) Microflora activities and decomposition. In: Brown J, Miller PC, Tieszen LL, Bunnell FL (eds) An arctic ecosystem: the coastal tundra at Barrow, Alaska. Dowden, Hutchinson & Ross, Inc., Stroudsburg, pp 291–334

Frolking S, Roulet N, Fuglestvedt J (2006) How northern peatlands influence the Earth's radiative budget: sustained methane emission versus sustained carbon sequestration. J Geophys Res 111:G01008. doi:10.1029/2005JG000091

Glenn AJ, Flanagan LB, Syed KH, Carlson PJ (2006) Comparison of net ecosystem CO_2 exchange in two peatlands in western Canada with contrasting dominant vegetation, *Sphagnum* and *Carex*. Agric For Meteorol 140:115–135

Gorham E (1991) Northern peatlands: role in the carbon balance and probable responses to climatic warming. Ecol Appl 1:182–195

Goulden ML, Munger JW, Fan S-M, Daube BC, Wofsy SC (1996) Measurements of carbon sequestration by long-term eddy covariance: methods and a critical evaluation of accuracy. Glob Change Biol 2:169–182

Griffis TJ, Rouse WR, Waddington JM (2000) Interannual variability of net ecosystem CO_2 exchange at a subarctic fen. Glob Biogeochem Cycles 14:1109–1121

Harazono Y, Mano M, Miyata A, Zulueta RC, Oechel WC (2003) Inter-annual carbon dioxide uptake of a wet sedge tundra ecosystem in the Arctic. Tellus B 55:215–231

Hargreaves KJ, Fowler D, Pitcairn CER, Aurela M (2001) Annual methane emission from Finnish mire estimated from eddy covariance campaign measurements. Theor Appl Climatol 70: 203–213

Hendriks DMD, Dolman AJ, van der Molen MK, van Huissteden J (2008) A compact and stable eddy covariance set-up for methane measurements using off-axis integrated cavity output spectroscopy. Atmos Chem Phys 8:431–443

Hirano T, Segah H, Harada T, Limin S, June T, Hirata R, Osaki M (2007) Carbon dioxide balance of a tropical peat swamp forest in Kalimantan, Indonesia. Glob Change Biol 13:412–425

Hooijer A, Silvius M, Wosten H, Page S (2006) PEAT-CO_2, Assessment of CO_2 emissions from drained peatlands in SE Asia. Delft Hydraulics report Q3943

Huemmrich KF, Black TA, Jarvis PG, McCaughey J, Hall FG (1999) High temporal resolution NDVI phenology from micrometeorological radiation sensors. J Geophys Res 104:27935–27944

Jackowicz-Korczyński M, Christensen TR, Bäckstrand K, Crill P, Friborg T, Mastepanov M, Ström L (2010) Annual cycle of methane emission from a subarctic peatland. J Geophys Res 115:G02009. doi:10.1029/2008jg000913

Joiner DW, Lafleur PM, McCaughey JH, Bartlett PA (1999) Interannual variability in carbon dioxide exchanges at a boreal wetland in the BOREAS northern study area. J Geophys Res 104:27663–27672

Kroon PS, Hensen A, Jonker HJJ, Zahniser MS, van 't Veen WH, Vermeulen AT (2007) Suitability of quantum cascade laser spectrometry for CH_4 and N_2O eddy covariance measurements. Biogeosciences 4:715–728

Kuhry P, Turunen J (2006) The postglacial development of boreal and subarctic peatlands. In: Wieder RK, Vitt DH (eds) Boreal peatland ecosystems. Springer, Berlin, pp 25–46

Kutzbach L, Wille C, Pfeiffer E-M (2007) The exchange of carbon dioxide between wet arctic tundra and the atmosphere at the Lena River Delta, Northern Siberia. Biogeosciences 4: 869–890

Kwon H-J, Oechel WC, Zulueta RC, Hastings SJ (2006) Effects of climate variability on carbon sequestration among adjacent wet sedge tundra and moist tussock tundra ecosystems. J Geophys Res 111:G03014. doi:10.1029/2005JG000036

Lafleur PM, Roulet NT, Admiral SW (2001) Annual cycle of CO_2 exchange at a bog peatland. J Geophys Res 106:3071–3081

Lafleur PM, Roulet NT, Bubier JL, Frolking S, Moore TR (2003) Interannual variability in the peatland-atmosphere carbon dioxide exchange at an ombrotrophic bog. Glob Biogeochem Cycles 17:1036. doi:10.1029/2002GB001983

Laine A, Sottocornola M, Kiely G, Byrne KA, Wilson D, Tuittila E-S (2006) Estimating net ecosystem exchange in a patterned ecosystem: example from blanket bog. Agric For Meteorol 138:231–243

Lappalainen E (ed) (1996) Global Peat Resources. International Peat Society, Jyväskylä, Finland, 368 pp

Laurila T, Asmi E, Aurela M, Uttal T, Makshtas A, Reshetnikov A, Ivakhov V, Hatakka J, Lihavainen H, Hyvärinen A-P, Viisanen Y, Laaksonen A, Taalas P (2010) Long-term measurements of greenhouse gases and aerosols at Siberian Arctic site, Tiksi. In: Abstracts of the international polar year conference, June 2010, Oslo. http://ipy-osc.no/abstract/382008

Limpens J, Berendse F, Blodau C, Canadell JG, Freeman C, Holden J, Roulet N, Rydin H, Schaepman-Strub G (2008) Peatlands and the carbon cycle: from local processes to global implications – a synthesis. Biogeosciences 5:1475–1491

Lohila A, Laurila T, Aro L, Aurela M, Tuovinen J-P, Laine J, Kolari P, Minkkinen K (2007) Carbon dioxide exchange above a 30-year-old Scots pine plantation established on organic soil cropland. Boreal Environ Res 12:141–157

Lohila A, Minkkinen K, Laine J, Savolainen I, Tuovinen J-P, Korhonen L, Laurila T, Tietäväinen H, Laaksonen A (2010) Forestation of boreal peatlands: impacts of changing albedo and greenhouse gas fluxes on radiative forcing. J Geophys Res 115:G04011. doi:10.1029/2010JG001327

Lund M, Lindroth A, Christensen TR, Ström L (2007) Annual CO_2 balance of a temperate bog. Tellus 59B:804–811

Maanavilja L, Riutta T, Aurela M, Pulkkinen M, Laurila T, Tuittila E-S (2011) Spatial variation in CO_2 exchange at a northern aapa mire. Biogeochemistry 104:325–345. doi: 10.1007/s10533-010-9505-7

Mäkiranta P, Hytönen J, Aro L, Maljanen M, Pihlatie M, Potila H, Shurpali N, Laine J, Lohila A, Martikainen PJ, Minkkinen K (2007) Soil greenhouse gas emissions from afforested organic soil croplands and cutaway peatlands. Boreal Environ Res 12:159–175

Massman WJ, Sommerfield RA, Mosier AR, Zeller KF, Hehn TJ, Rochelle SG (1997) A model investigation of turbulence-driven pressure pumping effects on the rate of diffusion of CO_2, N_2O, and CH_4 through layered snowpacks. J Geophys Res 102:18851–18863

Mitsch WJ, Gosselink JG, Anderson CJ, Zhang L (2009) Wetland ecosystems. Wiley, Hoboken, 295 pp

Moore TR, Knowles R (1990) Methane emissions from fen, bog and swamp peatlands in Quebec. Biogeochemistry 11:45–61

Neumann HH, den Hartog G, King KM, Chipanshi AC (1994) Carbon dioxide fluxes over a raised open bog at the Kinosheo Lake tower site during the Northern Wetlands Study (NOWES). J Geophys Res 99:1529–1538

Nieveen JP, Jacobs CMJ, Jacobs AFG (1998) Diurnal and seasonal variation of carbon dioxide exchange from a former true raised bog. Glob Change Biol 4:823–834

Nilsson M, Sagerfors J, Buffam I, Laudon H, Eriksson T, Grelle A, Klemedtsson L, Weslien P, Lindroth A (2008) Contemporary carbon accumulation in a boreal oligotrophic minerogenic mire – a significant sink after accounting for all C fluxes. Glob Change Biol 14:2317–2332

Ochsner TE, Sauer TJ, Horton R (2006) Field tests of the soil heat flux plate method and some alternatives. Agronomy J 98:1005–1014

Regina K, Nykänen H, Silvola J, Martikainen PJ (1996) Fluxes of nitrous oxide from boreal peatlands as affected by peatland type, water table level and nitrification capacity. Biogeochemistry 35:401–418

Repo ME, Susiluoto S, Lind SE, Jokinen S, Elsakov V, Biasi C, Virtanen T, Martikainen PJ (2009) Large N_2O emissions from cryoturbated peat soil in tundra. Nat Geosci 2:189–192

Rinne J, Riutta T, Pihlatie M, Aurela M, Haapanala S, Tuovinen J-P, Tuittila E-S, Vesala T (2007) Annual cycle of methane emission from a boreal fen measured by the eddy covariance technique. Tellus 59B:449–457

Riutta T, Laine J, Aurela M, Rinne J, Vesala T, Laurila T, Haapanala S, Pihlatie M, Tuittila E-S (2007) Spatial variation in plant community functions regulates carbon gas dynamics in a boreal fen ecosystem. Tellus 59B:838–852

Sachs T, Wille C, Boike J, Kutzbach L (2008) Environmental controls on ecosystem-scale CH_4 emission from polygonal tundra Lena river delta, Siberia. J Geophys Res 113:G00A03. doi:10.1029/2007JG000505

Sagerfors J, Lindroth A, Grelle A, Klemedtsson L, Weslien P, Nilsson M (2008) Annual CO_2 exchange between a nutrient-poor, minerotrophic, boreal mire and the atmosphere. J Geophys Res 113:G01001. doi:10.1029/2006JG000306

Saunders MJ, Jones MB, Kansiime F (2007) Carbon and water cycles in tropical papyrus wetlands. Wetl Ecol Manage 15:489–498

Schulze ED, Valentini R, Sanz MJ (2002) The long way from Kyoto to Marrakesh: implications of the Kyoto protocol negotiations for global ecology. Glob Change Biol 8:1–14

Shurpali NJ, Verma SB, Kim J (1995) Carbon dioxide exchange in a peatland ecosystem. J Geophys Res 100:14319–14326

Soegaard H, Nordstroem C (1999) Carbon dioxide exchange in a high-arctic fen estimated by eddy covariance measurements and modelling. Glob Change Biol 5:547–562

Sottocornola M, Kiely G (2005) An Atlantic blanket bog is a modest CO_2 sink. Geophys Res Lett 32:L23804. doi:10.1029/2005GLO24731

Suyker AE, Verma SB, Arkebauer TJ (1997) Season-long measurement of carbon dioxide exchange in a boreal fen. J Geophys Res 102:29021–29028

Syed KH, Flanagan LB, Carlson P, Glenn A, Gaalen DEV (2006) Environmental control of net ecosystem CO_2 exchange in a treed, moderately rich fen in northern Alberta. Agric For Meteorol 140:97–114
Turunen J, Tomppo E, Tolonen K, Reinikainen A (2002) Estimating carbon accumulation rates of undrained mires in Finland – application to boreal and subarctic regions. Holocene 12:79–90
van der Molen MK, van Huissteden J, Parmentier FJW, Petrescu AMR, Dolman AJ, Maximov TC, Kononov AV, Karsanaev SV, Suzdalov DA (2007) The growing season greenhouse gas balance of a continental tundra site in the Indigirka lowlands, NE Siberia. Biogeosciences 4:985–1003
Verma SB, Ullman FG, Billesbach DP, Clement RJ, Kim J (1992) Micrometeorogical measurements of methane flux in a northern peatland ecosystem. Bound Layer Meteorol 58:289–304
Vourlitis GL, Oechel WC (1997) Landscape-scale CO_2, water vapour, and energy flux of moist-wet coastal tundra ecosystems over two growing seasons. J Ecol 85:575–590
Vourlitis GL, Oechel WC (1999) Eddy covariance measurements of net CO_2 flux and energy fluxes of an Alaskan tussock tundra ecosystem. Ecology 80:686–701
Yan Y, Zhao B, Chen JQ, Guo HQ, Gu YJ, Wu QH, Li B (2008) Closing the carbon budget of estuarine wetlands with towerbased measurements and MODIS time series. Glob Change Biol 14:1690–1702
Zahniser MS, Nelson DD, McManus JB, Kebabian PK (1995) Measurement of trace gas fluxes using tunable diode laser spectroscopy. Philos Trans R Soc A 351:371–382
Zimov SA, Zimova GM, Daviodov SP, Daviodova AI, Voropaev YV, Voropaeva ZV, Prosiannikov SF, Prosiannikova OV, Semiletova IV, Semiletov IP (1993) Winter biotic activity and production of CO_2 in Siberian soils: a factor in the greenhouse effect. J Geophys Res 98:5017–5023

Chapter 15
Eddy Covariance Measurements over Lakes

Timo Vesala, Werner Eugster, and Anne Ojala

15.1 Introduction

We give an overview on the status of eddy covariance measurements over lake surfaces with a focus on CO_2 fluxes. Inland waters have a significant role in the sequestration, transport, and mineralization of organic carbon (Battin et al. 2009; Tranvik et al. 2009). Although inland waters are especially important in lateral transports of carbon, their direct carbon exchange with the atmosphere, so-called outgassing, has also been recognized to be a significant component in the global carbon budget (Tranvik et al. 2009; Bastviken et al. 2011). Lakes also store carbon (C) effectively in their sediments, but for instance in the boreal zone, annual CO_2 emissions are 17–43 times higher than the net sedimentation of C (Kortelainen et al. 2006). In forested catchments, the annual CO_2 efflux from lakes has been estimated to be up to 14% of the annual net ecosystem exchange (Hanson et al. 2004).

Lakes cover only about 3% of the Earth's surface (Downing et al. 2006), but in the boreal zone lakes cover on average 7% of land area, and furthermore in some parts in Finland (Raatikainen and Kuusisto 1990) and northern Canada (Spence et al. 2003) they occupy up to 20% and 30% of the landscape, respectively. Many of the water bodies are small (Fig. 15.1). The average number in Finland – the country in Europe which probably has the highest density of lakes per unit area – is 56 lakes per 100 km^2 (Raatikainen and Kuusisto 1990) and the number of lakes with

T. Vesala (✉)
Department of Physics, University of Helsinki, Helsinki, Finland
e-mail: timo.vesala@helsinki.fi

W. Eugster
Institute of Agricultural Sciences, ETH Zürich, Zürich, Switzerland
e-mail: werner.eugster@agrl.ethz.ch

A. Ojala
Department of Environmental Sciences, University of Helsinki, Helsinki, Finland
e-mail: Anne.Ojala@helsinki.fi

M. Aubinet et al. (eds.), *Eddy Covariance: A Practical Guide to Measurement and Data Analysis*, Springer Atmospheric Sciences, DOI 10.1007/978-94-007-2351-1_15,

Fig. 15.1 Lake Valkea-Kotinen (surface area of 0.041 km^2) in southern Finland is surrounded by a typical boreal coniferous forest (©Ilpo Hakala)

a surface area less than 0.01 km^2 is over 130,000. More importantly, the Arctic tundra is similarly occupied by numerous small ponds and lakes, the response of which to high-latitude warming and the resulting changes in CO_2 and CH_4 effluxes is very uncertain (Walter et al. 2007). MacIntyre et al. (2010) concludes that regional and global fluxes of greenhouse gases from lakes may be considerably larger than current estimates.

The present outgassing estimates are still provisional and probably underestimated (Alsdorf et al. 2007). The eddy covariance (EC) technique would be an indispensable tool for directly assessing the fluxes from lakes, rather than using chambers, although there is the expectation that agreement between methods can be found with appropriate chamber design (Cole et al. 2010). Beside chambers, gas exchange models are applied based on the measured CO_2 partial pressure difference between the air and water and parameterized bulk transfer coefficients, but the transfer coefficient is difficult to experimentally determine (MacIntyre et al. 2010). However, long-term EC flux measurement data are very scarce and much more data from lakes of different sizes, lake types (like water color), and meteorological conditions are urgently needed to assess the role of lakes in local, regional, and global carbon budgets. There is a negative relationship between lake size and the gas saturation (i.e., surface water CO_2 concentration relative to atmospheric equilibrium) and especially small lakes, are a relatively large source of CO_2 (Kelly et al. 2001; Kortelainen et al. 2006). To guarantee the best achievements, the EC-related work should be carried out in close collaboration of atmospheric physicists/meteorologists with limnologists and aquatic ecologists.

15.2 Existing Studies

Six articles have reported on EC measurements of CO_2 fluxes over lakes. Anderson et al. (1999) (AN) have used the method over a small woodland lake in Minnesota, USA; Morison et al. (2000) (MO) studied the productivity of a tropical *Echinochloa* grassland in high-water phase in Amazon floodplain; Eugster et al. (2003) (EU) determined exchange rates over an Arctic Alaskan and an isolated (no inlet and no outlet) mid-latitude Swiss lake; Vesala et al. (2006) (VE) reported fluxes for a full open-water period for a small boreal uppermost (no inlet) lake; Guerin and Abril (2007) (GU) investigated a tropical reservoir in French Guiana and in Jonsson et al. (2008) (JO) the studied lake was located in Sweden, north of the Arctic Circle, in the boreal forest zone; Huotari et al. (2011) (HU) presented a five-years flux record for the same lake studied by VE. The lengths of most of the records are rather short. In AN lake-atmosphere exchanges were measured over 5 weeks in spring, summer, and fall, over a 3-year period. MO reports 2 weeks for aquatic phase of the grassland and about 1 month for terrestrial low-water period when the study site cannot be regarded as an inland water body. EU covers three separate periods of few days, primarily looking at the process of outgassing rather than its long-term relevance for C budgets. During the first one (2 days), the instruments were mounted on Toolik Lake (Alaska) shore requiring the exclusion of data when the mean wind was from the land. The second period, also at Toolik Lake, covers 5 days, but the equipment was mounted in the center of the lake on a moored float. During the third period at Lake Soppensee (Switzerland), the measurements were carried out over 3 days, again with instruments mounted on a moored float. GU reports data only for 24 h measured a few hundred meters upstream of the dam of the studied reservoir. In JO the measurements were made over about 3 months and the shortest fetch to the shore was 350 m. The two longest available data sets published in a peer-reviewed journal are a full open-water period from April to November 2003 and five consecutive ice-free periods (2003–2007) from Finland, Lake Valkea-Kotinen in VE and HU, respectively.

Nordbo et al. (2011) have stressed the importance and scarcity of data on small lakes. The sizes on the reported studies excluding MO study are, in the order of increasing size, as follows: Lake Valkea-Kotinen (VE) is about 460 m long and 130 m wide (average depth (ad) 2.5 m); Lake Soppensee (EU) is about 800 m long and 400 m wide (ad 12 m); Williams Lake (AN) is ellipsoidal with 900-m major axis and 550-m minor axis (ad 5.2 m); Toolik Lake (EU) has a surface area of 1.5 km^2 (ad 7 m); Lake Merasjärvi (JO) has a surface area of 3.8 km^2 (ad 5.1 m); the average surface of the Petit-Saut dam reservoir at Sinnamary River (GU), owing to high and low water levels, is 300 km^2. An overview of long- (>5 months) and short-term sensible and latent heat flux measurements by EC over lake, tabulated by Nordbo et al. (2011) in their Table 1, lists beside the already mentioned studies (VE, EU, AN, JO, and GU) three long-term (Rouse et al. 2008; Liu et al. 2009) and six short-term energy flux records (Elo 2007; Panin et al. 2006; Beyrich et al. 2006; Assouline et al. 2008; Vercauteren et al. 2008; Salgado and Le Moigne 2010).

EC flux measurements of CH_4 have only recently been started, but no studies have been published in the peer-reviewed literature so far and this aspect will not be covered in this chapter.

The earliest article (AN) concludes: "In view of the uncertainty in predicting lake-atmosphere CO_2 transfer and its important global implications in air-water exchange, we strongly encourage other investigators to make comparative measurements of CO_2 flux in an effort to better understand and quantify the environment controls regulating air-water gas transfer in natural settings," in 1999. After 12 years, we still must agree with this statement.

15.3 Surface-Specific Siting Problems

CO_2 fluxes over lakes are typically much smaller than the fluxes over adjacent land, if there is vegetation growing. Depending on lake type and trophic conditions it may even be problematic to clearly define the outlines of a lake, that is, in many lakes littoral zone is an essential part of the ecosystem. If aquatic macrophytes such as reed and sedges are growing on the lake banks, special consideration must be given to these conditions, which should be considered as wetlands (Chap. 14) irrespective of the fact that they are part of a lake.

Even lakes with sharp margins do not necessarily behave similarly: limnological knowledge is necessary to distinguish stratified lakes from shallow lakes that do not show the typical stratification of temperatures with depth of lake. Shallow lakes are typically oxic, that is, the water body mixes entirely from bottom to top, and thus a direct linkage between any (microbial decomposition) processes on the lake bottom and the CO_2 concentration in the surface waters can be expected. Depending on lake depth, however, a time lag between what is going on at the lake bottom and what one finds in the surface waters should be considered. As turbulent mixing in the atmosphere is a continuous dynamic process, it takes a certain time period (depending on turbulent mixing conditions in the water body) to transport a unit of gas produced at the lake bottom to the surface waters.

One should also note that flux footprints (source areas) tend to be long over smooth lake surfaces due to low levels of mechanical turbulence (see Chap. 8 on Footprints) and measurements over small water bodies may face the problem with long enough fetches. However, Vesala et al. (2006) studied the small Lake Valkea-Kotinen and demonstrated that source areas can be relatively short because of the presence of turbulence generated by the surrounding forest, compared to a larger lake with an extended smooth surface. Lakes may have much less day versus night difference in footprints as other surfaces, since the surface tends to be cool during day, stabilizing the near-surface layer and the warm surface with respect to the air at night.

For proper interpretations of EC fluxes, one also needs to monitor various other variables in addition to the CO_2 flux. Anyone planning to set up a lake EC facility should consider a "shopping list" of sensors for basic meteorological variables.

The following list shows the ultimate desire of scientists wishing to understand the processes behind the outgassing from a lake. For a simple quantification of CO_2 effluxes alone, a reduced set of variables will also do in most cases. The ultimate list includes downwelling and upwelling radiation components (short-wave and long-wave components separately), inclinometer, buoy/platform orientation, air CO_2 concentration gradient, water temperature profile, sediment temperature (or water temperature close to the lake bottom), water velocities, water conductivity, water CO_2 concentration profile, dissolved inorganic carbon (DIC) and nitrogen (DIN), dissolved organic carbon (DOC) and nitrogen (DON), particulate organic matter (POM) for lakes with relevant inflows, dissolved oxygen profile/redox potential, pH, chlorophyll concentration, total nitrogen, total phosphorous, and sediment samplers. It would be desirable to measure fluxes also by chambers for intercomparison. The monitoring of CO_2, DIC, DOC, and POM from inlet and outlet water allows for a full carbon balance estimate.

15.3.1 Stratification of Lakes

The conditions are even more complicated in stratified lakes, as is shown schematically in Fig. 15.2. Stratified have a mixed layer (epilimnion) on top of stably stratified and denser water at depth (hypolimnion). The two layers are separated by a transition layer (metalimnion) where a steep density gradient is found. Since this density gradient is mostly a function of water temperatures, the separation between the layers is called a thermocline. However, in saline lakes, the density may be a function also of salt concentrations, and thus the separating layer may be called pycnocline. If organic substrates are available on the lake bottom, for example, from recent organic sediments and plant litter, or from melting permafrost in cold areas, then the decomposition products from these substrates may accumulate in the hypolimnion. Exchange across the thermocline may be a limiting factor, as is expressed with the small arrows in Fig. 15.2. Once the decomposition products are transferred to the epilimnion, they are in a more or less oxic environment and eventually are oxidized to CO_2 before they reach the surface waters. Since mixing in the epilimnion is typically much better (large arrows in Fig. 15.2) than the exchange across the thermocline, the conditions in the surface waters may rather reflect the conditions of exchange across the thermocline than the mixing conditions in the epilimnion.

15.3.2 Aqueous Chemistry of CO_2

A further important aspect that needs to be considered in the special case of CO_2 is its active chemistry in the water. Since CO_2 is highly soluble as carbonic acid (H_2CO_3), it does not necessarily outgas as CO_2 from the lake whenever it reaches

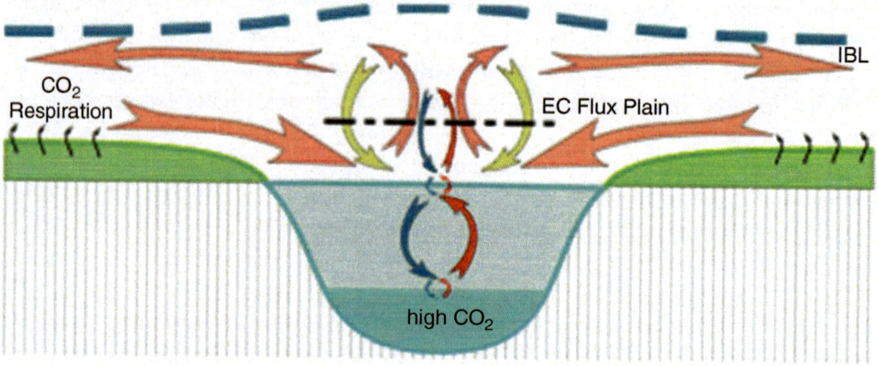

Fig. 15.2 Processes influencing the eddy covariance (EC) flux measurements above a lake surface at night. Because EC measurements cannot be performed directly at the air-water interface, the CO_2 exchange with the lake (*blue and red arrows*) at EC reference height (*black dash-dotted line*) is measured together with the exchange flux of CO_2-rich air from the land surrounding the lake (*pink and yellow arrows*) where CO_2 originates from respiration of soils and vegetation (*black arrows*). This local lake-breeze type circulation is expected to be restricted in its vertical extent by an internal boundary layer (IBL) (Reprinted from Eugster et al. (2003))

the surface waters. This is again shown with small arrows in Fig. 15.2, indicating that even with good mixing in the epilimnion there might be a rather small exchange flux across the lake surface. On the other hand, any change in the lake chemistry, namely the pH value of the waters, may quickly change the conditions for outgassing of CO_2 without a necessary driving force from the atmosphere. One factor affecting the surface water pH is photosynthetic activity.

In summary, eddy covariance flux measurements that are performed above the lake surface may accurately reflect current conditions for surface fluxes, but at the same time they may be very difficult to interpret because the main driving forces for the fluxes are not purely related to atmospheric turbulence, but to turbulent mixing in the lake as well (if not primarily due to the lake processes).

15.3.3 Land-Lake Interactions

An inherent characteristic of lakes is that they are surrounded by land, and in most cases the spatial extent of a lake is too small to strictly expect equilibrium conditions of the atmospheric turbulence with the local water surface. Using eddy covariance flux measurements, this is most easily tested by back-calculating the roughness length z_0 from momentum flux and stability measurements. The values of z_0 obtained in this way will most likely show much rougher conditions than what one would expect from a relatively smooth water surface. This is not a measurement error, but reflects the fact that especially the larger eddies with longer timescales are carrying a memory of the upwind surface roughness. The turbulent conditions do not quickly adapt to the smoother conditions of the lake surface as air travels from the rough surrounding terrestrial areas out to the lake surface (Jensen 1978). This effect is of course more pronounced with small lakes, but is not automatically ruled out on large lakes either.

Similarly, this land–lake interaction can also influence CO_2 concentrations and fluxes measured above the lake surface, as is shown with large arrows in Fig. 15.2. Consider daytime conditions, where the CO_2 uptake over a well-vegetated land surface surrounding the lake is 2 orders of magnitude stronger than the CO_2 uptake or release across the lake surface. Then low CO_2 concentrations originating from processes over the land may be advected to the lake, where standard eddy covariance data processing might mistake this as a local downward flux, namely if the lake water surface is cooler than the surrounding terrestrial vegetation, and advected air creeps in below the eddy covariance sensor as is depicted in Fig. 15.2. The specific aspects of local circulation and CO_2 advection to a lake from the surrounding forest has been addressed in detail by Sun et al. (1998).

The solution that was tried by Eugster et al. (2003) and Vesala et al. (2006) was to reduce the eddy covariance averaging time from 30 to 5 min. The argument is that atmospheric turbulence scales with z/u, the ratio of measurement height and horizontal wind speed, and thus a shorter averaging time eliminates part of the distal part of the flux footprint area from the integration of the fluxes (see Eugster et al. 2003 for more details). Under ideal neutral or unstable conditions one would expect that a 5-min flux average still yields 92% of the 30-min flux. Figure 15.3 from the Finnish Lake Valkea-Kotinen clearly showed that under daytime and evening conditions the fluxes were very similar for 5- and 30-min averaging periods, whereas in the early morning hours large differences were found. These differences were interpreted by the influence of respiration fluxes from the surrounding forest that could effectively be eliminated or at least reduced by choosing 5-min averages. Other filtering techniques may further improve the accuracy of flux measurements; it should however always be kept in mind that any filter applied to time series may introduce artifacts and thus always deserves a critical evaluation and should only be used with caution (see e.g., Stull 1988).

As Fig. 15.3 shows, the lake efflux of CO_2 during daytime was on the order of 0.3 μmol m^{-2} s^{-1} during which the expected CO_2 uptake from forest at such a locality would easily exceed 10–15 μmol m^{-2} s^{-1}, which is a 30–50 times stronger flux magnitude than the exchange across the lake surface. This indicates that high-precision closed-path gas analyzers are required to reliably resolve such fluxes. An open-path sensor for CO_2 concentration measurements would most likely be subject to Webb et al. (1980) density flux corrections and Burba heat flux corrections (Järvi et al. 2009; see also Sects. 4.1.4 and 4.1.5.2) that are of the same order of magnitude or even several times larger than the true net flux.

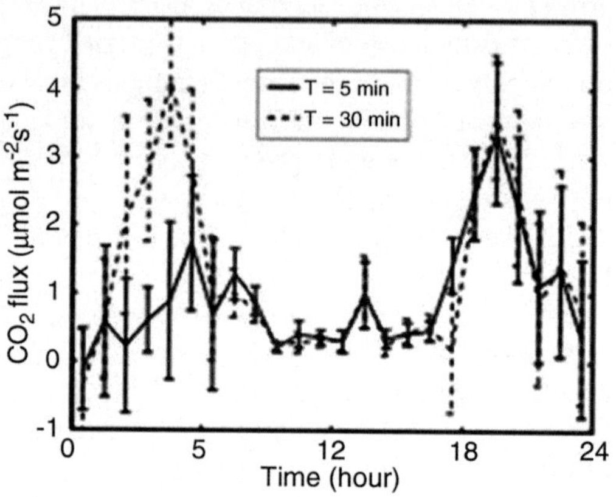

Fig. 15.3 Diurnal average curves of CO_2 fluxes from Lake Valkea-Kotinen (Finland) for July 2003, obtained with 30 min averaging (time series additionally detrended) and 5 min averaging (Reprinted from Vesala et al. (2006))

15.3.4 Quality Control Procedures

A critical note is necessary with respect to widely established empirical concepts to screen eddy covariance data (quality control) which may not be appropriate for lake flux studies. As already noted above, the spatial dimension of lakes is most probably small compared to the spatial and temporal scales of atmospheric turbulence. Thus, an integrative turbulence statistics criterion such as the σ_w/u_* criterion used by Foken and Wichura (1996) and Foken et al. (2004) (Sect. 4.3.2.2) may be misleading. First, of course, because of the slowness of adaptation of turbulent conditions to the smoother surface roughness over the lake (Jensen 1978), which suggests that σ_w/u_* measured over a lake is most likely strongly influenced by the upwind roughness surrounding the lake. Moreover, a lake of reasonable size is a huge energy store and can lead to atmospheric stratification that is the inverse of the one observed over the surrounding landscape. This is most pronounced during fall when lakes are still warm but the terrestrial surface cools considerably at night. Thus, over a lake it is not unlikely that during the night the surface is warmer than the air, thus leading to unstable stratification over the lake embedded in a more or less stably stratified atmosphere surrounding the lake. Hence, it is not easy to establish robust expectation values for σ_w/u_* under such conditions. If the lake is cooler than the air during daytime, then stable stratification over the lake may result within an otherwise unstably stratified atmosphere surrounding the lake. Since wind speeds and turbulence are however well developed during the daytime in the atmospheric boundary layer as a whole, the larger-scale conditions may still enhance atmospheric mixing of the air above the lake, thereby reducing any problems that are often associated with low u_* conditions over land surfaces during calm nights.

It should also be recalled that u_* is only a scaling parameter for mechanical turbulence, and it vanishes at the local free convection limit ($\zeta \ll -1$, e.g., Wyngaard et al. 1971). Local free convection is however not unlikely during nights over lakes, especially in fall (see above). Thus, an apparently low u_* value may be

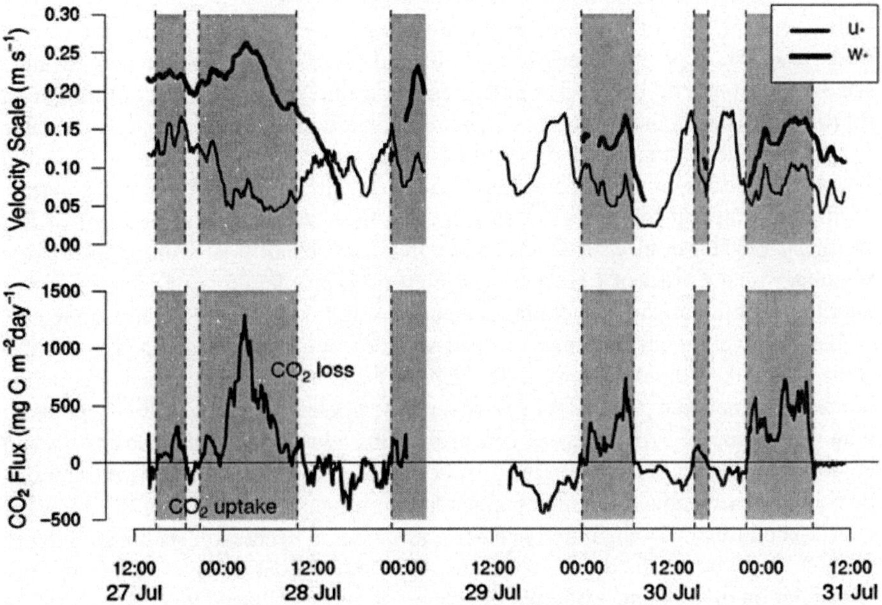

Fig. 15.4 Shear and convective velocity scales (*upper panel*) and gas fluxes (*lower panel*) from Toolik Lake, Alaska, during a 4-day period in July 1995. Maximum gas fluxes are not correlated with wind speed, as represented by u^*, but by whether or not heat was being lost from the surface layer ($w^* > 0$, *gray background*). Gas fluxes are less than 100 mg C m^{-2} day^{-1} at highest wind speeds, but increase above 500 mg C m^{-2} day^{-1} when buoyancy flux leads to convective overturning (Modified and updated from MacIntyre et al. (2001). Velocity scales u^* and w^* are expressed with respect to the atmosphere)

misinterpreted as low turbulent mixing, whereas in reality the turbulent mixing is mostly convection (over the lake) and not by mechanical mixing (shear flow). As the CO_2 efflux from lakes is most likely limited by processes in the water body, the effect of convection in the lake was investigated in a few cases. Figure 15.4 shows an example where the convective velocity scale w_* was found to be correlated with measured CO_2 fluxes, whereas gas fluxes were independent of u_*. In summary, great caution should be taken when transferring empirical concepts that may be adequate for eddy covariance flux measurements over forest surfaces to other surfaces, namely lake surfaces.

15.3.5 Mounting Instruments

A special challenge for eddy covariance flux measurements is the mounting of the instruments on a tower. There are basically two approaches: (1) install a tower with a solid foundation on the lake bottom, or (2) use a moored float to attach the instruments to. The first option has the advantage that it is sturdier than the second one. However, depending on lake currents and shear stress effects under stronger wind conditions, the lateral dynamic pressure on the tower construction might be a problem. The moored float is a dynamic construction which is less vulnerable to lake currents and associated lateral pressure, but it is never as sturdy as a solid tower. Depending on the dimensions of the float and the mooring there is a certain oscillation inherent to the measurements on a float. A detailed assessment of this factor by Eugster et al. (2003) has shown that for a small float with an oscillation frequency on the order of 1 Hz there was no serious problem for flux measurements, although additional variance contaminated the vertical wind speed component.

For lightweight instruments, a successful deployment of a damped buoy (designed by Mike Schurter, Eawag, Switzerland) was tested where a heavy concrete block was attached to a metal tube of several meters length that was then kept afloat with a set of buoys. At the upper end of the tube eddy covariance instruments can be attached. Depending on length of tube, the oscillation of the construction can be shifted to very low frequencies, since any motion is damped using this physical inertia-pendulum system. When heavier equipment is necessary, such a stabilizing pendulum can be attached to a conventionally moored float to reduce the frequency of oscillation of the float. Although Eugster et al. (2003) showed that this oscillation of the float is not overly critical for the flux measurements, the goal should still be to get highest accuracy for both covariances and variances of the measured time series. And this requires either a sturdy construction or a damping system that shifts the oscillation frequency of the float outside of the range of the relevant frequencies of fluxes and variances.

References

Alsdorf D, Bates P, Melack J et al (2007) Spatial and temporal complexity of the Amazon flood measured from space. Geophys Res Lett 34. doi:10.1029/2007GL029447

Anderson DE, Striegl RG, Stannard DI, Micherhuizen CM, McConnaughey TA, LaBaugh JW (1999) Estimating lake-atmosphere CO_2 exchange. Limnol Oceanogr 44:988–1001

Assouline S, Tyler SW, Tanny J, Cohen S, Bou-Zeid E, Parlange MB, Katul GG (2008) Evaporation from three water bodies of different sizes and climates: measurements and scaling analysis. Adv Water Res 31:160–172

Bastviken D, Tranvik LJ, Downing JA, Crill PM, Enrich-Prast A (2011) Freshwater methane emissions offset the continental carbon sink. Science 331:50

Battin TJ, Luyssaert S, Kaplan LA, Aufdenkampe AK, Richter A, Tranvik LJ (2009) The boundless carbon cycle. Nat Geosci 2:598–600

Beyrich F, Leps J-P, Mauder M, Bange J, Foken T, Hunke S, Lohse H, Lüdi A, Meijninger WML, Mironov D, Weisensee U, Zitterl P (2006) Area-averaged surface fluxes over the LITFASS region based on eddy-covariance measurements. Bound Layer Meteorol 121:33–65

Cole JJ, Bade DL, Bastviken D, Pace ML, Van de Bogert M (2010) Multiple approaches to estimating air–water gas exchange in small lakes. Limnol Oceanogr Methods 8:285–293

Downing JA, Prairie YT, Cole JJ, Duarte CM, Tranvik LJ, Striegl RG, McDowell WH, Kortelainen P, Caraco NF, Melack JM, Middelburg JJ (2006) The global abundance and size distribution of lakes, ponds, and impoundments. Limnol Oceanogr 51(5):2388–2397

Elo A-R (2007) The energy balance and vertical thermal structure of two small boreal lakes in summer. Boreal Environ Res 12:585–600

Eugster W, Kling G, Jonas T, McFadden JP, Wüest A, MacIntyre S, Chapin FS III (2003) CO_2 exchange between air and water in an arctic Alaskan and midlatitude Swiss lake: importance of convective mixing. J Geophys Res 108:4362–4380. doi:10.1029/2002JD002653

Foken T, Wichura B (1996) Tools for quality assessment of surface-based flux measurements. Agric For Meteorol 78:83–105

Foken T, Göckede M, Mauder M, Mahrt L, Amiro B, Munger W (2004) Post-field data quality control. In: Lee X, Massman W, Law B (eds) Handbook of micrometeorology. Kluwer Academic, Dordrecht, pp 181–208

Guerin F, Abril G (2007) Significance of pelagic aerobic methane oxidation in the methane and carbon budget of a tropical reservoir. J Geophys Res Biogeosci 112:G03006

Hanson PC, Pollard AI, Bade DL, Predick K, Carpenter SR, Foley JA (2004) A model of carbon evasion and sedimentation in temperate lakes. Glob Change Biol 10:1285–1298

Huotari J, Ojala A, Peltomaa E, Nordbo A, Launiainen S, Pumpanen J, Rasilo T, Hari P, Vesala T (2011) Long-term direct CO_2 flux measurements over a boreal lake: Five years of eddy covariance data. Geophys Res Lett 38. doi:10.1029/2011GL048753

Järvi L, Mammarella I, Eugster W, Ibrom A, Siivola E, Dellwik E, Keronen P, Burba G, Vesala T (2009) Comparison of net CO_2 fluxes measured with open- and closed-path infrared gas analyzers in urban complex environment. Boreal Environ Res 14:499–514

Jensen NO (1978) Change of surface roughness and the planetary boundary layer. Q J R Meteorol Soc 104:351–356

Jonsson A, Åberg J, Lindroth A, Jansson M (2008) Gas transfer rate and CO2 flux between an unproductive lake and the atmosphere in northern Sweden. J Geophys Res 113:G04006. doi:10.1029/2008JG000688

Kelly CA, Fee E, Ramlal PS, Rudd JWM, Hesslein RH, Anema C, Schindler EU (2001) Natural variability of carbon dioxide and net epilimnetic production in the surface waters of boreal lakes of different sizes. Limnol Oceanogr 46:1054–1064

Kortelainen P et al (2006) Sediment respiration and lake trophic state are important predictors of large CO_2 evasion from small boreal lakes. Glob Change Biol 12:1554–1567

Liu H, Zhang Y, Liu S, Jiang H, Sheng L, Williams QL (2009) Eddy covariance measurements of surface energy budget and evaporation in a cool season over southern open water in Mississippi. J Geophys Res 114:D04110. doi:10.1029/2008JD010891

MacIntyre S, Eugster W, Kling GW (2001) The critical importance of buoyancy flux for gas flux across the air-water interface. In: Donelan MA, Drennan WM, Saltzman ES, Wanninkhof R (eds) Gas transfer at water surfaces, Geophysical Monograph 127. American Geophysical Union, Washington, DC, pp 135–139

MacIntyre S, Jonsson A, Jansson M, Aberg J, Turney DE, Miller SD (2010) Buoyancy flux, turbulence, and the gas transfer coefficient in a stratified lake. Geophys Res Lett 37:L24604. doi:10.1029/2010GL044164

Morison JIL, Piedade MTF, Müller E, Long SP, Junk WJ, Jones MB (2000) Very high productivity of the C_4 aquatic grass *Echinochloa polystachya* in the Amazon floodplain confirmed by net ecosystem CO_2 flux measurements. Oecologia 125:400–411

Nordbo N, Launiainen S, Mammarella I, Leppäranta M, Huotari J, Ojanen A, Vesala T (2011) Long-term energy flux measurements and energy balance over a small boreal lake using eddy covariance technique. J Geophys Res 116. doi:10.1029/2010JD014542

Panin GN, Nasonov AE, Foken T, Lohse H (2006) On the parameterisation of evaporation and sensible heat exchange for shallow lakes. Theor Appl Climatol 85:123–129. doi: 10.1007/s00704-005-0185-5

Raatikainen M, Kuusisto E (1990) Suomen järvien lukumäärä ja pinta-ala (The number and surface area of the lakes in Finland). Terra 102:97–110 (In Finnish with English summary)

Rouse WR, Blanken PD, Bussières N, Oswald CJ, Schertzer WM, Spence C, Walker AE (2008) An investigation of the thermal and energy balance regimes of Great Slave and Great Bear Lakes. J Hydrometeorol 9(6):1318–1333

Salgado R, Le Moigne P (2010) Coupling of the FLake model to the Surfex externalized surface model. Boreal Environ Res 15:231–244

Spence C, Rouse WR, Worth D, Oswald C (2003) Energy budget processes of a small northern lake. J Hydrometeorol 4:694–701

Stull RB (1988) An introduction to boundary layer meteorology. Kluwer, Dordrecht, 310 pp

Sun J, Desjardins R, Marth L, MacPherson I (1998) Transport of carbon dioxide, water vapor, and ozone by turbulence and local circulations. J Geophys Res 103:25873–25885

Tranvik LJ, Downing JA, Cotner JB, Loiselle SA, Striegl RG, Ballatore TJ, Dillon P, Finlay K, Fortino K, Knoll LB, Kortelainen PL, Kuster T, Larsen S, Laurion I, Leech DM, McCallister SL, McKnight DM, Melack JM, Overholt E, Porter JA, Prairie Y, Renwick WH, Roland F, Sherman BS, Schindler DW, Sobek S, Tremblay A, Vanni MJ, Verschoor AM, van Wachenfeldt E, Weyhenmeyer GA (2009) Lakes and reservoirs as regulators of carbon cycling and climate. Limnol Oceanogr 54:2298–2314

Vercauteren N, Bou-Zeid E, Parlange MB, Lemmin U, Huwald H, Selker J, Meneveau C (2008) Subgrid-scale dynamics of water vapour, heat, and momentum over a lake. Bound Layer Meteorol 128:205–228. doi:10.1007/s10546-008-9287-9

Vesala T, Huotari J, Rannik Ü, Suni T, Smolander S, Sogachev A, Launiainen S Ojala A (2006) Eddy covariance measurements of carbon exchange and latent and sensible heat fluxes over a boreal lake for a full open-water period. J Geophys Res Atmos 111, D11101. doi:10.1029/2005JD006365

Walter KM, Smith LC, Chapin FS III (2007) Methane bubbling from northern lakes: present and future contributions to the global methane budget. Philos Trans R Soc A 365:1657–1676

Webb EK, Pearman GI, Leuning R (1980) Correction of flux measurements for density effects due to heat and water vapour transfer. Q J R Meteorol Soc 106:85–100

Wyngaard JC, Coté OR, Izumi Y (1971) Local free convection, similarity, and the budgets of shear stress and heat flux. J Atmos Sci 28:1171–1182

Chapter 16
Eddy Covariance Measurements Over Urban Areas

Christian Feigenwinter, Roland Vogt, and Andreas Christen

16.1 Introduction

Throughout the last two decades, numerous research projects applied the eddy covariance (EC) approach to urban ecosystems to directly measure turbulent fluxes between the urban surface and the atmosphere to quantify the exchange of energy, water vapor, greenhouse gases, air pollutants, and aerosols in connection with the assessment of (air pollutant) dispersion and of the urban energy, water, and carbon balances. Numerical models for dispersion, air pollution, and weather forecasting in cities rely on parameterization schemes for turbulence and surface exchange, which should take into account the implications that arise from the extremely rough surface of cities. Although, to date, the physics of dispersion and energy exchange taking place in the urban roughness sublayer (see Sect. 16.1.2) are mostly understood, it is challenging to parameterize and/or simplify them. Similarity theory is likely to fail in the lower part of the roughness sublayer, which is unfortunate because this is where urban residents live and hence it is the most important layer for forecasting products. Further, partitioning of the urban energy balance is strongly modified compared to rural ecosystems, due to the specific properties of the urban surface (3D geometry, roughness, impervious surfaces, anthropogenic heat injections) and due to complex source/sink distributions. The turbulent fluxes

C. Feigenwinter (✉) • R. Vogt
Institute of Meteorology, Climatology and Remote Sensing, University of Basel, Basel, Switzerland
e-mail: feigenwinter@metinform.ch; Roland.Vogt@unibas.ch

A. Christen
Department of Geography and Atmospheric Science Program, University of British Columbia, Vancouver, Canada
e-mail: andreas.christen@ubc.ca

M. Aubinet et al. (eds.), *Eddy Covariance: A Practical Guide to Measurement and Data Analysis*, Springer Atmospheric Sciences, DOI 10.1007/978-94-007-2351-1_16,

Table 16.1 Urban scales

Urban scale	Horizontal length scale	Built features	Meteorological scale
Building	10 × 10 m	Single-family house, high-rise	Micro
Canyon	30 × 40 m	Street, canyon	
Block	500 × 500 m	Block, factory	
Neighborhood (local climate zone)	5 × 5 km	City center, residential, industrial zone, etc.	Local
City	25 × 25 km	Urban area	Meso
Urban region	100 × 100 km	City plus its environs	

Adapted from Oke (2006a)

of energy and matter are also modified by human injections of heat, water, and carbon into the urban atmosphere by traffic, space heating, waste management, etc. In this sense, urban ecosystems have additional premises compared to nonurbanized ecosystems.

Many of the restrictions on EC measurements over very rough surfaces, like forests, also apply to urban surfaces. However, there are some differences mainly originating from the presence of a deep urban roughness sublayer, where the flow is significantly influenced by the presence of individual buildings/objects. In analogy to vegetation canopies, the assemblage of buildings, trees, and other objects in a city can be regarded as the "urban canopy" (Oke 1976).

16.1.1 Scales in Urban Climatology

Modifications of the land-atmosphere exchange by urban areas span over space and timescales of several orders of magnitude. Oke (2006a) presented concepts of urban scales that should be appreciated when applying the EC technique in urban environments (see Table 16.1 and Fig. 16.1):

(a) *Building, canyon, and block;* also *"microscale"*: dominated by irregular 3D structures (buildings, trees, roads, gardens, courtyards, plazas). Scale of choice for computational fluid dynamics (CFD) simulations. Restricted use of EC measurements, as it is challenging to attribute source areas.

(b) *Neighborhood scale;* also *"local scale"*: represented by a *Local Climate Zone* (see Sect. 16.1.4) made up of repetitive patches of surface cover, size and spacing of buildings, and human activity (residential, commercial, industry). Scale of choice for EC, as fluxes represent an integral response from a specific urban ecosystem.

(c) *City and urban region;* also *"mesoscale"*: represents typical city-wide pattern (e.g., the urban heat island UHI) and interactions between city and its rural counterpart. Note that a single station cannot represent this scale and representative pairs or networks of rural and urban EC measurements are required.

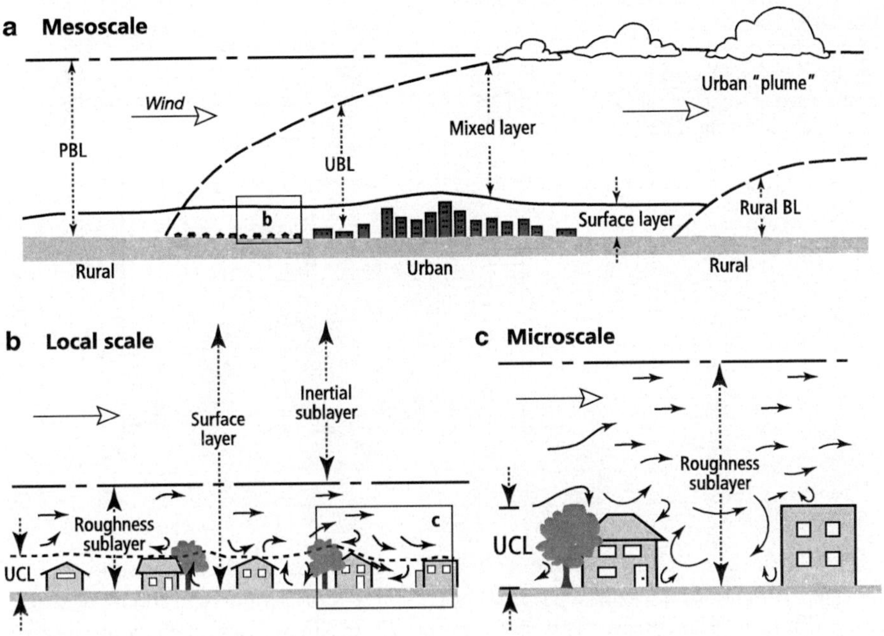

Fig. 16.1 Vertical (planetary boundary layer *PBL*, urban boundary layer *UBL*, urban canopy layer *UCL*) and horizontal scales (**a** mesoscale, **b** local scale, **c** microscale) in urban boundary layers (Adapted from Oke (2006b))

16.1.2 The Urban Atmosphere

The surface layer of an urban ecosystem can be conceptually divided into two sub-layers: the urban roughness sublayer (RSL) closer to the urban canopy responding to individual microscale elements, and the overlying inertial sublayer (ISL) due to the mixing of all sub-neighborhood scales in the RSL. The lowest part of the urban RSL from the ground up to the height of buildings z_h is called "urban canopy layer" (UCL). The restricted convective and radiative coupling between the airspace located in the canopy (e.g., street canyons) and the roughness sublayer above roofs allows the UCL to maintain its own climate. Microclimatic effects only persist for a short distance away from their source until they are blended, horizontally and vertically, by turbulence. In the horizontal, these effects may persist for a few hundred meters, while in the vertical, they are evident in the urban RSL, which extends from ground level to the blending height z_r that ranges from about 1.5 z_h over densely built-up areas up to 5 z_h in low density areas (Grimmond and Oke 1999). Exchange in the RSL is not only driven by turbulent exchange, but also by dispersive fluxes ("form-induced" fluxes) that – in contrast to permeable and irregular natural canopies – contribute significantly to the exchange in the UCL through stationary vortices arising from the flow around buildings and other large

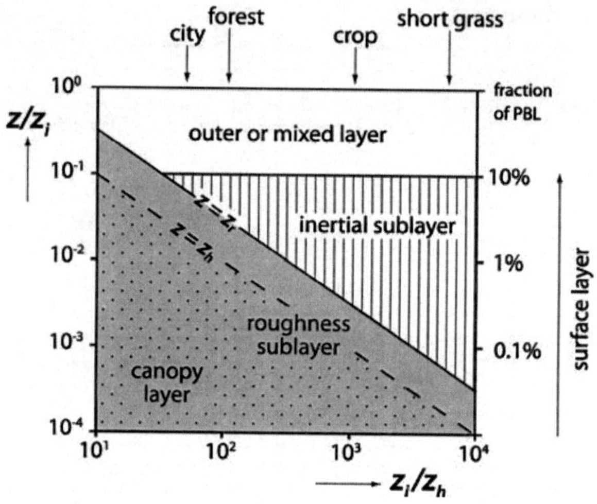

Fig. 16.2 The surface-specific different depths of *PBL* sublayers. z_i refers to the height of the PBL. (Adapted from Rotach (1999))

surface elements. EC measurements above z_r, that is, in the ISL, are supposed to measure a blended, spatially averaged signal that is representative of the local scale. Main differences in the vertical structure of the urban surface layer compared to nonurbanized ecosystem are the substantial depths of the RSL (Fig. 16.2) and the importance of nonturbulent exchanges (small-scale advection, dispersive fluxes). Note also in Fig. 16.2 that the ISL may even disappear at the expense of the RSL at urban sites with tall buildings.

16.1.3 Exchange Processes in the Urban Atmosphere

The main features that govern the vertical turbulent exchange of mass and scalars in the urban atmosphere can be characterized as follows (adapted from Roth 2000):

- *An intense shear layer* forms near the top of the canopy, *whose properties (turbulent kinetic energy, high turbulence intensities) differ systematically from those of the overlying inertial sublayer* due to wakes behind buildings and vortices of canyon flows.
- *Wake diffusion* behind roughness elements and *form drag* due to pressure differences across individual roughness elements (bluff-bodies) lead to high turbulent kinetic energy, efficient vertical and horizontal mixing, and create stationary vortices that can lead to significant dispersive fluxes.
- *3D organization of sources/sinks* (sunlit/shaded roofs, streets and walls, cold/hot spots, wet/dry spots, point and line sources of pollutants) results in a complex system of active surfaces and dissimilarities in the energy and mass transport due to high spatial variability.

- *Extreme surface heterogeneity* at all length-scales makes it nearly impossible to establish uniform, so local advection is very likely.
- *Organized motions,* manifested in ramp structures and sweeps and ejections, originate from a certain regularity of building structures that are very efficient in transporting heat.
- *Larger boundary-layer heights and reduced atmospheric stability* are created by the enhanced mechanical mixing and the UHI effect.

The urban surface layer is strongly influenced by the growth rate of internal boundary layers, whose properties are crucial to the location of the source areas for EC sensors. The flow structures and thermodynamic properties of internal boundary layers, generated by local scale surfaces, are adapted to the properties of their respective surface types. The growth rate of internal boundary layers depends on roughness and stratification. Since cities tend to neutral conditions, due to enhanced mechanical and thermal turbulence associated with large roughness and the heat island effect, we propose a typical height/fetch ratio in the range of 1:25–1:50. If surface properties inside the required fetch are not similar, then the measurements will not be representative of the local urban ecosystem. Fetch requirements can therefore be a significant restriction when choosing a site location.

It follows from the preceding discussion that a good characterization of the source area is of highest importance for the correct interpretation of the measurements. Footprint models (see Chap. 8) can provide reliable estimates of the turbulent transfer processes in the ISL, where Monin-Obukhov similarity theory (MOST) applies. Below this level, that is, in the RSL, complications arise due to the complex 3D geometry of buildings and the blockage and channeling of flow, which characterize the UCL, so the turbulent fluxes of momentum, energy, moisture, and pollutants are height dependent (Rotach 2001).

16.1.4 Characterization of the Urban Surface–Atmosphere Interface

The wide range of urban land covers and urban land uses presents a large variety of boundary conditions to the atmosphere, which manifest themselves in the specific properties of the urban surface relating to its roughness (the size, shape and separation of buildings, trees and other large structures), the radiative, thermal, and moisture characteristics and their spatial arrangement, and the pattern of emissions (e.g., carbon dioxide). These properties are organized uniquely in any given urban area. Recognition of this is of highest importance to understand the spatial and temporal variability of surface-atmosphere exchanges within cities (Grimmond et al. 2004) and has an important impact on the interpretation of the results from urban studies.

A useful tool to characterize urban areas and districts, and a necessity for the interpretation and comparability of meteorological measurements in cities, is the scheme of *Local Climate Zones (LCZ)* (Stewart and Oke 2009; Stewart 2009).

An essential first step is to evaluate the physical nature of the urban terrain in terms of transitions and inhomogeneity. The properties of radiation and airflow heavily depend on the dimensions of the roughness elements and their spatial arrangement, which can be characterized by some typical measures related to geometric features, surface fractions, and length scales. Partly based on such measures, Stewart and Oke (2009) present a field site classification of *Local Climate Zones* (LCZ) that includes a "Built Series." The main features of urban structure parameters and the "Built Series" of LCZ are presented in Fig. 16.3.

Typical relationships for urban roughness length z_0 and zero plane displacement height d derived from mean building height z_h are: $z_0/z_h = 0.1$ and $d/z_h = 0.5$, respectively (Grimmond et al. 1998). Measurements at a chosen location should represent the properties of the respective LCZ, which means that the source areas of the sensors are fully representative for the LCZ under investigation.

16.2 Conceptual Framework for Urban EC Measurements

Starting from the early 1980s, experimental attempts using EC technique in cities have been performed in conjunction with the phenomenon of the urban heat island (UHI) (e.g., Oke 1976) to better quantify the fluxes that contribute to the urban energy balance (UEB), that is, the sensible and latent heat flux and, indirectly, the storage heat flux, that is the most problematic flux to quantify in the UEB. This has led to the need for better parameterizations in numerical forecast models for urban areas and has motivated researchers to investigate the urban-specific structure of turbulence by the means of the EC technique. Most recently, more attention has been paid to the measurement of CO_2 fluxes in cities, and efforts are underway by the International Association for Urban Climate (IAUC) to organize and document urban flux sites into the URBAN FLUX NETWORK (accessible by the IAUC homepage on www.urban-climate.org).

A complete review of existing studies is beyond the scope of this chapter, however, several comprehensive reviews of urban turbulence studies that abstract the most important findings of the last decades are available. Roth (2000) provides an excellent "Review of turbulence over cities" based on stringently selected high-quality experimental results published up to the year 2000, Arnfield (2003) reviewed "Two decades of urban climate research," and Grimmond (2006) summarizes the "Progress in measuring and observing the urban atmosphere." Though the latter two are written from a more general point of view, all three give insight into the state of the art of EC measurements in urban environments. In addition, we refer to the BRIDGE report (Grimmond et al. 2010) for a comprehensive overview of existing studies related to energy, water, and CO_2 fluxes in cities.

BUILT SERIES

LOCAL CLIMATE ZONE (LCZ)		Sky view factor[1]	Aspect ratio[2]	Building surface fraction[3] (%)	Impervious surface fraction[4] (%)	Natural surface fraction[5] (%)	Height of roughness elements[6] z_H (m)	Terrain roughness class[7]	Anthropogenic heat flux density[8] Q_F (Wm^{-2})
B1: Compact highrise		0.2–0.5	>2	40–60	40–60	<10	>35	8	50–300
B2: Compact midrise		0.3–0.6	0.75–1.25	40–70	30–50	<15	8–25	6-7	<75
B3: Compact lowrise		0.3–0.5	1–1.5	50–70	20–30	<15	3–8	6	<75
B4: Open-set highrise		0.4–0.7	0.75–1.25	20–40	30–40	30–40	>30	7–8	<50
B5: Open-set midrise		0.8–0.9	0.3–0.5	20–40	20–40	20–40	8–25	5–6	<25
B6: Open-set lowrise		0.6–0.8	0.5–0.75	20–50	20–30	30–50	3–8	5–6	<25
B7: Extensive lowrise		>0.9	0.1–0.3	30–50	40–50	<20	3–10	5	<50
B8: Lightweight lowrise		0.3–0.5	1–1.5	50–80	<10	10–30	2–4	4–5	<5
B9: Sparsely developed		>0.9	0.1–0.2	10–20	<20	60–80	3–25	5–6	<10
B10: High-energy industrial		0.7–0.9	0.2–0.5	20–30	20–40	40–50	5–10	5–6	>300

1. Proportion of sky hemisphere visible from ground level.
2. Mean height-to-width ratio of street canyons (B1 thru B4, B6, B8) and building spacing (B5, B7, B9, B10).
3. Proportion of zone plan surface covered by buildings.
4. Proportion of zone plan surface covered by impervious materials.
5. Proportion of zone plan surface covered by natural materials.
6. Geometric average of building heights.
7. Based on Davenport *et al.* (2000) classification of effective terrain roughness.
8. Mean annual anthropogenic heat flux density. Varies significantly with latitude and season.

Fig. 16.3 The "Built Series" of local climate zones (Adapted from Stewart (2009))

16.2.1 Turbulence Characteristics

The results from Roth (2000) and the studies investigated therein show strong similarities in the integral statistics and (co)spectra of turbulent flows over both urban environment and plant canopies. Roth concludes that urban turbulence may be interpreted in the framework of plane mixing-layer flows (Raupach et al. 1996) with modifications to take into account wake turbulence. Work by Kastner-Klein and Rotach (2004), Kanda (2006), Moriwaki and Kanda (2006), Moriwaki et al. (2006) and Christen et al. (2009a) confirmed many of the analogies of urban RSL statistics to those found in plant-canopy flows. Similarly, organized structures have been shown to be very efficient in the transport of heat (Feigenwinter and Vogt 2005; Oikawa and Meng 1995; Christen et al. 2007) and, though to a lesser amount, water vapor and CO_2 (Moriwaki and Kanda 2006) in urban environments.

16.2.2 The Volume Balance Approach

In the following, we review specific aspects of the turbulent fluxes of sensible and latent heat and CO_2 in the context of a *volume balance approach* according to Fig. 16.4, which is the preferred concept for urban ecosystems due to their 3D nature.

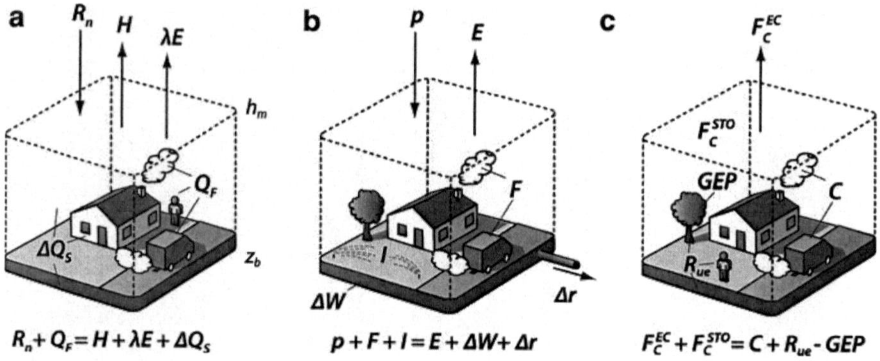

Fig. 16.4 Conceptual diagrams illustrating the volume balance approach for (**a**) the urban energy balance, (**b**) the urban water balance, and (**c**) the urban carbon balance. *All arrows* indicate the definition of positive flux densities. Note that advective fluxes are not considered. For a description of symbols see the text

16.2.2.1 Turbulent Heat Fluxes in the Context of Urban Energy Balance Studies

The urban surface energy balance is typically approached as a *volume budget* (Fig. 16.4). The upper boundary of the budget volume h_m is chosen to be above the roughness sublayer height z_r in the ISL, in order to provide horizontal homogeneity for EC and radiation measurements. The lower boundary z_b is located in the subsurface where annual substrate temperature variations approach zero (Fig. 16.4a):

$$R_n + Q_F = H + \lambda E + \Delta Q_S \tag{16.1}$$

Net all-wave radiation (R_n) and the turbulent flux densities of sensible and latent heat (H and λE) can be directly measured at h_m, the latter two terms by EC. The two additional terms account for the *anthropogenic heat flux density* Q_F and the *net storage heat flux density* ΔQ_S in the budget volume. Since these terms are unique to urban environments, they are briefly discussed.

The *anthropogenic heat flux density* Q_F is an extra input term to account for injections of sensible and latent heat to the atmosphere by human activity (combustion processes, heating/cooling of buildings, heat released by the human metabolism). Clearly those heat injections are not homogeneously distributed in the UCL, but often can be constrained both horizontally and vertically. Care has to be taken that EC systems measuring H and λE are located sufficiently distant from anthropogenic point and line heat sources (engines, ventilation systems, chimneys, etc.) but are still able to capture the integrative, local scale heat release by Q_F. Common methods for the estimation of Q_F are *inventory approaches* (top-down modeling of energy consumption statistics, traffic load data, etc.), *building energy models and traffic models* (bottom-up modeling) or by assuming energy balance closure and using *long-term* (e.g., annual, with ΔQ_S approaching zero) measurements of R_n, H, and λE. The annual residual term is then Q_F (Christen and Vogt 2004; Offerle et al. 2005; Pigeon et al. 2007a). Typical values of Q_F (see also Fig. 16.3) range between 10 and 15 W m^{-2} for suburban areas (LCZ B6 open-set lowrise), 20 and 30 W m^{-2} in more compact built-up areas (LCZ B3 compact lowrise), but can exceed 400 W m^{-2} in extreme cases such as reported for restricted areas, such as the central business district of Tokyo (Ichinose et al. 1999).

The *net storage heat flux density* ΔQ_S can be written as the sum of changes in the heat storage in the subsurface ground materials (ΔQ_{Sg}), in buildings (ΔQ_{Sb}), in vegetation (ΔQ_{Sv}) and in indoor and outdoor air (ΔQ_{Sa}). Both ΔQ_{Sv} and ΔQ_{Sa} are small compared to ΔQ_{Sg} and ΔQ_{Sb}. Since a direct measurement of ΔQ_S is impractical in an urban area, two empirical methods are commonly applied to estimate ΔQ_S (Roberts et al. 2006): First, the *energy-balance residual approach,* where all other terms are measured (Q_F might be estimated) and Eq. 16.1 is solved for ΔQ_S; second, the *thermal mass scheme* (TMS), which uses an array of temperature measurements (surface and intra-material) to estimate the rate of change of sensible heat of many representative materials and facets of the urban interface. Using weighting factors based on survey data, the measured rate of change in temperature of the "urban body" is used to calculate the energy released or removed from the interface. In comparison to non-urbanized ecosystems, ΔQ_S can be a substantial term (up to almost 60% of R_n) in the surface energy balance.

16.2.2.2 Evapotranspiration in the Context of Urban Water Balance Studies

In analogy to the urban energy balance, the urban water balance can be written as a *volume budget* (Fig. 16.4b, Oke 1987):

$$P_r + I + F = E + \Delta W + \Delta r \tag{16.2}$$

Similar to the water balance in non-urbanized ecosystems, all terms are expressed in mm per unit time. In the urban setting, this requires integration over a sufficiently large area or catchment to retrieve a representative water balance at the local scale. Inputs to the balancing volume are precipitation P_r, and anthropogenic water input by irrigation I and combustion processes F. Irrigation water supply I can be substantial, and can exceed monthly water input by precipitation in dry cities (e.g., by lawn sprinkling). F is typically small and can be approximated using direct CO_2 flux measurements (see Sect. 16.2.2.3). Outputs are evapotranspiration E, typically measured by EC at h_m, and runoff Δr. Runoff in urban systems is greatly enhanced relative to non-urbanized ecosystems, because impervious surface materials and infrastructure, such as sewer systems, promote rapid runoff. The only significant storage within the balancing volume occurs in the subsurface materials (ΔW) as other urban surface materials (roofs, walls) are typically relatively impervious and hence water storage is negligible.

16.2.2.3 CO_2 Fluxes in the Context of Urban Metabolism Studies

In analogy to the energy and water balance, the urban carbon balance is also approached as a *volume budget* with the upper boundary at h_m, although the lower boundary is the 3D surface in order to avoid transformations between CO_2 and organic carbon pools (Fig. 16.4c). In urban ecosystems, there is import of carbon as construction material, food, etc. and export of carbon in form of waste, litter, and lawn clippings (Churkina 2008), so it is more straightforward to focus on transformation processes between the atmosphere and the surface (respiration, combustion, photosynthesis) rather than tracking the complex carbon pools. This approaches the carbon balance simply as a mass balance of CO_2 in the UCL.

Using the eddy covariance approach, the integrative turbulent mass flux density of CO_2 (F_c^{EC}) is measured in μmol m^{-2} s^{-1} at h_m:

$$F_c^{EC} + F_c^{STO} = C + R_{ue} + GEP \tag{16.3}$$

The exchange at the surface between urban carbon pools and the atmosphere can be separated into *combustion processes C*, *urban ecosystem respiration* R_{ue}, and *photosynthesis GEP*. F_c^{STO} is the storage change over time in the indoor and outdoor air between the surface and the measurement level h_m (see also Sect. 16.4.1).

In contrast to non-urbanized ecosystems, CO_2 exchange between urban surfaces and the atmosphere is nearly always positive and often dominated by emissions from fossil-fuel combustion processes, *C*. Mass flux densities of fossil-fuel combustion can be conceptually separated into emissions from vehicles C_V and emissions from buildings C_B due to space heating and industrial processes. Similar to the anthropogenic heat release Q_F, sources of C_V and C_B are not distributed homogeneously in the vertical or horizontal and C_V is released from mobile sources at ground level, while C_B comes from point sources (venting systems, chimneys). C_V and C_B follow diurnal, weekday, and seasonal human activity cycles (traffic load, heating requirements).

Further, urban ecosystem respiration R_{ue} is not only the result of autotrophic and heterotrophic respirations in urban soils and vegetation R_{SV}, but also includes waste decomposition R_W and human respiration R_M. Respiration in soils and vegetation R_{SV} is promoted by intensive irrigation and fertilization, that are common in heavily managed residential ecosystems. Residential lawns can regularly reach respiration flux densities larger than 10 μmol m^{-2} s^{-1} (lawn) for well-watered plots under warm summertime conditions (Christen et al. 2009b). Also human respiration R_M can be a significant term in the carbon balance. Moriwaki and Kanda (2004) estimated for a densely populated urban area in Japan (118 inhabitants ha^{-1}) that human body respiration is 2.2 μmol m^{-2} s^{-1}, which contributes at their site to 38% to the total F_c^{EC} in summer and 17% in winter.

Urban vegetation (trees, lawns, gardens, etc.) also can be expected to show higher productivity and higher annual total carbon sequestration due to: (1) water availability by irrigation, (2) generally warmer and more conservative temperatures in urban ecosystems (urban heat island) that extend the vegetation period and reduce frost damage, and (3) fertilization by elevated atmospheric nitrogen deposition and elevated CO_2 concentration in cities (Trusilova and Churkina 2008). On the other hand, air pollution can also lead to significant physiological stress and damage and reduce *GEP* in areas where in particular O_3 concentrations are high.

16.2.3 Other Trace Gases and Aerosols

Few studies exist concerning eddy covariance measurements of other non-CO_2 trace gases and aerosols in cities. With the development and availability of new fast response instruments based on narrow band spectroscopy and mass spectrometry (see Sects. 2.4.4 and 10.3.2 also Chap. 2), it is very likely that in the context of air pollution and greenhouse gas emission reduction strategies, EC-based studies will become more common in the near future, since the technique provides an excellent tool to improve and evaluate emissions inventories, and to better understand urban atmospheric chemistry. Recent studies reported EC measurement of VOC fluxes (Velasco et al. 2005a, b, 2009; Langford et al. 2009), N_2O fluxes (Famulari et al. 2009) and aerosol fluxes including their composition (Dorsey et al. 2002; Longley et al. 2004; Donateo et al. 2006; Järvi et al. 2009b).

16.3 Challenges in the Siting of Urban EC Stations

Despite the heterogeneity of urban surfaces, valid and representative results can be obtained if attention is paid to some principles and concepts specific to urban areas. However, choosing an "ideal" location for EC measurement in cities is hardly possible; the limitations set by logistical and experimental difficulties encountered in urban environments are much higher even than they are for forests. Observations may therefore be limited to certain aspects of turbulent exchange in urban ecosystems since measurements campaigns with more than one measurement tower/location are the exception. Due to logistical, safety restrictions and the need to have public and regulatory acceptance, measurements often have to be performed from existing towers, which are restricted in height and/or with non-adequate fetch. A good companion is flexibility, because it is necessary to consider "nonstandard" exposures in terms of height, surfaces, buildings, and anthropogenic sources of heat, water vapor, and CO_2 (Oke 2006b).

EC measurements reported from urban ecosystems are collected driven by different motivations, and to address various spatial scales. Most commonly, EC systems are implemented to measure integrative fluxes from a "typical" LCZ. In that case, EC systems should be mounted on sufficiently tall towers near the top or above the RSL, but within the ISL of the internal boundary layer of the underlying LCZ. Given the various properties of urban districts it follows that the placement of towers and instruments must be adapted to the respective LCZ. The patchiness of LCZs in an urban landscape limits the vertical extent of the internal boundary layers z_{ib} that develop over each LCZ and form the upper boundary for LCZ-specific measurements. The lower boundary is the height of the RSL z_r so as to avoid the influence of individual roughness elements. The resulting narrow height-range $z_r < z < z_{ib}$, along with logistical limitations often require trade-offs between an acceptable level of RSL influence (e.g., measure in the upper RSL) and nonideal footprints (e.g., measure above z_{ib} for selected stabilities and wind directions but exclude certain cases from analysis). We therefore strongly recommend checking the estimates given by simple footprint models (refer to Chap. 8) before deciding on placement of a flux tower. Note that source areas for turbulent fluxes and radiation sensors are unlikely to match, so special care must be taken so as to gather energy balance measurements that represent the same LCZ.

In addition, some studies use tall towers to investigate regional fluxes from urban-rural landscapes, and special systems have been used to measure the variability of fluxes within the RSL/UCL. EC systems within the RSL can provide localized measurements (e.g., turbulent kinetic energy TKE, statistical moments), but their interpretation is severely limited because many underlying assumptions of the EC theory are not fulfilled (including horizontal homogeneity, vertical direction of relevant flux densities, negligible dispersive fluxes, and negligible advection). It is of great importance to avoid (if not explicitly desired) zones that are characterized by streamlines that have been perturbed by flows around isolated high-rise buildings, roof geometry, street canyons, and flow regimes caused by varying canyon aspect ratios.

16.4 Implications of the Peculiarities of the Urban Boundary Layer on EC Measurements

One of the great advantages of urban sites compared to flux towers in non-urbanized ecosystems is the usually excellent documentation of the urban environment. Aerial photographs, emission inventories, high-resolution maps and 3D surface models, census data, traffic statistics, etc. are typically available from city authorities. It is strongly recommended to make extensive use of those data sets before, during, and after a campaign. With the additional application of footprint models, the source area of an urban flux tower can be described in detail and that allows geostatistical interpretation of the results, the definition of rejection criteria, and passive experimental control.

In the following, the most important derivations related to the EC technique if compared to nonurbanized ecosystems are addressed.

16.4.1 Advection and Storage

While fluxes from grassland, crops, and forests are supposed to be representative of a specific ecosystem, urban fluxes rather represent a specific LCZ, if at all. Since spatial heterogeneity of the roughness elements and of the source/sink distributions is the rule in urban ecosystems; therefore, advective and storage fluxes have to be considered within the concept of the volume balance approach discussed in Sect. 16.2.2.

Advection occurs at three scales in cities. First, at the microscale, advection is expected to be the rule within the UCL, for example, for sensible heat between shadowed and sunlit patches, for latent heat between wet and dry patches, and for air pollutants between high-emission patches (e.g., streets) and passive patches (e.g., courtyards). Microscale advection, however, is not a concern to local scale EC measurements in the ISL, as turbulence blends those effects and the EC system on top of the RSL responds to the integral effects of a microscale patchwork. However, microscale advection also can mean that nonlinear interactions could exist, for example, evapotranspiration is boosted in a small-scale patchwork of dry and wet patches compared to a single large dry patch separated from a large wet patch, even though the relative fractions of wet to dry surfaces might be similar in both cases. Second, at the local scale, advective fluxes are present due to the too close proximity of urban parks, water bodies, and between built-up areas of different density. If the study is not specifically interested in such fluxes, source areas that include several LCZs and steep topography should be avoided. Third, mesoscale advection occurs between the city as a whole and the surrounding rural environment ("urban breeze"), or, for coastal cities, due to the presence of sea breezes (Pigeon et al. 2007b). Further, the surrounding topography may also induce anabatic and katabatic flows similar to those found in connection with other ecosystems (e.g., mountain-valley and/or slope wind systems, drainage flows). In practice, advection is rarely measured in urban field experiments and similar difficulties to those reported from nonurbanized ecosystems (Aubinet et al. 2010) may arise in interpreting the impact of advective fluxes in urban environments.

The considerable height aboveground where EC systems are operated in urban studies can result in non-negligible vertical flux divergences in the air volume below measurement height h_{m}. A flux divergence of heat and concentrations over time within the balancing volume is explicitly considered in the budget equations and as part of the storage terms (see Sect. 16.2.2 and Fig. 16.4). However, in many applications, instantaneous emissions and uptake at the surface become significant, that is, at the 3D ground–building–air interface, not at h_{m}. Flux densities measured at h_{m} by eddy covariance $\overline{w'c_s{}'}|_{h_m}$ can be "reduced" to a spatial average flux density at the interface, F_0. This is achieved using representative measurements (typically vertical profiles) of the change of concentrations of heat and mass in the air volume over time $\left\langle \overline{\partial c_s / \partial t} \right\rangle$, in analogy to common practice in nonurban ecosystems with tall canopies (see also Eq. 1.24, Sect. 1.4.2):

$$F_s = \overline{w'c_s{}'}|_{h_m} + \int_0^{h_m} \Lambda_a \left\langle \frac{\overline{\partial c_s}}{\partial t} \right\rangle dz \tag{16.4}$$

In addition to forest ecosystems, the additional term Λ_{a} must be introduced; it is the volume fraction of outdoor air in the total balancing volume at a particular height z. In the UCL, buildings occupy a significant fraction of the total volume, so Λ_{a} can be as small as 30% in dense urban neighborhoods (in forests, the volume occupied by trees is negligible). Strictly speaking, buildings contain airspace as well but "indoor" airspace is mechanically and thermally decoupled from the outdoor atmosphere and the interface of interest is typically the building shell. The vertical profile of Λ_{a} can be extracted from 3D building data sets, or simply be estimated as a single number in the UCL based on the plan area fraction λ_{P} of buildings.

Table 16.2 summarizes typical values of measured flux divergences of sensible heat ΔQ_{Sa} and carbon dioxide ΔS_a in the air volume below h_{m} compared to the magnitude of the fluxes measured by an EC system on a 30 m tower in a dense European city center (Vogt et al. 2005). Storage changes were calculated using a

Table 16.2 Average values (June 15–July 15, 2002, Basel-Sperrstrasse) of flux densities measured at the top of an urban 30 m tower and the effect of storage change in air volume below the measurement height

	Sensible heat flux			CO_2 flux		
	Tower	Storage	Corrected	Tower	Storage	Corrected
	$H_{(hm)}$	ΔQ_{Sa}	$H_{(0)}$	$F_{C(h_m)}$	ΔS_a	$F_{C(0)}$
Time	(W m^{-2})	(W m^{-2})	(W m^{-2})	(μmol m^{-2} s^{-1})	(μmol m^{-2} s^{-1})	(μmol m^{-2} s^{-1})
03:00	+21	−3	+18	+5.3	+1.2	+6.5
09:00	+114	+7	+121	+14.8	−3.4	+11.4
15:00	+222	+2	+224	+14.4	+0.0	+14.4
21:00	+23	−5	+18	+11.7	+2.5	+14.2

profile of six thermometers, and ten CO_2 gas multiplexer inlets at various heights and taking into account the vertical profile of Λ_a. Table 16.2 underlines that for sensible heat flux, during the day, flux divergence below h_m is small ($\Delta Q_{Sa} < 5\%$ of H), but more relevant during the night (up to 30% of H). Note that during the night, measured H at tower top stays positive, but the air in the UCL below cools down. For the CO_2 flux, storage in the air below measurement level is even more relevant, in particular in the morning when the onset of thermal mixing "flushes" out CO_2 enriched air from the street canyons, creating an overestimated CO_2 flux at the tower top. For longer time periods, such as daily or yearly totals, storage in the air volume can be neglected for all fluxes.

Finally, note that in our underlying conceptual framework (Fig. 16.4 and Eqs. 16.1, 16.2, 16.3 and 16.4) advection is completely neglected, but in a real urban setting, horizontal advection processes are likely to occur on several scales. So ultimately, the concentration change over time in the volume below h_m could also be the result of horizontal exchange processes.

Another urban-specific challenge arises from the fact that Eq. 16.4 includes the horizontally averaged concentration change $\left\langle \overline{\partial c_s} / \partial t \right\rangle$ at various heights. While air within forest canopies can be expected to be reasonably well mixed (horizontally), we often encounter horizontally disconnected airspaces in the lower UCL, for example, inner courtyards can be separated from street canyons, and show different concentrations and changes over time. Under situations with horizontally isolated airspaces, a single profile is inadequate to quantify the storage change, and ideally, several, horizontally separated measurements are required.

16.4.2 Flow Distortion

Average streamlines that are not parallel to the ground are a problem for flux measurements (Finnigan et al. 2003). Normally, this problem can, at least partially, be overcome by appropriate rotation and calibration procedures (see Sects. 3.2.4 and 7.3.3.2). However, flow distortion in urban environments can also arise from

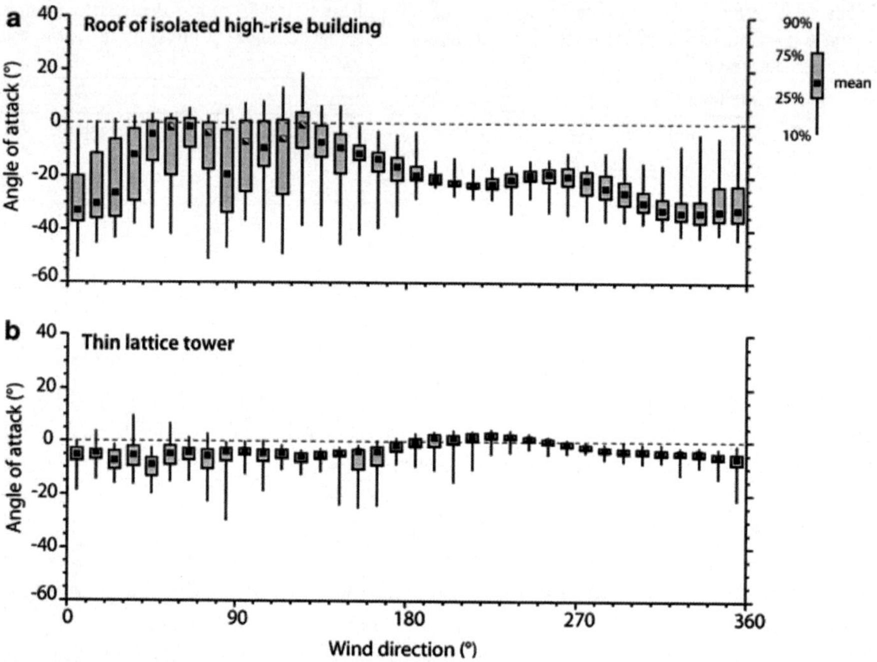

Fig. 16.5 Flow distortion from two selected urban EC measurement sites expressed as average vertical angle of attack: System (**a**) is located 3 m above the roof top of a 120 m high-rise building, (**b**) system is operated on a 20 m lattice tower on the top of a 20 m building

flow separation and deflection around and over local and distant buildings in an urban canopy. In contrast to plant canopies that are mostly permeable and porous, buildings are bluff bodies that are impermeable and inflexible. Therefore, buildings create strong dynamical pressure differences across their facets that in turn leads to significant vertical wind components, decreased mean wind, and enhanced turbulent kinetic energy. For isolated buildings, a displacement of the mean streamlines can be detected several building heights above the roof and significant wake effects can be still found 10–15 times the building's height downwind (e.g., Oke 1987, Fig. 7.6).

As a consequence, EC measurements near dynamical pressure gradients, that is, near exposed walls and on roofs of buildings, are to be avoided (Oke 2006b). For logistical reasons, it may seem inviting to use isolated high-rise buildings as platforms for EC systems in urban areas, but in nearly all cases they are inappropriate. Figure 16.5a shows the angle of attack versus wind direction from an EC system mounted on a small 3 m mast on the edge of the roof of an isolated high-rise building (building height: 120 m, building width: ~40 m) and gives an idea of the strongly deformed wind in the region directly at the roof's edge. Although the EC system is located six times above the average height of the buildings in this neighborhood ($h_m/z_h \sim 6$) and hence expected to be well within the ISL, flux measurements at this location are impossible. The second example in Fig. 16.5b shows the same graph for an EC system on a 20 m lattice tower on top of a building considered "typical" for the given LCZ (building height: 20 m, building width: ~20 m). Although the vertical location of this EC system is at only $h_m/z_h = 2$, the measurement location shows much less flow distortion. In summary, not only is the height of the EC measurements a determinant of the location of urban flux measurements, but also flow distortion can be a severe limitation in the choice of appropriate platforms and measurement locations.

16.4.3 Night Flux Problem, Gap Filling, and QC/QA

The typical premises that lead to the night flux problem (Chap. 5), – that is, low u_* and a stably stratified and decoupled canopy layer – are rarely found in urban atmospheres. The significant roughness of all urban surface forms produces mechanical turbulence which together with the release of stored and anthropogenic heat promotes thermal turbulence, and this produces a well-mixed ISL day and night. Hence, an underestimation of fluxes during nighttime is considered not as critical in cities as it is in a forest ecosystem. Figure 16.6 shows the nighttime frequency for dynamical stability classes in the ISL at five simultaneously operated sites in and around the city of Basel, Switzerland. The frequency of nighttime stable situations decreases from 60% at the rural sites to only 10% in the city center. Unstable situations dominate the nighttime atmosphere at the urban sites due to significant storage and anthropogenic heat releases (Christen and Vogt 2004).

It follows from this discussion that gap filling of urban EC data is mainly restricted to statistical methods (Chap. 6), since the models for respiration and light response have to be adapted to the specific urban conditions and/or are of lesser importance for certain LCZs. Quality-control tests, as described in Chap. 4, may result in a large number of rejected data, since these tests are heavily based on MOST, which is subject to fail in the urban RSL, and nonstationarity is likely to be increased, in particular during daytime, due to increased thermal convection. However, because there is currently no urban-specific QC/QA framework available, it is recommended that the data be tested using the procedures described in Chap. 4 as a first step. Some restrictions may be eased in a further step.

16.4.4 Service and Maintenance of Instruments

In addition to the particular problems arising from the very nature of a city, as described in previous sections, contamination of instruments (transducers, IRGA windows) due to aerosols is the most crucial issue. This implies increased attention to service and maintain the site and the instruments in particular. Advantages and disadvantages of open- and closed-path systems have been previously discussed in

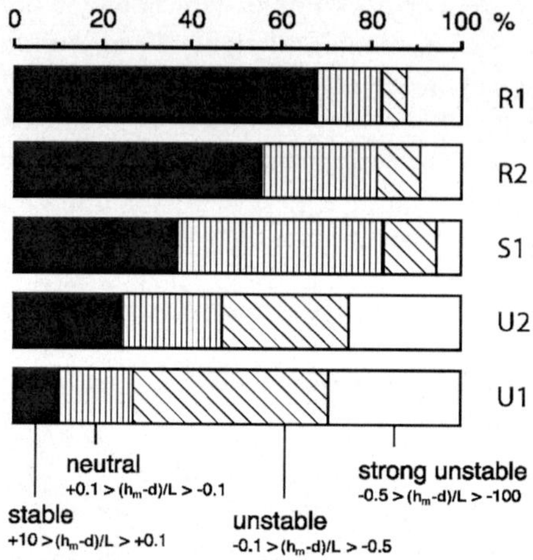

Fig. 16.6 Frequencies of different stability classes for nighttime cases (22–04 h, CET) measured simultaneously at five EC towers in Basel, Switzerland, between 10 June and 10 July 2002 (Modified from Christen and Vogt (2004))

this book (Sect. 2.4.4) and also apply for urban flux towers and instrumentation (Järvi et al. 2009a). However, for both systems, additional care should be taken. For closed-path systems we recommend an interval of a maximum of 1 week for the replacement of the inlet filters, and this may need to be reduced to a few days in heavy polluted environments. The same rule applies to the cleaning interval of open-path sensors. Internal sampling cells of closed-path system IRGAs, though protected by air filters, also need increased attention when exposed to urban polluted air.

16.5 Summary and Conclusions

Using the EC methodology for urban areas is still not a plug-and-play application. With some efforts, however, adequate results from EC measurements in urban areas can be obtained if attention is paid to certain urban-specific peculiarities. Siting of an urban flux tower is much more crucial as it is for sites in non-urbanized ecosystems. It is of highest importance that the researcher is aware of the site-specific influences on the flux measurements, because the "ideal" urban flux site does not exist. Extensive knowledge of the source area characteristics and careful analysis of the flow distortion by the close surroundings of the site are inevitable for a proper interpretation of the measurements. Keeping this in mind and applying the usual data processing chain, flux measurements by the EC technique can be a valuable tool for the characterization of part of the urban metabolism, that is, energy and mass fluxes. Originally restricted to fluxes of energy, water, and carbon, new instruments made the EC technique also attractive to measure fluxes of particle matter, VOC and N_2O, which in turn is helpful for the characterization of the chemistry of the urban atmosphere in the context of air pollution studies.

The instrumentation of an urban and a "nonurban" flux tower is essentially identical and the same advantages and disadvantages for open- and closed-path sensors apply. Some attention should be paid to the increased contamination of instruments and air filters due to the higher air pollution; apart from that, urban flux towers are serviced in the same manner as flux towers in non-urbanized ecosystems.

Acknowledgments The research leading to these results has received funding from the European Community's 7th Framework Programme (FP7/2007-2013) under grant agreement ne211345. The authors thank Tim Oke for valuable scientific discussion.

References

Arnfield AJ (2003) Two decades of urban climate research: a review of turbulence, exchanges of energy and water, and the urban heat island. Int J Climatol 23:1–16

Aubinet M, Feigenwinter C, Heinesch B, Bernhofer C, Canepa E, Lindroth A, Montagnani L, Rebmann C, Sedlak P, Van Gorsel E (2010) Direct advection measurements do not help to solve the night-time CO2 closure problem: evidence from three different forests. Agric For Meteorol 150:655–664

Christen A, Vogt R (2004) Energy and radiation balance of a central European city. Int J Climatol 24:1395–1421

Christen A, van Gorsel E, Vogt R (2007) Coherent structures in urban roughness sublayer turbulence. Int J Climatol 27:1955–1968

Christen A, Vogt R, Rotach MW (2009a) The budget of turbulent kinetic energy in the urban roughness sublayer. Bound Layer Meteorol 131:193–223

Christen A, Coops N, Crawford B, Liss K, Oke, TR (2009b) The role of soils and lawns in urban-atmosphere exchange of carbon dioxide. In: 7th International Conference on Urban Climate, Yokahama, Japan

Churkina G (2008) Modeling the carbon cycle of urban systems. Ecol Model 216(2):107–113

Donateo A, Contini D, Belosi F (2006) Real time measurements of PM2.5 concentrations and vertical turbulent fluxes using an optical detector. Atmos Environ 40:1346–1360

Dorsey JR, Nemitz E, Gallagher MW, Fowler D, Williams PI, Bower KN, Beswick KM (2002) Direct measurements and parameterisation of aerosol flux, concentration and emission velocity above a city. Atmos Environ 36:791–800

Famulari D, Nemitz E, Di Marco C, Phillips GJ, Thomas R, House E, Fowler D (2009) EC measurements of nitrous oxide fluxes above a city. Agric For Meteor 150:786–793. doi:10.1016/j.agrformet.2009.08.003

Feigenwinter C, Vogt R (2005) Detection and analysis of coherent structures in urban turbulence. Theor Appl Climatol 81:219–230. doi:10.1007/s00704-004-0111-2

Finnigan JJ, Clement R, Malhi Y, Leuning R, Cleugh HA (2003) A re-evaluation of long-term flux measurement techniques – Part I: averaging and coordinate rotation. Bound Layer Meteorol 107:1–48

Grimmond CSB (2006) Progress in measuring and observing the urban atmosphere. Theor Appl Climatol 84:3–22. doi:10.1007/s00704-005-0140-5

Grimmond CSB, Oke TR (1999) Aerodynamic properties of urban areas derived from analysis of urban form. J Appl Meteorol 38:1262–1292

Grimmond CSB, King TS, Roth M, Oke TR (1998) Aerodynamic roughness of urban areas derived from wind observations. Bound Layer Meteorol 89:1–24

Grimmond CSB, Salmond JA, Oke TR, Offerle B, Lemonsu A (2004) Flux and turbulence measurements at a densely built-up site in Marseille: heat, mass (water and carbon dioxide), and momentum. J Geophys Res 109:D24101. doi:10.1029/2004JD004936

Grimmond CSB, Young D, Lietzke B, Vogt R, Marras S, Spano D (2010) Inventory of current state of empirical and modelling knowledge of energy, water and carbon sinks, sources and fluxes. BRIDGE-Collaborative Project. FP7 contract 211345, pp. 108. Available online on: http://www.bridge-fp7.eu/images/reports/BRIDGE%20D.2.1.pdf

Ichinose T, Shimodozono K, Hanaki K (1999) Impact of anthropogenic heat on urban climate in Tokyo. Atmos Environ 33:3897–3909

Järvi L, Mammarella I, Eugster W, Ibrom A, Siivola E, Dellwik E, Keronen P, Burba G, Vesala T (2009a) Comparison of net CO2 fluxes measured with open- and closed-path infrared gas analyzers in an urban complex environment. Boreal Environ Res 14:499–514

Järvi L, Rannik Ü, Mammarella I, Sogachev A, Aalto PP, Keronen P, Siivola E, Kumala M, Vesala T (2009b) Annual particle flux observations over a heterogeneous urban area. Atmos Chem Phys 9:7847–7856

Kanda M (2006) Large eddy simulations on the effects of surface geometry of building arrays on turbulent organized structures. Bound Layer Meteorol 118:151–168

Kastner-Klein P, Rotach MW (2004) Mean flow and turbulence characteristics in an urban roughness layer. Bound Layer Meteorol 111:55–84

Langford B, Davison B, Nemitz E, Hewitt CN (2009) Mixing ratios and eddy covariance flux measurements of volatile organic compounds from an urban canopy Manchester, (UK). Atmos Chem Phys 9:1971–1987

Longley D, Gallagher MW, Dorsey JR, Flynn M, Bowr KN, Allen JD (2004) Street canyon aerosol pollutant transport measurements. Sci Total Environ 334–335:327–336

Moriwaki R, Kanda M (2004) Seasonal and diurnal fluxes of radiation, heat, water vapor and CO_2 over a suburban area. J Appl Meteorol 43:1700–1710

Moriwaki R, Kanda M (2006) Local and global similarity in turbulent transfer of heat, water vapour, and CO_2 in the dynamic convective sublayer over a suburban area. Bound Layer Meteorol 120:163–179. doi:10.1007/s10546-005-9034-4

Moriwaki R, Kanda M, Harumi N (2006) CO_2 build-up within a suburban canopy layer in winter night. Atmos Environ 40:1394–1407. doi:10.1016/j.atmosenv.2005.10.059

Offerle B, Grimmond CSB, Fortuniak K (2005) Heat storage and anthropogenic heat flux in relation to the energy balance of a central European city centre. Int J Climatol 25:1405–1419

Oikawa S, Meng Y (1995) Turbulence characteristics and organized motions in a suburban roughness sublayer. Bound Layer Meteorol 74:289–312

Oke TR (1976) The distinction between canopy and boundary-layer urban heat islands. Atmosphere 14:268–277

Oke T (1987) Boundary layer climates. Methuen, London

Oke TR (2006a) Towards better scientific communication in urban climate. Theor Appl Climatol 84:179–190

Oke TR (2006b) Initial guidance to obtain representative meteorological observations at urban sites. Instrument and Observing Methods (IOM), report No.81, WMO/TD. No. 1250. World Meteorological Organization, Geneva

Pigeon G, Legain D, Durand P, Masson V (2007a) Anthropogenic heat release in an old European agglomeration (Toulouse, France). Int J Climatol 27:1969–1981

Pigeon G, Lemonsu A, Grimmond CSB, Durand P, Thouron O, Masson V (2007b) Divergence of turbulent fluxes in the surface layer: case of a coastal city. Bound Layer Meteorol 124:269–290

Raupach MR, Finnigan JJ, Brunet Y (1996) Coherent eddies and turbulence in vegetation canopies: the mixing-layer analogy. Bound Layer Meterorol 78:351–382

Roberts SM, Oke TR, Grimmond CSB, Voogt JA (2006) Comparison of four methods to estimate urban heat storage. J Appl Meteorol Climatol 45:1766–1781

Rotach MW (1999) On the influence of the urban roughness sublayer on turbulence and dispersion. Atmos Environ 33:4001–4008

Rotach MW (2001) Simulation of urban-scale dispersion using a Lagrangian stochastic dispersion model. Bound Layer Meteorol 99:379–410

Roth M (2000) Review of atmospheric turbulence over cities. Q J R Meteorol Soc 126:941–990

Stewart I (2009) Classifying urban climate field sites by "Local Climate Zones". Urban Climate News 34, 8–11. www.urban-climate.org/IAUC034.pdf

Stewart I, Oke TR (2009) Newly developed 'thermal climate zones' for defining and measuring urban heat island magnitude in the canopy layer. Preprints T.R. Oke symposium & eighth symposium on urban environment, Phoenix, Paper J8.3. American Meteorological Society, Boston, MA

Trusilova K, Churkina G (2008) The response of the terrestrial biosphere to urbanization: land cover conversion, climate, and urban pollution. Biogeosciences 5:1505–1515

Velasco E, Lamb B, Pressley S, Allwine E, Westberg H, Jobson BT, Alexander M, Prazeller P, Molina L, Molina M (2005a) Flux measurements of volatile organic compounds from an urban landscape. Geophys Res Lett 32:L20802. doi:10.1029/2005GL023356

Velasco E, Pressley S, Allwine E, Westberg H, Lamb B (2005b) Measurements of CO_2 fluxes from the Mexico City urban landscape. Atmos Environ 39:7433–7446

Velasco E et al (2009) EC flux measurements of pollutant gases in urban Mexico City. Atmos Chem Phys 9:7325–7342

Vogt R, Christen A, Rotach MW, Roth M, Satyanarayana ANV (2005) Temporal dynamics of CO_2 fluxes and profiles over a Central European city. Theor Appl Climatol 84:117–126. doi:10.1007/s00704-005-0149-9

Chapter 17
Database Maintenance, Data Sharing Policy, Collaboration

Dario Papale, Deborah A. Agarwal, Dennis Baldocchi, Robert B. Cook, Joshua B. Fisher, and Catharine van Ingen

"*If I have seen further*," Sir Isaac Newton wrote to Robert Hooke in 1676, "it is by standing on the shoulders of giants."[1] What Newton was implying was that he was able to do more, understand more, and further advance science as a whole because he was able to build on the advancements of his predecessors. If these "giants" had

[1]The phrase was, in fact, based on that of Bernard of Chartres five centuries earlier. (d. 1130): "We are like dwarfs sitting on the shoulders of giants. We see more than they do, indeed farther..." (*"Nous sommes des nains juchés sur des épaules de géant. Nous voyons ainsi davantage et plus loin qu'eux, non parce que notre vue est plus aigüe ou notre taille plus haute, mais parce qu'ils nous portent en l'air et nous élèvent de toute leur hauteur gigantesque."*). Gimpel, J., 1961. The Cathedral Builders. Grove Press, New York.

D. Papale (✉)
DIBAF, University of Tuscia, Viterbo, Italy
e-mail: darpap@unitus.it

D.A. Agarwal
Lawrence Berkeley National Laboratory, Berkeley, CA, USA
e-mail: DAAgarwal@lbl.gov

D. Baldocchi
Department of Environmental Science, Policy and Management, University of California, Berkeley, CA, USA
e-mail: baldocchi@berkeley.edu

R.B. Cook
Oak Ridge National Laboratory, Oak Ridge, TN, USA
e-mail: cookrb@ornl.gov

J.B. Fisher
Jet Propulsion Laboratory, California Institute of Technology, Pasadena, CA, USA

C. van Ingen
Microsoft Research, San Francisco, CA, USA
e-mail: vaningen@microsoft.com

M. Aubinet et al. (eds.), *Eddy Covariance: A Practical Guide to Measurement and Data Analysis*, Springer Atmospheric Sciences, DOI 10.1007/978-94-007-2351-1_17,

not shared their "shoulders," then Newton would have been limited in his ability to "see"; in other words, if his predecessors had not shared their work – namely their science and data – then Newton would not have been able to make as many scientific contributions as he did.

Scientific questions of today are now more global than ever before. The answers to these questions are buried within multiple disciplines and across a diverse range of scientists and institutions. The expanse and complexity of data required by researchers often exceed the means of a single scientist. Data sharing in the form of its distributed collection and analysis is increasingly common. Collective research now takes place in what may be called "collaboratories" or in "centers without walls" (Clery 2006).

Creating effective artifacts, which enable scientists to collaborate on data analyses, continues to be a significant challenge for today's science activities. It is rare that providing a file system abstraction on distributed data enables acceleration of scientific discoveries. By explicitly identifying and addressing the different requirements for data contributors, data curators, and data consumers, we can create a data management architecture which enables the creation of datasets that evolve over time with growing and changing data, data annotations, participants, and use rules.

This involves also a crucial contribution by the teams and people collecting the data, that in addition to carefully acquire and process the measurements and to be ready to share their measurements within the scientific community, need to follow general rules that help to make their data well documented and safely stored and to maximize visibility to their works and sites.

In this chapter, we provide examples of the types of functions and capabilities typically provided within the data management systems, focusing in particular on databases structures and characteristics, data practices, and data user services. Finally, the importance and advantages of collective efforts like data sharing for synthesis activities and the relative data policy options are discussed and analyzed.

17.1 Data Management

The eddy covariance (EC) technique produces a vast amount of data, from the 10 Hz measurements to the aggregated 30 min fluxes and ancillary data. In addition, differently from other centralized activities like, for example, the remote sensing data acquisition centers, the eddy covariance community is heterogeneous and it requires additional efforts to get the community working together using comparable measurements. To achieve this goal it is crucial to have a structure operating in this direction and the EC fluxes community is served by several interoperating data management facilities. Each of these facilities contributes to the overall data management, coherence, and usability of the network of sites measuring EC fluxes. The first data management layer is provided by the flux tower itself and the team that manages the tower. The raw data generated from measurements at the tower are quality checked and archived by the team responsible for the tower. The other data management centers include the regional networks and the FLUXNET global network. Each of these data management centers has a role in providing the overall flux data archiving and user services.

17.1.1 Functions

A typical data management system for EC fluxes and meteorological data provides one or more of the functions of the full data management system including robust archiving of data, generation of standardized data products, authenticated access to users, additional data products, and documentation for the data. The data management system as a whole becomes the focus of collaboration among the EC fluxes measurement scientists and a means to interact with the users of the EC flux data. As a whole, the combined data management systems for shared scientific datasets should exhibit the following properties:

- *Archive:* Carbon, energy, water, and other gas flux measurements data are exceedingly valuable and careful archiving of the raw data and generated products should be an integral part of the overall data management system.
- *Quality:* Data quality indicators to identify potential problems with the data stored as parameters of the data in the database. Identified data quality problems should also be addressed and corrected when possible. In addition to the quality checks applied by the data providers the data management center should apply additional controls to identify suspicious or erroneous data that need to be flagged and communicated to both the users and the data providers, providing an additional and independent quality control tool to the latters. The correction and processing methods implemented in the database must be always updated and advanced in agreement with the new scientific and technical findings.
- *Secure:* Most scientific data require access control and accountability (e.g., to determine *post facto* who has accessed the data) for a variety of reasons. For example, even when the policy allows anyone to access the data, requiring a registration and subsequent authentication step before access is granted allows access to be tracked (e.g., so that the impact of the data might be quantified via number of unique data accesses). The overall system must meet the collective security requirements (policy and mechanism) of the data providers and the users.
- *Scalable:* The network has grown dramatically in the last 10 years (Fig. 17.1), and it is expected to continue to grow in particular in the less covered areas of the globe. The system must be ready to answer to changes along a number of dimensions: size of dataset managed, size of meta-data managed, and number of active participants (authors, curators, publishers, and consumers). In addition, the data management system should be efficient in its behavior: stability and robustness, ability to serve the data to consumers, ability for contributors to upload new data/meta-data, and ability for curators to determine what requested

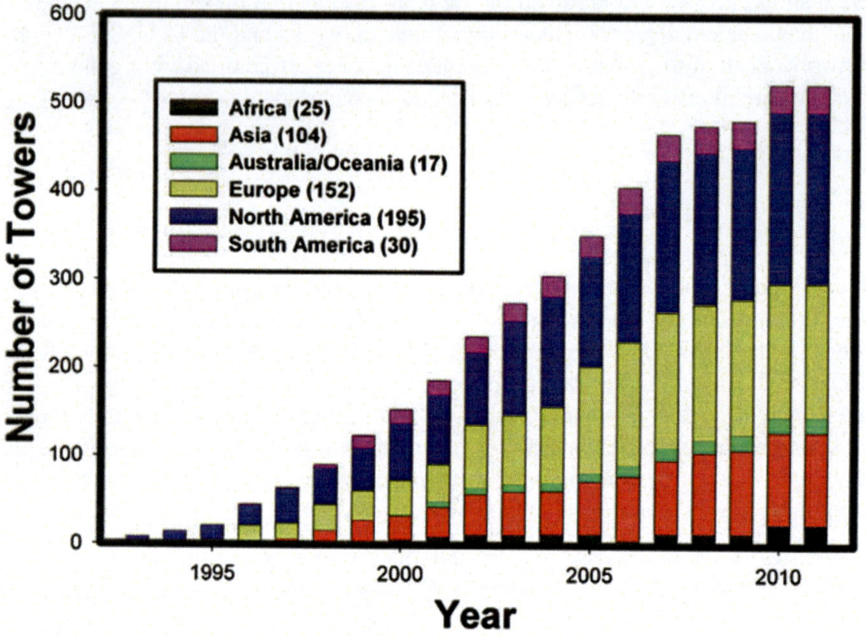

Fig. 17.1 Growth of FLUXNET by continent since 1993 (updated to March 2011). Data are from registrations to the FLUXNET ORNL database (www.fluxnet.ornl.gov, visit the website for updated information)

modifications are pending are just some examples of the services that need to be scalable and ready to answer new requests.

- *Usable:* Users must be able to easily find and obtain the data they need along with the relevant meta-data for their scientific explorations, explanations of the processes used to generate the data, and quality metrics for the data. Collectively, users need to be able to explore the data both based on keywords and on application-specific properties of the data – for example, "locate all scientific output (papers, derived datasets, etc.) directly or indirectly based on observations from SensorX in the range January 1, 1999 through June 30, 2002."
- *No special-purpose software for data access:* Ideally, the participants should not be required to learn new software packages to fully participate in the data analysis collaboration. The most attractive approach to this requirement is to ensure that users can fully participate using only a Web browser or spreadsheet program.
- *Provenance:* The data and meta-data that are held by the data management system is connected via potentially multiple set of relationships. For example, a potential consumer of a particular set of data might ask a question about the data in a particular blog, which might generate an answer that explicitly references another piece of data or meta-data. The data management system must be able to keep track of such histories and origins of data and meta-data, and such

provenance must be efficiently integrated into the rest of the data management system (such as the search capability).

- *Notifications:* The users of the system should not be expected to directly engage the data management system in order to determine what has changed since the last time they visited the system. That is, the users of the system should be able to register their interest in a variety of types of additions/modifications (e.g., data revisions/additions, meta-data revisions/additions, new users of a particular class) and be able to receive notifications via their choice of a variety of mechanisms (e.g., e-mail or SMS). In essence, the system should enable subscription to and *push* of information to the users of the system.
- *Additional processing:* The system should provide a number of derived variables calculated centrally using the raw data (in this case raw data are the calculated fluxes) ensuring the same methodology and calculation scheme. These derived products could include quality assurance and quality control flags (Chap. 4), gap-filled datasets (Chap. 6), and calculated additional variables like GPP and R_{eco} from partitioning (Chap. 9) or data-derived products such as water and radiation use efficiency, potential evapotranspiration, surface conductance, etc. In addition, it could be important to implement links with other data sources relevant for the users like meteorological networks, remote sensing products, and climate models results.
- *Track usage:* Papers and other scientific results derived from the data should be traceable, for example, creating a public list of products where each of the datasets included in the database are involved or cited such that the impact of a site's data for a particular time period can be identified by the data owner and funding agencies.

17.1.2 Flux Tower Repositories

The long-term value and quality of the EC data depends first and foremost on the quality of the measurements and care of the data provided by the individual flux tower teams. The flux tower teams have a critical role in the overall data management. Archiving of the raw measurement data by the tower team helps to ensure that the data are not lost and available for future recalculations using new methods or corrections. The use of standardized names and units for variables and keeping the site identifier and site name consistent throughout the life of the measurement site significantly reduces confusion and processing mistakes in particular during the reprocessing of older data. This applies to all of the data from the site not just the EC and meteorological data.

It is important to organize and archive the information about the data gathered together with meta-data such as measurement setup, instruments makes, models and serial numbers, instruments locations in the tower (height) or in the footprint (location, depth in the soil), calibration dates and methods, maintenance and disturbance information for each sensor, methodologies used to calculate fluxes and correct the data. The same is valid for all the biological disturbances and management information about the site that needs to be stored together with the information about the methodologies and people that collected the measurements and information. These data are critical to the utility of the EC measurements. Ideally data should be archived in at least two locations with one being off-site (in many cases the regional network will serve as the second off-site archive of the data). Regular testing of the archiving system increases the likelihood that data can be recovered in the case of an emergency.

17.1.3 Regional Repositories

The first examples of regional databases dedicated to eddy covariance measurements were proposed at the end of the 1990s in the context of the two major regional networks AmeriFlux and EuroFlux-CarboEurope. The two databases were initially relatively simple and without many functions. But they have evolved during the last 10 years, adding new functionality and collaboration with the aim to coordinate and standardize the services.

Coordinated and interconnected regional databases are more efficient than a single global network. First, political and cultural positions are better managed by local coordinators, who are more connected and have deeper understanding of local conditions than outside coordinators. Second, the lack of global-scale funding agencies currently interested in fully supporting a global database system for EC and related data, is somewhat offset by the interest of regional and continental funding availability to support regional databases (e.g., DOE for USA or EU for Europe). Therefore, harmonization of regional databases is needed to maintain and improve the interoperability and inter-database standardization. FLUXNET is a global initiative that works with regional networks and helps to create a network of networks that share processing options, standards, and policies to enable global-scale studies (Fig. 17.2).

There are many different regional databases that coordinate networks ranging from a few tens of sites to more than 100 stations. The main EC databases are AmeriFlux (http://public.ornl.gov/ameriflux/index.html), CarboEurope and CarboAfrica (http://www.europe-fluxdata.eu), Fluxnet-Canada (http://fluxnet.ccrp.ec.gc.ca/e_about.htm), and Asiaflux (https://db.cger.nies.go.jp/asiafluxdb/).

17.1.3.1 One Example: The European Eddy Covariance Flux Database System

The European network of EC sites started in 1996 with 16 forest sites in the EuroFlux-EU project and has grown to more than 140 sites since then, thanks to several other EU-funded projects like TCOS-Siberia, GreenGrass, CarboMont, CarboEuroflux, CarboEurope-IP, IMECC, CarboExtreme, and GHG-Europe. Paral-

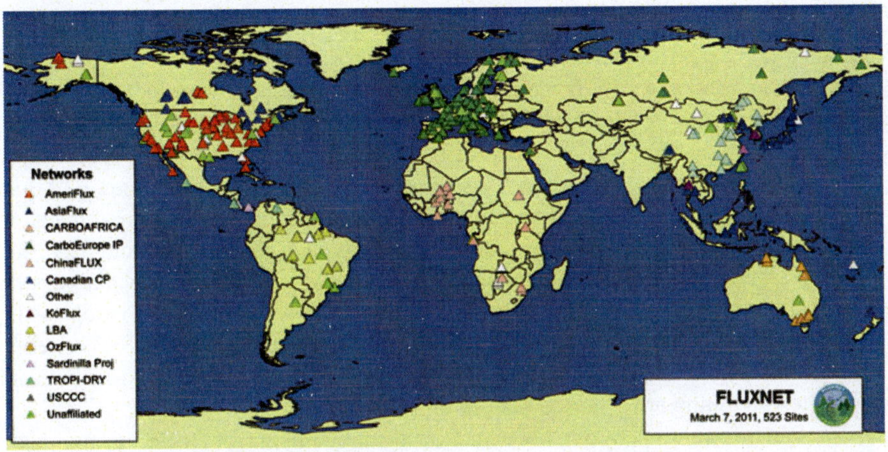

Fig. 17.2 The regional networks and FLUXNET. Continuously updated version is available at the FLUXNET ORNL webpage (www.fluxnet.ornl.gov)

lel to the network, the database system has been developed to host, quality control, standardize, and distribute the data acquired at the different sites. The database, based on SQL and .NET, is currently located in Viterbo (Italy) at the University of Tuscia and the server has a double backup system and is connected to a set of workstations for data processing. To ensure standardization and promote and help data exchange between the different projects, the different projects share the same database structure and processing so that data products are presented in the same way (same format, same quality flags meaning, same variables names and units, same processing scheme); sites involved in more than one project do not need to upload and curate the data in different interfaces (see www.europe-fluxdata.eu). At the same time, the project part of the general database maintains specific interfaces that give visibility, simplify the intra-project coordination and can host additional project-specific datasets.

In 2005, the European Eddy Covariance fluxes database introduced a number of new tools that improved the suite of services offered to data owners and data users and changed the philosophy of the database, moving for a data repository to a real data management system. A standard data processing was introduced, including the implementation of the QAQC described in Papale et al. (2006), the u_* filtering in Reichstein et al. 2005, two gap-filling alternatives (MDS – Reichstein et al. 2005 and ANN – Papale and Valentini 2003), flux partitioning (Reichstein et al. 2005), and the dataset was provided in a standard format to simplify the use and automate the reading. At the time when this chapter was prepared, the suite of processing options implemented in the database was under review to include new QAQC tools and a new partitioning method (Lasslop et al. 2010). The data processing will likely continue to evolve and improve over time and it is one of the most important characteristics of a database system to be always ready to implement new processing schemes proposed in the literature also because these are often difficult to be applied individually by each single site manager.

Two other main services introduced were the versioning and the PIs Information system (PIs are the Principal Investigators, the site responsible and data owners). The versioning system was introduced to track new versions or new processing schemes applied to already existing and used datasets. In fact, development in the calculation and processing methods or errors discovered late in the data are the main causes of new versions of datasets that have been already downloaded and used. In these cases, it is important to inform all the users that a new version is available and to track all the changes that have occurred to the data from the first version released. Only with this information is it possible to fully understand the differences between successive studies based on the same sites but using different datasets. More details about the versioning are reported later in this chapter.

The PIs Information System is a section of the database dedicated to the data providers. Here the PIs can upload the data, track the status of the data processing, define the data access and data use policies to be applied to the data, and see the list of the users that downloaded their sites measurements. In addition, an e-mail service automatically informs the PIs about who requested access to their data and analysis planned. Providing these information, the PIs are on one side more comfortable in sharing the data since they know who accessed the data and, if interested in collaborating with the study, can directly contact the data users to propose joint activities; on the other hand, they can have a quantification of how much their sites are interesting and useful for the scientific community and for this reason important to maintain active and to be funded.

17.1.4 The FLUXNET Initiative and Database

FLUXNET, which is a network of networks, is a collaboration between the regional networks and independent measurement sites which results in a global EC fluxes measurement network. This global network is brought together to enable global-scale synthesis activities. The main role of FLUXNET is to establish contacts between regional networks and EC people around the world and to maintain an updated inventory identifying which sites are active, their period of data taking, and what is measured at a site to facilitate exchanges and coordination between sites. FLUXNET also promotes synthesis activities through meetings and workshops where standardized datasets and derived products are used in scientific analyses at global scale. FLUXNET receives new data when a new dataset is gathered in preparation for a global synthesis effort (not continuously). The aim of each dataset collection effort is to produce the new FLUXNET dataset collecting preprocessed data from all the existing sites, through the regional databases or directly from independent sites. Typically, a FLUXNET Synthesis workshop is held in conjunction with the data collection effort to enable users to discuss the dataset's scientific potential, propose new cross-site synthesis analyses, and start these analyses. The first FLUXNET Synthesis workshop was held at the Marconi Conference Center in 2000. In preparation for the Marconi workshop, a dataset was gathered that contained 97 site years of data from 40 sites located primarily in the Americas and Europe. The dataset was quality checked and gap filled using a standardized methodology. The synthesis analysis efforts using this Marconi dataset resulted in 11 synthesis papers published in a special issue of Agricultural Forest and Meteorology in December 2002.

In the case of the Marconi dataset, all the synthesis teams knew the measurement site scientists personally, so communication and trust among the group members was relatively easy to establish. The data download and communication functions were handled manually (ftp, e-mail, and phone) for the Marconi dataset. The resulting dataset contained only the flux-met data from each site. The synthesis teams obtained any needed ancillary data for the sites directly from the measurement teams.

The second FLUXNET synthesis workshop was held in La Thuile, Italy, in February, 2007. As a result of the workshop and continuing efforts since the workshop, the La Thuile dataset was released to synthesis teams in September 2007. The delay in releasing the data was due to both the time needed to complete the gathering and processing of the data and the time needed to develop a portal to support sharing of the dataset. The released La Thuile dataset contains over 960 site years of data from more than 180 scientists working at over 250 measurement sites around the world. Four years later, there were more than 110 synthesis teams writing papers using these data (see http://www.fluxdata.org for the current list of published papers).

As the body of EC flux measurements increases, their value to the broader scientific community increases as well. For example, the FLUXNET LaThuile dataset was an order of magnitude bigger than any carbon flux measurement dataset that had been available before and is enabling cross-site, regional, ecosystem, and global-scale analyses. The LaThuile dataset was also more valuable than previous FLUXNET datasets because advanced standard processing methods were applied and the ancillary data were included as part of the dataset.

The amount of data and users involved in the LaThuile-2007 FLUXNET synthesis activity, together with the progress in the computing and IT structures available, led also to a radical change of the database structure with respect to the Marconi initiative. A database structure accessible through a web interface was developed, and it is currently the reference access point for the FLUXNET synthesis studies (www.fluxdata.org). This database is maintained updated and synchronized with the FLUXNET meta-database and the regional databases. A new data collection and processing is ongoing with the aim to release a new FLUXNET dataset in 2012. This new dataset will be significantly larger than the LaThuile dataset and with additional processing and uncertainty estimation tool included.

17.2 Data Practices

17.2.1 Contributing Data and Reporting Protocols

Data are generally submitted by the flux tower team to the regional databases, which are responsible for the next level of quality control and processing and also for the data transmission to the FLUXNET database for synthesis activities. This two-step process allows more precise control of the data and policies but can introduce differences between the regional networks and FLUXNET. For this reason coordination is important.

Protocols for data transmission to the regional networks are generally provided by the different databases and are often different. This does not affect the ability to share data between networks because it is one of the roles of the regional databases to import the data in the original template and export the data after quality control and quality assurance in a standard exchange format. From a centralized standpoint, one common data reporting template and protocol for all the sites is desirable. However, it is difficult to agree on protocols that can be applied in completely different ecosystems and climatic regions. Despite these difficulties, several large coordination efforts are ongoing. Some results have already been achieved; the Biological, Ancillary, Disturbances, Management (BADM) template (see www.fluxdata.org), originally developed in AmeriFlux, has been adopted by other networks (Europe, Africa). A period of suggestion of modifications and improvements to the original AmeriFlux template allowed inclusion of data types and information specific to other regions and originally missing from the template. This template is now used across the networks to report all the measurements and information that have low time resolution (daily to annual) but that are fundamental for a correct data interpretation. The BADM template is becoming an international standard in the FLUXNET community and is available as an Excel workbook to enable tower teams to fill in the information off-line and then submit the completed template. However, the data collected by the BADM template cover a wide range of data types and complexities. Some examples include plant species percents and site disturbance information which each require the entry of a text string that comes from a control vocabulary (species or disturbance type) and a value (percent or year) along with other relevant information. Although the template provides an easier mechanism for data input, the Excel format does not enable the input to be checked as it is entered. If the users do not follow the instructions for data input carefully, a great deal of manual work is required at the time of data ingest to correct the template.

Web-based database form interfaces can also provide services to directly update or submit ancillary data. This method has the advantage of fast submission and registration of the data. It also enables direct tracking of the origin of data and the ability to apply simple rules regarding the values reported to be sure that they can be imported correctly. An example is the submission of a text string where a value was expected: if entering for example the variable "disturbance_year" (the year when a disturbance happened) is reported as [two thousand], or [2000–2001], or [January 2001], or [03-01-2001], a web interface can immediately warn the submitter that it needs to be corrected. The correct value to report would be [2001] and other information about uncertainty in the year or exact day is entered as related variables or as a comment or annotation. A web interface for data entry improves the likelihood that the data can be imported correctly and will not require manual corrections.

Despite this improved control over data submission, data submitted using web-form interfaces are still affected by reporting errors. To enable these errors to be detected, a temporary table can be used to store newly submitted information until a curator for the site has verified it. The curators are experts who check the submission and ask the PI for clarification if the values are suspicious. A common example is an error in the units that makes the value "possible" but may be not "probable." Once a value has been confirmed, the curator accepts it and it is moved into the database.

17.2.2 Common Naming/Units/Reporting/Versioning

FLUXNET is a collaboration where data products are made by individual networks as well as cross-network synthesis groups. Data products include different processing methods of fluxes and meteorological time series field data, derived variables like light use efficiency, water use efficiency, or drought indices, ancillary data such as site classification, disturbance, and management history, or biomass characteristics and remotely sensing data. Today, these data products are usually produced as a collection of files. As the community grows and cross-site and cross-network synthesis studies have become more important, the need for standardization across the data has also grown.

17.2.2.1 Enabling Cross-site Analysis: Site Identifier, Variables, and Units

A prerequisite of cross-site analysis of the data is an ability to compare the data from different sites and regions. The first step in enabling analysis of the data is unique site identifiers: a site's identifier does not change unless the site changes significantly the location (i.e., different footprint, see Chap. 8) and different sites have different identifiers. This is typically accomplished through assigning codes for the identifiers for each of the sites, associated with the geographic coordinates. The use of identifiers allows the human readable name of the site to change without affecting the unique identifier (although it is not recommended that the site name change unless necessary).

A cross-site analysis also needs the data itself to be comparable across the sites. This would require in theory exactly the same instruments, setup, and methodologies to acquire and process the data implemented at all the sites. Although there are efforts ongoing in this direction with the organization of highly standardized measurements networks (see, e.g., ICOS, www.icos-infrastructure.eu), the knowledge of the sources of uncertainties related to the EC measurements processing (Chap. 7) allows the use of not fully standardized data for synthesis activities. This, however, requires the following of some basic rules such as reporting of common variables across the sites with associated meta-data to describe system and processing but also that the data be reported by all the sites with the same variable names and units. Although ideally the original data should be reported using standardized names and units that are agreed across the regional networks and FLUXNET, this is sometimes difficult due to the heterogeneity related to relevant measurements in different environments, different levels of detail, and different units adopted in different countries. A compromise is that the regional databases should harmonize variable names and units inside the network and implement conversion tools to meet the FLUXNET standards. The updated list of the standardized variable names and units are available in the regional databases and FLUXNET web pages and must be consulted during the dataset preparation before submission.

17.2.2.2 Data Releases

One of the aims for the near future is to update continuously (or frequently) the FLUXNET datasets with new and updated data. In this scenario, the flux-met data will arrive in large yearly batches and the ancillary data would be continuously updated. With a dataset in continuous evolution, the data users need to be able to indicate a version of the data that they used in performing their analysis. There are a wide number of models we can draw from in designing a data release strategy. Agencies such as USGS process their data once from collection to quality checking and release. Released data do not change. NASA uses a collection abstraction for their releases. They continually add new data to a collection as they are received and processed. NASA reprocesses the entire dataset if there are revised processing calibrations or algorithms. Reprocessed data are released as a new collection. The fluxdata.org site has adopted a strategy similar to the NASA collections and releases a frozen version of the data before each major update to the flux-met data (particularly when all the data has been reprocessed, which is currently the case). A data release spans sites and contains files with the same data types in the same format and units. For example, a data release might consist of flux-met data files with half-hourly and daily aggregations for a set of site years and quality flags indicating the results of the centralized quality controls.

All files need to be versioned, which is effectively a serial number for the file and increments monotonically at any change to the data or file format. If the same file is used to create more than one data release, the version is unchanged. For example, data files for inactive sites will remain unchanged across different successive data releases unless the data are reprocessed using a new version of the data processing.

Data releases are either classified as *frozen* or *latest.*

- A frozen data release does not change. Users of a frozen data release have the guarantee that the same analysis will give the same answer. When possible, scientists should use a frozen data release for publication, particularly for synthesis studies. Frozen data releases are made by halting all changes to an existing latest release.
- A latest data release may change at any time. New files may be added, new file versions may replace existing files, files may be removed, and processing algorithms may change.

Networks create data releases at any time. These data releases may be published to only network members, for a synthesis set such as LaThuile, or to other collaborations. A release may not contain all sites in the network either due to quality bars or due to the targeted science.

When freezing a cross-network synthesis set (e.g., a FLUXNET dataset like for the LaThuile data collection – www.fluxdata.org – or the Marconi synthesis dataset – Falge et al. 2005), all of the contributing network and unaffiliated data releases should also be frozen. In other words, when a synthesis data release x is frozen, it should be based on frozen data releases from each network as well as the unaffiliated holdings of the original data used to create that synthesis data product. This may require the networks or unaffiliated sites to freeze an existing latest data release when they would not normally do so but the advantage is that the data release is then traceable throughout the system (the network knows exactly which data went into the release as does the site).

Best practices include:

- Each network makes available at least one frozen data release and the latest data release for any data products published by that network. For example, AmeriFlux would publish both the original data used to create the cross-network synthesis version x (the last time data were frozen by that network) as well as a latest release.
- Descriptive summary documentation should be provided to explain the differences between data releases, though it is not necessary to detail the change in every single point. Knowing what data have not changed is often as interesting as knowing what data have changed. Differences in annual values or variability may also be helpful.
- Data releases should include fluxes, meteorological data, and ancillary data. The combination enables richer science.

17.2.2.3 File Naming

The name of the file downloaded by the user, exchanged between regional networks or submitted for FLUXNET synthesis datasets preparation could include important information about the content. In this paragraph one possibility will be discussed and presented to better explain and illustrate the possibilities of standard file naming. The goal is to create file names that can easily be understood by humans, clearly identify the version and source of the data, identify the type of data and processing used, and clearly identify the set this data was published into. All this information could and should also be included in the meta-data and data information files, but having a summary in the file name would help to have an overview of the data characteristics.

File names could include, in addition to a code to identify the site and a number to identify the year, information about:

- Data type, like fluxes, meteorological, biological, and ancillary data, and remote sensing cutouts.
- Data version, with a monotonically increasing number that will identify successive versions of the same data (e.g., due to changes in the processing or errors in the previous data). This number changes only when file content is different from the previous version.
- Processing applied, if the data are as submitted to the database or have been processed with additional QAQC, gap filling, partitioning.
- Time resolution of the measurements, in particular for aggregated data from daily to yearly.
- Regional networks or synthesis activity that produced the data, useful for users that are working in the context of specific continental or project activities.

For example, a file name could be structured as

AAAA_LL_PPP_T_CC-SSS_YYYY_vvv.<extension>

where AAAA identifies the network (e.g., CEUR for CarboEurope, AMFL for AmeriFlux, CAFL for Canada); LL, the processing level (e.g., L1 for raw data, L2 processed data, L3 quality controlled, L4 gap filled . . .); PPP, the data type (e.g., FLX for fluxes, MET for meteo, ANC for ancillary, SLR for soil respiration . . .); T, the time resolution (e.g., H for half-hourly, D for daily, W for weekly . . .); CC-SSS, the site code with CC indicating the country, and SSS, the site code; YYYY, the year; and vvv, the version number.

17.2.3 Ancillary Data Collection

In 2007, ancillary data reporting protocols were developed in the context of a FLUXNET synthesis activity (BADM template, see www.fluxdata.org), but the formatting diversity and heterogeneity of these data still makes consistency difficult. Reporting of data ranges, approximate values, and qualifiers in lieu of simple numeric values is a common practice, so all ancillary data should be stored as text fields to preserve accuracy information and support non-numeric entries while conversion to numbers is possible during the data quality evaluation and control.

The submission date, user, and method should also be captured when ancillary data are submitted and this provenance information stored along with the values. In addition, all past values and their provenance information are kept so that the history of the values provided for a variable can be reconstructed. This enables ancillary data views that correspond to the database state at any particular point in time to be reconstructed.

Fluxdata.org provides an example of web interfaces and protocols for reporting values. The challenge is to capture ancillary data in the portal and check the submission format of the data when they are collected by the tower and analysis teams.

17.3 Data User Services

Flux data have the potential to benefit a wide range of scientific analyses. This broad usage has the potential to significantly enhance the impact and value of the data, but it can only be achieved if the data are available and usable. The data standardization, versioning, quality assessment, and curation components mentioned earlier in this chapter are precursors to data reusability. In this section, we discuss some of the other services and capabilities and data sharing rules and motivations.

17.3.1 Data Products: The Example of fluxdata.org

Large-scale synthesis studies are becoming more common. These synthesis studies are often conducted by science teams that are geographically distributed and on datasets that are global in scale. A broad array of collaboration and data analytics tools are now available to support these science teams. However, building tools that scientists actually use requires a lot of work. In this section we will refer to an example of collaboration between the regional networks and FLUXNET communities that led to the development of the fluxdata.org portal to support data analysis by users. The fluxdata.org infrastructure provides advanced data organization, mining, and analysis features through utilization of a database to organize the data. Cross-site data reports and On-Line Analytical Processing (OLAP) data cubes enable browsing of the data. It is an example of the types of user services and products that can be provided to the user community. We discuss the types of functions and users that an EC flux portal should support.

17.3.1.1 Users and Use Cases

A first important step in the construction of a database infrastructure is the definition of the users and related needs. In fluxdata.org, there are four primary types of users and associated use cases for a carbon flux data portal. These are:

- Analysis scientists (data users) – site selection, dataset information data download, analysis support, and paper writing support
- Measurement site scientists (data providers) – information about proposed and published papers using their data, data download, and data submission/update
- Regional flux networks (data curators) – data correction, checking, and update, coordination of the regional contributions
- The public – proposed and published papers information, dataset information, and funding information

Although it is tempting to think of each of the above user groups as distinct, this is not the case. Multiple groups share many of the use cases. Measurement site scientists are typically involved in synthesis activities and regional networks.

The primary use cases are identified below:

- Synthesis site selection – evaluate criteria that will determine which sites are suited to an analysis. Typically, most of the site selection process is done using high-level aggregated data about the sites like annual sums or averages or percentages of high-quality data and ancillary and meta-information about the site.
- Dataset information – ability to quickly answer simple questions about the dataset such as which sites are included, which years of what data, where are the sites located, who is the measurement scientist in charge of a site, and what are the measurement system characteristics.
- Data download – ability to browse and download fluxes, meteorological and ancillary data for sites. Provide different levels of access to data according with the specific data access policies.
- Data update – submit updates to ancillary data and new ancillary data and track their provenance.
- Paper writing support – enable communication with measurement site scientists and gathering of citations and acknowledgments for data.
- Proposed and published paper information – access to proposed and published papers, paper progress, and paper site year usage information.
- Data curation – inform curator of submitted changes and provide an opportunity for a person familiar with the site, the *curator,* to sanity check data submissions.

Fluxes and meteorological data submission is not listed above. These data, which are the core of the database, are collected by the regional databases and transferred to fluxdata.org periodically. This organization gives the possibility to maintain a direct link with and between the regional networks. The scientists responsible for the measurements work with the regional database that imports, check and process the data and help also the fluxdata.org, giving distributed responsibilities for data gathering and processing.

In the next sections, we will discuss some of the elements needed in a user portal to support analysis usage of the carbon flux data.

17.3.1.2 The Public Access Area

The public area of the portal is accessed by all users without any restrictions or identification. The public area contains all information about the dataset and collaboration that can be made openly available. This public information is designed to be accessed by all users so the content is not replicated in other areas of the portal. The aim of this section is to present the activities to potential users, potential data contributors, and to the agencies. The public area of the portal typically contains:

- Characteristics and locations of the measurement sites with information about the science teams running the measurement sites and funding acknowledgments of the measurement sites. This information is ideally presented using interactive maps and reports. Examples include an interactive mashup of tower locations, and reports containing the average annual values of any site that can be made publicly available.
- Lists of the variables measured at sites including the explanations and availability/years of those variables along with explanations of the derived variables, gap-filling techniques, aggregation method, units, and quality markers.
- Consistent data versioning information that enables users to identify specific versions of the data for repeatability and traceability.
- Measurement site pages each listing all public information about the site and data from the site including pictures of the site if available.
- A news feed providing regular updates, announcements of changes to the dataset, and information about new functionality.
- The analysis teams' membership, papers, progress, and lists of the sites involved in each analysis, if available.
- Data fair use and publication guidelines.
- Instructions about how to participate in the activities by both sharing new data and using data for scientific analysis.

The public access area helps users get oriented. It also allows potential new users to evaluate the expected utility of the dataset before requesting access to use the data.

17.3.1.3 The Authorized User Support Area

Access to the rest of the portal should be controlled through authentication of users and by tracking user access activity. This enables the registration of data download activities, that is, important information for the sites responsible to know the number and activity of users interested in their sites and to use this information to support the existence of the site or modify the data sharing policy. In addition, having the list of users that downloaded each specific dataset gives the possibility to contact the users in case of new data versions or possible errors discovered after download. As the amount of data available increases automated data analysis and synthesis support infrastructure is helpful to users. Usually analysis teams will be less familiar with the data than the measurement site teams. Providing users with enhanced data products such as a quality-controlled and gap-filled dataset, and calculations of derived variables such as gross and net production are highly valuable. A critical element of providing this information is the accompanying methodology explanations so that the user can correctly interpret the values.

Another critical element needed by data users is the ancillary information about the site that enables interpretation and use of the data for a broad set of analyses. This information is collected using the BADM protocols mentioned earlier. Since these data tend to be collected by a wide array of individuals, they are more difficult to bring together and methods for centrally collecting, storing, updating, and presenting the ancillary site data can be an important function of the portal.

Functions ideally available in this area of a portal include:

- Download of flux-met data in standardized formats and with QA/QC data identifiers.
- Browsing and download of compilations of ancillary data for a site and across sites.
- Access for analysis teams to update the status of their analysis (allows measurement site scientists to track progress), update the list of measurement sites used in the analysis, exchange e-mail with measurement site scientists and inform them when a paper is published.
- Access reports containing annual aggregates of site flux, meteorological, and ancillary data as well as cross-site compilations of that same data and data quality indicators. Such compilations of site information enable analysis teams to narrow down their site selections without having to download and analyze the data itself.
- Notify data contributors of data usage and enable communication.
- Quick-look tools to visualize the data (e.g., simple plots) to better preselect the variables and sites of interest.

There are two typical options for notification of data usage to data contributors. The first is for an automated e-mail to be sent to the contributor of the data each time a user downloads data. A second option is to provide the analysis teams with mechanisms to identify when they are using data from a site and then provide displays which allow measurement scientists to see what analyses have indicated usage of their data. Even if the automated e-mail on download is provided, a means of indicating sites in use by an analysis can enable ongoing communication and tracking of the contributing data to a resulting publication. The scale of the datasets and the need to enable the building of trust between participants mean it was no longer possible to rely on informal and manual mechanisms to manage the communication between analysis teams and the measurement site teams.

17.3.1.4 Measurement Site Scientist Support Functions

Although typically a data portal is created to serve data users, there are many valuable capabilities it can also provide to the data contributors. Site scientists should have access to the data from their site through the portal so they can check the data processing applied in the database and the version of the data they are actually sharing. Functions available in this area of the portal should also include the following:

- Download the flux-met data for the site.
- Display all ancillary data collected for the site.
- Submit new ancillary data and update existing ancillary data.
- Search which synthesis activities have specified they are using the site's data.
- List and contact the users that have downloaded the site's data.
- Surface data releases and accompanying documentation.
- List the papers published that use their site's data.
- Specify specific papers related to the sites that should be consulted and cited if relevant.
- Specify specific acknowledgments for their sites that could be added in the publications.

17.4 Data Sharing and Policy of Uses

17.4.1 Data Sharing Motivation

The governance of shared data is an issue with practical implications. Limitations to sharing can block advancements. For example, if every part of a car is patented by a different person, then it becomes decreasingly likely that the car would ever be fully built or functional – for example, everyone might come together except for the person who owns the patents to the wheels. In fact, this "anticommons" has been a demonstrated problem in the field of biotechnology (English and Schweik 2007; Heller 1998; Heller and Eisenberg 1998). Problems may occur when cooperation or collective action ceases before a product achieves its full potential. For shared data, similar problems may be foreseeable in that a global scale scientific question that relies on multiple data sources to answer may not be answerable if data sharing is problematic.

Data sharing is not easy. An analog comes from sharing physical natural resources. Garret Hardin's "Tragedy of the Commons" (Hardin 1968) is simple to understand because at first glance it makes sense: people consume a limited resource until it is depleted because if they miss out, then someone else will not. However, there are numerous cases of shared resources governed sustainably. How is it that these shared resources do not go the way of the tragedy? Elinor Ostrom, who won the Nobel Prize in Economics in 2009, has dedicated her research career to this question. She found that certain features in a shared resource commons are consistent throughout all the sustainable cases she examined; when those features (or, as she calls them, *design principles*) are not present, then the commons is less likely to succeed sustainably (Ostrom 1990). There are eight design principles: (1) clearly defined boundaries, (2) appropriation rules related to local conditions, (3) collective-choice arrangements, (4) monitoring, (5) graduated sanctions, (6) conflict-resolution mechanisms, (7) minimal recognition of rights to organize, and (8) nested enterprises.

Nonetheless, these principles apply to natural resources that are exhaustible (e.g., can be deleted or eliminated) – data are an intellectual property resource that is not necessarily exhaustible. However, intellectual property can be misused too. Data have been considered in the literature as intellectual property, with attention to public access when data are in the public domain (i.e., publicly funded) (Drazen 2002; Hess and Ostrom 2003; Hughes 1988; Litman 1990; May 2000; Posey et al. 1995; Rai 1999; Reichman and Samuelson 1997). Data may be hoarded by data producers or taken without permission by data users (first-use rights may be compromised), and poor analyses can taint a perfectly good dataset. The finite aspect of data may not be the data themselves, but the publications and/or acknowledgments. Recent research that combined common property theory with intellectual property theory focusing in part on FLUXNET data sharing showed that cases of data sharing without conflict or irresolvable dispute incorporated more of Ostrom's design principles than did those characterized by conflict (Fisher and Fortmann 2010).

Scientifically, generally the more data, the merrier: statistical power increases and spatial representativeness can increase (though not always, and actually may bias some regions more than others) with more data. The variety and distribution of biome types, age classes, disturbance regimes, climatic controls, atmospheric coupling, among all the other myriad of ecosystem complexity components, increase with more data sharing. Perhaps more importantly, the number of eyes on the analysis increases with data sharing, which lends different perspectives, theoretical backgrounds, biases, cultural understandings, and ideas on how the world functions, or at least how each ecosystem functions. If our grand objective is to understand how the world works, then we cannot operate in isolation: we must work with the world.

Data can be made available for broader use through individual tower web sites, regional network sites, and the FLUXNET fluxdata.org site. Clear rules governing the use of the carbon-flux datasets should be laid out in a data fair use policy readily accessible to the users. Each team that wants to use the data must abide by the data fair use policy. The data fair use policy defines proper usage, coauthorship, citation, and acknowledgment behaviors. In particular, it defines required actions to be taken by the data users before publication of any results.

17.4.2 *Data Policy of Use*

In the business world, the central aim is to maximize monetary profit. In the academic/scientific world, the currency is publications. Publications advance careers, establish priority, and generally aim to impact scholarly communication and society (Suber 2007). Understanding the rules governing acknowledgment, citation, and authorship is of the upmost importance in academia and science.

The fundamental intellectual property rights question for scientific data is: Who has the rights to the data produced (May 2000)? For example, a *professor* who is supported by a *federal grant*, employed by and uses resources from a *university*, hires a *lab technician*, works with a post doc funded from a *private organization*, and publishes in a scientific *journal* might be required to share or give up a number of rights to the work produced. These same types of agents and agencies may surface again when a *data user* uses the generated work for further analyses. These agents and agencies grow exponentially when *international collaborators* become involved. Generally, these agents and agencies support data ownership in the public domain, though they are subject to different levels of intellectual property laws in addition to Federal statutes, contractual rights and duties, and limits within state-funded programs.

The rules that govern data ownership operate at three scales: macro, meso, and micro (Fisher and Fortmann 2010). At the macro-scale in the USA, for example, data can be copyrighted,[2] although some researchers are interested in copyrighting their work so that they can distribute their work freely (termed a "copyleft") (Heffan 1997). Meso-scale governance of data sharing and ownership occurs among institutions like universities, national academies, and organizations such as FLUXNET (Fienberg et al. 1985; Fisher and Fortmann 2010). Micro-scale rules operate at the personal level from individual understandings (or conflicts) to unwritten relationships and norms (Rai 1999).

It could be useful also to distinguish between data access policy and data use policy. They are clearly connected but not overlapping. A user could be interested to access the data without the intention to use them in publications or presentation. Examples could be just personal use or to verify published results. Giving open and direct access to the data increases transparency and visibility for the data providers and is also in line with recommendation of the Global Climate Observing System (GCOS) and the Group on Earth Observation's (GEO) guidelines and data principles.

Data use policies are instead the list of rules, steps, and requests that regulate the use of data, including the way to recognize and acknowledge the work of the data providers and database managers and could include requests of citations, acknowledgment, or coauthorship.

---

[2]For example, the "Digital Millennium Copyright Act" (H.R. 2281), which updated the Copyright Act (Title 17 of the US Code) to include digital data; the "Public Domain Enhancement Act" (H.R. 2601); the "Public Access to Science Act" (H.R. 2613); the "Consumer Access to Information Act of 2004" (H.R. 3872); and the "Digital Media Consumers Rights Act of 2005" (H.R. 1201).

The critical question then is "When should you give (or be given) coauthorship versus a citation versus an acknowledgment?" This is constantly thought about and debated for good reason: there is no consistent answer. Linking to the fact that ownership alone (see above) is as complicated as a food web, it may now be understandable why assigning the proper credit is also as sticky as a spider's web. Expectations, norms, and practices vary widely across disciplines, institutions, and countries. These inconsistencies are not necessarily without good reason. For example, in some scientific communities it might be expected that simply sharing data is insufficient for coauthorship, but some sort of "significant intellectual contribution" is needed for coauthorship. Whereas in other communities or regions, for example, in a developing country, where data collection and publication can be a much more difficult and lengthy process logistically and due to language and other barriers, simply sharing data warrants coauthorship. The concern by the latter case is that a Westerner can easily take data from a developing country, and publish the results much more quickly – and first. If that were to occur, then the scientist in the developing country would be less incentivized to share data.

Even within a country such as the USA, the rules are unclear. For example, the International Committee of Medical Journal Editors recommends that authors meet three conditions: (1) substantial contributions to conception and design, or acquisition of data, or analysis and interpretation of data; (2) drafting the article or revising it critically for important intellectual content; and (3) final approval of the version to be published. However, even within this explicit definition is undefined terminology. What constitutes "substantial," "critically," or "important"? In the Merriam-Webster Dictionary, "author" is defined as "one that originates or creates," or "initiates," or "brings into existence." While the person who shared the data may not have been the one to initiate or bring the paper into existence, the paper's existence would be changed without the data.

As scientists perhaps we seek that unifying first-principles, physics-based model to robustly and universally define authorship versus acknowledgment. Although the search for such might well be alchemic, many journals like *Nature* and *PNAS* ask authors to clearly identify their contribution to the paper in designing the research, performing the research, and contributing to analysis and writing. Designing authorship rules could follow the design principles of Elinor Ostrom (1990). We need clearly defined boundaries between who goes where, but these boundaries need to be sensitive to local conditions constructed collectively, rather than as one group imposing them on all others. Or, we could simply end the debate and hand down a set of hard-and-fast rules that go along the lines of: (1) coauthorship if "significant intellectual input" as defined by the primary author, (2) citation where the data/idea were first mentioned/published, (3) acknowledgment – make as big as possible without getting ridiculous. However, there would be people unhappy with these rules and situations where these rules poorly dictate what to do. Nonetheless, we could use this as a starting point – a lump of clay now dropped on the table, waiting to be sculpted by the community that would use and appreciate it. Working at a global level like in the FLUXNET community means that different positions, views, cultural behaviors, attitudes, and barriers need to be considered, analyzed, and possibly synthesized in a common agreed data use policy that would act as a first step to build a more connected and integrated global flux community. When the first examples of eddy covariance data synthesis activities were published more than 10 years ago, coauthorship was also used as a tool to get the people involved and ready to share their data that at that time were new and rare. Nowadays the coauthorship meaning is changing in the direction of a more direct participation in the scientific message preparation and discussion and directions to find a compromise with the data providers' acknowledgment expectations.

Examples of data policy requirements may require that the analysis team must contact the scientists at the measurement sites used in the analysis and inform them of the data usage, confirm permission to use the data, request additional information needed for the analysis, invite participation in the analysis effort, and obtain proper citations and acknowledgments for the data.

Data for a measurement site scientist represents a potential revenue stream in that it enables analyses that can be carried out using the data. For the data contributors to databases and synthesis activities like in the LaThuile initiative, the potential reward is that they get coauthorship, acknowledgment or citation of their data but the risk is that the data contribution will not be acknowledged or will be misinterpreted. The measurement site scientists' conditions for sharing their data with synthesis activities are in general to receive proper "credit" for their data contributions, have an opportunity to explain peculiarities of their data, and ensure that their own local analysis efforts not be "scooped" by the synthesis activity. Regarding the request of proper "credit" there are different opinions and positions in the community, from the request to have an opportunity to contribute to the papers and become coauthor to an open policy where the only request is to be informed about the data use and publications that include their sites.

In the case of the LaThuile FLUXNET synthesis activity in 2007, there are three defined data policies (available at www.fluxdata.org) that try to cover the different views and positions of the sites' managers participating in the activity; each data contributor has the possibility to decide under which policy he or she wants to share his or her measurements. In addition, a steering committee that includes representatives of the measurement site scientists, regional networks, and synthesis teams manages the synthesis activities, ensuring the respect of the policies and trying to solve possible conflicts. A data use policy system structured in this way may look (and in fact it is) complicated and not completely open. However, an open and free data sharing, if it is not imposed directly by the funding agencies, is possible only by creating a community where participants trust the other participants' correctness and see the advantage of sharing data and having papers published where their data are included. The objective of the LaThuile FLUXNET 2007 policies system was to start the building of such community and the increasing number of participants that decided to share more openly their data is a good indication for the near future.

17.4.3 Additional Credit Possibilities

The primary scientific end product of the flux data, besides the database itself, is the publications generated by using the data in analyses. Measurement site scientists are generally interested in seeing their data used but also want to receive "credit" for having supplied the data, ensure that the appropriate funding sources have been acknowledged, and provide appropriate references for the paper to cite. In the past it was common to add as coauthor the data providers and their staff, even if their contribution to the paper preparation was limited to providing the measurements. The requests from the measurement site scientists were handled through a personal conversation between each data provider and the paper writing team and often the simpler way was to send a draft almost finished with an open coauthors list. Now the approach to this issue is changing and it is a common and generally accepted rule to add as coauthors only the scientists that contributed intellectually and substantially to the scientific message presented in the paper. In the case of publications that analyze data from a small number of sites, this communication between lead author of the paper and data providers is still straightforward and may easily lead to coauthorship, in particular if the data are fundamental for the paper's message and the measurements responsible help their interpretation. However, there is an increasing number of analyses which rely on a large number of global carbon-flux sites and/or incorporate data from many other sources and contributors. Providing the opportunity for coauthorship and intellectual input opportunities to all data contributors in these cases is not always feasible because it would mean to receive often opposite point of view and suggestion is not easy to reconcile. In addition, there are more and more scientific analyses that make use of EC products in complex models or analysis (e.g., in data assimilation systems) where the underlying methodology and statistical analysis is often so specialized and focused that it is difficult to fully understand for many of the scientist that are working in different fields. In these cases, also offering coauthorship could be problematic since it is important for all the coauthors to fully understand methods and results of the papers where they are involved. At the same time it is however important that the data contributors' rights are preserved and full acknowledgment and credit is ensured to their work.

This same issue has been faced by many other scientific fields and in particular high-energy physics which in the case of large experiments have thousands of scientists involved in each experiment. These groups typically converge on a formal set of guidelines for authorship and create a group authorship designator which is included in all publications. In addition, they typically converge on a single acknowledgment that can be used to acknowledge funding. In the case of large-scale experiments, these mechanisms are critical since otherwise, identifying individual contributions is too difficult. Flux datasets are a bit different since each individual site year has a clear set of contributors and funders. What is needed is a hybrid approach to the problem that incorporates aspects of the authorship mechanisms used in large experiments but that also enables tracking to individual contributions.

One possibility that has been already discussed in the past is to assign a Digital Object Identifier (DOI) (see also www.doi.org) to each dataset. The DOI is a digital identifier for any object of intellectual property and provides a means of persistently identifying a dataset on a digital network and associating it with related current data like authors, owner, characteristics, location, and all the other information relevant to describe and characterize the data. The system is currently used by most of the scientific journals to identify each single paper and could be easily implemented also for the data. However, the advantage of such a system for the data contributor is also linked to a change in the evaluation schemes adopted by funding agencies but also single institutions that should start to consider identifying datasets by DOI as a high-level product. The DOI release could be for example delegated to the regional databases and linked to a minimum level of quality, documentation, and completeness to be assigned, increasing the value of such identifier. An example of a data collection identified by a DOI is the Marconi Synthesis dataset (Falge et al. 2005).

References

Clery D (2006) Can grid computing help us work together? Science 313(5786):433–434

Drazen JA (2002) Who owns the data in a clinical trial? Sci Eng Ethics 8(3):407–411

English R, Schweik CM (2007) Identifying success and tragedy of FLOSS commons: a preliminary classification of Sourceforge.net projects. Paper presented at 29th international conference on software engineering workshops, Minneapolis

Falge E, Aubinet M, Bakwin P, Baldocchi D, Berbigier P, Bernhofer C, Black A, Ceulemans R, Davis K, Dolman A, Goldstein A, Goulden M, Granier A, Hollinger D, Jarvis P, Jensen N, Pilegaard K, Katul G, Kyaw Tha Paw P, Law B, Lindroth A, Loustau D, Mahli Y, Monson R, Moncrieff P, Moors E, Munger W, Meyers T, Oechel W, Schulze E, Thorgeirsson H, Tenhunen J, Valentini R, Verma S, Vesala T, Wofsy S (2005) FLUXNET Marconi Conference Gap-Filled Flux and Meteorology Data, 1992–2000. Data set. Available on-line [http//www.daac.ornl.gov] from Oak Ridge National Laboratory Distributed Active Archive Center, Oak Ridge, Tennessee, U.S.A. doi:10.3334/ORNLDAAC/811

Fienberg SE, Martin ME, Straf ML (1985) Sharing research data. Committee on National Statistics, National Research Council, Washington, DC, 240 pp

Fisher JB, Fortmann LP (2010) Governing the data commons: policy, practice, and the advancement of science. Inf Manage 47:237–245

Hardin G (1968) The tragedy of the commons. Science 162(3859):1243–1248

Heffan IV (1997) Copyleft: licensing collaborative works in the digital age. Stanford Law Rev 49(6):1487–1521

Heller MA (1998) The tragedy of the anticommons: property in the transition from Marx to markets. Harv Law Rev 111(3):621–688

Heller MA, Eisenberg RS (1998) Can patents deter innovation? The anticommons in biomedical research. Science 280:698–701

Hess C, Ostrom E (2003) Ideas, artifacts, and facilities: information as a common-pool resource. Law Contemp Probl 66:111

Hughes J (1988) The philosophy of intellectual property. Georget Law J 77(287):296–314

Lasslop G, Reichstein M, Papale D, Richardson AD, Arneth A, Barr A, Stoy P, Wohlfahrt G (2010) Separation of net ecosystem exchange into assimilation and respiration using a light response curve approach: critical issues and global evaluation. Glob Change Biol 16:187–208. ISSN:1354–1013, doi: 10.1111/j.1365-2486.2009.02041.x

Litman J (1990) The public domain. Emory Law J 965:975
May C (2000) A global political economy of intellectual property rights: the new enclosures? Routledge, New York
Ostrom E (1990) Governing the commons: the evolution of institutions for collective action. Cambridge University Press, Cambridge
Papale D, Valentini R (2003) A new assessment of European forests carbon exchanges by eddy fluxes and artificial neural network ki. Glob Change Biol 9:525–535, ISSN: 1354–1013
Papale D, Reischtein M, Aubinet M, Canfora E, Bernhoher C, Longdoz B, Kutsch W, Rambal S, Valentini R, Vesala T, Yakir D (2006) Towards a standardized processing of Net Ecosystem Exchange measured with eddy covariance technique: algorithms and uncertainty estimation. Biogeosciences 3:571–583, ISSN: 1726–4170
Posey DA, Dutfield G, Plenderleith K (1995) Collaborative research and intellectual property-rights. Biodivers Conserv 4(8):892–902
Rai AK (1999) Regulating scientific research: intellectual property rights and the norms of science. Northwest Univ Law Rev 94(1):77–152
Reichman JH, Samuelson P (1997) Intellectual property rights in data? Vanderbilt Law Rev 51(1):49–166
Reichstein M, Falge E, Baldocci D, Papale D, Aubinet M, Berbigier P, Bernhofer C, Buchmann N, Gilmanov T, Granier A, Grunwald T, Havrankova K, Ilvesniemi H, Janous D, Knohl A, Laurila T, Lohila A, Loustau D, Matteucci G, Meyers T, Miglietta F, Ourcival J-M, Pumpanen J, Rambal S, Rotenberg E, Sanz M, Tenhunen J, Seufert G, Vaccari F, Vesala T, Yakir D, Valentini R (2005) On the separation of net ecosystem exchange into assimilation and ecosystem respiration: review and improved algorithm. Glob Change Biol 11:1424–1439
Suber P (2007) Creating an Intellectual Commons through Open Access. In: Hess C, Ostrom E (eds) Understanding knowledge as a commons: from theory to practice. The MIT Press, Cambridge

Symbol Index

Symbol	Meaning
A_{SFi}	Sapwood area (m^2 of sapwood $m^{-2}$of soil)
Bo	Bowen ratio
c, c_1, $c_2 and$ c_3	Sound velocity and its components measured along each sonic anemometer axis
c_p	Air specific heat
c_s	Molar concentration of component s [mole m^{-3}]
C	CO2 fluxes resulting from combustion processes in urban areas
C_{ss}	Spectral density of a signal χ_s
C_{ws}, C_{wc}, C_{wu}, $C_{w\theta}$	Cospectral density of w and χ_s,χ_c, u, θ
d	Zero plane displacement height
d_s	Path length of scalar sensor (e.g. IRGA or sonic thermometer)
d_{ss}	Lateral separation distance between the two sensors
d_{pl}	Sonic anemometer path length
D_s	Molecular diffusivity of scalar s
e	Vapour pressure
E_0	Activation energy (in Lloyd and Taylor equation)
E_{SF}	Sap flow density of the stand
E_{tot}, E_{plant}, E_{int}, E_{soil}	Evapotranspiration and its components : plant transpiration, evaporation of intercepted water and from soil
f	Frequency
f_{os}	Half power frequency (for component s)
f_s	Sampling frequency
$\mathcal{F}_S(f)$, $\mathcal{F}_S^*(f)$	Fourier transform of the signal and its complex conjugate
F	Water vapour flux due to combustion (urban areas)
F_c, F_s, F_v	Net ecosystem exchange of CO2, of component s, of water vapour, expressed in mass units

(continued)

M. Aubinet et al. (eds.), *Eddy Covariance: A Practical Guide to Measurement and Data Analysis*, Springer Atmospheric Sciences, DOI 10.1007/978-94-007-2351-1,

(continued)

Symbol	Meaning
$F_c^{EC}, F_s^{EC}, F_v^{EC}$	Turbulent flux of CO2, of component s, of water vapour, expressed in mass units
$F_{s,mol}^{EC}, F_{c,mol}^{EC}, F_{v,mol}^{EC}$	Turbulent flux of any scalar s, of CO2, of water vapour, expressed in molar units
F_{manure}, $F_{animal-products}$, $F_{erosion}$, F_{leach}, $F_{harvest}$, F_{fire}, F_{VOC}	Different CO_2 fluxes exchanged at the ecosystem scale in grasslands
g	Gravity acceleration
G	Soil heat flux
G_s	Apparatus transfer functions due to sensor loss
GEP	Gross ecosystem productivity
H	Sensible heat flux
H_s	Buoyancy flux
h, h_c	Canopy height
h_m	Measurement height
I	Water flux due to irrigation
K_ζ	Molecular diffusivity of quantity ζ
L	Obukhov length
L_{self}	Filter inductance
L_t	Tube length in closed path systems
m_d	Dry air molar mass
m_v	Water vapor molar mass
n	Dimensionless frequency
NBP	Net biome productivity
NECB	Net ecosystem carbon balance
NEE	Net ecosystem exchange
Og_{ws}	Ogive
P_r	Precipitation
p	Atmospheric pressure
$\overline{p}_i$	Air pressure measured within the gas analyser
PPFD	Photosynthetically active photon flux density
Q	Air flow in closed path tube
Q_F	Anthropogenic heat flux density
$Q(\hat{x})$	Source emission rate/sink strength in the surface-vegetation volume.
Q_{10}	Temperature sensitivity parameter
r_h	Relative humidity
r_t	Tube radius in closed path systems
R	Universal gas constant
R_{eco}	Ecosystem respiration
R_{ue}	Urban Ecosystem Respiration
Re	Reynolds number
R_g	Global radiation
R_n	Net radiation
REW	Relative water content
R_{01}, R_{12}, R_{23}, R_{03}	Rotation matrices

(continued)

(continued)

Symbol	Meaning
R_{10}	Ecosystem base respiration at 10 C
S_s	Component s source/sink strength
S_i	Component s source/sink strength (in momentum equation)
SWC	Soil water content
t	Time
t_l	Time lag
T_{ws}	Total apparatus transfer function
T_{ss}, T_{pw}, T_{ps}, T_{ta}	Apparatus transfer functions due to lateral separation, wind averaging, scalar averaging and tube attenuation.
u	Horizontal wind velocity component parallel to average wind velocity
$\vec{u}$	Wind velocity vector
u_i	Wind velocity component (in momentum equation)
u_f	Flow velocity in closed path tubes
u_{SFi}	Sap flux density ($m^3$of water $m^{-2}$of sapwood s^{-1})
u_*	Friction velocity
u_{*crit}	Friction velocity threshold
v	Lateral component of velocity
V	Voltage output of measurement device
$\boldsymbol{V}(t)$	Trajectory velocities as a function of time (lagrangian footprint model)
VAI	Vegetation area index
w	Vertical wind velocity component
WTD	Water table depth
$\overline{w'u'}$	Eddy covariance term
x	Horizontal Cartesian coordinate, parallel to the average wind velocity
x_f	Distance of the fetch
$\boldsymbol{X}(t)$	Trajectory co-ordinates as a function of time (lagrangian footprint model)
y	Horizontal Cartesian coordinate, perpendicular to the average velocity
z	Vertical Cartesian coordinate
z_h, z_i, z_r	Height of urban canopy, of PBL, of roughness sublayer

Greek Symbols

α, β, γ	Euler angles
α, β, γ	Parameters of the NEE to light response (respectively, quantum efficiency, saturation value and dark respiration)
γ	Ratio of constant pressure and constant volume heat capacities
χ_s	Mass mixing ratio of component s[kg kg^{-1}]
$\chi_{s,m}$	Molar mixing ratio of component s [mole mole^{-1}]
$\hat{\chi}$	Measured value of variable χ
Δ	Laplacian operator $\left(\frac{\partial^2}{\partial x^2} + \frac{\partial^2}{\partial y^2} + \frac{\partial^2}{\partial z^2}\right)$
ΔQ_S	Net storage heat flux density (in urban areas)
Δr	Water run off (urban areas)
ΔW	Water storage in subsurface material (in urban areas)
δ	Systematic error on flux measurements
δ_{cal}	Calibration error
δ_a, δ_N, $\delta_{R,}$	Isotopic composition of air, respiration and NEE
δ_s / F_s^{EC}	Relative spectral error
Δp	Pressure drop in closed path chambers
Δ_{canopy}	Isotopic discrimination by photosynthesis
ε	Random error on flux measurements
$\overline{\varepsilon}$	Mean dissipation rate of turbulent kinetic energy
ϕ	Concentration or Flux footprint function
κ	von Karmann constant
λ	Latent heat of vaporization for water
λE	Latent heat flux
ν	Kinematic viscosity of the air
η	Quantity being measured at location x (*footprint analysis*)
θ	Air temperature
$\overline{\theta}_i$	Air temperature measured within the gas analyser
θ_{ref}	Reference temperature (in Lloyd and Taylor equation)
θ_s, θ_v	Sonic temperature, virtual temperature
θ_{NRH}	Degree of curvature of non rectangular hyperbola
ρ, ρ_s, ρ_c, ρ_d, ρ_v	Density of total air, of scalar s, of CO2, of dry air, of water vapour
σ	Ratio of water vapor and dry air densities
σ	Standard deviation
σ_θ ,$\sigma_{\theta s}$	Standard deviation of temperature, of sonic temperature
τ	Momentum flux
τ_R	System response time
μ	Ratio of molar masses of dry air and water vapor
ζ	Stability parameter
ζ	Any variable
ξ	Any variable

Other Symbols

∂	Partial derivative operator
$\overline{}$	Reynolds averaging operator (by extension : time average)
$\langle\rangle$	Ensemble average (Chap 8) or spatial average (Chap 16)
$\widehat{}$	True value

Subscripts

BA	Block average
c	Refers to CO2
d	Refers to dry air
DR	Double rotation
LD	Linear detrending
M	Mitscherlich equation
NRH	Non rectangular hyperbola
PF	Planar fit
RH	Rectangular hyperbola
s	Refers to any scalar or sonic temperature
v	Refers to water vapour

Abbreviations and Acronyms

ABL	Atmospheric boundary layer
AF	Autoregressive filtering
ANN	Artificial neural networks
BA	Block average
CFD	Computational fluid dynamics
DBH	Diameter at breast height
DEC	Disjunct eddy covariance
DI	Disturbance indicator
DIC	Dissolved inorganic carbon
DIN	Dissolved inorganic nitrogen
DOC	Dissolved organic carbon
DR	Double rotation
EC	Eddy covariance
EW	Extractable soil water
GHG	Greenhouse gases
HeNe	Helium neon
IRGA	Infrared gas analyser
ISL	Inertial sublayer
ISR	Intermediate storage reservoir
LAI	Leaf area index
LD	Linear detrending
LES	Large eddy simulation
LS	Lagrangian stochastics
LUT	Look-up table
MDS	Marginal Distribution Sampling
MDV	Mean Diurnal Variation method
MOST	Monin-Obukhov similarity theory
NDVI	Normalised differential vegetation index
NEON	National earth observation network

(continued)

M. Aubinet et al. (eds.), *Eddy Covariance: A Practical Guide to Measurement and Data Analysis*, Springer Atmospheric Sciences, DOI 10.1007/978-94-007-2351-1,

(continued)

NLR	Non-linear regressions
PBL	Planetary boundary layer
PDF	Probability distribution function
PF	Planar fit
POM	Particulate organic matter
PTR-MS	Proton transfer reaction mass spectrometer
RSL	Roughness sublayer
SAT	Sonic anemometer-thermometer
TDL	Tunable diode laser
TDR	Time-domain reflectometer
TMS	Thermal mass scheme
UBL	Urban boundary layer
UCL	Urban canopy layer
UEB	Urban energy balance
UHI	Urban heat island
VOC	Volatile organic compounds
VPD	Vapor pressure deficit
WPL	Webb Pearman and Leuning (dilution correction)

Index

M. Aubinet et al. (eds.), *Eddy Covariance: A Practical Guide to Measurement and Data Analysis*, Springer Atmospheric Sciences, DOI 10.1007/978-94-007-2351-1,

I

K

L

M

N

O